Viral Infection and Apoptosis

Special Issue Editor
Marc Kvansakul

MDPI • Basel • Beijing • Wuhan • Barcelona • Belgrade

MDPI

Special Issue Editor
Marc Kvansakul
La Trobe University
Australia

Editorial Office
MDPI AG
St. Alban-Anlage 66
Basel, Switzerland

This edition is a reprint of the Special Issue published online in the open access journal *Viruses* (ISSN 1999-4915) in 2017 (available at: http://www.mdpi.com/journal/viruses/special_issues/viruses_and_apoptosis).

For citation purposes, cite each article independently as indicated on the article page online and as indicated below:

Lastname, F.M.; Lastname, F.M. Article title. *Journal Name*. **Year**. *Article number*, page range.

First Edition 2018

Image courtesy of Marc Kvansakul

ISBN 978-3-03842-655-4 (Pbk)
ISBN 978-3-03842-654-7 (PDF)

Table of Contents

About the Special Issue Editor

Marc Kvansakul earned a B.Sc in Biochemistry from Imperial College London, followed by a PhD with Prof Erhard Hohenester at Imperial College London examining protein-protein interactions in the extracellular matrix. Marc then moved to the Walter and Eliza Hall Institute in Melbourne/ Australia to work with Prof Peter Colman to investigate the structure and function of cell death regulators encoded by DNA viruses. In 2010, Marc accepted an independent position at La Trobe University joining the La Trobe Institute for Molecular Science, where he is currently an Associate Professor and serves as Deputy Head of the Department of Biochemistry and Genetics.

viruses

MDPI

Editorial
Viral Infection and Apoptosis

Marc Kvansakul

Department of Biochemistry and Genetics, La Trobe Institute for Molecular Science, La Trobe University, Melbourne, VIC 3086, Australia; m.kvansakul@latrobe.edu.au; Tel.: +61-3-94792263

Received: 21 November 2017; Accepted: 22 November 2017; Published: 23 November 2017

Viruses are master molecular manipulators, and evolved to thrive and survive in all species. Key to their continuing success has been the ability to subvert host cell defence systems to ensure viral survival, replication and proliferation. Amongst the diverse arsenal of defence mechanisms deployed by multicellular hosts are those that rely on rapid activation of programmed cell death or apoptosis pathways, to trigger premature cell death of infected host cells. Apoptosis of an infected host cell has been identified as a powerful mechanism to curtail viral spread, and consequently, viruses have evolved sophisticated molecular strategies to subvert host cell apoptotic defences.

This special issue is devoted to the interplay of viruses and host cell apoptotic signalling pathways. Many viruses manipulate apoptosis for their own purpose, and Zhou et al. review the interplay of viruses with death receptor mediated apoptosis [1]. This contribution is complemented by a companion review focussed on the role of the intrinsic or Bcl-2 mediated pathway of apoptosis in host–virus interactions by Kvansakul et al. [2], as well as by a research article examining the structure and function of a canarypoxvirus encoded Bcl-2 homolog from Anasir et al. [3]. The ability of other poxviruses to modulate apoptosis signalling is discussed by Nichols et al. [4]. From the large group of DNA viruses that have been shown to subvert apoptosis signalling, Epstein–Barr virus is examined by Fitzsimmons & Kelly [5], with another prominent herpesvirus, cytomegalovirus, being reviewed by Brune & Andoniou [6]. This set of reviews is completed by an overview from Dixon and colleagues on the African swine fever virus-encoded modulators of apoptosis [7]. Influenza virus interference with apoptosis is reviewed by Shim et al. [8], which is complemented by a research article from Othumpangat et al. [9] describing the role of miRNA-4776 in regulating host cell NF-κB signalling. Moving onto RNA viruses, the interplay between flaviviruses and cell death pathways is reviewed by Okamoto et al. [10], whereas HIV induced bystander apoptosis is reviewed by Garg & Joshi [11]. Another important RNA virus, hepatitis B virus, is reviewed by Lin & Zhang [12], with a final research article describing the establishment of an Avian infectious bronchitis cell culture system by Han et al. [13].

I hope that the research articles and reviews that were assembled for this special issue will shed light for both experts and interested bystanders on the complex and sometimes unexpected interplay between viruses and host cell apoptosis signalling pathways, and inspires researchers to investigate the many unresolved questions about the ways and means utilized by viruses to manipulate life and death of an infected cell.

References

1. Zhou, X.; Jiang, W.; Liu, Z.; Liu, S.; Liang, X. Virus infection and death receptor-mediated apoptosis. *Viruses* **2017**, *9*, 316. [CrossRef] [PubMed]
2. Kvansakul, M.; Caria, S.; Hinds, M.G. The Bcl-2 family in host-virus interactions. *Viruses* **2017**, *9*, 290. [CrossRef] [PubMed]
3. Anasir, M.I.; Baxter, A.A.; Poon, I.K.H.; Hulett, M.D.; Kvansakul, M. Structural and functional insight into canarypox virus CNP058 mediated regulation of apoptosis. *Viruses* **2017**, *9*, 305. [CrossRef] [PubMed]

4. Nichols, D.B.; de Martini, W.; Cottrell, J. Poxviruses utilize multiple strategies to inhibit apoptosis. *Viruses* **2017**, *9*, 215. [CrossRef] [PubMed]
5. Fitzsimmons, L.; Kelly, G.L. EBV and apoptosis: The viral master regulator of cell fate? *Viruses* **2017**, *9*, 339. [CrossRef] [PubMed]
6. Brune, W.; Andoniou, C.E. Die another day: Inhibition of cell death pathways by cytomegalovirus. *Viruses* **2017**, *9*, 249. [CrossRef] [PubMed]
7. Dixon, L.K.; Sanchez-Cordon, P.J.; Galindo, I.; Alonso, C. Investigations of pro- and anti-apoptotic factors affecting african swine fever virus replication and pathogenesis. *Viruses* **2017**, *9*, 241. [CrossRef] [PubMed]
8. Shim, J.M.; Kim, J.; Tenson, T.; Min, J.Y.; Kainov, D.E. Influenza virus infection, interferon response, viral counter-response, and apoptosis. *Viruses* **2017**, *9*, 223. [CrossRef] [PubMed]
9. Othumpangat, S.; Bryan, N.B.; Beezhold, D.H.; Noti, J.D. Upregulation of miRNA-4776 in influenza virus infected bronchial epithelial cells is associated with downregulation of NFKBIB and increased viral survival. *Viruses* **2017**, *9*, 94. [CrossRef] [PubMed]
10. Okamoto, T.; Suzuki, T.; Kusakabe, S.; Tokunaga, M.; Hirano, J.; Miyata, Y.; Matsuura, Y. Regulation of apoptosis during flavivirus infection. *Viruses* **2017**, *9*, 243. [CrossRef] [PubMed]
11. Garg, H.; Joshi, A. Host and viral factors in HIV-mediated bystander apoptosis. *Viruses* **2017**, *9*, 237. [CrossRef] [PubMed]
12. Lin, S.; Zhang, Y.J. Interference of apoptosis by hepatitis B virus. *Viruses* **2017**, *9*, 230.
13. Han, X.; Tian, Y.; Guan, R.; Gao, W.; Yang, X.; Zhou, L.; Wang, H. Infectious bronchitis virus infection induces apoptosis during replication in chicken macrophage HD11 cells. *Viruses* **2017**, *9*, 198. [CrossRef] [PubMed]

Review

Virus Infection and Death Receptor-Mediated Apoptosis

Xingchen Zhou †, Wenbo Jiang †, Zhongshun Liu, Shuai Liu and Xiaozhen Liang *

Key Laboratory of Molecular Virology & Immunology, Institut Pasteur of Shanghai, University of Chinese Academy of Sciences, Chinese Academy of Sciences, Shanghai 200031, China; xczhou@ips.ac.cn (X.Z.); wbjiang@ips.ac.cn (W.J.); zsliu@ips.ac.cn (Z.L.); sliu@ips.ac.cn (S.L.)
* Correspondence: xzliang@ips.ac.cn; Tel.: +86-021-5492-3096
† These authors contributed equally to this work.

Received: 21 September 2017; Accepted: 25 October 2017; Published: 27 October 2017

Abstract: Virus infection can trigger extrinsic apoptosis. Cell-surface death receptors of the tumor necrosis factor family mediate this process. They either assist persistent viral infection or elicit the elimination of infected cells by the host. Death receptor-mediated apoptosis plays an important role in viral pathogenesis and the host antiviral response. Many viruses have acquired the capability to subvert death receptor-mediated apoptosis and evade the host immune response, mainly by virally encoded gene products that suppress death receptor-mediated apoptosis. In this review, we summarize the current information on virus infection and death receptor-mediated apoptosis, particularly focusing on the viral proteins that modulate death receptor-mediated apoptosis.

Keywords: virus infection; death receptor; extrinsic apoptosis; host immune response

1. Introduction of Virus-Mediated Apoptosis

Apoptosis, necroptosis, and pyroptosis are the three major ways of programed cell death (PCD) following virus infection [1,2]. Among them, apoptosis is the most extensively investigated PCD during viral infection. Apoptosis elicited by virus infection has both negative and positive influence on viral replication. Host cells eliminate virally infected cells via apoptosis, which aborts virus infection. On the other hand, some viruses take advantage of inducing apoptosis as a way to release and disseminate progeny viruses [3–5]. In both cases, it requires certain viral products to block or delay apoptosis in order to produce sufficient progeny and establish successful viral replication [2,6]. A number of important reviews have provided valuable insights into viruses and apoptosis. This review specifically focuses on viral modulation of death receptor-mediated apoptosis.

2. Intrinsic and Extrinsic Apoptosis Pathways

Apoptosis can be triggered by two distinct signaling pathways, namely the intrinsic and extrinsic pathways [7–9]. The intrinsic apoptotic pathway is elicited by a wide range of intracellular stress conditions, including cytokine deprivation, DNA damage, oxidative stress, cytosolic Ca^{2+} overload and endoplasmic reticulum stress. These heterogeneous apoptotic signals converge to trigger one pivotal event—mitochondrial outer membrane permeabilization (MOMP), which promotes the release of cytochrome c and other mitochondrial factors into the cytosol, ultimately leading to the generation of initiator and effector caspases and subsequent cell death [7,10]. MOMP is mainly controlled by the B-cell lymphoma 2 (BCL-2) family of proapoptotic proteins. The three BCL-2 homology domains (BH3)-only proteins BCL-2-interacting mediator (BIM) and BH3-interacting domain death agonist (BID) can transiently interact with BCL-2-associated X protein (BAX) or BCL-2 antagonist or killer (BAK) upon activation induced by intrinsic apoptosis signals, leading to the activation and conformational changes of BAX and BAK [11,12]. Activated BAX and BAK allow the formation of high-order homo-oligomers and stable insertion into the outer mitochondrial membranes, promoting MOMP [13,14]. The extrinsic apoptotic pathway is activated by extracellular stress stimulation that is sensed and triggered through activation of death receptors of the tumor necrosis factor (TNF) family, including TNF receptor 1 (TNF-R1), Fas (also called CD95 or Apo-1 or TNFRSF6), TNF-related apoptosis-inducing ligand (TRAIL) receptors (TRAIL-R1 and TRAIL-R2, also known as DR4 and DR5), DR3 and DR6 [7]. Death receptors are type-I transmembrane proteins and are characterized by an extracellular ligand-binding region containing 1-5 cysteine-rich domains, a membrane-spanning region, and a 60- to 80-amino acid cytoplasmic death domain [15–17]. Ligand binding to the death receptor on the cell surface leads to signal transduction through the formation of the death-inducing signaling complex (DISC), which mediates the subsequent apoptotic signal transduction [18,19]. Unlike the intrinsic pathway, apoptosis via death receptor-mediated extrinsic pathway does not always require mitochondria. In type I cells without the involvement of intrinsic pathway, the extrinsic apoptotic pathway results in the activation of caspase-8, which can directly induce the activation of caspase-3 and -7, leading to apoptosis [12]. However, both pathways generate similar effector caspases that serve to amplify the initial death signal [20]. In type II cells, the extrinsic pathway can also link to the intrinsic apoptotic pathway via caspase-8 cleavage of BH3-only protein BID [8]. Following death receptor stimulation, activated caspase-8 cleaves BID into 15 kDa truncated form tBID. tBID then triggers MOMP and cytochrome C release, initiating effector caspase activation and apoptosis [21].

Signaling through both Fas and TRAIL-R1/-R2 leads to the oligomerization of death receptors and intracellular assembly of DISC composing of dead receptor, the adaptor molecule FADD (FAS-associated with a death domain), procaspase-8, procaspase-10 and cellular FADD-like interleukin-1β converting enzyme (FLICE)-like inhibitory protein (c-FLIP). c-FLIP has two protein isoforms, c-FLIP long (c-FLIP$_L$) and c-FLIP short (c-FLIP$_S$) [22]. c-FLIP isoforms control the activation of caspase cascade that emanates from caspase-8 and then initiate the apoptotic program or nonapoptotic caspase-8 signaling [23–25]. In contrast to Fas and TRAIL death receptors, TNF-R1-mediated apoptotic signaling is more complex. TNF-α binding to TNF-R1 recruits TNFR-associated protein with death domain (TRADD) as an adaptor protein which subsequently recruits FADD, TNF-associated factor-2 (TRAF-2), receptor-interacting protein (RIP), and RIP-associated interleukin-1 beta-converting enzyme homolog (ICH-1)/cell death protein-3 (CED-3)-homologous protein with a death domain (RAIDD). FADD then binds and activates caspase-8 and -10, leading to apoptosis [17,26].

3. Viral Induction and Viral Activators of Death Receptor-Mediated Apoptosis

Death receptor-mediated apoptosis represents an efficient mechanism by which the virus can induce cell death and disseminate progeny, which plays an important role in viral pathogenesis and provides a potential therapeutic target. Regulation of death receptor-mediated apoptosis by the virus is mainly through upregulation of death receptors or their ligand on the cell surface of the infected individuals and increased sensitivity of the cells to death receptor-mediated apoptosis (Figure 1). Many viruses encode viral proteins to regulate death receptor-mediated apoptosis in a variety of different ways (Table 1).

Table 1. Viral proteins that modulate death receptor-mediated apoptosis.

Virus	Viral Activator	Mediated Signaling	References
HIV-1	Tat	Fas, TRAIL-R	[27–29]
	Env	Fas	[30]
	Vpu	Fas	[31]
	gp120	Fas, TNF-R	[28,32,33]
	gp160	Fas	[34]
HBV	HBX	TRAIL-R	[35,36]
HCV	Core protein	Fas, TNF-R, TRAIL-R	[37–40]
EBV	LMP2A	Fas	[41]
	LMP1	Fas	[42]
HPV	E2	Fas, TNF-R	[43]
	E7	TNF-R	[44]
HTLV-1	Tax	TRAIL-R	[45]
Lyssavirus	Matrix protein	TRAIL-R	[46]
Virus	**Viral Inhibitor**	**Mediated Signaling**	**References**
Fibroma virus	TNFR2 ortholog	TNF-R	[47,48]
Vaccinia virus	CrmE	TNF-R	[49]
	SPI-1	Fas	[50]
	SPI-2	Fas, TNF-R	[50–54]
Myxoma virus	T2	TNF-R	[55]
Cowpox virus	CrmA	Fas, TNF-R	[56]
HSV-1	gD	Fas–	[57]
	Ribonucleotide reductase R1	Fas	[58]
HSV-2	Ribonucleotide reductase R1	Fas	[58]
HCMV	IE2	Fas, TRAIL-R	[59]
	vMIA	Fas	[60]
	UL36	Fas	[61]
MCMV	M36	Fas	[62]
	M45	TNF-R	[63]
KSHV	v-FLIP	Fas, TRAIL-R	[64,65]
MCV	MC159	Fas, TNF-R, TRAIL-R	[65–67]
EHV2	E8	Fas, TNF-R	[66]
HVS	v-FLIP	Fas, TRAIL-R	[65]
BHV-4	v-FLIP	Fas, TRAIL-R	[65]
EBV	EBER	Fas	[68,69]
	BHRF1	Fas, TNF-R, TRAIL-R	[69,70]
	BZLF1	TNF-R	[71]
	LMP1	Fas, TRAIL-R	[72,73]
MHV68	M11	Fas, TNF-R	[74]
HCV	Core protein	Fas, TNF-R	[75–77]
	E2	Fas	[78]
	NS5A	TNF-R	[79]
HBV	Core protein	Fas	[80]
ADV	E3-10.4K/14.5K complex	Fas	[81]
	E3-6.7K/10.4K/14.5K complex	TRAIL-R	[82]
	E3-RID complex	Fas, TNF-R, TRAIL-R	[83]
	E3-14.7K	Fas, TNF-R	[84]
HPV	E5	Fas, TRAIL-R	[85,86]
	E6	TNF-R, TRAIL-R	[87–90]
	E7	Fas, TNF-R	[91,92]
EBOV	Glycoprotein	Fas	[93]
MARV	Glycoprotein	Fas	[93]
HIV-1	Nef	Fas	[94]
	Tat	TRAIL-R	[95]
HTLV-1	Tax	Fas	[96,97]
HTLV-2	Tax	Fas	[98]

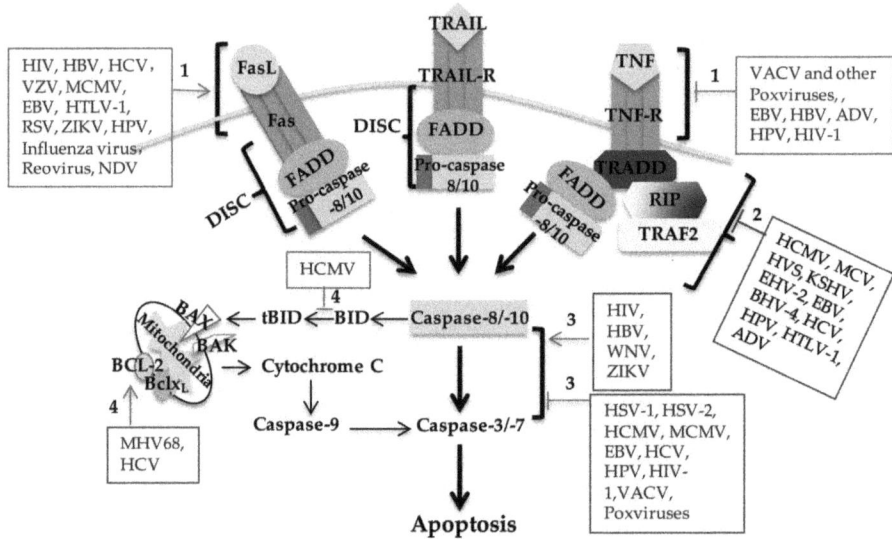

Figure 1. Viral modulation of death-receptor mediated apoptosis. Death receptors Fas, TRAIL-R and TNF-R form DISC upon binding to their ligands, activate caspase cascade and subsequently initiate extrinsic apoptosis. Caspase-8 activation can cleavage BID to tBID and link to mitochondria mediated intrinsic apoptosis pathway. Virus infection regulates death receptor-mediated extrinsic apoptosis mainly through virally encoded proteins. The regulatory mechanisms involve: (1) regulating the expression and function of death receptors/ligands; (2) interfering DISC formation and function; (3) regulating caspase activities; (4) regulating the expression and function of pro-apoptotic and anti-apoptotic proteins. Black arrow represents signal induction; grey arrow represents signal induced by viruses; grey T bar represents signal inhibited by viruses.

3.1. Human Immunodeficiency Virus (HIV)

Apoptosis mediated by death receptors plays an important role during HIV-1 infection. An increased expression of Fas antigen in CD4+ and CD8+ T lymphocytes occurs in patients infected with HIV-1, rendering those cells sensitive to FasL/Fas system-mediated apoptosis and contributing to T lymphocyte depletion in HIV-infected individuals [99–105]. HIV-1 infection also upregulates FasL expression in macrophage and mediates apoptosis and depletion of T lymphocytes [106].

Regulation of TRAIL expression and TRAIL-mediated apoptosis during HIV-1 infection has been well demonstrated. HIV-1 infection induces expression of TRAIL and DR5 and leads to TRAIL-mediated apoptosis in primary CD4+ T cells, which is regulated by IFN-α that is produced by HIV-1-stimulated plasmacytoid dendritic cells (pDCs) [107,108]. Upregulation of TRAIL in primary macrophages during HIV-1 infection occurs and mediates apoptosis in bystander T cells and neuronal cells [27,109,110]. An elevated level of plasma TRAIL was observed in HIV-infected patients and preferentially provokes apoptosis of HIV-1-infected monocyte-derived macrophages and partially mediates CD4+ T-cell apoptosis [111,112]. Additionally, HIV infection results in TRAIL expression and TRAIL-mediated apoptosis in memory B cells, leading to the loss of memory B cells [113]. As such, Fas- or TRAIL-mediated signaling could be exploited for the development of therapeutic target aimed at the prevention of T cell death in AIDS and preventive HIV vaccine.

HIV-1-encoded proteins modulate death receptor-mediated apoptosis in different cell types. HIV-1 Tat, Vpu, gp120 and gp160 proteins sensitize T cells to Fas-mediated apoptosis with different regulatory mechanisms, possibly contributing to T-cell depletion in AIDS [28,31,34]. HIV-1 gp120 accelerates the apoptosis of human lamina propria T cells induced by Fas-mediated activation which is related to increased induction of FasL mRNA [32], whereas gp160 enhancement of Fas-mediated apoptosis involves the activation of caspase-3 and requires calmodulin binding to the C-terminal binding domain of gp160 [34]. The Env protein of CCR5 tropic HIV strains activates Fas and caspase-8 as well as triggers FasL production, which eventually results in CD4+ T cell apoptosis [30]. Furthermore, HIV-1 Tat upregulates TRAIL in peripheral blood mononuclear cells (PBMCs) and primary macrophages, leading to apoptosis and depletion of uninfected bystander cells [27,29]. A recent report indicates that soluble gp120 shed from HIV-1-infected cells and virus infection itself induces TNF-α expression on macrophages, and upregulates TNF-receptor 2 (TNF-R2) expression on the surface of CD8+ T cells. However, whether T-cell death occurs when these two cell populations interact is unexpected, because reports of apoptosis by TNF-R2 signaling are rare [33].

3.2. Hepatitis Viruses

In chronic hepatitis C virus (HCV) infection, enhanced hepatocyte apoptosis and upregulation of the death receptors and death-inducing ligands have been described [114–116]. Fas expression on PBMCs of HCV-infected patients increases significantly compared with the cells from normal subjects [117]. HCV infection sensitizes human hepatocytes to TRAIL-induced apoptosis in a caspase 9-dependent manner through upregulating DR4 and DR5 [118,119]. Fas- and TRAIL-mediated apoptosis of hepatocytes triggered by viral infection appears to correlate with liver pathology and contributes to fibrogenesis [114,120]. Hepatitis B virus (HBV) replication can also enhance TRAIL-mediated apoptosis in human hepatocytes, in part, by HBV-encoded antigen (HBxAg)-dependent upregulation of TRAIL-R1/DR4 [121].

The pro- and anti-apoptotic roles of HCV proteins are controversial and dependent on the experimental system used [122]. HCV core protein increases the sensitivity of Jurkat T cells to Fas-mediated apoptosis by binding to the cytoplasmic domain of Fas and potentially enhancing the downstream signaling event of Fas-mediated apoptosis [37]. The core protein induces apoptosis in a target T cell expressing Fas, which is mediated by FasL that is upregulated in hepatoblastoma cell line [38]. It also enhances TNF-induced apoptosis by binding to the cytoplasmic domain of TNF-R1 [39]. Additionally, it increases TRAIL-mediated apoptotic cell death in hepatocellular carcinoma cell line, which is dependent on the activation of mitochondria apoptosis signaling pathway [40]. The impact of HBV viral products on death receptor-mediated apoptosis is less clear. HBV X protein (HBX) has been shown to increase DR5 expression through NF-κB pathway and sensitize TRAIL-induced apoptosis in hepatocytes by inhibiting the E3 ubiquitin ligase A20. A20 negatively regulates caspase-8 cleavage and activation through mediating RIP1 polyubiquitination [35,36].

3.3. Herpesviruses

Fas antigen expression significantly increases on PBMCs obtained from varicella-zoster virus (VZV) seropositive donors after culture with VZV antigen. The cultured cells undergo Fas-mediated apoptosis, suggesting a potential role of Fas-mediated apoptosis in the elimination of lymphocytes activated by VZV infection [123]. Another report shows that VZV-induced apoptosis activates caspase-8 in human melanoma cells [124]. Murine cytomegalovirus (MCMV) infection increases Fas expression and Fas-mediated apoptosis, leading to reduced number of hematopoietic progenitor cells and contributing to CMV-induced myelosuppression [125], whereas latent infection of myeloid progenitors by human CMV (HCMV) are refractory to Fas-mediated killing through the cellular IL-10/PEA-15 pathway, and HCMV infection in fibroblasts suppresses Fas expression and protects the cells against Fas-mediated apoptosis through de novo virus-encoded gene expression [126,127]. Epstein–Barr virus (EBV)-infected cells release Fas ligand in exosomal fractions and induce FasL-mediated extrinsic pathway in a number of different cell types including B cells, T cells and epithelial cells [128]. EBV also induces Fas expression in CD4+ T cells and FasL expression in B cells and macrophage, which leads to EBV-stimulated T cells undergoing apoptosis [129]. Both EBV latent membrane protein 1 (LMP1) and protein 2A (LMP2A) sensitize the infected B cells to Fas-mediated apoptosis through the increase of Fas expression, susceptible to elimination by the immune system [41,42].

3.4. Other RNA Viruses

Influenza virus infection activates Fas gene expression and induces apoptosis of infected cells [130–132]. Furthermore, influenza virus infection induces co-expression of Fas and FasL on the surface of infected cells, which causes apoptosis when the infected cells come into contact with each other [133]. Additionally, influenza virus infection increases TRAIL and receptor DR5 expression which plays an important role in the virus clearance by the immune response [134]. Respiratory syncytial virus (RSV) infection in the epithelial cells and other primary airway cells induces extrinsic cell death through an increase of Fas expression and upregulation of TRAIL and its receptors DR4 and DR5 [135,136]. Similarly, reovirus-induced apoptosis is also mediated by the increase of TRAIL release and expression of DR4 and DR5 [137]. Reovirus infection sensitizes different types of cancer cell lines to TRAIL-mediated apoptosis in a caspase 8-dependent manner or through inhibition of NF-κB activation [138,139]. Newcastle disease virus (NDV) infection triggers upregulation of TNF-α and TRAIL which initiate extrinsic apoptosis [140]. Chandipura virus induces neuronal death through the Fas-mediated extrinsic apoptotic pathway [141]. One report shows that dengue virus-induced apoptosis involves in FasL/Fas pathway in vascular endothelial cells [142]. West Nile virus (WNV) infection activates death-receptor-mediated apoptosis in the brains of infected animals through upregulation of caspase activity, which in turn contributes to WNV-induced neuronal injury and pathogenesis [143]. Zika virus (ZIKV) infection of neuronal cells can increase TNF-α expression and activate caspase-3/-7, -8 and -9, which might contribute to ZIKV-induced neuronal cell death and neurotoxicity [144,145]. Both Fas- and TNF-α-mediated cell death signaling play a role in Ebola virus (EBOV)-induced lymphocyte apoptosis, which might contribute to lymphopenia in the infected patients [146–148]. Neurovirulent strain of Sindbis virus infection induces TNF-α-mediated apoptosis in PC-12 cells [149].

3.5. Other DNA Viruses and Retroviruses

Human papillomavirus (HPV) E2 protein induces apoptosis mediated by FasL and TNF-α in HPV-positive and negative cervical cancer cell lines through interacting with c-FLIP and abrogating the apoptosis-inhibitory function of c-FLIP [43]. HPV E7 expression in genital keratinocytes can also sensitize the cells to TNF-mediated apoptosis [44]. Human T-cell leukemia virus-I (HTLV-1) Tax oncoprotein stimulates NF-κB-dependent expression of TRAIL mRNA and induces TRAIL-mediated T cell death [45]. Likewise, lyssavirus, which is a member of the Rhabdoviridae family, induces TRAIL-dependent apoptosis in neuroblastoma cells through the release of a soluble, active form of TRAIL by encoded matrix protein [46].

4. Viral Inhibitors of Death Receptor-Mediated Apoptosis

Elimination of infected cells via death receptor-mediated apoptosis is one of the defense mechanisms against virus infection. Induction of early cell death would severely limit virus production and reduce or eliminate the spread of progeny virus in the host. Thus, many viruses have evolved many different strategies to interfere with death receptor signaling and prevent apoptosis through virally encoded antiapoptotic factors (Table 1), thereby allowing for the production and spread of progeny virus. Some viruses express death receptor orthologs and specifically target death receptors to inhibit apoptosis. The secreted TNF-R2 ortholog of Shope fibroma virus (rabbit poxvirus) can neutralize TNF as a soluble decoy receptor, which is one of the first-described evasion strategies [47,48]. The poxvirus-encoded TNFR ortholog T2 protein and vaccinia virus (VACV)-encoded TNFR ortholog CrmE inhibit TNF-mediated apoptosis of infected cells [49,55]. HCMV also contains a TNFR ortholog encoded by the UL144 gene, but its functional significance remains obscure [150]. Most viral proteins block death receptor-mediated apoptosis mainly through regulation of death receptors or their ligand expression, interaction with apoptotic signaling molecules and interfering with signaling pathways (Figure 1).

4.1. Herpesviruses

Herpesviruses have been most instructive for viral inhibitors of death receptor-mediated apoptosis [151]. Herpes simplex virus-1 (HSV-1) glycoprotein D (gD) exhibits NF-κB-dependent protection against Fas-mediated apoptosis in U937 monocytoid cells, which is associated with decreased levels of caspase-8 activity and upregulation of antiapoptotic proteins [57]. The ribonucleotide reductase R1 subunits of HSV-1 and HSV-2 protect cells against FasL-induced apoptosis by interacting with caspase-8 [58]. The HCMV protein IE2 induces the expression of c-FLIP in human retinal pigment epithelial cells and contributes to protection from Fas- and TRAIL-mediated apoptosis [59], whereas HCMV-encoded viral mitochondria-localized inhibitor of apoptosis (vMIA), a product of the viral *UL37* gene, inhibits Fas-mediated apoptosis at a point downstream of caspase-8 activation and Bid cleavage [60]. HCMV UL36 and MCMV homologous protein M36 inhibit Fas-mediated apoptosis through prevention of caspase-8 activation by binding to pro-caspase-8 [61,62]. MCMV-encoded M45 blocks TNF-induced apoptosis through the binding of M45 to the TNFR adaptor protein RIP1 in a manner that is independent of caspase activation [63]. Additionally, M45 also inhibits TNF-α-dependent necrosis by targeting RIP3 and disrupting RIP1–RIP3 interaction [152].

Like poxvirus molluscum contagiosum virus (MCV)-encoded MC159 protein which is a viral FLICE-inhibitory protein (v-FLIP) with two death effector domains and inhibits both Fas- and TNFR-mediated apoptosis [66], several gamma-herpesviruses including herpesvirus saimiri (HVS), Kaposi sarcoma-associated virus (KSHV), equine herpesvirus 2 (EHV-2) and bovine herpesvirus 4 (BHV-4) also encode the v-FLIP. These v-FLIP proteins protect against apoptosis induced by Fas, TNF-R1, and TRAIL-R through interaction with FADD and prevention of procaspase-8 maturation [64–67]. EBV-encoded small nonpolyadenylated RNA (EBER) protein confers resistance to Fas-mediated apoptosis by blocking protein kinase PKR activity in intestine 407 cells [68,69]. EBV-encoded BHRF1 protein with distant homology to BCL-2 inhibits TNF- and Fas-mediated apoptosis in a cell type-specific manner; the protective mechanism of BHRF1 against apoptosis resembles that of BCL-2 and Bcl-XL as it inhibits activation of cytosolic phospholipase A2 and caspase-3 [70]. However, BHRF1 inhibits TRAIL-induced apoptosis in BJAB cells by functioning downstream of Bid cleavage and upstream of mitochondrial damage [69]. EBV BZLF1 prevents TNF-α activation of target genes and TNF-α-induced apoptosis by downregulating TNFR1 [71]. EBV LMP1 expression confers partial resistance to Fas-mediated apoptosis by reducing caspase activity in BJAB cells [72], and it inhibits TRAIL-mediated apoptosis through activation of PI3K/Akt and expression of c-FLIP in nasopharyngeal carcinoma cells [73]. The murine gammaherpesvirus-68 (MHV68) M11 encodes a BCL-2 ortholog which inhibits Fas- and TNF-α-mediated apoptosis [74].

4.2. Hepatitis Viruses

HCV also encodes several proteins that antagonize host cell death signals. Although HCV core protein sensitizes Jurkat T cells to Fas-mediated apoptosis, it inhibits Fas-mediated apoptosis via NF-κB activation in particular HepG2 cell lines, suggesting its cell type-specific function [75]. The core protein blocks TNF-α-mediated apoptosis through inhibition of caspase-8 activation by sustaining c-FLIP expression and proteolytic cleavage of the death substrate poly (SDP-ribose) polymerase [76,77]. HCV E2 protein activates phosphorylation of IkBα, increases the expression of antiapoptotic BCL-2 family proteins, and confers Raji cells and primary human B lymphocytes protection against Fas-mediated apoptosis [78]. HCV non-structural protein 5A (NS5A) impairs TNF-mediated apoptosis by interfering the association between TRADD and FADD [79]. HBV core protein prevents Fas-mediated apoptosis by regulation of Fas and FasL expression [80].

4.3. Adenoviruses

The E3 region of adenoviruses (ADV) encodes several proteins that modulate death receptors on the cell surface and death receptor-mediated apoptosis. The E3-10.4K/14.5K complex selectively mediates loss of Fas surface expression and blocks Fas-induced apoptosis of virus-infected cells [81], whereas the E3 proteins, 6.7K, 10.4K and 14.5K complex, can induce downregulation of TRAIL-R1 and TRAIL-R2 from the cell surface and block the infected cells from TRAIL-mediated apoptosis [82]. The E3 receptor internalization and degradation (RID) complex prevents apoptotic cell death initiated through dead receptors including TNF-R1, TRAIL-R1, and Fas [83]. Adenovirus type 5 encoded 14.7 kDa inhibits Fas-mediated apoptosis through interaction with FLICE and TNF-mediated apoptosis by inhibiting TNF-R1 internalization and DISC formation [84].

4.4. Human Papillomaviruses

High-risk HPV type 16 (HPV16) and 18 (HPV18) play a pivotal role in the pathophysiology of cervical cancer. Like other viruses, HPV has also developed strategies to block host-mediated apoptosis and regulate the survival of infected cells [153]. Some evidence suggests that the oncoproteins of HPV and E5 can inhibit death receptor signaling pathway by different mechanisms [85,86]. E5 inhibits Fas-induced apoptosis, in part, by decreasing the cell surface expression of the Fas receptor whereas E5 inhibits TRAIL signaling by interfering with the formation of TRAIL DISC and subsequent cleavage of procaspases-8 and -3, as well as of PARP [85]. The E6 oncoprotein of HPV can inhibit TNF-mediated apoptosis through interacting with the death domain of the TNF-R1 and blocking TNF-R1 interaction with TRADD in mouse fibroblasts, human monocytes/histocytes, and osteosarcoma cells [88–90]. The E6 protein can also protect TRAIL-induced apoptosis by facilitating the degradation of FADD and caspase-8 [87]. The E7 oncoprotein of HPV inhibits TNF-mediated apoptosis in keratinocytes by upregulation of antiapoptotic protein c-IAP2 [92]. The mechanism of E7 in delaying Fas-mediated apoptosis and preventing TNF-mediated apoptosis is also involved in the suppression of caspase-8 activation [91].

4.5. Other Viruses

Glycoproteins of EBOV and Marburg virus (MARV) suppress Fas-mediated apoptosis in Hela cells [93]. HIV-1 Nef expression confers resistance against Fas-mediated apoptosis through inhibition of caspase-3 and caspase-8 activation [94], whereas HIV-1 Tat protects Jurkat T cells from TRAIL-mediated apoptosis [95]. HTLV-1 transactivator protein Tax inhibits Fas-mediated apoptosis by induction of c-FLIP through activation of NF-κB [96,97]. HTLV-2 Tax protein also inhibits Fas-mediated apoptosis, but the mechanism remains unclear [98]. Poxviruses encode conserved serine protease inhibitors (serpins) which inhibit caspase-8 activity and Fas- and TNF-mediated apoptosis, such as CrmA protein of cowpox virus, SPI-2 of rabbitpox, vaccinia, variola and ectromelia viruses, and SPI-1 protein of vaccinia virus [50–54,56].

5. Consequence of Death Receptor-Mediated Apoptosis during Viral Infection

For many viruses, induction of apoptosis during lytic infection or at late stages of infection may be an important step for the dissemination of progeny virus to neighboring cells while also evading host immune inflammatory and immune responses. With some viruses, inhibition of apoptosis in virus-infected cells can prevent premature death of the host cell and impair virus production, which enables the establishment of viral latency and facilitates persistent infection, contributing to the avoidance of immune surveillance by the host. Therefore, in certain circumstance, either induction or inhibition of death receptor-mediated apoptosis could assist viral infection and contribute to viral pathogenesis.

For the host, death receptors can be mediators of the innate immune response to viral infection. The murine and human TRAIL promoters contain interferon regulatory elements and can be activated by interferons, and thus TRAIL is one of the earliest genes induced by interferons [154,155]. Many innate immune cells increase TRAIL expression by proinflammation cytokines like interferons that are produced during viral infection. TRAIL-mediated apoptosis thus could play a role in the clearance of virus-infected cells by innate immune cells, especially natural killer (NK) cells. NK cells express the TNF family of cytokines and mediate cytotoxicity through the TRAIL/TRAIL-R signaling and granzyme/perforin mechanisms [155]. TRAIL expression on NK cells can be induced by other cytokines and has been shown to involve in the killing of activated NK cells against virus-infected cells [156]. For instance, IFN-α- or IL26-induced TRAIL expression on NK cells is associated with antiviral cytotoxicity of NK cells and the control of HCV infection in chronic HCV-infected patients [157,158]. Similarly, IFN-α/β-induced modulation of the TRAIL/TRAIL-R system enhances the NK cell-mediated apoptotic killing of murine cells infected with encephalomyocarditis virus [155]. Besides, NK cells can eliminate virus-specific T cells through TRAIL-mediated apoptosis. Such as, NK cells rapidly eliminate HBV-specific T cells which display high-level expression of TRAIL-R2 in patients with chronic hepatitis B and activated CD4+ T cells in the salivary gland during chronic MCMV infection [159,160]. However, some viral proteins can antagonize NK-mediated killing through modulation of TRAIL/TRAIL-R system. HCMV glycoprotein UL141 binds to TRAIL-R2 and thus protects virus-infected cells from TRAIL and TRAIL-dependent NK cell-mediated killing [161,162]. MCMV m166 open reading frame inhibits expression of TRAIL-DR in infected cells and thus thwarts NK-mediated killing [163]. Apart from its important role in NK cell killing activity, TRAIL-mediated apoptosis is also involved in the cytotoxicity of pDCs. Measles virus and influenza virus can induce TRAIL expression on the surface of pDC and enable the cytotoxic killing of pDC against TRAIL-sensitive target cells [164,165]. One study reports that HIV-1 viremia is associated with the upregulation of TRAIL-R1 on activated CD4+ T cells which become susceptible to TRAIL-dependent pDC-mediated killing [166].

In addition to the role in the cytotoxic activity of innate immune NK cells and pDC cells, death receptor-mediated apoptosis plays an important role in the cytotoxic T cell killing during viral infection. It is well demonstrated that some virus-specific cytotoxic T lymphocytes (CTLs) use the FasL/Fas-dependent lytic mechanism to kill virus-infected or bystander cells, such as lymphocytic choriomeningitis virus (LCMV)-infected cell lysis by LCMV-specific CD4+ CTL [167], MHC class I-restricted killing of neurons by LCMV-specific CD8+ T lymphocytes [168], Ag-bearing cell killing and non-Ag-bearing bystander cell killing by HCV-specific CTLs [169,170], and growth inhibition of EBV- or MHV68-infected B cells by virus-specific CTLs [171–173]. In addition, Fas- and TRAIL-mediated apoptosis regulate clearance of influenza A virus (IAV) by IAV-specific CD8+ T cells [174,175]. Conversely, Fas-mediated apoptosis can also cause the elimination of some virus-specific CTLs, such as HIV-, HCV- and EBV-specific CTLs [176–178]. The sensitivity of CTLs to Fas-induced apoptosis is of particular importance for the virus as it impairs the capability of virus-specific CTLs to kill virus-infected cells, thus resulting in the escape of virally infected cells from the CTL response.

Death receptors also mediate apoptosis-independent processes during viral infection. For instance, FasL/Fas system participates in the induction of inflammatory response during virus infection. This has been mainly demonstrated in the context of HSV-2 infection, during which it regulates inflammation in vaginal tissue via the Fas/FasL pathway [179–181]. This content is not within the focus of this review and would not be further discussed here.

6. Concluding Remarks

Death receptor-mediated apoptosis represents a complex and co-evolved mechanism used by the virus and the host, which contributes to viral pathogenesis and host immune surveillance. The infected host cell uses it as part of the antiviral response, whereas the virus appears to balance apoptotic and anti-apoptotic effect to facilitate viral infection. With respect to the potential use of death receptor-mediated apoptosis in the treatment of viral diseases, therapeutic strategies to enhance death receptor-mediated apoptotic clearance of virus-infected cells may be beneficial in some viral infections, whereas in viral infections in which pathogenesis and propagation are enhanced by apoptosis, inhibition of death receptor-mediated apoptosis may be the therapeutic goal. Furthermore, death receptor-mediated apoptosis plays a critical role in the control of virus-infected cells by NK cells, pDCs, and CTLs, which could be the basis for the development of targeted immune control of virus infection. Future studies will need to elucidate in more detail the mechanisms of death receptor-mediated apoptosis by which those immune cells mediate antiviral function. Viral products involved in the induction and suppression of death receptor-mediated apoptosis provide critical insights into cellular apoptotic processes, which could be useful in treating viral diseases. Understanding the mechanism of virally induced death receptor-mediated apoptosis is vital because of its involvement in the pathophysiology of diseases and therapeutic intervention. Given the multifaceted role of death receptor-mediated apoptosis, further preclinical and clinical studies are required in order to determine its specific usage in the treatment of viral diseases.

Acknowledgments: This work was supported by the grants from National Natural Science Foundation of China (81371825), Chinese Academy of Sciences "100 talents" program (2060299) and National Key R & D Program of China (2016YFA0502100).

Conflicts of Interest: The authors have no conflicts of interest.

References

1. Danthi, P. Viruses and the diversity of cell death. *Annu. Rev. Virol.* **2016**, *3*, 533–553. [CrossRef] [PubMed]
2. Teodoro, J.G.; Branton, P.E. Regulation of apoptosis by viral gene products. *J. Virol.* **1997**, *71*, 1739–1746. [PubMed]
3. Galluzzi, L.; Brenner, C.; Morselli, E.; Touat, Z.; Kroemer, G. Viral control of mitochondrial apoptosis. *PLoS Pathog.* **2008**, *4*, e1000018. [CrossRef] [PubMed]
4. Everett, H.; McFadden, G. Apoptosis: An innate immune response to virus infection. *Trends Microbiol.* **1999**, *7*, 160–165. [CrossRef]
5. Shen, Y.; Shenk, T.E. Viruses and apoptosis. *Curr. Opin. Genet. Dev.* **1995**, *5*, 105–111. [CrossRef]
6. Roulston, A.; Marcellus, R.C.; Branton, P.E. Viruses and apoptosis. *Annu. Rev. Microbiol.* **1999**, *53*, 577–628. [CrossRef] [PubMed]
7. Galluzzi, L.; Vitale, I.; Abrams, J.M.; Alnemri, E.S.; Baehrecke, E.H.; Blagosklonny, M.V.; Dawson, T.M.; Dawson, V.L.; El-Deiry, W.S.; Fulda, S.; et al. Molecular definitions of cell death subroutines: Recommendations of the nomenclature committee on cell death 2012. *Cell Death Differ.* **2012**, *19*, 107–120. [CrossRef] [PubMed]
8. Ichim, G.; Tait, S.W. A fate worse than death: Apoptosis as an oncogenic process. *Nat. Rev. Cancer* **2016**, *16*, 539–548. [CrossRef] [PubMed]
9. Elmore, S. Apoptosis: A review of programmed cell death. *Toxicol. Pathol.* **2007**, *35*, 495–516. [CrossRef] [PubMed]
10. Di Pietro, R.; Zauli, G. Emerging non-apoptotic functions of tumor necrosis factor-related apoptosis-inducing ligand (TRAIL)/Apo2L. *J. Cell. Physiol.* **2004**, *201*, 331–340. [CrossRef] [PubMed]
11. Chipuk, J.E.; Bouchier-Hayes, L.; Green, D.R. Mitochondrial outer membrane permeabilization during apoptosis: The innocent bystander scenario. *Cell Death Differ.* **2006**, *13*, 1396–1402. [CrossRef] [PubMed]
12. Tait, S.W.; Green, D.R. Mitochondria and cell death: Outer membrane permeabilization and beyond. *Nat. Rev. Mol. Cell Biol.* **2010**, *11*, 621–632. [CrossRef] [PubMed]
13. Dewson, G.; Kratina, T.; Sim, H.W.; Puthalakath, H.; Adams, J.M.; Colman, P.M.; Kluck, R.M. To trigger apoptosis, Bak exposes its BH3 domain and homodimerizes via BH3: Groove interactions. *Mol. Cell* **2008**, *30*, 369–380. [CrossRef] [PubMed]
14. George, N.M.; Evans, J.J.; Luo, X. A three-helix homo-oligomerization domain containing BH3 and BH1 is responsible for the apoptotic activity of Bax. *Genes Dev.* **2007**, *21*, 1937–1948. [CrossRef] [PubMed]
15. Papoff, G.; Hausler, P.; Eramo, A.; Pagano, M.G.; Di Leve, G.; Signore, A.; Ruberti, G. Identification and characterization of a ligand-independent oligomerization domain in the extracellular region of the CD95 death receptor. *J. Biol. Chem.* **1999**, *274*, 38241–38250. [CrossRef] [PubMed]

16. Siegel, R.M.; Frederiksen, J.K.; Zacharias, D.A.; Chan, F.K.; Johnson, M.; Lynch, D.; Tsien, R.Y.; Lenardo, M.J. Fas preassociation required for apoptosis signaling and dominant inhibition by pathogenic mutations. *Science* **2000**, *288*, 2354–2357. [CrossRef] [PubMed]

17. Yoon, J.H.; Gores, G.J. Death receptor-mediated apoptosis and the liver. *J. Hepatol.* **2002**, *37*, 400–410. [CrossRef]

18. Martin-Villalba, A.; Llorens-Bobadilla, E.; Wollny, D. CD95 in cancer: Tool or target? *Trends Mol. Med.* **2013**, *19*, 329–335. [CrossRef] [PubMed]

19. Peter, M.E.; Hadji, A.; Murmann, A.E.; Brockway, S.; Putzbach, W.; Pattanayak, A.; Ceppi, P. The role of CD95 and CD95 ligand in cancer. *Cell Death Differ.* **2015**, *22*, 549–559. [CrossRef] [PubMed]

20. Shepard, B.D.; Badley, A.D. The biology of TRAIL and the role of TRAIL-based therapeutics in infectious diseases. *Antiinfect. Agents Med. Chem.* **2009**, *8*, 87–101. [CrossRef] [PubMed]

21. Kantari, C.; Walczak, H. Caspase-8 and bid: Caught in the act between death receptors and mitochondria. *Biochim. Biophys. Acta* **2011**, *1813*, 558–563. [CrossRef] [PubMed]

22. Irmler, M.; Thome, M.; Hahne, M.; Schneider, P.; Hofmann, K.; Steiner, V.; Bodmer, J.L.; Schroter, M.; Burns, K.; Mattmann, C.; et al. Inhibition of death receptor signals by cellular FLIP. *Nature* **1997**, *388*, 190–195. [CrossRef] [PubMed]

23. Strasser, A.; Jost, P.J.; Nagata, S. The many roles of FAS receptor signaling in the immune system. *Immunity* **2009**, *30*, 180–192. [CrossRef] [PubMed]

24. Peter, M.E.; Krammer, P.H. The CD95(APO-1/Fas) DISC and beyond. *Cell Death Differ.* **2003**, *10*, 26–35. [CrossRef] [PubMed]

25. Hughes, M.A.; Powley, I.R.; Jukes-Jones, R.; Horn, S.; Feoktistova, M.; Fairall, L.; Schwabe, J.W.; Leverkus, M.; Cain, K.; MacFarlane, M. Co-operative and hierarchical binding of c-FLIP and Caspase-8: A unified Model defines how c-FLIP Isoforms differentially control cell fate. *Mol. Cell* **2016**, *61*, 834–849. [CrossRef] [PubMed]

26. Wajant, H.; Pfizenmaier, K.; Scheurich, P. Tumor necrosis factor signaling. *Cell Death Differ.* **2003**, *10*, 45–65. [CrossRef] [PubMed]

27. Zhang, M.; Li, X.; Pang, X.; Ding, L.; Wood, O.; Clouse, K.; Hewlett, I.; Dayton, A.I. Identification of a potential HIV-induced source of bystander-mediated apoptosis in T cells: Upregulation of trail in primary human macrophages by HIV-1 tat. *J. Biomed. Sci.* **2001**, *8*, 290–296. [CrossRef] [PubMed]

28. Westendorp, M.O.; Frank, R.; Ochsenbauer, C.; Stricker, K.; Dhein, J.; Walczak, H.; Debatin, K.M.; Krammer, P.H. Sensitization of T cells to CD95-mediated apoptosis by HIV-1 Tat and gp120. *Nature* **1995**, *375*, 497–500. [CrossRef] [PubMed]

29. Yang, Y.; Tikhonov, I.; Ruckwardt, T.J.; Djavani, M.; Zapata, J.C.; Pauza, C.D.; Salvato, M.S. Monocytes treated with human immunodeficiency virus Tat kill uninfected CD4(+) cells by a tumor necrosis factor-related apoptosis-induced ligand-mediated mechanism. *J. Virol.* **2003**, *77*, 6700–6708. [CrossRef] [PubMed]

30. Algeciras-Schimnich, A.; Vlahakis, S.R.; Villasis-Keever, A.; Gomez, T.; Heppelmann, C.J.; Bou, G.; Paya, C.V. CCR5 mediates Fas- and caspase-8 dependent apoptosis of both uninfected and HIV infected primary human CD4 T cells. *AIDS* **2002**, *16*, 1467–1478. [CrossRef] [PubMed]

31. Casella, C.R.; Rapaport, E.L.; Finkel, T.H. Vpu increases susceptibility of human immunodeficiency virus type 1-infected cells to fas killing. *J. Virol.* **1999**, *73*, 92–100. [PubMed]

32. Boirivant, M.; Viora, M.; Giordani, L.; Luzzati, A.L.; Pronio, A.M.; Montesani, C.; Pugliese, O. HIV-1 gp120 accelerates Fas-mediated activation-induced human lamina propria T cell apoptosis. *J. Clin. Immunol.* **1998**, *18*, 39–47. [CrossRef] [PubMed]

33. Herbein, G.; Mahlknecht, U.; Batliwalla, F.; Gregersen, P.; Pappas, T.; Butler, J.; O'Brien, W.A.; Verdin, E. Apoptosis of CD8+ T cells is mediated by macrophages through interaction of HIV gp120 with chemokine receptor CXCR4. *Nature* **1998**, *395*, 189–194. [PubMed]

34. Micoli, K.J.; Pan, G.; Wu, Y.; Williams, J.P.; Cook, W.J.; McDonald, J.M. Requirement of calmodulin binding by HIV-1 gp160 for enhanced FAS-mediated apoptosis. *J. Biol. Chem.* **2000**, *275*, 1233–1240. [CrossRef] [PubMed]

35. Zhang, H.; Huang, C.; Wang, Y.; Lu, Z.; Zhuang, N.; Zhao, D.; He, J.; Shi, L. Hepatitis B Virus X protein sensitizes TRAIL-induced Hepatocyte apoptosis by inhibiting the E3 ubiquitin ligase A20. *PLoS ONE* **2015**, *10*, e0127329. [CrossRef] [PubMed]

36. Kong, F.; You, H.; Zhao, J.; Liu, W.; Hu, L.; Luo, W.; Hu, W.; Tang, R.; Zheng, K. The enhanced expression of death receptor 5 (DR5) mediated by HBV X protein through NF-κB pathway is associated with cell apoptosis induced by (TNF-α related apoptosis inducing ligand) TRAIL in hepatoma cells. *Virol. J.* **2015**, *12*, 192. [CrossRef] [PubMed]

37. Hahn, C.S.; Cho, Y.G.; Kang, B.S.; Lester, I.M.; Hahn, Y.S. The HCV core protein acts as a positive regulator of Fas-mediated apoptosis in a human lymphoblastoid T cell line. *Virology* **2000**, *276*, 127–137. [CrossRef] [PubMed]

38. Ruggieri, A.; Murdolo, M.; Rapicetta, M. Induction of FAS ligand expression in a human hepatoblastoma cell line by HCV core protein. *Virus Res.* **2003**, *97*, 103–110. [CrossRef] [PubMed]

39. Zhu, N.; Khoshnan, A.; Schneider, R.; Matsumoto, M.; Dennert, G.; Ware, C.; Lai, M.M. Hepatitis C virus core protein binds to the cytoplasmic domain of tumor necrosis factor (TNF) receptor 1 and enhances TNF-induced apoptosis. *J. Virol.* **1998**, *72*, 3691–3697. [PubMed]

40. Chou, A.H.; Tsai, H.F.; Wu, Y.Y.; Hu, C.Y.; Hwang, L.H.; Hsu, P.I.; Hsu, P.N. Hepatitis C virus core protein modulates TRAIL-mediated apoptosis by enhancing Bid cleavage and activation of mitochondria apoptosis signaling pathway. *J. Immunol.* **2005**, *174*, 2160–2166. [CrossRef] [PubMed]

41. Incrocci, R.; Hussain, S.; Stone, A.; Bieging, K.; Alt, L.A.; Fay, M.J.; Swanson-Mungerson, M. Epstein-barr virus latent membrane protein 2A (LMP2A)-mediated changes in Fas expression and Fas-dependent apoptosis: Role of Lyn/Syk activation. *Cell. Immunol.* **2015**, *297*, 108–119. [CrossRef] [PubMed]

42. Le Clorennec, C.; Youlyouz-Marfak, I.; Adriaenssens, E.; Coll, J.; Bornkamm, G.W.; Feuillard, J. EBV latency III immortalization program sensitizes B cells to induction of CD95-mediated apoptosis via LMP1: Role of NF-κB, STAT1, and p53. *Blood* **2006**, *107*, 2070–2078. [CrossRef] [PubMed]

43. Wang, W.; Fang, Y.; Sima, N.; Li, Y.; Li, W.; Li, L.; Han, L.; Liao, S.; Han, Z.; Gao, Q.; et al. Triggering of death receptor apoptotic signaling by human papillomavirus 16 E2 protein in cervical cancer cell lines is mediated by interaction with c-FLIP. *Apoptosis* **2011**, *16*, 55–66. [CrossRef] [PubMed]

44. Stoppler, H.; Stoppler, M.C.; Johnson, E.; Simbulan-Rosenthal, C.M.; Smulson, M.E.; Iyer, S.; Rosenthal, D.S.; Schlegel, R. The E7 protein of human papillomavirus type 16 sensitizes primary human keratinocytes to apoptosis. *Oncogene* **1998**, *17*, 1207–1214. [CrossRef] [PubMed]

45. Rivera-Walsh, I.; Waterfield, M.; Xiao, G.; Fong, A.; Sun, S.C. NF-κB signaling pathway governs TRAIL gene expression and human T-cell leukemia virus-I Tax-induced T-cell death. *J. Biol. Chem.* **2001**, *276*, 40385–40388. [CrossRef] [PubMed]

46. Kassis, R.; Larrous, F.; Estaquier, J.; Bourhy, H. Lyssavirus matrix protein induces apoptosis by a TRAIL-dependent mechanism involving caspase-8 activation. *J. Virol.* **2004**, *78*, 6543–6555. [CrossRef] [PubMed]

47. Benedict, C.A.; Norris, P.S.; Ware, C.F. To kill or be killed: Viral evasion of apoptosis. *Nat. Immunol.* **2002**, *3*, 1013–1018. [CrossRef] [PubMed]

48. Smith, C.A.; Davis, T.; Anderson, D.; Solam, L.; Beckmann, M.P.; Jerzy, R.; Dower, S.K.; Cosman, D.; Goodwin, R.G. A receptor for tumor necrosis factor defines an unusual family of cellular and viral proteins. *Science* **1990**, *248*, 1019–1023. [CrossRef] [PubMed]

49. Reading, P.C.; Khanna, A.; Smith, G.L. Vaccinia virus CrmE encodes a soluble and cell surface tumor necrosis factor receptor that contributes to virus virulence. *Virology* **2002**, *292*, 285–298. [CrossRef] [PubMed]

50. Macen, J.L.; Garner, R.S.; Musy, P.Y.; Brooks, M.A.; Turner, P.C.; Moyer, R.W.; McFadden, G.; Bleackley, R.C. Differential inhibition of the Fas- and granule-mediated cytolysis pathways by the orthopoxvirus cytokine response modifier A/SPI-2 and SPI-1 protein. *Proc. Natl. Acad. Sci. USA* **1996**, *93*, 9108–9113. [CrossRef] [PubMed]

51. Turner, S.J.; Silke, J.; Kenshole, B.; Ruby, J. Characterization of the ectromelia virus serpin, SPI-2. *J. Gen. Virol.* **2000**, *81*, 2425–2430. [CrossRef] [PubMed]

52. Dobbelstein, M.; Shenk, T. Protection against apoptosis by the vaccinia virus SPI-2 (B13R) gene product. *J. Virol.* **1996**, *70*, 6479–6485. [PubMed]

53. Veyer, D.L.; Carrara, G.; Maluquer de Motes, C.; Smith, G.L. Vaccinia virus evasion of regulated cell death. *Immunol. Lett.* **2017**, *186*, 68–80. [CrossRef] [PubMed]

54. Brooks, M.A.; Ali, A.N.; Turner, P.C.; Moyer, R.W. A rabbitpox virus serpin gene controls host range by inhibiting apoptosis in restrictive cells. *J. Virol.* **1995**, *69*, 7688–7698. [PubMed]

55. Schreiber, M.; Sedger, L.; McFadden, G. Distinct domains of M-T2, the myxoma virus tumor necrosis factor (TNF) receptor homolog, mediate extracellular TNF binding and intracellular apoptosis inhibition. *J. Virol.* **1997**, *71*, 2171–2181. [PubMed]

56. Srinivasula, S.M.; Ahmad, M.; Fernandes-Alnemri, T.; Litwack, G.; Alnemri, E.S. Molecular ordering of the Fas-apoptotic pathway: The Fas/APO-1 protease Mch5 is a CrmA-inhibitable protease that activates multiple Ced-3/ICE-like cysteine proteases. *Proc. Natl. Acad. Sci. USA* **1996**, *93*, 14486–14491. [CrossRef] [PubMed]

57. Medici, M.A.; Sciortino, M.T.; Perri, D.; Amici, C.; Avitabile, E.; Ciotti, M.; Balestrieri, E.; De Smaele, E.; Franzoso, G.; Mastino, A. Protection by herpes simplex virus glycoprotein D against Fas-mediated apoptosis - Role of nuclear factor κB. *J. Biol. Chem.* **2003**, *278*, 36059–36067. [CrossRef] [PubMed]

58. Dufour, F.; Sasseville, A.M.; Chabaud, S.; Massie, B.; Siegel, R.M.; Langelier, Y. The ribonucleotide reductase R1 subunits of herpes simplex virus types 1 and 2 protect cells against TNFα- and FasL-induced apoptosis by interacting with caspase-8. *Apoptosis* **2011**, *16*, 256–271. [CrossRef] [PubMed]

59. Chiou, S.H.; Yang, Y.P.; Lin, J.C.; Hsu, C.H.; Jhang, H.C.; Yang, Y.T.; Lee, C.H.; Ho, L.L.; Hsu, W.M.; Ku, H.H.; et al. The immediate early 2 protein of human cytomegalovirus (HCMV) mediates the apoptotic control in HCMV retinitis through up-regulation of the cellular FLICE-inhibitory protein expression. *J. Immunol.* **2006**, *177*, 6199–6206. [CrossRef] [PubMed]

60. Goldmacher, V.S.; Bartle, L.M.; Skaletskaya, A.; Dionne, C.A.; Kedersha, N.L.; Vater, C.A.; Han, J.W.; Lutz, R.J.; Watanabe, S.; Cahir McFarland, E.D.; et al. A cytomegalovirus-encoded mitochondria-localized inhibitor of apoptosis structurally unrelated to BCL-2. *Proc. Natl. Acad. Sci. USA* **1999**, *96*, 12536–12541. [CrossRef] [PubMed]

61. Skaletskaya, A.; Bartle, L.M.; Chittenden, T.; McCormick, A.L.; Mocarski, E.S.; Goldmacher, V.S. A cytomegalovirus-encoded inhibitor of apoptosis that suppresses caspase-8 activation. *Proc. Natl. Acad. Sci. USA* **2001**, *98*, 7829–7834. [CrossRef] [PubMed]

62. McCormick, A.L.; Skaletskaya, A.; Barry, P.A.; Mocarski, E.S.; Goldmacher, V.S. Differential function and expression of the viral inhibitor of caspase 8-induced apoptosis (vICA) and the viral mitochondria-localized inhibitor of apoptosis (vMIA) cell death suppressors conserved in primate and rodent cytomegaloviruses. *Virology* **2003**, *316*, 221–233. [CrossRef] [PubMed]

63. Mack, C.; Sickmann, A.; Lembo, D.; Brune, W. Inhibition of proinflammatory and innate immune signaling pathways by a cytomegalovirus RIP1-interacting protein. *Proc. Natl. Acad. Sci. USA* **2008**, *105*, 3094–3099. [CrossRef] [PubMed]

64. Belanger, C.; Gravel, A.; Tomoiu, A.; Janelle, M.E.; Gosselin, J.; Tremblay, M.J.; Flamand, L. Human herpesvirus 8 viral FLICE-inhibitory protein inhibits Fas-mediated apoptosis through binding and prevention of procaspase-8 maturation. *J. Hum. Virol.* **2001**, *4*, 62–73. [PubMed]

65. Thome, M.; Schneider, P.; Hofmann, K.; Fickenscher, H.; Meinl, E.; Neipel, F.; Mattmann, C.; Burns, K.; Bodmer, J.L.; Schroter, M.; et al. Viral FLICE-inhibitory proteins (FLIPs) prevent apoptosis induced by death receptors. *Nature* **1997**, *386*, 517–521. [CrossRef] [PubMed]

66. Bertin, J.; Armstrong, R.C.; Ottilie, S.; Martin, D.A.; Wang, Y.; Banks, S.; Wang, G.H.; Senkevich, T.G.; Alnemri, E.S.; Moss, B.; et al. Death effector domain-containing herpesvirus and poxvirus proteins inhibit both Fas- and TNFR1-induced apoptosis. *Proc. Natl. Acad. Sci. USA* **1997**, *94*, 1172–1176. [CrossRef] [PubMed]

67. Thurau, M.; Everett, H.; Tapernoux, M.; Tschopp, J.; Thome, M. The TRAF3-binding site of human molluscipox virus FLIP molecule MC159 is critical for its capacity to inhibit Fas-induced apoptosis. *Cell Death Differ.* **2006**, *13*, 1577–1585. [CrossRef] [PubMed]

68. Nanbo, A.; Yoshiyama, H.; Takada, K. Epstein-Barr virus-encoded poly(A)- RNA confers resistance to apoptosis mediated through Fas by blocking the PKR pathway in human epithelial intestine 407 cells. *J. Virol.* **2005**, *79*, 12280–12285. [CrossRef] [PubMed]

69. Kawanishi, M. Epstein-Barr virus BHRF1 protein protects intestine 407 epithelial cells from apoptosis induced by tumor necrosis factor α and anti-Fas antibody. *J. Virol.* **1997**, *71*, 3319–3322. [PubMed]

70. Foghsgaard, L.; Jaattela, M. The ability of BHRF1 to inhibit apoptosis is dependent on stimulus and cell type. *J. Virol.* **1997**, *71*, 7509–7517. [PubMed]

71. Morrison, T.E.; Mauser, A.; Klingelhutz, A.; Kenney, S.C. Epstein-Barr virus immediate-early protein BZLF1 inhibits tumor necrosis factor α-induced signaling and apoptosis by downregulating tumor necrosis factor receptor 1. *J. Virol.* **2004**, *78*, 544–549. [CrossRef] [PubMed]

72. Snow, A.L.; Lambert, S.L.; Natkunam, Y.; Esquivel, C.O.; Krams, S.M.; Martinez, O.M. EBV can protect latently infected B cell lymphomas from death receptor-induced apoptosis. *J. Immunol.* **2006**, *177*, 3283–3293. [CrossRef] [PubMed]

73. Li, S.S.; Yang, S.; Wang, S.; Yang, X.M.; Tang, Q.L.; Wang, S.H. Latent membrane protein 1 mediates the resistance of nasopharyngeal carcinoma cells to TRAIL-induced apoptosis by activation of the PI3K/Akt signaling pathway. *Oncol. Rep.* **2011**, *26*, 1573–1579. [PubMed]

74. Wang, G.H.; Garvey, T.L.; Cohen, J.I. The murine gammaherpesvirus-68 M11 protein inhibits Fas- and TNF-induced apoptosis. *J. Gen. Virol.* **1999**, *80 Pt 10*, 2737–2740. [CrossRef] [PubMed]

75. Marusawa, H.; Hijikata, M.; Chiba, T.; Shimotohno, K. Hepatitis C virus core protein inhibits Fas- and tumor necrosis factor α-mediated apoptosis via NF-κB activation. *J. Virol.* **1999**, *73*, 4713–4720. [PubMed]

76. Ray, R.B.; Meyer, K.; Steele, R.; Shrivastava, A.; Aggarwal, B.B.; Ray, R. Inhibition of tumor necrosis factor (TNF-α)-mediated apoptosis by hepatitis C virus core protein. *J. Biol. Chem.* **1998**, *273*, 2256–2259. [CrossRef] [PubMed]

77. Kim, H.; Ray, R. Evasion of TNF-α-mediated apoptosis by hepatitis C virus. *Methods Mol. Biol.* **2014**, *1155*, 125–132. [PubMed]

78. Chen, Z.H.; Zhu, Y.Z.; Ren, Y.L.; Tong, Y.M.; Hua, X.; Zhu, F.H.; Huang, L.B.; Liu, Y.; Luo, Y.; Lu, W.; et al. Hepatitis C virus Protects human B lymphocytes from Fas-mediated apoptosis via E2-CD81 engagement. *PLoS ONE* **2011**, *6*, e18933. [CrossRef] [PubMed]

79. Majumder, M.; Ghosh, A.K.; Steele, R.; Zhou, X.Y.; Phillips, N.J.; Ray, R.; Ray, R.B. Hepatitis C virus NS5A protein impairs TNF-mediated hepatic apoptosis, but not by an anti-FAS antibody, in transgenic mice. *Virology* **2002**, *294*, 94–105. [CrossRef] [PubMed]

80. Liu, W.; Lin, Y.T.; Yan, X.L.; Ding, Y.L.; Wu, Y.L.; Chen, W.N.; Lin, X. Hepatitis B virus core protein inhibits Fas-mediated apoptosis of hepatoma cells via regulation of mFas/FasL and sFas expression. *FASEB J.* **2015**, *29*, 1113–1123. [CrossRef] [PubMed]

81. Shisler, J.; Yang, C.; Walter, B.; Ware, C.F.; Gooding, L.R. The adenovirus E3-10.4K/14.5K complex mediates loss of cell surface Fas (CD95) and resistance to Fas-induced apoptosis. *J. Virol.* **1997**, *71*, 8299–8306. [PubMed]

82. Benedict, C.A.; Norris, P.S.; Prigozy, T.I.; Bodmer, J.L.; Mahr, J.A.; Garnett, C.T.; Martinon, F.; Tschopp, J.; Gooding, L.R.; Ware, C.F. Three adenovirus E3 proteins cooperate to evade apoptosis by tumor necrosis factor-related apoptosis-inducing ligand receptor-1 and -2. *J. Biol. Chem.* **2001**, *276*, 3270–3278. [CrossRef] [PubMed]

83. McNees, A.L.; Garnett, C.T.; Gooding, L.R. The adenovirus E3 RID complex protects some cultured human T and B lymphocytes from Fas-induced apoptosis. *J. Virol.* **2002**, *76*, 9716–9723. [CrossRef] [PubMed]

84. Schneider-Brachert, W.; Tchikov, V.; Merkel, O.; Jakob, M.; Hallas, C.; Kruse, M.L.; Groitl, P.; Lehn, A.; Hildt, E.; Held-Feindt, J.; et al. Inhibition of TNF receptor 1 internalization by adenovirus 14.7K as a novel immune escape mechanism. *J. Clin. Investig.* **2006**, *116*, 2901–2913. [CrossRef] [PubMed]

85. Kabsch, K.; Alonso, A. The human papillomavirus type 16 E5 protein impairs TRAIL- and FasL-mediated apoptosis in HaCaT cells by different mechanisms. *J. Virol.* **2002**, *76*, 12162–12172. [CrossRef] [PubMed]

86. Lagunas-Martinez, A.; Madrid-Marina, V.; Gariglio, P. Modulation of apoptosis by early human papillomavirus proteins in cervical cancer. *Biochim. Biophys. Acta* **2010**, *1805*, 6–16. [CrossRef] [PubMed]

87. Garnett, T.O.; Filippova, M.; Duerksen-Hughes, P.J. Accelerated degradation of FADD and procaspase 8 in cells expressing human papilloma virus 16 E6 impairs TRAIL-mediated apoptosis. *Cell Death Differ.* **2006**, *13*, 1915–1926. [CrossRef] [PubMed]

88. Filippova, M.; Song, H.; Connolly, J.L.; Dermody, T.S.; Duerksen-Hughes, P.J. The human papillomavirus 16 E6 protein binds to tumor necrosis factor (TNF) R1 and protects cells from TNF-induced apoptosis. *J. Biol. Chem.* **2002**, *277*, 21730–21739. [CrossRef] [PubMed]

89. Duerksen-Hughes, P.J.; Yang, J.; Schwartz, S.B. HPV 16 E6 blocks TNF-mediated apoptosis in mouse fibroblast LM cells. *Virology* **1999**, *264*, 55–65. [CrossRef] [PubMed]

90. Filippova, M.; Filippov, V.A.; Kagoda, M.; Garnett, T.; Fodor, N.; Duerksen-Hughes, P.J. Complexes of human papillomavirus type 16 E6 proteins form pseudo-death-inducing signaling complex structures during tumor necrosis factor-mediated apoptosis. *J. Virol.* **2009**, *83*, 210–227. [CrossRef] [PubMed]

91. Thompson, D.A.; Zacny, V.; Belinsky, G.S.; Classon, M.; Jones, D.L.; Schlegel, R.; Munger, K. The HPV E7 oncoprotein inhibits tumor necrosis factor α-mediated apoptosis in normal human fibroblasts. *Oncogene* **2001**, *20*, 3629–3640. [CrossRef] [PubMed]

92. Yuan, H.; Fu, F.; Zhuo, J.; Wang, W.; Nishitani, J.; An, D.S.; Chen, I.S.; Liu, X. Human papillomavirus type 16 E6 and E7 oncoproteins upregulate c-IAP2 gene expression and confer resistance to apoptosis. *Oncogene* **2005**, *24*, 5069–5078. [CrossRef] [PubMed]

93. Noyori, O.; Nakayama, E.; Maruyama, J.; Yoshida, R.; Takada, A. Suppression of Fas-mediated apoptosis via steric shielding by filovirus glycoproteins. *Biochem. Biophys. Res. Commun.* **2013**, *441*, 994–998. [CrossRef] [PubMed]

94. Yoon, K.; Jeong, J.G.; Kim, S. Stable expression of human immunodeficiency virus type 1 Nef confers resistance against Fas-mediated apoptosis. *AIDS Res. Hum. Retrov.* **2001**, *17*, 99–104. [CrossRef] [PubMed]

95. Gibellini, D.; Re, M.C.; Ponti, C.; Maldini, C.; Celeghini, C.; Cappellini, A.; La Placa, M.; Zauli, G. HIV-1 Tat protects CD4+ Jurkat T lymphoblastoid cells from apoptosis mediated by TNF-related apoptosis-inducing ligand. *Cell. Immunol.* **2001**, *207*, 89–99. [CrossRef] [PubMed]

96. Krueger, A.; Fas, S.C.; Giaisi, M.; Bleumink, M.; Merling, A.; Stumpf, C.; Baumann, S.; Holtkotte, D.; Bosch, V.; Krammer, P.H.; et al. HTLV-1 Tax protects against CD95-mediated apoptosis by induction of the cellular FLICE-inhibitory protein (c-FLIP). *Blood* **2006**, *107*, 3933–3939. [CrossRef] [PubMed]

97. Okamoto, K.; Fujisawa, J.; Reth, M.; Yonehara, S. Human T-cell leukemia virus type-I oncoprotein Tax inhibits Fas-mediated apoptosis by inducing cellular FLIP through activation of NF-κB. *Genes Cells* **2006**, *11*, 177–191. [CrossRef] [PubMed]

98. Zehender, G.; Varchetta, S.; de Maddalena, C.; Colasante, C.; Riva, A.; Meroni, L.; Moroni, M.; Galli, M. Resistance to Fas-mediated apoptosis of human T-cell lines expressing human T-lymphotropic virus type-2 (HTLV-2) Tax protein. *Virology* **2001**, *281*, 43–50. [CrossRef] [PubMed]

99. Aries, S.P.; Schaaf, B.; Muller, C.; Dennin, R.H.; Dalhoff, K. Fas (CD95) expression on CD4+ T cells from HIV-infected patients increases with disease progression. *J. Mol. Med.* **1995**, *73*, 591–593. [CrossRef] [PubMed]

100. Sloand, E.M.; Young, N.S.; Kumar, P.; Weichold, F.F.; Sato, T.; Maciejewski, J.P. Role of Fas ligand and receptor in the mechanism of T-cell depletion in acquired immunodeficiency syndrome: Effect on CD4+ lymphocyte depletion and human immunodeficiency virus replication. *Blood* **1997**, *89*, 1357–1363. [PubMed]

101. Baumler, C.B.; Bohler, T.; Herr, I.; Benner, A.; Krammer, P.H.; Debatin, K.M. Activation of the CD95 (APO-1/Fas) system in T cells from human immunodeficiency virus type-1-infected children. *Blood* **1996**, *88*, 1741–1746. [PubMed]

102. McCloskey, T.W.; Oyaizu, N.; Kaplan, M.; Pahwa, S. Expression of the Fas antigen in patients infected with human immunodeficiency virus. *Cytometry* **1995**, *22*, 111–114. [CrossRef] [PubMed]

103. Gehri, R.; Hahn, S.; Rothen, M.; Steuerwald, M.; Nuesch, R.; Erb, P. The Fas receptor in HIV infection: Expression on peripheral blood lymphocytes and role in the depletion of T cells. *AIDS* **1996**, *10*, 9–16. [CrossRef] [PubMed]

104. Boudet, F.; Lecoeur, H.; Gougeon, M.L. Apoptosis associated with ex vivo down-regulation of BCL-2 and up-regulation of Fas in potential cytotoxic CD8+ T lymphocytes during HIV infection. *J. Immunol.* **1996**, *156*, 2282–2293. [PubMed]

105. Katsikis, P.D.; Wunderlich, E.S.; Smith, C.A.; Herzenberg, L.A.; Herzenberg, L.A. Fas antigen stimulation induces marked apoptosis of T lymphocytes in human immunodeficiency virus-infected individuals. *J. Exp. Med.* **1995**, *181*, 2029–2036. [CrossRef] [PubMed]

106. Badley, A.D.; McElhinny, J.A.; Leibson, P.J.; Lynch, D.H.; Alderson, M.R.; Paya, C.V. Upregulation of Fas ligand expression by human immunodeficiency virus in human macrophages mediates apoptosis of uninfected T lymphocytes. *J. Virol.* **1996**, *70*, 199–206. [PubMed]

107. Herbeuval, J.P.; Grivel, J.C.; Boasso, A.; Hardy, A.W.; Chougnet, C.; Dolan, M.J.; Yagita, H.; Lifson, J.D.; Shearer, G.M. CD4+ T-cell death induced by infectious and noninfectious HIV-1: Role of type 1 interferon-dependent, TRAIL/DR5-mediated apoptosis. *Blood* **2005**, *106*, 3524–3531. [CrossRef] [PubMed]

108. Herbeuval, J.P.; Hardy, A.W.; Boasso, A.; Anderson, S.A.; Dolan, M.J.; Dy, M.; Shearer, G.M. Regulation of TNF-related apoptosis-inducing ligand on primary CD4+ T cells by HIV-1: Role of type I IFN-producing plasmacytoid dendritic cells. *Proc. Natl. Acad. Sci. USA* **2005**, *102*, 13974–13979. [CrossRef] [PubMed]

109. Miura, Y.; Koyanagi, Y.; Mizusawa, H. TNF-related apoptosis-inducing ligand (TRAIL) induces neuronal apoptosis in HIV-encephalopathy. *J. Med. Dent. Sci.* **2003**, *50*, 17–25. [PubMed]

110. Miura, Y.; Misawa, N.; Maeda, N.; Inagaki, Y.; Tanaka, Y.; Ito, M.; Kayagaki, N.; Yamamoto, N.; Yagita, H.; Mizusawa, H.; et al. Critical contribution of tumor necrosis factor-related apoptosis-inducing ligand (TRAIL) to apoptosis of human CD4+ T cells in HIV-1-infected hu-PBL-NOD-SCID mice. *J. Exp. Med.* **2001**, *193*, 651–660. [CrossRef] [PubMed]

111. Kim, N.; Dabrowska, A.; Jenner, R.G.; Aldovini, A. Human and simian immunodeficiency virus-mediated upregulation of the apoptotic factor TRAIL occurs in antigen-presenting cells from AIDS-susceptible but not from AIDS-resistant species. *J. Virol.* **2007**, *81*, 7584–7597. [CrossRef] [PubMed]

112. Huang, Y.; Erdmann, N.; Peng, H.; Herek, S.; Davis, J.S.; Luo, X.; Ikezu, T.; Zheng, J. TRAIL-mediated apoptosis in HIV-1-infected macrophages is dependent on the inhibition of Akt-1 phosphorylation. *J. Immunol.* **2006**, *177*, 2304–2313. [CrossRef] [PubMed]

113. Van Grevenynghe, J.; Cubas, R.A.; Noto, A.; DaFonseca, S.; He, Z.; Peretz, Y.; Filali-Mouhim, A.; Dupuy, F.P.; Procopio, F.A.; Chomont, N.; et al. Loss of memory B cells during chronic HIV infection is driven by Foxo3a- and TRAIL-mediated apoptosis. *J. Clin. Investig.* **2011**, *121*, 3877–3888. [CrossRef] [PubMed]

114. Mundt, B.; Kuhnel, F.; Zender, L.; Paul, Y.; Tillmann, H.; Trautwein, C.; Manns, M.P.; Kubicka, S. Involvement of TRAIL and its receptors in viral hepatitis. *FASEB J.* **2003**, *17*, 94–96. [CrossRef] [PubMed]

115. Calabrese, F.; Pontisso, P.; Pettenazzo, E.; Benvegnu, L.; Vario, A.; Chemello, L.; Alberti, A.; Valente, M. Liver cell apoptosis in chronic hepatitis C correlates with histological but not biochemical activity or serum HCV-RNA levels. *Hepatology* **2000**, *31*, 1153–1159. [CrossRef] [PubMed]

116. Mita, E.; Hayashi, N.; Iio, S.; Takehara, T.; Hijioka, T.; Kasahara, A.; Fusamoto, H.; Kamada, T. Role of Fas ligand in apoptosis induced by hepatitis C virus infection. *Biochem. Biophys. Res. Commun.* **1994**, *204*, 468–474. [CrossRef] [PubMed]

117. Taya, N.; Torimoto, Y.; Shindo, M.; Hirai, K.; Hasebe, C.; Kohgo, Y. Fas-mediated apoptosis of peripheral blood mononuclear cells in patients with hepatitis C. *Br. J. Haematol.* **2000**, *110*, 89–97. [CrossRef] [PubMed]

118. Lan, L.; Gorke, S.; Rau, S.J.; Zeisel, M.B.; Hildt, E.; Himmelsbach, K.; Carvajal-Yepes, M.; Huber, R.; Wakita, T.; Schmitt-Graeff, A.; et al. Hepatitis C virus infection sensitizes human hepatocytes to TRAIL-induced apoptosis in a caspase 9-dependent manner. *J. Immunol.* **2008**, *181*, 4926–4935. [CrossRef] [PubMed]

119. Deng, Z.; Yan, H.; Hu, J.; Zhang, S.; Peng, P.; Liu, Q.; Guo, D. Hepatitis C virus sensitizes host cells to TRAIL-induced apoptosis by up-regulating DR4 and DR5 via a MEK1-dependent pathway. *PLoS ONE* **2012**, *7*, e37700. [CrossRef] [PubMed]

120. Silberstein, E.; Ulitzky, L.; Lima, L.A.; Cehan, N.; Teixeira-Carvalho, A.; Roingeard, P.; Taylor, D.R. HCV-mediated Apoptosis of hepatocytes in culture and viral pathogenesis. *PLoS ONE* **2016**, *11*, e0155708. [CrossRef] [PubMed]

121. Janssen, H.L.; Higuchi, H.; Abdulkarim, A.; Gores, G.J. Hepatitis B virus enhances tumor necrosis factor-related apoptosis-inducing ligand (TRAIL) cytotoxicity by increasing TRAIL-R1/death receptor 4 expression. *J. Hepatol.* **2003**, *39*, 414–420. [CrossRef]

122. Fischer, R.; Baumert, T.; Blum, H.E. Hepatitis C virus infection and apoptosis. *World J. Gastroenterol.* **2007**, *13*, 4865–4872. [CrossRef] [PubMed]

123. Ito, M.; Watanabe, M.; Ihara, T.; Kamiya, H.; Sakurai, M. Fas antigen and BCL-2 expression on lymphocytes cultured with cytomegalovirus and varicella-zoster virus antigen. *Cell. Immunol.* **1995**, *160*, 173–177. [CrossRef]

124. Brazeau, E.; Mahalingam, R.; Gilden, D.; Wellish, M.; Kaufer, B.B.; Osterrieder, N.; Pugazhenthi, S. Varicella-zoster virus-induced apoptosis in MeWo cells is accompanied by down-regulation of BCL-2 expression. *J. Neurovirol.* **2010**, *16*, 133–140. [CrossRef] [PubMed]

125. Mori, T.; Ando, K.; Tanaka, K.; Ikeda, Y.; Koga, Y. Fas-mediated apoptosis of the hematopoietic progenitor cells in mice infected with murine cytomegalovirus. *Blood* **1997**, *89*, 3565–3573. [PubMed]

126. Poole, E.; Lau, J.C.; Sinclair, J. Latent infection of myeloid progenitors by human cytomegalovirus protects cells from FAS-mediated apoptosis through the cellular IL-10/PEA-15 pathway. *J. Gen. Virol.* **2015**, *96*, 2355–2359. [CrossRef] [PubMed]

127. Seirafian, S.; Prod'homme, V.; Sugrue, D.; Davies, J.; Fielding, C.; Tomasec, P.; Wilkinson, G.W. Human cytomegalovirus suppresses Fas expression and function. *J. Gen. Virol.* **2014**, *95*, 933–939. [CrossRef] [PubMed]

128. Ahmed, W.; Philip, P.S.; Attoub, S.; Khan, G. Epstein-Barr virus-infected cells release Fas ligand in exosomal fractions and induce apoptosis in recipient cells via the extrinsic pathway. *J. Gen. Virol.* **2015**, *96*, 3646–3659. [CrossRef] [PubMed]

129. Tanner, J.E.; Alfieri, C. Epstein-barr virus induces Fas (CD95) in T cells and Fas ligand in B cells leading to T-cell apoptosis. *Blood* **1999**, *94*, 3439–3447. [PubMed]

130. Wada, N.; Matsumura, M.; Ohba, Y.; Kobayashi, N.; Takizawa, T.; Nakanishi, Y. Transcription stimulation of the Fas-encoding gene by nuclear factor for interleukin-6 expression upon influenza virus infection. *J. Biol. Chem.* **1995**, *270*, 18007–18012. [CrossRef] [PubMed]

131. Takizawa, T.; Matsukawa, S.; Higuchi, Y.; Nakamura, S.; Nakanishi, Y.; Fukuda, R. Induction of programmed cell death (apoptosis) by influenza virus infection in tissue culture cells. *J. Gen. Virol.* **1993**, *74*, 2347–2355. [CrossRef] [PubMed]

132. Takizawa, T.; Fukuda, R.; Miyawaki, T.; Ohashi, K.; Nakanishi, Y. Activation of the apoptotic Fas antigen-encoding gene upon influenza virus infection involving spontaneously produced β-interferon. *Virology* **1995**, *209*, 288–296. [CrossRef] [PubMed]

133. Fujimoto, I.; Takizawa, T.; Ohba, Y.; Nakanishi, Y. Co-expression of Fas and Fas-ligand on the surface of influenza virus-infected cells. *Cell Death Differ.* **1998**, *5*, 426–431. [CrossRef] [PubMed]

134. Ishikawa, E.; Nakazawa, M.; Yoshinari, M.; Minami, M. Role of tumor necrosis factor-related apoptosis-inducing ligand in immune response to influenza virus infection in mice. *J. Virol.* **2005**, *79*, 7658–7663. [CrossRef] [PubMed]

135. O'Donnell, D.R.; Milligan, L.; Stark, J.M. Induction of CD95 (Fas) and apoptosis in respiratory epithelial cell cultures following respiratory syncytial virus infection. *Virology* **1999**, *257*, 198–207. [CrossRef] [PubMed]

136. Kotelkin, A.; Prikhod'ko, E.A.; Cohen, J.I.; Collins, P.L.; Bukreyev, A. Respiratory syncytial virus infection sensitizes cells to apoptosis mediated by tumor necrosis factor-related apoptosis-inducing ligand. *J. Virol.* **2003**, *77*, 9156–9172. [CrossRef] [PubMed]

137. Clarke, P.; Meintzer, S.M.; Gibson, S.; Widmann, C.; Garrington, T.P.; Johnson, G.L.; Tyler, K.L. Reovirus-induced apoptosis is mediated by TRAIL. *J. Virol.* **2000**, *74*, 8135–8139. [CrossRef] [PubMed]

138. Clarke, P.; Meintzer, S.M.; Moffitt, L.A.; Tyler, K.L. Two distinct phases of virus-induced nuclear factor κB regulation enhance tumor necrosis factor-related apoptosis-inducing ligand-mediated apoptosis in virus-infected cells. *J. Biol. Chem.* **2003**, *278*, 18092–18100. [CrossRef] [PubMed]

139. Clarke, P.; Meintzer, S.M.; Spalding, A.C.; Johnson, G.L.; Tyler, K.L. Caspase 8-dependent sensitization of cancer cells to TRAIL-induced apoptosis following reovirus-infection. *Oncogene* **2001**, *20*, 6910–6919. [CrossRef] [PubMed]

140. Liao, Y.; Wang, H.X.; Mao, X.; Fang, H.; Wang, H.; Li, Y.; Sun, Y.; Meng, C.; Tan, L.; Song, C.; et al. RIP1 is a central signaling protein in regulation of TNF-α/TRAIL mediated apoptosis and necroptosis during Newcastle disease virus infection. *Oncotarget* **2017**, *8*, 43201–43217. [PubMed]

141. Ghosh, S.; Dutta, K.; Basu, A. Chandipura virus induces neuronal death through Fas-mediated extrinsic apoptotic pathway. *J. Virol.* **2013**, *87*, 12398–12406. [CrossRef] [PubMed]

142. Liao, H.; Xu, J.; Huang, J. FasL/Fas pathway is involved in dengue virus induced apoptosis of the vascular endothelial cells. *J. Med. Virol.* **2010**, *82*, 1392–1399. [CrossRef] [PubMed]

143. Clarke, P.; Leser, J.S.; Quick, E.D.; Dionne, K.R.; Beckham, J.D.; Tyler, K.L. Death receptor-mediated apoptotic signaling is activated in the brain following infection with West Nile virus in the absence of a peripheral immune response. *J. Virol.* **2014**, *88*, 1080–1089. [CrossRef] [PubMed]

144. Olmo, I.G.; Carvalho, T.G.; Costa, V.V.; Alves-Silva, J.; Ferrari, C.Z.; Izidoro-Toledo, T.C.; da Silva, J.F.; Teixeira, A.L.; Souza, D.G.; Marques, J.T.; et al. Zika virus Promotes neuronal cell death in a non-cell autonomous manner by triggering the release of neurotoxic factors. *Front. Immunol.* **2017**, *8*, 1016. [CrossRef] [PubMed]

145. Souza, B.S.; Sampaio, G.L.; Pereira, C.S.; Campos, G.S.; Sardi, S.I.; Freitas, L.A.; Figueira, C.P.; Paredes, B.D.; Nonaka, C.K.; Azevedo, C.M.; et al. Zika virus infection induces mitosis abnormalities and apoptotic cell death of human neural progenitor cells. *Sci. Rep.* **2016**, *6*, 39775. [CrossRef] [PubMed]

146. Bradfute, S.B.; Swanson, P.E.; Smith, M.A.; Watanabe, E.; McDunn, J.E.; Hotchkiss, R.S.; Bavari, S. Mechanisms and consequences of ebolavirus-induced lymphocyte apoptosis. *J. Immunol.* **2010**, *184*, 327–335. [CrossRef] [PubMed]

147. Impagliazzo, A.; Milder, F.; Kuipers, H.; Wagner, M.V.; Zhu, X.; Hoffman, R.M.; van Meersbergen, R.; Huizingh, J.; Wanningen, P.; Verspuij, J.; et al. A stable trimeric influenza hemagglutinin stem as a broadly protective immunogen. *Science* **2015**, *349*, 1301–1306. [CrossRef] [PubMed]

148. Reed, D.S.; Hensley, L.E.; Geisbert, J.B.; Jahrling, P.B.; Geisbert, T.W. Depletion of peripheral blood T lymphocytes and NK cells during the course of ebola hemorrhagic Fever in cynomolgus macaques. *Viral Immunol.* **2004**, *17*, 390–400. [CrossRef] [PubMed]

149. Sarid, R.; Ben-Moshe, T.; Kazimirsky, G.; Weisberg, S.; Appel, E.; Kobiler, D.; Lustig, S.; Brodie, C. vFLIP protects PC-12 cells from apoptosis induced by Sindbis virus: Implications for the role of TNF-α. *Cell Death Differ.* **2001**, *8*, 1224–1231. [CrossRef] [PubMed]

150. Benedict, C.A.; Butrovich, K.D.; Lurain, N.S.; Corbeil, J.; Rooney, I.; Schneider, P.; Tschopp, J.; Ware, C.F. Cutting edge: A novel viral TNF receptor superfamily member in virulent strains of human cytomegalovirus. *J. Immunol.* **1999**, *162*, 6967–6970. [PubMed]

151. Sedy, J.R.; Spear, P.G.; Ware, C.F. Cross-regulation between herpesviruses and the TNF superfamily members. *Nat. Rev. Immunol.* **2008**, *8*, 861–873. [CrossRef] [PubMed]

152. Upton, J.W.; Kaiser, W.J.; Mocarski, E.S. Virus inhibition of RIP3-dependent necrosis. *Cell Host Microbe* **2010**, *7*, 302–313. [CrossRef] [PubMed]

153. Garnett, T.O.; Duerksen-Hughes, P.J. Modulation of apoptosis by human papillomavirus (HPV) oncoproteins. *Arch. Virol.* **2006**, *151*, 2321–2335. [CrossRef] [PubMed]

154. Gong, B.; Almasan, A. Genomic organization and transcriptional regulation of human Apo2/TRAIL gene. *Biochem. Biophys. Res. Commun.* **2000**, *278*, 747–752. [CrossRef] [PubMed]

155. Sato, K.; Hida, S.; Takayanagi, H.; Yokochi, T.; Kayagaki, N.; Takeda, K.; Yagita, H.; Okumura, K.; Tanaka, N.; Taniguchi, T.; et al. Antiviral response by natural killer cells through TRAIL gene induction by IFN-α/β. *Eur. J. Immunol.* **2001**, *31*, 3138–3146. [CrossRef]

156. Smyth, M.J.; Takeda, K.; Hayakawa, Y.; Peschon, J.J.; van den Brink, M.R.; Yagita, H. Nature's TRAIL—On a path to cancer immunotherapy. *Immunity* **2003**, *18*, 1–6. [CrossRef]

157. Stegmann, K.A.; Bjorkstrom, N.K.; Veber, H.; Ciesek, S.; Riese, P.; Wiegand, J.; Hadem, J.; Suneetha, P.V.; Jaroszewicz, J.; Wang, C.; et al. Interferon-α-induced TRAIL on natural killer cells is associated with control of hepatitis C virus infection. *Gastroenterology* **2010**, *138*, 1885–1897. [CrossRef] [PubMed]

158. Miot, C.; Beaumont, E.; Duluc, D.; Le Guillou-Guillemette, H.; Preisser, L.; Garo, E.; Blanchard, S.; Hubert Fouchard, I.; Creminon, C.; Lamourette, P.; et al. IL-26 is overexpressed in chronically HCV-infected patients and enhances TRAIL-mediated cytotoxicity and interferon production by human NK cells. *Gut* **2015**, *64*, 1466–1475. [CrossRef] [PubMed]

159. Peppa, D.; Gill, U.S.; Reynolds, G.; Easom, N.J.; Pallett, L.J.; Schurich, A.; Micco, L.; Nebbia, G.; Singh, H.D.; Adams, D.H.; et al. Up-regulation of a death receptor renders antiviral T cells susceptible to NK cell-mediated deletion. *J. Exp. Med.* **2013**, *210*, 99–114. [CrossRef] [PubMed]

160. Schuster, I.S.; Wikstrom, M.E.; Brizard, G.; Coudert, J.D.; Estcourt, M.J.; Manzur, M.; O'Reilly, L.A.; Smyth, M.J.; Trapani, J.A.; Hill, G.R.; et al. TRAIL+ NK cells control CD4+ T cell responses during chronic viral infection to limit autoimmunity. *Immunity* **2014**, *41*, 646–656. [CrossRef] [PubMed]

161. Nemcovicova, I.; Benedict, C.A.; Zajonc, D.M. Structure of human cytomegalovirus UL141 binding to TRAIL-R2 reveals novel, non-canonical death receptor interactions. *PLoS Pathog.* **2013**, *9*, e1003224. [CrossRef] [PubMed]

162. Smith, W.; Tomasec, P.; Aicheler, R.; Loewendorf, A.; Nemcovicova, I.; Wang, E.C.; Stanton, R.J.; Macauley, M.; Norris, P.; Willen, L.; et al. Human cytomegalovirus glycoprotein UL141 targets the TRAIL death receptors to thwart host innate antiviral defenses. *Cell Host Microbe* **2013**, *13*, 324–335. [CrossRef] [PubMed]

163. Verma, S.; Loewendorf, A.; Wang, Q.; McDonald, B.; Redwood, A.; Benedict, C.A. Inhibition of the TRAIL death receptor by CMV reveals its importance in NK cell-mediated antiviral defense. *PLoS Pathog.* **2014**, *10*, e1004268. [CrossRef] [PubMed]

164. Achard, C.; Guillerme, J.B.; Bruni, D.; Boisgerault, N.; Combredet, C.; Tangy, F.; Jouvenet, N.; Gregoire, M.; Fonteneau, J.F. Oncolytic measles virus induces tumor necrosis factor-related apoptosis-inducing ligand (TRAIL)-mediated cytotoxicity by human myeloid and plasmacytoid dendritic cells. *Oncoimmunology* **2017**, *6*, e1261240. [CrossRef] [PubMed]

165. Chaperot, L.; Blum, A.; Manches, O.; Lui, G.; Angel, J.; Molens, J.P.; Plumas, J. Virus or TLR agonists induce TRAIL-mediated cytotoxic activity of plasmacytoid dendritic cells. *J. Immunol.* **2006**, *176*, 248–255. [CrossRef] [PubMed]

166. Stary, G.; Klein, I.; Kohlhofer, S.; Koszik, F.; Scherzer, T.; Mullauer, L.; Quendler, H.; Kohrgruber, N.; Stingl, G. Plasmacytoid dendritic cells express TRAIL and induce CD4+ T-cell apoptosis in HIV-1 viremic patients. *Blood* **2009**, *114*, 3854–3863. [CrossRef] [PubMed]

167. Zajac, A.J.; Quinn, D.G.; Cohen, P.L.; Frelinger, J.A. Fas-dependent CD4+ cytotoxic T-cell-mediated pathogenesis during virus infection. *Proc. Natl. Acad. Sci. USA* **1996**, *93*, 14730–14735. [CrossRef] [PubMed]

168. Medana, I.M.; Gallimore, A.; Oxenius, A.; Martinic, M.M.; Wekerle, H.; Neumann, H. MHC class I-restricted killing of neurons by virus-specific CD8+ T lymphocytes is effected through the Fas/FasL, but not the perforin pathway. *Eur. J. Immunol.* **2000**, *30*, 3623–3633. [CrossRef]

169. Ando, K.; Hiroishi, K.; Kaneko, T.; Moriyama, T.; Muto, Y.; Kayagaki, N.; Yagita, H.; Okumura, K.; Imawari, M. Perforin, Fas/Fas ligand, and TNF-α pathways as specific and bystander killing mechanisms of hepatitis C virus-specific human CTL. *J. Immunol.* **1997**, *158*, 5283–5291. [PubMed]

170. Gremion, C.; Grabscheid, B.; Wolk, B.; Moradpour, D.; Reichen, J.; Pichler, W.; Cerny, A. Cytotoxic T lymphocytes derived from patients with chronic hepatitis C virus infection kill bystander cells via Fas-FasL interaction. *J. Virol.* **2004**, *78*, 2152–2157. [CrossRef] [PubMed]

171. Wilson, A.D.; Redchenko, I.; Williams, N.A.; Morgan, A.J. CD4+ T cells inhibit growth of Epstein-Barr virus-transformed B cells through CD95-CD95 ligand-mediated apoptosis. *Int. Immunol.* **1998**, *10*, 1149–1157. [CrossRef] [PubMed]

172. Tan, L.; Zhang, C.; Dematos, J.; Kuang, L.; Jung, J.U.; Liang, X. CD95 Signaling Inhibits B Cell receptor-mediated γherpesvirus replication in apoptosis-resistant B lymphoma cells. *J. Virol.* **2016**, *90*, 9782–9796. [CrossRef] [PubMed]

173. Topham, D.J.; Cardin, R.C.; Christensen, J.P.; Brooks, J.W.; Belz, G.T.; Doherty, P.C. Perforin and Fas in murine gammaherpesvirus-specific CD8(+) T cell control and morbidity. *J. Gen. Virol.* **2001**, *82*, 1971–1981. [CrossRef] [PubMed]

174. Brincks, E.L.; Gurung, P.; Langlois, R.A.; Hemann, E.A.; Legge, K.L.; Griffith, T.S. The magnitude of the T cell response to a clinically significant dose of influenza virus is regulated by TRAIL. *J. Immunol.* **2011**, *187*, 4581–4588. [CrossRef] [PubMed]

175. Brincks, E.L.; Katewa, A.; Kucaba, T.A.; Griffith, T.S.; Legge, K.L. CD8 T cells utilize TRAIL to control influenza virus infection. *J. Immunol.* **2008**, *181*, 4918–4925. [CrossRef] [PubMed]

176. Mueller, Y.M.; De Rosa, S.C.; Hutton, J.A.; Witek, J.; Roederer, M.; Altman, J.D.; Katsikis, P.D. Increased CD95/Fas-induced apoptosis of HIV-specific CD8(+) T cells. *Immunity* **2001**, *15*, 871–882. [CrossRef]

177. Liu, Z.X.; Govindarajan, S.; Okamoto, S.; Dennert, G. Fas-mediated apoptosis causes elimination of virus-specific cytotoxic T cells in the virus-infected liver. *J. Immunol.* **2001**, *166*, 3035–3041. [CrossRef] [PubMed]

178. Contini, P.; Ghio, M.; Merlo, A.; Brenci, S.; Filaci, G.; Indiveri, F.; Puppo, F. Soluble HLA class I/CD8 ligation triggers apoptosis in EBV-specific CD8+ cytotoxic T lymphocytes by Fas/Fas-ligand interaction. *Hum. Immunol.* **2000**, *61*, 1347–1351. [CrossRef]

179. Krzyzowska, M.; Shestakov, A.; Eriksson, K.; Chiodi, F. Role of Fas/FasL in regulation of inflammation in vaginal tissue during HSV-2 infection. *Cell Death Dis.* **2011**, *2*, e132. [CrossRef] [PubMed]

180. Krzyzowska, M.; Baska, P.; Orlowski, P.; Zdanowski, R.; Winnicka, A.; Eriksson, K.; Stankiewicz, W. HSV-2 regulates monocyte inflammatory response via the Fas/FasL pathway. *PLoS ONE* **2013**, *8*, e70308. [CrossRef] [PubMed]

181. Krzyzowska, M.; Baska, P.; Grochowska, A.; Orlowski, P.; Nowak, Z.; Winnicka, A. Fas/FasL pathway participates in resolution of mucosal inflammatory response early during HSV-2 infection. *Immunobiology* **2014**, *219*, 64–77. [CrossRef] [PubMed]

![viruses logo]

Review

The Bcl-2 Family in Host-Virus Interactions

Marc Kvansakul [1,*], Sofia Caria [1] and Mark G. Hinds [2,*]

[1]   Department of Biochemistry and Genetics, La Trobe Institute for Molecular Science, La Trobe University, Melbourne, VIC 3086, Australia; s.caria@latrobe.edu.au

[2]   Department of Chemistry and Physics, La Trobe Institute for Molecular Science, La Trobe University, Melbourne, VIC 3086, Australia

*   Correspondence: m.kvansakul@latrobe.edu.au (M.K.); m.hinds@latrobe.edu.au (M.G.H.);
    Tel.: +61-3-94792263 (M.K.)

Received: 22 September 2017; Accepted: 3 October 2017; Published: 6 October 2017

Abstract: Members of the B cell lymphoma-2 (Bcl-2) family are pivotal arbiters of mitochondrially mediated apoptosis, a process of fundamental importance during tissue development, homeostasis, and disease. At the structural and mechanistic level, the mammalian members of the Bcl-2 family are increasingly well understood, with their interplay ultimately deciding the fate of a cell. Dysregulation of Bcl-2-mediated apoptosis underlies a plethora of diseases, and numerous viruses have acquired homologs of Bcl-2 to subvert host cell apoptosis and autophagy to prevent premature death of an infected cell. Here we review the structural biology, interactions, and mechanisms of action of virus-encoded Bcl-2 proteins, and how they impact on host-virus interactions to ultimately enable successful establishment and propagation of viral infections.

Keywords: Bcl-2; apoptosis; autophagy; structural biology; poxvirus; herpesvirus; asfarvirus; iridovirus; adenovirus; host-pathogen interactions

1. Introduction

From the observation of specific changes in cell morphology upon cellular suicide, and ending in engulfment of the cell by phagocytes, Kerr et al. in 1972 concluded that there must be a genetically programmed form of cell death responsible for cell deletion, that they called apoptosis [1]. From these early origins, it is now recognised that apoptosis is one of a spectrum of programmed cell death (PCD) pathways that includes not only apoptosis but also autophagy, necroptosis and more specialised forms of PCD such as pyroptosis, ferroptosis, anoikis, entosis, pathanatos, netosis, and cornification [2]. The genetic and molecular basis of these different pathways are still being determined.

Correct apoptosis regulation is key to homeostasis, and regulates the clearance of cells that are no longer required in development, and are damaged, dangerous, or infected [3,4]. Apoptosis is an important regulator of the immune response, and pathogens have acquired genes that both prevent the cell from initiating apoptosis during their replicative stage, and initiating apoptosis to release their progeny. Using molecular mimicry of host proteins, pathogens have evolved mechanisms to overcome host cell defences. Such host-pathogen interactions are poorly understood, but several of these pathways are regulated by the activity of proteins of the B-cell lymphoma-2 (Bcl-2) family, a group of about 20 proteins, and numerous studies have been performed to further understand and characterise its mechanism [5–9]. Ultimately, the caspase cascade is activated [10], enabling disassembly of the apoptotic cell followed by its phagocyte-mediated engulfment and elimination via lysosomes [11].

Bcl-2 proteins arose early in metazoan evolution [12,13], and are characterised by the existence of short conserved sequence regions, the Bcl-2 homology (BH) motifs [14] (Figure 1). Two phylogenetically [15] and structurally [8] separate groups of proteins constitute the Bcl-2 family, one group consists of intrinsically disordered proteins (IDP) [7,16], while the other group have a globular α-helical fold structure known as the "Bcl-2 fold" [8]. The former group are IDPs and are exclusively pro-apoptotic, while the latter folded group are either pro-survival or pro-apoptotic. Both Bcl-2 family groups bear BH motifs. The pro-apoptotic BH3-only proteins bear only the BH3 motif (Figure 1), and in mammals includes the proteins Bim, Bad, Bmf, Hrk, Puma, Bik, and Noxa. The BH3-only proteins acquire secondary structure upon binding to their folded Bcl-2 targets, neutralising or activating this second group of the Bcl-2 family. The Bcl-2 fold generates a hydrophobic groove that accommodates a BH3 motif. The Bcl-2 family in mammals includes the pro-survival members Bcl-2, Bcl-x$_L$, Mcl-1, Bcl-w, Bcl-B, and A1/Bfl-1 [17] while the pro-apoptotic group includes Bax and Bak, and a member Bok that as yet, has a less-well defined role [18]. The pro-apoptotic group all bear the BH3-motif, whereas the pro-survival proteins do not always bear this motif [8]. Bid bears only a BH3-motif, but has a folded α-helical structure [19–21], and is activated by caspase cleavage in the α1-α2 loop to form truncated Bid (tBid), the C-terminal fragment that bears the BH3-motif. It is not clear how the caspase cleaved Bid (cBid) decomposes to form N-Bid, the N-terminal fragment, and tBid, the C-terminal fragments, as cBid is stable [19]. Though tBid retains some secondary structure [22], like the other BH3-only proteins, tBid is intrinsically disordered [22,23].

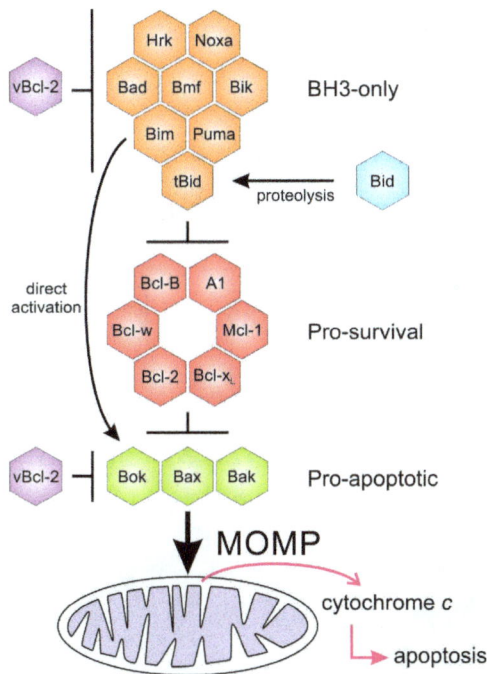

Figure 1. Pathways for Bcl-2 protein action. In mammals, a tripartite mechanism regulated by the Bcl-2 family controls the integrity of the mitochondrial outer membrane (MOM). Activation of Bax and Bak leads to MOM permeabilisation (MOMP), and escape of factors such as cytochrome c from the mitochondrial intermembrane space to initiate the caspase cascade that is the defining step in apoptosis. BH3-only proteins either activate Bax and/or Bak either by removal of their inhibition by pro-survival proteins, or by direct interaction. The BH3-only protein Bid is activated by proteolytic cleavage that releases its BH3-motif for interaction. Viral Bcl-2 (vBcl-2) orthologues can act on the BH3-only proteins, or directly block the action of Bax and Bak to prevent apoptosis initiation. Activation steps are shown as arrows and inhibition as bars.

Bcl-2 proteins are not only pivotal in higher organisms such as mammals, worms and flies, but have been identified in evolutionary ancient species such as sponges [24,25] and hydra [26]. Key molecular and mechanistic features appear to be well preserved, with the identification of pro-survival Bcl-2 and pro-apoptotic Bak in sponges [27], along with representatives of pro-survival Bcl-2, as well as BH3-only proteins and Bak like proteins in hydra [28]. The key role of the membrane is preserved, and interactions between the pro-survival Bcl-2 and Bax-like proteins are conserved in hydra, with the Bcl-2 orthologue HyBcl-2 co-localising with HyBax on the periphery of mitochondria [29,30]. The Bcl-2 family probably forms part of a primitive immune response for cnidarians, and Bax has been shown to be upregulated in response to disease in the coral *Acropora hyacinthus* [31]. The demosponge *Geodia cydonium* Bcl-2 orthologue, BHP2, is the most ancient Bcl-2 protein to be described at a molecular level to date [32]. Structurally, the Bcl-2 fold and the BH3 motif-in-groove interaction is conserved, as shown for the sponge Bcl-2 BHP2 (Figure 2), although subtle differences allow BHP2 to discriminate between most human pro-apoptotic Bcl-2 proteins to be selective for its sponge counterparts [32].

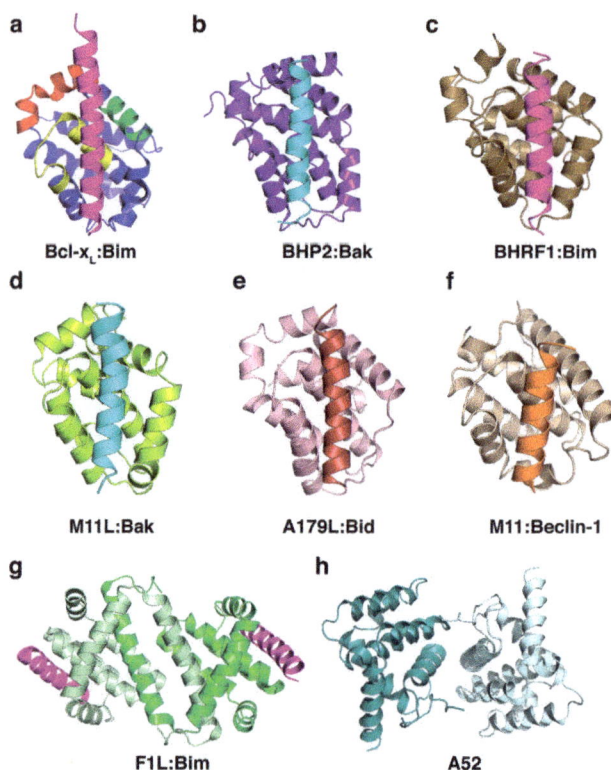

Figure 2. Structures of Bcl-2 family members. (**A**) Human Bcl-x$_L$:Bim complex [33] (PDB ID 1PQ1); (**B**) *G. cydonium* BHP2:LB-Bak-2 complex [32] (PDB ID 5TWA); (**C**) EBV BHRF1:Bim complex [34] (PDB ID 2WH6); (**D**) Myxomavirus M11L:Bak complex [35] (PDB ID 2JBY); (**E**) African swine fever virus A179L:Bid complex [36] (PDB ID 5UA4); (**F**) Murine γ-herpesvirus 68 M11:Beclin-1 complex [37] (PDB ID 3BL2); (**G**) Vaccinia virus F1L:Bim complex [38] (PDB ID 4D2M); (**H**) Vaccinia virus A52 [39] (PDB ID 2VVW).

It is now emerging that the apoptotic machinery is closely associated with other cellular regulatory pathways such as autophagy, the unfolded protein response, and endoplasmic reticulum (ER) stress signalling [40]. Viral subversion of the cellular Unfolded Protein Response (UPR) is a mechanism that is increasingly recognised as being fundamental for host immunity [41]. The UPR is a multimodal response to perturbed ER function ("ER stress") that results from unfolded proteins accumulating in the ER faster than they are able to be folded, leading to shut down of translation, an increase in the rate of protein folding, activation of degradation pathways of the ubiquitin-proteasome or autophagy, and ultimately apoptosis if the stress is unrelieved. Thus, ER stress, autophagy, and apoptosis are all tightly linked and regulated by viruses.

The gatekeepers of mitochondrial integrity are the pro-apoptotic proteins Bax and Bak that have overlapping roles [42]. Bax and Bak are necessary for instigation of apoptosis; however, the details of their mode of action are still disputed. In mammals, after apoptotic stimuli, cytosolic Bax migrates to the mitochondrial outer membrane (MOM) to generate pores in the mitochondrial outer membrane that allows the escape of apoptogenic factors that have activated the caspase cascade (Figure 1). The apoptotic programme is not conserved in all aspects; for example, apoptotis in *Caenorhabditis elegans* differs from that in mammals. In *C. elegans*, there is a single folded Bcl-2 protein (CED-9) present in this organism that is associated with mitochondria [43], and it interacts with the caspase activator CED-4 to inhibit apoptosis. The CED-9:CED-4 interaction is antagonised by the BH3-only protein, Egl-1, to release CED-4 and activate the caspase CED-3 and the caspase cascade.

2. An Expanding Family of Viral Bcl-2 Orthologues has been Discovered

The importance of apoptosis and the Bcl-2 proteins in immune cell regulation and innate immunity responses has created an evolutionary pressure for viruses to acquire the genes for the pro-survival Bcl-2 proteins [44]. There are a multitude of large DNA viruses that mimic pro-survival Bcl-2 (vBcl-2) proteins, hijacking the intrinsic apoptotic pathway for their benefit; these are summarised in Table 1.

Table 1. Pro-survival Bcl-2 proteins encoded by viruses.

Virus-Encoded Pro-Survival Bcl-2	Reference
γ-herpesviruses 68 M11	[45]
Adenovirus E1B19K	[46]
Epstein-Barr virus BHRF1	[34,47]
Epstein-Barr virus BALF1	[34,47]
Kaposi's sarcoma-associated herpesvirus Ks-Bcl-2	[48,49]
Turkey herpesvirus vnr-13	[50]
African swine fever virus A179L	[36,51]
Grouper iridovirus GIV66	[52,53]
Myxoma virus M11L	[35,54]
Vaccinia virus F1L	[55–57]
Variola virus F1L	[58]
Ectromelia virus EMV025	[59]
Sheeppox virus SPPV14	[60]
Deerpox virus DPV022	[61,62]
Fowlpox virus FPV029	[63,64]
Canarypox CNP058	[65]
Lumpy skin disease virus LD17	[60]
Orfvirus ORFV125	[66]

3. Membrane Interactions

The accumulation and oligomerisation of Bax and Bak at the intracellular membrane is the key event and the point of no return in apoptosis, and it is the least well understood at a molecular level [67]. vBcl-2 orthologues also play a role here, with many localising to intracellular membranes such as the mitochondrial membrane, ER and nuclear envelope in the host cell. The presence of a putative hydrophobic transmembrane (TM) region for many of the Bcl-2 family indicates the importance of this interaction, though the exact molecular mechanisms remain ill-defined, Bax and Bak accumulate on the mitochondrial surface and ultimately lead to its disruption and leakage. Viral control over membrane disruption is thus crucial to maintaining the host cell viability for replication [68].

The folded Bcl-2 proteins are partitioned between the cytosol and intracellular membranes and trafficking between the two environments occurs. Differences in cytosol-membrane partitioning are dependent on their rate of translocation, which is in turn dependent on their TM regions [69]. Bak and Bcl-2 are predominantly membrane-associated [70,71], while others, including Bax and Bcl-x_L are predominantly cytosolic but become membrane integrated after an apoptotic stimulus [72,73]. Trafficking of Bax and Bak between the membrane and cytosol is a process dependent on Bcl-x_L and Bcl-2 [69,74–76]. In addition to the requirement for the TM region the interaction between the pro-survival and proteins requires an exposed BH3-motif on Bax and Bak, a process that necessarily requires a conformational change from their solution conformation where the key residues of the BH3-motif are buried [77]. This suggests an interaction of the BH3 motif of Bax or Bak in the groove of the pro-survival protein, although this was one of the first interactions observed in the Bcl-2 family [78] there remains a deficit of structural data on this interaction.

BH3-only proteins are also associated with intracellular membranes (see [7] for a discussion), some such as Bik bear hydrophobic C-terminal regions suggestive of membrane interacting proteins. The interaction with BH3-only proteins releases the TM region from the BH3-binding groove in pro-survival proteins [79,80], potentially releasing it for membrane binding. Full biological activity of pro-survival Bcl-2 is not observed if the C-terminal region is truncated from these molecules even though binding to their BH3-targets is improved [79]. The same behaviour has been observed for viral Bcl-2 orthologues, where C-terminal deletions reduce the pro-survival activity (See Opgenorth et al. 1992 [81] for M11L truncation). Similarly, deletion of the C-terminal TM region of Bax impairs its membrane localisation and biological activity [82]. A combination of biophysical, biochemical and genetic studies have shown that there is multiple redundancy in the interactions [17,83,84]. Several models have been put forward for BH3-only protein PCD activation but most BH3-only proteins are able to activate Bax or Bak [85]. Thus, the membrane interaction is critical to the pro-survival or pro-apoptotic activity of the Bcl-2 family and further complicates an already complex multilevel-redundant regulation mechanism for mammalian apoptosis.

Many vBcl-2 orthologues, including the first viral Bcl-2 orthologue found, adenovirus E1B 19K, though without an obvious hydrophobic C-terminal region, are closely associated with intracellular membranes, the ER and nuclear envelope [86] and the association with membranes is required for its function [87]. Frog virus 3 Bcl-2 orthologue 97R localises to the ER and deletion of the C-terminal 29 residues inactivates the protein [88]. The African swine fever virus Bcl-2 orthologue A179L, is closely associated with viral factories, and though it lacks an obvious TM region, it is associated with the ER and mitochondrial membranes. However, the mutant G85A A179L loses its ability to keep cells alive, but also associates with ER membranes [89], probably indicating that a competent BH3-binding is required for ER association, as this mutant destroys binding to the BH3-only proteins [36]. EBV BHRF1 is associated with membranes [47]. Combined, these features attest to the importance of membrane association or integration to the activity of the folded Bcl-2 proteins, including those encoded by viruses.

Though many questions remain about the exact nature of the molecular assemblies of Bax and Bak that disrupt the MOM, a more consistent mechanism is now emerging where Bax and Bak undergo a series of conformational changes to form high-order aggregates to create the membrane disrupting pores [90]. However, in solution Bax is a monomeric and relatively rigid protein with little evidence of conformational mobility [77,91], a finding consistent with fluorescence cross correlation studies showing that Bax associates with mitochondria prior to oligomerisation [92] into ring-like pores [93,94]. In a defined system consisting of only cBid, Bax and Bcl-x_L, Bleicken et al. showed that the interactions between these apoptotic regulators is spatially regulated. When embedded in the membrane, Bax is proposed to form a positive feed-back loop recruiting Bax, and Bcl-x_L inhibits this process by preventing Bax oligomer growth and translocating membrane Bax to the cytosol [92]. Accumulating evidence suggests that the TM regions are intermolecular interaction sites. Andreau-Fernandez et al. showed that the TM region of Bax interacts with the TM regions of Bcl-2 and

Bcl-x$_L$ [95]. Earlier structural studies on Bcl-w where a near-full-length sequence was well behaved [80], showed that like Bax [77] the C-terminal tail lies in the BH3-binding groove. Furthermore, the presence of the TM region in Bcl-x$_L$ [79,96] and Bcl-w [80] reduces their affinity for BH3-motifs. In the case of Bcl-x$_L$ is dimeric when the TM region is present [92] and monomeric in its absence, as shown by structural studies [97]. Nuclear Magnetic Resonance (NMR) investigation of the Bcl-x$_L$:membrane interaction showed that it has an α-helical C-terminal tail that anchors the folded globular Bcl-2 domain head to the membrane [98]. BH3-ligand displacement of the C-terminal residues of the α9 residues from the groove of Bcl-w renders them unstructured in solution [80], and a likely mechanism to drive Bcl-w to the membrane [79]. Biochemical studies showed that like Bcl-x$_L$, Bak has a transmembrane C-terminal anchor [99]. Combined, these studies indicate that the C-terminal region of the Bcl-2 fold are not by-standers, and play an important role in not only membrane targeting and anchoring, but also the protein-protein interactions of these molecules. The observation that the viral pro-survival proteins mimic these membrane localisation and activities suggests the importance of modulating PCD in infected cells.

Other models for the membrane oligomerisation include an initial dimerisation of Bax or Bak [100] prior to their oligomerisation at the membrane surface by unfolding an interaction. In this model, it is proposed that membrane rupture is caused by disordered clustering of Bak or Bax dimers. A "hit and run" mechanism has been suggested, where an initial weak interaction induces subsequent conformational changes in Bax or Bak. A second site has been proposed for binding BH3-only proteins on Bax, though it is a low affinity interaction that initiates apoptosis [101]. Structural investigation of detergent treated Bax in the presence of BH3-motifs gave a symmetrical Bax dimer with the BH3 bound in the groove; this was proposed to be the active from of Bax that further oligomerises to form pores [102]. Structures of Bim and Bid BH3-motifs in the groove of domain swapped dimer have been determined [103], similar domain-swapped dimers have been observed for Bcl-x$_L$, that also retain the ability to bind BH3 motifs [104]. A caveat on these studies is that they were performed with C-terminally truncated Bax and may not reflect membrane interactions in their entirety [92]. Further studies will be required to elucidate the exact mechanisms.

4. Viral Bcl-2-mediated Subversion of Programmed Cell Death

Considering the importance of Bcl-2 proteins in regulating apoptosis, as well as autophagy in higher organisms [11,84], it is unsurprising that numerous viruses have acquired sequence, functional and structural homologs of Bcl-2 to subvert host apoptosis as well as autophagy signalling for their own ends. Prevention of premature host cell death during the initial stages has been shown to be critical for successful infection of EBV [68] and demonstrates the pivotal role that disarming of host apoptotic defences plays in preventing clearance of virus to enable successful infection and propagation. However, whilst prevention of premature host cell apoptosis is highly desirable, ultimately, viruses also rely on host cell apoptosis at a later stage to aid viral dissemination, for example an avian reovirus triggers apoptosis to enable optimal release and dissemination of viral progeny [105].

The earliest identified virus encoded Bcl-2 homologs were E1B 19K from adenovirus and BHRF1 from EBV which both display substantial sequence identity (18% and 16% identical to human Bcl-2 respectively) to mammalian Bcl-2 and contain the hallmark BH1 and BH2 motifs. Functional studies of E1B 19K determined that it is a potent inhibitor of apoptosis that is induced by stimuli including Fas ligand, TNFα, and adenoviral infection. Mechanistically, E1B 19K engages Bax [106], Bak [107] and Bik [108], and is functionally interchangeable with Bcl-2 during adenovirus infection and transformation [46].

One mechanism for modulating apoptosis is to manipulate the BH3-only and Bax family through interaction in the binding groove (Figure 2). This interaction has now been extensively studied and binding affinities measured (Table 2), but the implications of binding specific BH3-motif bearing proteins is not always clear. For example, Bim is a universal Bcl-2 binder and binds all mammalian pro-survival proteins with relatively high affinity, yet is not bound by all vBcl-2 proteins (for example the variola virus F1L). Some viral Bcl-2 proteins are capable of binding nearly all pro-apoptotic proteins (e.g., A179L, FPV039), while others have a much more specific ligand range (Table 2). The specificities of the viral Bcl-2 proteins for their BH3-targets have been determined (Table 2), and in general, these interactions are of lower affinity than the pro-survival Bcl-2, probably reflecting a balancing act by the virus, as they need to block apoptosis during their replicative stage; however apoptosis is necessary for the escape of viral progeny on maturation. The cell type specificity of the virus also probably plays a role in deciding which host Bcl-2 proteins are inhibited, and this is an area for further investigation.

5. *Herpesviridae*-Encoded Bcl-2 Homologs

Many members of the *herpesviridae* encode Bcl-2 like proteins. Epstein-Barr virus (or human herpesvirus 4) is a large DNA virus belonging to the *γ-herpesviridae* and harbours two Bcl-2 homologs, BHRF1 and BALF1. BHRF1 was shown to be an enhancer of cell survival [47]. Biochemical and structural studies revealed that BHRF1 adopts a Bcl-2 fold [34,109] and is bound to the BH3-only proteins Bim, Bid, and Puma, as well as Bak and Bax [34,110] (Table 2). Mechanistically, BHRF1 was shown to rely on the sequestration of Bim [111] and Bak [34] to inhibit apoptosis, and to confer chemoresistance in a Burkitt lymphoma mouse model, similar to Bcl-2 [34]. BHRF1 was also shown to be constitutively overexpressed in a sub-set of EBV transformed B-cells, thus rendering them resistant to apoptosis [112]. The function of a second EBV-encoded Bcl-2 homolog, BALF1, remains controversial. Initial data suggested that BALF1 acts as a pro-survival Bcl-2 protein [113], however a second report showed that BALF1 is pro-apoptotic and inhibits the other EBV-encoded pro-survival protein BHRF1 [114]. Subsequently, others reported that both BHRF1 and BALF1 are required for successful EBV-induced B-cell transformation [68]. The identification of BHRF1 in transformed B-cells sparked interest in developing antagonists against BHRF1 for targeted cancer therapy, and the feasibility of such an approach was recently demonstrated via the use of an engineered protein that bound BHRF1 with picomolar affinity [115]. No small molecule antagonists for BHRF1 have been reported yet; however their development is underway [116].

Kaposi's sarcoma herpesvirus (KSHV or human herpesvirus 8) is also a large DNA virus and a member of the *γ-herpesviridae*. KSHV encodes a Bcl-2 homolog, Ks-Bcl-2 [117] that adopts a Bcl-2 fold [49] and is able to bind Bim, Bid, Bik, Bmf, Hrk, Noxa, and Puma [110] (Table 2). Conflicting data exist for binding of Bax and Bak, with one report indicating that neither bind Ks-Bcl-2 [48], whereas a subsequent study revealed a high affinity interaction for Bak (50 nM) and moderate affinity for Bax (980 nM) [109]. During viral infection, Ks-Bcl-2 appears to play a pivotal role in completion of the lytic cycle, as a Ks-Bcl-2 deletion virus of KSHV does not complete the lytic replication cycle. Interestingly, the Ks-Bcl-2-related ORF16 from rhesus rhadinovirus is able to functionally replace Ks-Bcl-2 during the lytic cycle, in contrast to other endogenous mammalian Bcl-2 proteins, such as Bcl-x_L or other herpesvirus-encoded vBcl-2 proteins including M11 and vMIA [118]. Another rhadinovirus, herpesvirus saimiri, also encodes a Bcl-2 homolog named ORF16 [119], which was shown to be anti-apoptotic and bound Bak and Bax in pull-down assays.

Murine γ-herpesvirus 68 encodes M11 [120] and was identified as an inhibitor of Fas and TNF induced apoptosis [45,121], but biochemical studies demonstrated binding to Bim, Bid, Bmf, Noxa, Puma, and Hrk, as well as Bak and Bax via the canonical ligand binding groove of its Bcl-2 fold [37]. Thus, M11 can inhibit the major mitochondrial pathways to apoptosis, however functional studies [37,122] indicate that the mitochondrial pathway may not be the primary target for M11 (see below).

Cytomegalovirus (CMV or human herpesvirus 5) is a large DNA virus belonging to the *β-herpesviridae*. CMV encodes proteins that directly target host pro-apoptotic proteins Bax and Bak, but appear to be neither sequence nor structural homologs of Bcl-2. Human CMV encodes vMIA, which has been shown to inhibit Bax [123] and Bak oligomerisation [124,125]. Interestingly, the interaction of vMIA with Bax does not involve the canonical Bcl-2 ligand binding groove. Unexpectedly, vMIA bound to an alternative binding site distinct from the canonical BH3 binding groove in Bax, which was mapped using NMR to define an interaction site comprising primarily of the loops connecting α1–α2, α3–α4, and α5–α6. Furthermore, the vMIA-Bax interaction was of high affinity with a K_d of 22 nM [126]. Whilst human CMV vMIA appears to be able to neutralise both Bax and Bak, in mouse CMV Bax and Bak are neutralised by two proteins with single specificity [127]. MCMV-encoded m38.5 has been shown to be mitochondrially localised, and to inhibit Bax activation [128,129], whereas Bak inhibition is achieved via m41.1 [130,131].

Amongst the α-herpesviruses, a virus-encoded homolog of the endogenous turkey pro-survival Bcl-2 protein NR13 [132], Bcl-B [133], Boo, Diva, or NRH [134] in mammals, has a Bcl-2 fold [135,136] and has been termed vnr-13 [50]. Though little is known about vnr-13, it was shown to localise to the outer mitochondrial membrane, and inhibit apoptosis after serum deprivation [50].

Table 2. Affinities (in nM) of different pro-survival Bcl-2 proteins for peptides spanning the BH3 motif of endogenous pro-apoptotic Bcl-2 family members or Beclin-1 (measurements taken from: [34–36,57,58,60,62,64,110,137–142]).

	Poxviral Bcl-2					
Pro-death	SPPV14	M11L	MVA_F1L	VAR_F1L	DPV022	FPV039
Bad	>2000	>1000	NB	NB	NB	653
Bid	341	100	NB	3200	NB	2
Bik	>2000	>1000	NB	NB	NB	30
Bim	26	5	250	NB	340	10
Bmf	67	100	NB	NB	NB	16
Hrk	63	>1000	NB	NB	NB	24
Noxa	>2000	>1000	NB	NB	NB	28
Puma	65	>1000	NB	NB	NB	24
Bak	46	50	4300	2640	6930	76
Bax	32	75	1850	960	4040	76
Beclin-1	n/a	n/a	n/a	n/a	NB	n/a
	Asfarviral Bcl-2		Herpesviral Bcl-2			
	A179L	BHRF1	Ks-Bcl-2	M11	N1L	
Bad	258	>2000	>1000	NB	>1000	
Bid	26	109	112	232	152	
Bik	190	>2000	>1000	NB	n/a	
Bim	6	18	29	131	72	
Bmf	254	>2000	>1000	300	n/a	
Hrk	1487	>1000	>1000	719	n/a	
Noxa	1575	>2000	>1000	132	n/a	
Puma	31	70	69	370	n/a	
Bak	29	150	<50	76.3	71	
Bax	26	1,400	980	690	n/a	
Beclin-1	n/a	n/a	n/a	40.2	n/a	
	Human Bcl-2					Sponge Bcl-2
	Bcl-2	Bcl-w	Bcl-x_L	Mcl-1	A1	BHP2
Bad	16	30	5.3	>100,000	15,000	NB
Bid	6800	40	82	2100	1	NB
Bik	850	12	43	1700	58	NB
Bim	2.6	4.3	4.6	2.4	1	NB
Bmf	3	9.8	9.7	1100	180	NB
Hrk	320	49	3.7	370	46	3760
Noxa	>100,000	>100,000	>100,000	24	20	NB
Puma	3.3	5.1	6.3	5	1	NB
Bak	>1000	500	50	10	3	66
Bax	100	58	130	12	n/a	NB
Beclin-1	n/a	n/a	2300	n/a	n/a	n/a

MVA = modified vaccinia virus Ankara, VAR = variole virus, n/a = not available, NB = no binding.

6. *Poxviridae*-Encoded Bcl-2 Homologs

The *poxviridae* encompass a number of families that encode for Bcl-2 proteins. Vaccinia virus is a large DNA virus, the prototypical member of the orthopoxviruses, and encodes F1L, a potent inhibitor of intrinsic apoptosis [55,56] that displays no recognisable sequence identity to Bcl-2. Structural studies revealed that F1L adopts an unusual Bcl-2 fold featuring a domain-swapped dimer configuration, in marked contrast to mammalian pro-survival Bcl-2 proteins, which are all monomeric [57,143]. Furthermore, F1L only bound a highly restricted subset of pro-apoptotic Bcl-2 including Bim [56,144] and Bak [145,146] (Table 2), and was shown to inhibit Bak activation by functionally replacing Mcl-1 during infection [147]. Although F1L is also able to inhibit Bax-mediated apoptosis [144], this activity is likely via an indirect mechanism as F1L does not engage Bax in a cellular context. Like other Bcl-2 family proteins F1L is localised to mitochondrial membranes through its C-terminal residues, and this region is necessary for full pro-survival activity [148]. Mechanistically, the interaction of F1L with Bim was identified as the primary mechanism underlying F1L-mediated inhibition of apoptosis in the context of a live viral infection [143]. Interestingly, the F1L homolog in variola virus, the causative agent of smallpox and another member of the *orthopoxviridae*, appears to utilise a different mechanism for apoptosis inhibition, despite adopting a near identical structure and sequence [58]. Unlike its vaccinia virus counterpart, variola virus F1L only binds Bid, Bak, and Bax, and not Bim (Table 2), and only inhibits Bax-mediated apoptosis [58]. A homolog of vaccinia virus F1L found in another orthopoxvirus, Ectromelia virus EMV025, was also shown to be anti-apoptotic by inhibiting Bax and Bak activation by directly engaging Bim and Bak [59].

Myxomavirus is a member of the *leporipoxviridae* and encodes for M11L, another potent inhibitor of intrinsic apoptosis lacking detectable sequence similarity with Bcl-2 [149]. M11L is able to engage several host pro-apoptotic Bcl-2 proteins including Bak [150], Bax, Bim, and Bid [35] (Table 2). Structural studies showed that M11L adopts a compact, monomeric Bcl-2 fold [35,151] where the canonical ligand binding groove is used to engage pro-apoptotic Bcl-2 proteins [35]. Interestingly, functional studies revealed that M11L primarily acts by sequestering Bak and Bax [35], in contrast to vaccinia virus F1L which acts primarily via Bim sequestration [143].

Orf virus is a parapoxvirus and encodes a readily identifiable Bcl-2 homolog, ORFV125, that potently inhibits intrinsic apoptosis [66]. Functional studies revealed that ORFV125 interacts with several BH3-only proteins including Bim, Puma, Hrk, Bik, and Noxa as well as active Bax but not Bak [152].

Among the *avipoxviridae*, both fowlpoxvirus FPV039 and canarypoxvirus CNP058 have been shown to suppress apoptosis. FPV039 inhibits apoptosis [153] after overexpression of all BH3-only proteins [153], and was shown to adopt a Bcl-2 fold and engage all major host pro-apoptotic Bcl-2 proteins [64] (Table 2). Interestingly, the closely related canarypoxvirus CNP058 also inhibits apoptosis in transfected cells, but engaged a different subset of pro-apoptotic Bcl-2 proteins, largely with weaker affinities than FPV039 [64].

Other Bcl-2 proteins-encoded by poxviruses include sheeppoxvirus SPPV14 and deerpoxvirus DPV022. SPPV14 displayed a broader spectrum of pro-apoptotic Bcl-2 interactions by binding Bim, Bid, Bmf, Hrk, Puma, as well as Bax and Bak [60] (Table 2). In contrast, DPV022 only engaged Bim, Bax and Bak [62] (Table 2). Intriguingly, DPV022 also adopted a domain-swapped Bcl-2 fold similar to those observed for vaccinia and variola virus F1L [57,58,143], suggesting that this particular topology for Bcl-2 proteins may be more widely found in nature.

7. *Asfarviridae* and *iridoviridae*-Encoded Bcl-2 Homologs

African swine fever virus is a large double stranded DNA virus, the only member of the *asfarviridae*, and encodes A179L [154]. A179L adopts a Bcl-2 fold [36], and displays extreme promiscuity by binding all host pro-apoptotic Bcl-2 proteins [36,155]. A179L localised to mitochondria [89] and potently inhibits apoptosis in cell culture assays [51]. Amongst the *iridoviridae*, grouper iridovirus was shown to encode GIV66, which inhibited apoptosis in a grouper kidney cell culture model [156]. However, the structural and functional basis of GIV66-mediated apoptosis inhibition has not been established.

8. Other Functional Roles of Viral Bcl-2 Homologs

Although the vast majority of virus-encoded Bcl-2 proteins primarily interfere with host cell intrinsic apoptosis signalling by targeting endogenous pro-apoptotic host Bcl-2 proteins, a number of studies have revealed that vBcl-2 proteins also harbour other activities. Several vBcl-2 proteins have been shown to inhibit autophagy. Murine γ-herpesvirus 68-encoded M11 utilises the canonical Bcl-2 ligand binding groove to bind the BH3 motif of Beclin-1, a key autophagy regulator [37,122]. Another member of the *herpesviridae*, KSHV, also targets Beclin-1 using Ks-Bcl-2 [157]. However, this ability to engage Beclin-1 is not limited to the *herpesviridae*, with African swine fever A179L displaying autophagy inhibitory activity in addition to anti-apoptotic activity [89]. Adenoviral E1B 19K was also shown to bind Beclin-1, and thus inhibit autophagy [158]. However, to date, no autophagy inhibitor has been identified amongst the *poxviridae*.

Another virus-encoded Bcl-2 protein with multiple functionalities is the vaccinia virus F1L. In addition to the anti-apoptotic activity mediated by sequestering Bim, F1L was also shown to inhibit inflammasome activation [159] via an unusual unstructured N-terminal extension prior to the Bcl-2 fold [38]. In addition to the ability to mediate inflammasome activation, the N-terminus of F1L was also proposed to act as a caspase-9 inhibitor [160,161], however, a subsequent study suggested that the F1L N-terminus is not involved in apoptosis inhibition [38].

F1L is not the only vaccinia virus-encoded Bcl-2 protein with dual functionality. N1L was shown to adopt a Bcl-2 fold (albeit lacking a TM anchoring region) and inhibit both intrinsic apoptosis by targeting several pro-apoptotic Bcl-2 proteins as well as modulating NF-κB signalling. Interestingly, N1L also adopts a Bcl-2 fold with dimeric topology; however, dimerisation is not achieved via a domain swap as seen in F1L and DPV022, but rather via a novel interface centering on the α1 and α6 helices [138,162]. Furthermore, the ability to manipulate both apoptosis and NF-κB signalling is mediated via two discrete sites on N1L [163].

In addition to N1L, several other NF-κB inhibitors that adopt Bcl-2 folds have now been identified in vaccinia virus. These include B14 [39] andA52 [39], as well as A46 [164,165], A49 [166] and K7 [167]. Whilst all four proteins inhibit NF-κB, they are distinguished by substantial differences in mechanism, cellular activity, and structure. Similar to N1L, B14 and A52 form dimers utilising an interface involving α1 and α6 helices, with small but significant differences in the orientation of monomeric chains with each other within the dimers amongst the three proteins [39]. Furthermore, A46 also forms dimers, however, this involves an interface formed by α4 and α6 helices of the C-terminal Bcl-2 like domain. Intriguingly, A46 harbours an additional N-terminal domain that mediates tetramerisation of A46, thus adding an additional layer of quaternary structure-based regulation [168]. In contrast, A49 and K7 are monomeric in solution. Unlike N1L, B14, A52 and K7 do not bind pro-apoptotic Bcl-2 proteins. However, K7 harbours dual functionality that is similar to N1L, and in addition to inhibition of NF-κB, it also binds to the human DEAD-box RNA helicase DDX3 to inhibit induction of the IFN-β promoter [169,170].

9. Concluding Remarks

Virus-encoded Bcl-2 proteins have demonstrated the remarkable adaptability of the Bcl-2 fold, and its ability to modulate signalling that involves several cell death-associated pathways via multiple mechanisms. Whilst mammalian pro-survival Bcl-2 proteins display several distinct rules of engagement for their interactions with pro-apoptotic Bcl-2, the picture is not as clear amongst the virus-encoded homologs of Bcl-2. In mammals, key interactions between pro-survival and pro-apoptotic Bcl-2 are typically characterised by high affinities, with some such as Bcl-x_L:Bim interactions straying into picomolar affinities. The caveat with the binding studies is that they have been performed with C-terminally truncated molecules, and probably overestimate true affinities [79,80]. Furthermore, all mammalian pro-survival Bcl-2 proteins target Bim, the sole universal pro-apoptotic BH3-only protein [135] and either Noxa or Bad, but not both [139]. In contrast, key interactions between virus-encoded pro-survival proteins tend to display weaker affinities, typically with dissociation constants (K_d) in the nanomolar range, although the dimeric poxvirus-encoded pro-survival proteins F1L [57,58] and DPV022 [62] only display low nanomolar or micromolar K_d values. It remains to be determined if these markedly weaker affinities are related to the different oligomeric state of the virus-encoded proteins. Furthermore, the Bad/Noxa dyad does not apply to virus-encoded pro-survival Bcl-2, with fowlpoxvirus FPV039 [64] and ASFV A179L [36] binding both Noxa and Bad. Indeed, both of these proteins bind all major pro-apoptotic Bcl-2 proteins, another feature not previously observed amongst their mammalian counterparts. Lastly, Bim is not a universal target amongst virus-encoded Bcl-2, with variola virus F1L showing no affinity for Bim, and instead displaying weak binding to Bid [58] whilst being a potent inhibitor of Bax-mediated apoptosis in cellular assays. Overall the virus-encoded Bcl-2 pro-survival proteins display weaker affinities for host pro-apoptotic proteins. This could suggest that only small perturbations of the overall balance between pro-survival and pro-apoptotic proteins in a host cell are sufficient to impede apoptosis progression; however, the functional relevance of these lower affinities remains to be clarified.

When considering mechanisms and the long-standing debate on the precise mechanism of action of cellular pro-survival Bcl-2, current models and mechanisms have not been entirely resolved [84]. BH3-only proteins antagonise the pro-survival Bcl-2 family that in turn keep Bax and Bak in check, but they can also activate Bax and Bak (Figure 1). Direct binding to Bax and Bak has been observed though the affinities are generally low, for example, Bim binds full length Bax with a K_d of 3.1 μM compared with K_d in low nanomolar ranges for the pro-survival proteins [171]. While the exact details of the membrane pore generated by Bak and Bax remain under investigation, it is clear that the interactions at the membrane are crucial to the apoptotic response in mammals, and viruses also appear to exploit this. vMIA [172] and other viral Bcl-2 proteins translocate to the MOM, and a possible role for them would be to inhibit pore formation by either preventing pore growth through sequestration or retrotranslocation of the components into the cytosol as in the case of Bcl-x_L [74,92]. It is becoming apparent for the best understood virus-encoded Bcl-2 proteins that, in contrast to mammalian apoptosis, multiple mechanisms of action exist, though many of these need to be clarified with quantitative structure and binding studies that are complemented by live viral infection models. Whereas myxomavirus M11L was shown to act by sequestering Bax and Bak [35], vaccinia virus F1L was shown to only require neutralisation of Bim in a viral infection setting [143]. For EBV BHRF1, a combination of neutralisation of Bim [111] and Bak [34] was required.

When considering the role of membranes in Bcl-2 activity, and the observation that it is Bak and Bax accumulation at the membrane that is critical for intrinsic apoptosis to proceed, the association of virus-encoded Bcl-2 proteins with membranes is not unexpected. Nearly all apoptosis inhibitory vBcl-2 proteins harbour transmembrane anchoring regions to direct their subcellular localisation, chiefly to the MOM. Interestingly, different vBcl-2 proteins appear to inhibit different stages of Bax activation and translocation to the outer mitochondrial membrane [173]. E1B 19K and BHRF1, are examined for their ability to block Bax activation at different steps and thereby reveal the timing of mitochondrial changes during apoptosis. BHRF1 inhibited Bax activation but not upstream of apoptotic signalling events, whereas E1B19K permitted the initial stages of Bax activation to proceed, but prevented the subsequent oligomerisation of Bax. Furthermore, CMV-encoded m38.5 and vMIA appear to block Bax downstream of translocation to mitochondria, when Bax has already undergone structural changes [125].

These data suggest that no universal mechanism exists that enables virus-encoded Bcl-2 to subvert premature host cell apoptosis, and that the precise mechanism reflects the unique circumstances under which viral infection takes place. In particular, the mechanism of action may be heavily influenced by the initial site of contact and tissue type. A pertinent example is CMV-encoded m38.5, with an m38.5 deletion virus showing no overt signs of impaired replication in visceral organs, whereas in salivary glands a 10–100 fold difference was observed [128]. This suggests that particular tissues may be more prone to infection; however, this aspect and the observation that expression patterns of pro-apoptotic Bcl-2 proteins vary amongst tissues has not been adequately addressed for the vast majority of viruses. Interestingly, viruses can manipulate the host cell apoptosis program on many levels, with evidence emerging that viruses can manipulate the caspase cascade [174] and the endogenous levels of the host Bcl-2 family members [175]. These findings add an additional layer of complexity to the quest of identifying the precise molecular mechanism of action of vBcl-2 proteins, and ultimately suggest that more sophisticated approaches may be required to answer these questions.

Acknowledgments: This work was supported by the National Health and Medical Research Council Australia (Project Grant APP1007918 to Marc Kvansakul) and the Australian Research Council (Fellowship FT130101349 to Marc Kvansakul).

Conflicts of Interest: The authors declare no conflict of interest.

References

1. Kerr, J.F.; Wyllie, A.H.; Currie, A.R. Apoptosis: A basic biological phenomenon with wide-ranging implications in tissue kinetics. *Br. J. Cancer* **1972**, *26*, 239–257. [CrossRef] [PubMed]
2. Galluzzi, L.; Vitale, I.; Abrams, J.M.; Alnemri, E.S.; Baehrecke, E.H.; Blagosklonny, M.V.; Dawson, T.M.; Dawson, V.L.; El-Deiry, W.S.; Fulda, S.; et al. Molecular definitions of cell death subroutines: Recommendations of the Nomenclature Committee on Cell Death 2012. *Cell Death Differ.* **2012**, *19*, 107–120. [CrossRef] [PubMed]
3. Strasser, A.; Cory, S.; Adams, J.M. Deciphering the rules of programmed cell death to improve therapy of cancer and other diseases. *EMBO J.* **2011**, *30*, 3667–3683. [CrossRef] [PubMed]
4. Hardwick, J.M.; Soane, L. Multiple functions of BCL-2 family proteins. *Cold Spring Harb. Perspect. Biol.* **2013**, *5*. [CrossRef] [PubMed]
5. Youle, R.J.; Strasser, A. The BCL-2 protein family: Opposing activities that mediate cell death. *Nat. Rev. Mol. Cell Biol.* **2008**, *9*, 47–59. [CrossRef] [PubMed]
6. Kvansakul, M.; Hinds, M.G. The Bcl-2 family: Structures, interactions and targets for drug discovery. *Apoptosis* **2015**, *20*, 136–150. [CrossRef] [PubMed]
7. Kvansakul, M.; Hinds, M.G. The structural biology of BH3-only proteins. *Methods Enzymol.* **2014**, *544*, 49–74. [PubMed]
8. Kvansakul, M.; Hinds, M.G. Structural biology of the Bcl-2 family and its mimicry by viral proteins. *Cell Death Dis.* **2013**, *4*, e909. [CrossRef] [PubMed]
9. Delbridge, A.R.; Grabow, S.; Strasser, A.; Vaux, D.L. Thirty years of BCL-2: Translating cell death discoveries into novel cancer therapies. *Nat. Rev. Cancer* **2016**, *16*, 99–109. [CrossRef] [PubMed]
10. Fuentes-Prior, P.; Salvesen, G.S. The protein structures that shape caspase activity, specificity, activation and inhibition. *Biochem. J.* **2004**, *384*, 201–232. [CrossRef] [PubMed]
11. Marino, G.; Niso-Santano, M.; Baehrecke, E.H.; Kroemer, G. Self-consumption: The interplay of autophagy and apoptosis. *Nat. Rev. Mol. Cell Biol.* **2014**, *15*, 81–94. [CrossRef] [PubMed]
12. Lanave, C.; Santamaria, M.; Saccone, C. Comparative genomics: The evolutionary history of the Bcl-2 family. *Gene* **2004**, *333*, 71–79. [CrossRef] [PubMed]
13. Zmasek, C.M.; Godzik, A. Evolution of the animal apoptosis network. *Cold Spring Harb. Perspect. Biol.* **2013**, *5*, a008649. [CrossRef] [PubMed]
14. Aouacheria, A.; Rech de Laval, V.; Combet, C.; Hardwick, J.M. Evolution of Bcl-2 homology motifs: Homology versus homoplasy. *Trends Cell Biol.* **2013**, *23*, 103–111. [CrossRef] [PubMed]
15. Aouacheria, A.; Brunet, F.; Gouy, M. Phylogenomics of life-or-death switches in multicellular animals: Bcl-2, BH3-Only, and BNip families of apoptotic regulators. *Mol. Biol. Evol.* **2005**, *22*, 2395–2416. [CrossRef] [PubMed]
16. Rautureau, G.J.; Day, C.L.; Hinds, M.G. Intrinsically disordered proteins in Bcl-2 regulated apoptosis. *Int. J. Mol. Sci.* **2010**, *11*, 1808–1824. [CrossRef] [PubMed]
17. Shamas-Din, A.; Kale, J.; Leber, B.; Andrews, D.W. Mechanisms of action of Bcl-2 family proteins. *Cold Spring Harb. Perspect. Biol.* **2013**, *5*. [CrossRef] [PubMed]
18. Ke, F.; Voss, A.; Kerr, J.B.; O'Reilly, L.A.; Tai, L.; Echeverry, N.; Bouillet, P.; Strasser, A.; Kaufmann, T. BCL-2 family member BOK is widely expressed but its loss has only minimal impact in mice. *Cell Death Differ.* **2012**, *19*, 915–925. [CrossRef] [PubMed]

19. Chou, J.J.; Li, H.; Salvesen, G.S.; Yuan, J.; Wagner, G. Solution structure of BID, an intracellular amplifier of apoptotic signaling. *Cell* **1999**, *96*, 615–624. [CrossRef]
20. McDonnell, J.M.; Fushman, D.; Milliman, C.L.; Korsmeyer, S.J.; Cowburn, D. Solution structure of the proapoptotic molecule BID: A structural basis for apoptotic agonists and antagonists. *Cell* **1999**, *96*, 625–634. [CrossRef]
21. Billen, L.P.; Shamas-Din, A.; Andrews, D.W. Bid: A Bax-like BH3 protein. *Oncogene* **2008**, *27*, S93–S104. [CrossRef] [PubMed]
22. Wang, Y.; Tjandra, N. Structural insights of tBid, the caspase-8-activated Bid, and its BH3 domain. *J. Biol. Chem.* **2013**, *288*, 35840–35851. [CrossRef] [PubMed]
23. Yao, Y.; Bobkov, A.A.; Plesniak, L.A.; Marassi, F.M. Mapping the interaction of pro-apoptotic tBID with pro-survival BCL-XL. *Biochemistry* **2009**, *48*, 8704–8711. [CrossRef] [PubMed]
24. Wiens, M.; Diehl-Seifert, B.; Muller, W.E. Sponge Bcl-2 homologous protein (BHP2-GC) confers distinct stress resistance to human HEK-293 cells. *Cell Death Differ.* **2001**, *8*, 887–898. [CrossRef] [PubMed]
25. Wiens, M.; Miller, W.E.G. Cell death in Porifera: Molecular players in the game of apoptotic cell death in living fossils. *Can. J. Zool.* **2006**, *84*, 307–321. [CrossRef]
26. David, C.N.; Schmidt, N.; Schade, M.; Pauly, B.; Alexandrova, O.; Bottger, A. Hydra and the evolution of apoptosis. *Integr. Comp. Biol.* **2005**, *45*, 631–638. [CrossRef] [PubMed]
27. Wiens, M.; Belikov, S.I.; Kaluzhnaya, O.V.; Schroder, H.C.; Hamer, B.; Perovic-Ottstadt, S.; Borejko, A.; Luthringer, B.; Muller, I.M.; Muller, W.E. Axial (apical-basal) expression of pro-apoptotic and pro-survival genes in the lake baikal demosponge *Lubomirskia baicalensis*. *DNA Cell Biol.* **2006**, *25*, 152–164. [CrossRef] [PubMed]
28. Lasi, M.; Pauly, B.; Schmidt, N.; Cikala, M.; Stiening, B.; Kasbauer, T.; Zenner, G.; Popp, T.; Wagner, A.; Knapp, R.T.; et al. The molecular cell death machinery in the simple cnidarian Hydra includes an expanded caspase family and pro- and anti-apoptotic Bcl-2 proteins. *Cell Res.* **2010**, *20*, 812–825. [CrossRef] [PubMed]
29. Muller-Taubenberger, A.; Vos, M.J.; Bottger, A.; Lasi, M.; Lai, F.P.; Fischer, M.; Rottner, K. Monomeric red fluorescent protein variants used for imaging studies in different species. *Eur. J. Cell Biol.* **2006**, *85*, 1119–1129. [CrossRef] [PubMed]
30. Bottger, A.; Alexandrova, O. Programmed cell death in Hydra. *Semin. Cancer Biol.* **2007**, *17*, 134–146. [CrossRef] [PubMed]
31. Ainsworth, T.D.; Knack, B.; Ukani, L.; Seneca, F.; Weiss, Y.; Leggat, W. In situ hybridisation detects pro-apoptotic gene expression of a Bcl-2 family member in white syndrome-affected coral. *Dis. Aquat. Organ.* **2015**, *117*, 155–163. [CrossRef] [PubMed]
32. Caria, S.; Hinds, M.G.; Kvansakul, M. Structural insight into an evolutionarily ancient programmed cell death regulator - the crystal structure of marine sponge BHP2 bound to LB-Bak-2. *Cell Death Dis.* **2017**, *8*, e2543. [CrossRef] [PubMed]
33. Liu, X.; Dai, S.; Zhu, Y.; Marrack, P.; Kappler, J.W. The structure of a Bcl-xL/Bim fragment complex: Implications for Bim function. *Immunity* **2003**, *19*, 341–352. [CrossRef]
34. Kvansakul, M.; Wei, A.H.; Fletcher, J.I.; Willis, S.N.; Chen, L.; Roberts, A.W.; Huang, D.C.; Colman, P.M. Structural basis for apoptosis inhibition by Epstein-Barr virus BHRF1. *PLoS Pathog.* **2010**, *6*, e1001236. [CrossRef] [PubMed]
35. Kvansakul, M.; van Delft, M.F.; Lee, E.F.; Gulbis, J.M.; Fairlie, W.D.; Huang, D.C.; Colman, P.M. A structural viral mimic of prosurvival Bcl-2: A pivotal role for sequestering proapoptotic Bax and Bak. *Mol. Cell* **2007**, *25*, 933–942. [CrossRef] [PubMed]
36. Banjara, S.; Caria, S.; Dixon, L.K.; Hinds, M.G.; Kvansakul, M. Structural Insight into African Swine Fever Virus A179L-Mediated Inhibition of Apoptosis. *J. Virol.* **2017**, *91*. [CrossRef] [PubMed]
37. Ku, B.; Woo, J.S.; Liang, C.; Lee, K.H.; Hong, H.S.; E, X.; Kim, K.S.; Jung, J.U.; Oh, B.H. Structural and biochemical bases for the inhibition of autophagy and apoptosis by viral BCL-2 of murine gamma-herpesvirus 68. *PLoS Pathog.* **2008**, *4*, e25. [CrossRef] [PubMed]
38. Caria, S.; Marshall, B.; Burton, R.L.; Campbell, S.; Pantaki-Eimany, D.; Hawkins, C.J.; Barry, M.; Kvansakul, M. The N Terminus of the Vaccinia Virus Protein F1L Is an Intrinsically Unstructured Region That Is Not Involved in Apoptosis Regulation. *J. Biol. Chem.* **2016**, *291*, 14600–14608. [CrossRef] [PubMed]

39. Graham, S.C.; Bahar, M.W.; Cooray, S.; Chen, R.A.; Whalen, D.M.; Abrescia, N.G.; Alderton, D.; Owens, R.J.; Stuart, D.I.; Smith, G.L.; et al. Vaccinia virus proteins A52 and B14 Share a Bcl-2-like fold but have evolved to inhibit NF-κB rather than apoptosis. *PLoS Pathog.* **2008**, *4*, e1000128. [CrossRef] [PubMed]

40. Pihan, P.; Carreras-Sureda, A.; Hetz, C. BCL-2 family: Integrating stress responses at the ER to control cell demise. *Cell Death Differ.* **2017**, *24*, 1478–1487. [CrossRef] [PubMed]

41. Janssens, S.; Pulendran, B.; Lambrecht, B.N. Emerging functions of the unfolded protein response in immunity. *Nat. Immunol.* **2014**, *15*, 910–919. [CrossRef] [PubMed]

42. Lindsten, T.; Ross, A.J.; King, A.; Zong, W.X.; Rathmell, J.C.; Shiels, H.A.; Ulrich, E.; Waymire, K.G.; Mahar, P.; Frauwirth, K.; et al. The combined functions of proapoptotic Bcl-2 family members Bak and Bax are essential for normal development of multiple tissues. *Mol. Cell* **2000**, *6*, 1389–1399. [CrossRef]

43. Chen, F.; Hersh, B.M.; Conradt, B.; Zhou, Z.; Riemer, D.; Gruenbaum, Y.; Horvitz, H.R. Translocation of C. elegans CED-4 to nuclear membranes during programmed cell death. *Science* **2000**, *287*, 1485–1489. [CrossRef] [PubMed]

44. Neumann, S.; El Maadidi, S.; Faletti, L.; Haun, F.; Labib, S.; Schejtman, A.; Maurer, U.; Borner, C. How do viruses control mitochondria-mediated apoptosis? *Virus Res.* **2015**, *209*, 45–55. [CrossRef] [PubMed]

45. Wang, G.H.; Garvey, T.L.; Cohen, J.I. The murine gammaherpesvirus-68 M11 protein inhibits Fas- and TNF-induced apoptosis. *J. Gen. Virol.* **1999**, *80*, 2737–2740. [CrossRef] [PubMed]

46. Chiou, S.K.; Tseng, C.C.; Rao, L.; White, E. Functional complementation of the adenovirus E1B 19-kilodalton protein with Bcl-2 in the inhibition of apoptosis in infected cells. *J. Virol.* **1994**, *68*, 6553–6566. [PubMed]

47. Henderson, S.; Huen, D.; Rowe, M.; Dawson, C.; Johnson, G.; Rickinson, A. Epstein-Barr virus-coded BHRF1 protein, a viral homologue of Bcl-2, protects human B cells from programmed cell death. *Proc. Natl. Acad. Sci. USA* **1993**, *90*, 8479–8483. [CrossRef] [PubMed]

48. Cheng, E.H.; Nicholas, J.; Bellows, D.S.; Hayward, G.S.; Guo, H.G.; Reitz, M.S.; Hardwick, J.M. A Bcl-2 homolog encoded by Kaposi sarcoma-associated virus, human herpesvirus 8, inhibits apoptosis but does not heterodimerize with Bax or Bak. *Proc. Natl. Acad. Sci. USA* **1997**, *94*, 690–694. [CrossRef] [PubMed]

49. Huang, Q.; Petros, A.M.; Virgin, H.W.; Fesik, S.W.; Olejniczak, E.T. Solution structure of a Bcl-2 homolog from Kaposi sarcoma virus. *Proc. Natl. Acad. Sci. USA* **2002**, *99*, 3428–3433. [CrossRef] [PubMed]

50. Aouacheria, A.; Banyai, M.; Rigal, D.; Schmidt, C.J.; Gillet, G. Characterization of *vNR-13*, the first alphaherpesvirus gene of the *bcl-2* family. *Virology* **2003**, *316*, 256–266. [CrossRef] [PubMed]

51. Brun, A.; Rivas, C.; Esteban, M.; Escribano, J.M.; Alonso, C. African swine fever virus gene A179L, a viral homologue of bcl-2, protects cells from programmed cell death. *Virology* **1996**, *225*, 227–230. [CrossRef] [PubMed]

52. Lu, L.; Zhou, S.Y.; Chen, C.; Weng, S.P.; Chan, S.M.; He, J.G. Complete genome sequence analysis of an iridovirus isolated from the orange-spotted grouper, Epinephelus coioides. *Virology* **2005**, *339*, 81–100. [CrossRef] [PubMed]

53. Tsai, C.T.; Ting, J.W.; Wu, M.H.; Wu, M.F.; Guo, I.C.; Chang, C.Y. Complete genome sequence of the grouper iridovirus and comparison of genomic organization with those of other iridoviruses. *J. Virol.* **2005**, *79*, 2010–2023. [CrossRef] [PubMed]

54. Graham, K.A.; Opgenorth, A.; Upton, C.; McFadden, G. Myxoma virus M11L ORF encodes a protein for which cell surface localization is critical in manifestation of viral virulence. *Virology* **1992**, *191*, 112–124. [CrossRef]

55. Wasilenko, S.T.; Stewart, T.L.; Meyers, A.F.; Barry, M. Vaccinia virus encodes a previously uncharacterized mitochondrial-associated inhibitor of apoptosis. *Proc. Natl. Acad. Sci. USA* **2003**, *100*, 14345–14350. [CrossRef] [PubMed]

56. Fischer, S.F.; Ludwig, H.; Holzapfel, J.; Kvansakul, M.; Chen, L.; Huang, D.C.; Sutter, G.; Knese, M.; Hacker, G. Modified vaccinia virus Ankara protein F1L is a novel BH3-domain-binding protein and acts together with the early viral protein E3L to block virus-associated apoptosis. *Cell Death Differ.* **2006**, *13*, 109–118. [CrossRef] [PubMed]

57. Kvansakul, M.; Yang, H.; Fairlie, W.D.; Czabotar, P.E.; Fischer, S.F.; Perugini, M.A.; Huang, D.C.; Colman, P.M. Vaccinia virus anti-apoptotic F1L is a novel Bcl-2-like domain-swapped dimer that binds a highly selective subset of BH3-containing death ligands. *Cell Death Differ.* **2008**, *15*, 1564–1571. [CrossRef] [PubMed]

58. Marshall, B.; Puthalakath, H.; Caria, S.; Chugh, S.; Doerflinger, M.; Colman, P.M.; Kvansakul, M. Variola virus F1L is a Bcl-2-like protein that unlike its vaccinia virus counterpart inhibits apoptosis independent of Bim. *Cell Death Dis.* **2015**, *6*, e1680. [CrossRef] [PubMed]

59. Mehta, N.; Taylor, J.; Quilty, D.; Barry, M. Ectromelia virus encodes an anti-apoptotic protein that regulates cell death. *Virology* **2015**, *475*, 74–87. [CrossRef] [PubMed]

60. Okamoto, T.; Campbell, S.; Mehta, N.; Thibault, J.; Colman, P.M.; Barry, M.; Huang, D.C.; Kvansakul, M. Sheeppox virus SPPV14 encodes a Bcl-2-like cell death inhibitor that counters a distinct set of mammalian proapoptotic proteins. *J. Virol.* **2012**, *86*, 11501–11511. [CrossRef] [PubMed]

61. Banadyga, L.; Lam, S.C.; Okamoto, T.; Kvansakul, M.; Huang, D.C.; Barry, M. Deerpox virus encodes an inhibitor of apoptosis that regulates Bak and Bax. *J. Virol.* **2011**, *85*, 1922–1934. [CrossRef] [PubMed]

62. Burton, D.R.; Caria, S.; Marshall, B.; Barry, M.; Kvansakul, M. Structural basis of Deerpox virus-mediated inhibition of apoptosis. *Acta Crystallogr. D Biol. Crystallogr.* **2015**, *71*, 1593–1603. [CrossRef] [PubMed]

63. Banadyga, L.; Gerig, J.; Stewart, T.; Barry, M. Fowlpox virus encodes a Bcl-2 homologue that protects cells from apoptotic death through interaction with the proapoptotic protein Bak. *J. Virol.* **2007**, *81*, 11032–11045. [CrossRef] [PubMed]

64. Anasir, M.I.; Caria, S.; Skinner, M.A.; Kvansakul, M. Structural basis of apoptosis inhibition by the fowlpox virus protein FPV039. *J. Biol. Chem.* **2017**, *292*, 9010–9021. [CrossRef] [PubMed]

65. Tulman, E.R.; Afonso, C.L.; Lu, Z.; Zsak, L.; Kutish, G.F.; Rock, D.L. The genome of canarypox virus. *J. Virol.* **2004**, *78*, 353–366. [CrossRef] [PubMed]

66. Westphal, D.; Ledgerwood, E.C.; Hibma, M.H.; Fleming, S.B.; Whelan, E.M.; Mercer, A.A. A novel Bcl-2-like inhibitor of apoptosis is encoded by the parapoxvirus ORF virus. *J. Virol.* **2007**, *81*, 7178–7188. [CrossRef] [PubMed]

67. Uren, R.T.; Iyer, S.; Kluck, R.M. Pore formation by dimeric Bak and Bax: An unusual pore? *Philos. Trans. R. Soc. Lond. B Biol. Sci.* **2017**, *372*. [CrossRef] [PubMed]

68. Altmann, M.; Hammerschmidt, W. Epstein-Barr virus provides a new paradigm: A requirement for the immediate inhibition of apoptosis. *PLoS Biol.* **2005**, *3*, e404. [CrossRef] [PubMed]

69. Todt, F.; Cakir, Z.; Reichenbach, F.; Emschermann, F.; Lauterwasser, J.; Kaiser, A.; Ichim, G.; Tait, S.W.; Frank, S.; Langer, H.F.; et al. Differential retrotranslocation of mitochondrial Bax and Bak. *EMBO J.* **2015**, *34*, 67–80. [CrossRef] [PubMed]

70. Zong, W.X.; Li, C.; Hatzivassiliou, G.; Lindsten, T.; Yu, Q.C.; Yuan, J.; Thompson, C.B. Bax and Bak can localize to the endoplasmic reticulum to initiate apoptosis. *J. Cell Biol.* **2003**, *162*, 59–69. [CrossRef] [PubMed]

71. Akao, Y.; Otsuki, Y.; Kataoka, S.; Ito, Y.; Tsujimoto, Y. Multiple subcellular localization of Bcl-2: Detection in nuclear outer membrane, endoplasmic reticulum membrane, and mitochondrial membranes. *Cancer Res.* **1994**, *54*, 2468–2471. [PubMed]

72. Wolter, K.G.; Hsu, Y.T.; Smith, C.L.; Nechushtan, A.; Xi, X.G.; Youle, R.J. Movement of Bax from the cytosol to mitochondria during apoptosis. *J. Cell Biol.* **1997**, *139*, 1281–1292. [CrossRef] [PubMed]

73. Hsu, Y.T.; Wolter, K.G.; Youle, R.J. Cytosol-to-membrane redistribution of Bax and Bcl-X(L) during apoptosis. *Proc. Natl. Acad. Sci. USA* **1997**, *94*, 3668–3672. [CrossRef] [PubMed]

74. Edlich, F.; Banerjee, S.; Suzuki, M.; Cleland, M.M.; Arnoult, D.; Wang, C.; Neutzner, A.; Tjandra, N.; Youle, R.J. Bcl-x(L) retrotranslocates Bax from the mitochondria into the cytosol. *Cell* **2011**, *145*, 104–116. [CrossRef] [PubMed]

75. Todt, F.; Cakir, Z.; Reichenbach, F.; Youle, R.J.; Edlich, F. The C-terminal helix of Bcl-x(L) mediates Bax retrotranslocation from the mitochondria. *Cell Death Differ.* **2013**, *20*, 333–342. [CrossRef] [PubMed]

76. Schellenberg, B.; Wang, P.; Keeble, J.A.; Rodriguez-Enriquez, R.; Walker, S.; Owens, T.W.; Foster, F.; Tanianis-Hughes, J.; Brennan, K.; Streuli, C.H.; et al. Bax exists in a dynamic equilibrium between the cytosol and mitochondria to control apoptotic priming. *Mol. Cell* **2013**, *49*, 959–971. [CrossRef] [PubMed]

77. Suzuki, M.; Youle, R.J.; Tjandra, N. Structure of Bax: Coregulation of dimer formation and intracellular localization. *Cell* **2000**, *103*, 645–654. [CrossRef]

78. Oltvai, Z.N.; Milliman, C.L.; Korsmeyer, S.J. Bcl-2 heterodimerizes in vivo with a conserved homolog, Bax, that accelerates programmed cell death. *Cell* **1993**, *74*, 609–619. [CrossRef]

79. Wilson-Annan, J.; O'Reilly, L.A.; Crawford, S.A.; Hausmann, G.; Beaumont, J.G.; Parma, L.P.; Chen, L.; Lackmann, M.; Lithgow, T.; Hinds, M.G.; et al. Proapoptotic BH3-only proteins trigger membrane integration of prosurvival Bcl-w and neutralize its activity. *J. Cell Biol.* **2003**, *162*, 877–887. [CrossRef] [PubMed]

80. Hinds, M.G.; Lackmann, M.; Skea, G.L.; Harrison, P.J.; Huang, D.C.; Day, C.L. The structure of Bcl-w reveals a role for the C-terminal residues in modulating biological activity. *EMBO J.* **2003**, *22*, 1497–1507. [CrossRef] [PubMed]

81. Opgenorth, A.; Graham, K.; Nation, N.; Strayer, D.; McFadden, G. Deletion analysis of two tandemly arranged virulence genes in myxoma virus, M11L and myxoma growth factor. *J. Virol.* **1992**, *66*, 4720–4731. [PubMed]

82. Nechushtan, A.; Smith, C.L.; Hsu, Y.T.; Youle, R.J. Conformation of the Bax C-terminus regulates subcellular location and cell death. *EMBO J.* **1999**, *18*, 2330–2341. [CrossRef] [PubMed]

83. Tait, S.W.; Green, D.R. Mitochondrial regulation of cell death. *Cold Spring Harb. Perspect. Biol.* **2013**, *5*. [CrossRef] [PubMed]

84. Czabotar, P.E.; Lessene, G.; Strasser, A.; Adams, J.M. Control of apoptosis by the BCL-2 protein family: Implications for physiology and therapy. *Nat. Rev. Mol. Cell Biol.* **2014**, *15*, 49–63. [CrossRef] [PubMed]

85. Hockings, C.; Anwari, K.; Ninnis, R.L.; Brouwer, J.; O'Hely, M.; Evangelista, M.; Hinds, M.G.; Czabotar, P.E.; Lee, E.F.; Fairlie, W.D.; et al. Bid chimeras indicate that most BH3-only proteins can directly activate Bak and Bax, and show no preference for Bak versus Bax. *Cell Death Dis.* **2015**, *6*, e1735. [CrossRef] [PubMed]

86. White, E.; Blose, S.H.; Stillman, B.W. Nuclear envelope localization of an adenovirus tumor antigen maintains the integrity of cellular DNA. *Mol. Cell Biol.* **1984**, *4*, 2865–2875. [CrossRef] [PubMed]

87. Rao, L.; Modha, D.; White, E. The E1B 19K protein associates with lamins in vivo and its proper localization is required for inhibition of apoptosis. *Oncogene* **1997**, *15*, 1587–1597. [CrossRef] [PubMed]

88. Ring, B.A.; Ferreira Lacerda, A.; Drummond, D.J.; Wangen, C.; Eaton, H.E.; Brunetti, C.R. Frog virus 3 open reading frame 97R localizes to the endoplasmic reticulum and induces nuclear invaginations. *J. Virol.* **2013**, *87*, 9199–9207. [CrossRef] [PubMed]

89. Hernaez, B.; Cabezas, M.; Munoz-Moreno, R.; Galindo, I.; Cuesta-Geijo, M.A.; Alonso, C. A179L, a new viral Bcl2 homolog targeting Beclin 1 autophagy related protein. *Curr. Mol. Med.* **2013**, *13*, 305–316. [CrossRef] [PubMed]

90. Cosentino, K.; Garcia-Saez, A.J. Bax and Bak Pores: Are We Closing the Circle? *Trends Cell Biol.* **2017**, *27*, 266–275. [CrossRef] [PubMed]

91. Barnes, C.A.; Mishra, P.; Baber, J.L.; Strub, M.P.; Tjandra, N. Conformational Heterogeneity in the Activation Mechanism of Bax. *Structure* **2017**, *25*, 1310–1316.e1313. [CrossRef] [PubMed]

92. Bleicken, S.; Hantusch, A.; Das, K.K.; Frickey, T.; Garcia-Saez, A.J. Quantitative interactome of a membrane Bcl-2 network identifies a hierarchy of complexes for apoptosis regulation. *Nat. Commun.* **2017**, *8*, 73. [CrossRef] [PubMed]

93. Grosse, L.; Wurm, C.A.; Bruser, C.; Neumann, D.; Jans, D.C.; Jakobs, S. Bax assembles into large ring-like structures remodeling the mitochondrial outer membrane in apoptosis. *EMBO J.* **2016**, *35*, 402–413. [CrossRef] [PubMed]

94. Salvador-Gallego, R.; Mund, M.; Cosentino, K.; Schneider, J.; Unsay, J.; Schraermeyer, U.; Engelhardt, J.; Ries, J.; Garcia-Saez, A.J. Bax assembly into rings and arcs in apoptotic mitochondria is linked to membrane pores. *EMBO J.* **2016**, *35*, 389–401. [CrossRef] [PubMed]

95. Andreu-Fernandez, V.; Sancho, M.; Genoves, A.; Lucendo, E.; Todt, F.; Lauterwasser, J.; Funk, K.; Jahreis, G.; Perez-Paya, E.; Mingarro, I.; et al. Bax transmembrane domain interacts with prosurvival Bcl-2 proteins in biological membranes. *Proc. Natl. Acad. Sci. USA* **2017**, *114*, 310–315. [CrossRef] [PubMed]

96. Yao, Y.; Fujimoto, L.M.; Hirshman, N.; Bobkov, A.A.; Antignani, A.; Youle, R.J.; Marassi, F.M. Conformation of BCL-XL upon Membrane Integration. *J. Mol. Biol.* **2015**, *427*, 2262–2270. [CrossRef] [PubMed]

97. Muchmore, S.W.; Sattler, M.; Liang, H.; Meadows, R.P.; Harlan, J.E.; Yoon, H.S.; Nettesheim, D.; Chang, B.S.; Thompson, C.B.; Wong, S.L.; et al. X-ray and NMR structure of human Bcl-xL, an inhibitor of programmed cell death. *Nature* **1996**, *381*, 335–341. [CrossRef] [PubMed]

98. Yao, Y.; Nisan, D.; Fujimoto, L.M.; Antignani, A.; Barnes, A.; Tjandra, N.; Youle, R.J.; Marassi, F.M. Characterization of the membrane-inserted C-terminus of cytoprotective BCL-XL. *Protein Expr. Purif.* **2016**, *122*, 56–63. [CrossRef] [PubMed]

99. Iyer, S.; Bell, F.; Westphal, D.; Anwari, K.; Gulbis, J.; Smith, B.J.; Dewson, G.; Kluck, R.M. Bak apoptotic pores involve a flexible C-terminal region and juxtaposition of the C-terminal transmembrane domains. *Cell Death Differ.* **2015**, *22*, 1665–1675. [CrossRef] [PubMed]

100. Dewson, G.; Kratina, T.; Czabotar, P.; Day, C.L.; Adams, J.M.; Kluck, R.M. Bak activation for apoptosis involves oligomerization of dimers via their α6 helices. *Mol. Cell* **2009**, *36*, 696–703. [CrossRef] [PubMed]

101. Gavathiotis, E.; Suzuki, M.; Davis, M.L.; Pitter, K.; Bird, G.H.; Katz, S.G.; Tu, H.C.; Kim, H.; Cheng, E.H.; Tjandra, N.; et al. BAX activation is initiated at a novel interaction site. *Nature* **2008**, *455*, 1076–1081. [CrossRef] [PubMed]

102. Czabotar, P.E.; Westphal, D.; Dewson, G.; Ma, S.; Hockings, C.; Fairlie, W.D.; Lee, E.F.; Yao, S.; Robin, A.Y.; Smith, B.J.; et al. Bax crystal structures reveal how BH3 domains activate Bax and nucleate its oligomerization to induce apoptosis. *Cell* **2013**, *152*, 519–531. [CrossRef] [PubMed]

103. Robin, A.Y.; Krishna Kumar, K.; Westphal, D.; Wardak, A.Z.; Thompson, G.V.; Dewson, G.; Colman, P.M.; Czabotar, P.E. Crystal structure of Bax bound to the BH3 peptide of Bim identifies important contacts for interaction. *Cell Death Dis.* **2015**, *6*, e1809. [CrossRef] [PubMed]

104. O'Neill, J.W.; Manion, M.K.; Maguire, B.; Hockenbery, D.M. BCL-XL dimerization by three-dimensional domain swapping. *J. Mol. Biol.* **2006**, *356*, 367–381. [CrossRef] [PubMed]

105. Rodriguez-Grille, J.; Busch, L.K.; Martinez-Costas, J.; Benavente, J. Avian reovirus-triggered apoptosis enhances both virus spread and the processing of the viral nonstructural muNS protein. *Virology* **2014**, *462–463*, 49–59. [CrossRef] [PubMed]

106. Han, J.; Sabbatini, P.; Perez, D.; Rao, L.; Modha, D.; White, E. The E1B 19K protein blocks apoptosis by interacting with and inhibiting the p53-inducible and death-promoting Bax protein. *Genes Dev.* **1996**, *10*, 461–477. [CrossRef] [PubMed]

107. Farrow, S.N.; White, J.H.; Martinou, I.; Raven, T.; Pun, K.T.; Grinham, C.J.; Martinou, J.C.; Brown, R. Cloning of a Bcl-2 homologue by interaction with adenovirus E1B 19K. *Nature* **1995**, *374*, 731–733. [CrossRef] [PubMed]

108. Han, J.; Wallen, H.D.; Nunez, G.; White, E. E1B 19,000-molecular-weight protein interacts with and inhibits CED-4-dependent, FLICE-mediated apoptosis. *Mol. Cell Biol.* **1998**, *18*, 6052–6062. [CrossRef] [PubMed]

109. Huang, Q.; Petros, A.M.; Virgin, H.W.; Fesik, S.W.; Olejniczak, E.T. Solution structure of the BHRF1 protein from Epstein-Barr virus, a homolog of human Bcl-2. *J. Mol. Biol.* **2003**, *332*, 1123–1130. [CrossRef] [PubMed]

110. Flanagan, A.M.; Letai, A. BH3 domains define selective inhibitory interactions with BHRF-1 and KSHV BCL-2. *Cell Death Differ.* **2008**, *15*, 580–588. [CrossRef] [PubMed]

111. Desbien, A.L.; Kappler, J.W.; Marrack, P. The Epstein-Barr virus Bcl-2 homolog, BHRF1, blocks apoptosis by binding to a limited amount of Bim. *Proc. Natl. Acad. Sci. USA* **2009**, *106*, 5663–5668. [CrossRef] [PubMed]

112. Kelly, G.L.; Long, H.M.; Stylianou, J.; Thomas, W.A.; Leese, A.; Bell, A.I.; Bornkamm, G.W.; Mautner, J.; Rickinson, A.B.; Rowe, M. An Epstein-Barr virus anti-apoptotic protein constitutively expressed in transformed cells and implicated in burkitt lymphomagenesis: The Wp/BHRF1 link. *PLoS Pathog.* **2009**, *5*, e1000341. [CrossRef] [PubMed]

113. Marshall, W.L.; Yim, C.; Gustafson, E.; Graf, T.; Sage, D.R.; Hanify, K.; Williams, L.; Fingeroth, J.; Finberg, R.W. Epstein-Barr virus encodes a novel homolog of the bcl-2 oncogene that inhibits apoptosis and associates with Bax and Bak. *J. Virol.* **1999**, *73*, 5181–5185. [PubMed]

114. Bellows, D.S.; Howell, M.; Pearson, C.; Hazlewood, S.A.; Hardwick, J.M. Epstein-Barr virus BALF1 is a BCL-2-like antagonist of the herpesvirus antiapoptotic BCL-2 proteins. *J. Virol.* **2002**, *76*, 2469–2479. [CrossRef] [PubMed]

115. Procko, E.; Berguig, G.Y.; Shen, B.W.; Song, Y.; Frayo, S.; Convertine, A.J.; Margineantu, D.; Booth, G.; Correia, B.E.; Cheng, Y.; et al. A computationally designed inhibitor of an Epstein-Barr viral Bcl-2 protein induces apoptosis in infected cells. *Cell* **2014**, *157*, 1644–1656. [CrossRef] [PubMed]

116. Caria, S.; Chugh, S.; Nhu, D.; Lessene, G.; Kvansakul, M. Crystallization and preliminary X-ray characterization of Epstein-Barr virus BHRF1 in complex with a benzoylurea peptidomimetic. *Acta Crystallogr. Sect. F Struct. Biol. Cryst. Commun.* **2012**, *68*, 1521–1524. [CrossRef] [PubMed]

117. Sarid, R.; Sato, T.; Bohenzky, R.A.; Russo, J.J.; Chang, Y. Kaposi's sarcoma-associated herpesvirus encodes a functional Bcl-2 homologue. *Nat. Med.* **1997**, *3*, 293–298. [CrossRef] [PubMed]

118. Gallo, A.; Lampe, M.; Gunther, T.; Brune, W. The Viral Bcl-2 Homologs of Kaposi's Sarcoma-Associated Herpesvirus and Rhesus Rhadinovirus Share an Essential Role for Viral Replication. *J. Virol.* **2017**, *91*. [CrossRef] [PubMed]

119. Nava, V.E.; Cheng, E.H.; Veliuona, M.; Zou, S.; Clem, R.J.; Mayer, M.L.; Hardwick, J.M. Herpesvirus saimiri encodes a functional homolog of the human Bcl-2 oncogene. *J. Virol.* **1997**, *71*, 4118–4122. [PubMed]

120. Virgin, H.W., 4th; Latreille, P.; Wamsley, P.; Hallsworth, K.; Weck, K.E.; Dal Canto, A.J.; Speck, S.H. Complete sequence and genomic analysis of murine γherpesvirus 68. *J. Virol.* **1997**, *71*, 5894–5904. [PubMed]

121. Roy, D.J.; Ebrahimi, B.C.; Dutia, B.M.; Nash, A.A.; Stewart, J.P. Murine γherpesvirus M11 gene product inhibits apoptosis and is expressed during virus persistence. *Arch. Virol.* **2000**, *145*, 2411–2420. [CrossRef] [PubMed]

122. Sinha, S.; Colbert, C.L.; Becker, N.; Wei, Y.; Levine, B. Molecular basis of the regulation of Beclin 1-dependent autophagy by the γ-herpesvirus 68 Bcl-2 homolog M11. *Autophagy* **2008**, *4*, 989–997. [CrossRef] [PubMed]

123. Goldmacher, V.S.; Bartle, L.M.; Skaletskaya, A.; Dionne, C.A.; Kedersha, N.L.; Vater, C.A.; Han, J.W.; Lutz, R.J.; Watanabe, S.; Cahir McFarland, E.D.; et al. A cytomegalovirus-encoded mitochondria-localized inhibitor of apoptosis structurally unrelated to Bcl-2. *Proc. Natl. Acad. Sci. USA* **1999**, *96*, 12536–12541. [CrossRef] [PubMed]

124. Karbowski, M.; Norris, K.L.; Cleland, M.M.; Jeong, S.Y.; Youle, R.J. Role of Bax and Bak in mitochondrial morphogenesis. *Nature* **2006**, *443*, 658–662. [CrossRef] [PubMed]

125. Norris, K.L.; Youle, R.J. Cytomegalovirus proteins vMIA and m38.5 link mitochondrial morphogenesis to Bcl-2 family proteins. *J. Virol.* **2008**, *82*, 6232–6243. [CrossRef] [PubMed]

126. Ma, J.; Edlich, F.; Bermejo, G.A.; Norris, K.L.; Youle, R.J.; Tjandra, N. Structural mechanism of Bax inhibition by cytomegalovirus protein vMIA. *Proc. Natl. Acad. Sci. USA* **2012**, *109*, 20901–20906. [CrossRef] [PubMed]

127. Cam, M.; Handke, W.; Picard-Maureau, M.; Brune, W. Cytomegaloviruses inhibit Bak- and Bax-mediated apoptosis with two separate viral proteins. *Cell Death Differ.* **2010**, *17*, 655–665. [CrossRef] [PubMed]

128. Manzur, M.; Fleming, P.; Huang, D.C.; Degli-Esposti, M.A.; Andoniou, C.E. Virally mediated inhibition of Bax in leukocytes promotes dissemination of murine cytomegalovirus. *Cell Death Differ.* **2009**, *16*, 312–320. [CrossRef] [PubMed]

129. Arnoult, D.; Skaletskaya, A.; Estaquier, J.; Dufour, C.; Goldmacher, V.S. The murine cytomegalovirus cell death suppressor m38.5 binds Bax and blocks Bax-mediated mitochondrial outer membrane permeabilization. *Apoptosis* **2008**, *13*, 1100–1110. [CrossRef] [PubMed]

130. Fleming, P.; Kvansakul, M.; Voigt, V.; Kile, B.T.; Kluck, R.M.; Huang, D.C.; Degli-Esposti, M.A.; Andoniou, C.E. MCMV-mediated inhibition of the pro-apoptotic Bak protein is required for optimal in vivo replication. *PLoS Pathog.* **2013**, *9*, e1003192. [CrossRef] [PubMed]

131. Handke, W.; Luig, C.; Popovic, B.; Krmpotic, A.; Jonjic, S.; Brune, W. Viral inhibition of BAK promotes murine cytomegalovirus dissemination to salivary glands. *J. Virol.* **2013**, *87*, 3592–3596. [CrossRef] [PubMed]

132. Lee, R.M.; Gillet, G.; Burnside, J.; Thomas, S.J.; Neiman, P. Role of Nr13 in regulation of programmed cell death in the bursa of Fabricius. *Genes Dev.* **1999**, *13*, 718–728. [CrossRef] [PubMed]

133. Ke, N.; Godzik, A.; Reed, J.C. Bcl-B, a novel Bcl-2 family member that differentially binds and regulates Bax and Bak. *J. Biol. Chem.* **2001**, *276*, 12481–12484. [CrossRef] [PubMed]

134. Aouacheria, A.; Arnaud, E.; Venet, S.; Lalle, P.; Gouy, M.; Rigal, D.; Gillet, G. Nrh, a human homologue of Nr-13 associates with Bcl-Xs and is an inhibitor of apoptosis. *Oncogene* **2001**, *20*, 5846–5855. [CrossRef] [PubMed]

135. Rautureau, G.J.; Yabal, M.; Yang, H.; Huang, D.C.; Kvansakul, M.; Hinds, M.G. The restricted binding repertoire of Bcl-B leaves Bim as the universal BH3-only prosurvival Bcl-2 protein antagonist. *Cell Death Dis.* **2012**, *3*, e443. [CrossRef] [PubMed]

136. Rautureau, G.J.; Day, C.L.; Hinds, M.G. The structure of Boo/Diva reveals a divergent Bcl-2 protein. *Proteins* **2010**, *78*, 2181–2186. [CrossRef] [PubMed]

137. Ku, B.; Woo, J.S.; Liang, C.; Lee, K.H.; Jung, J.U.; Oh, B.H. An insight into the mechanistic role of Beclin 1 and its inhibition by prosurvival Bcl-2 family proteins. *Autophagy* **2008**, *4*, 519–520. [CrossRef] [PubMed]

138. Aoyagi, M.; Zhai, D.; Jin, C.; Aleshin, A.E.; Stec, B.; Reed, J.C.; Liddington, R.C. Vaccinia virus N1L protein resembles a B cell lymphoma-2 (Bcl-2) family protein. *Protein Sci.* **2007**, *16*, 118–124. [CrossRef] [PubMed]

139. Chen, L.; Willis, S.N.; Wei, A.; Smith, B.J.; Fletcher, J.I.; Hinds, M.G.; Colman, P.M.; Day, C.L.; Adams, J.M.; Huang, D.C. Differential targeting of prosurvival Bcl-2 proteins by their BH3-only ligands allows complementary apoptotic function. *Mol. Cell* **2005**, *17*, 393–403. [CrossRef] [PubMed]

140. Smits, C.; Czabotar, P.E.; Hinds, M.G.; Day, C.L. Structural plasticity underpins promiscuous binding of the prosurvival protein A1. *Structure* **2008**, *16*, 818–829. [CrossRef] [PubMed]

141. Willis, S.N.; Chen, L.; Dewson, G.; Wei, A.; Naik, E.; Fletcher, J.I.; Adams, J.M.; Huang, D.C. Proapoptotic Bak is sequestered by Mcl-1 and Bcl-xL, but not Bcl-2, until displaced by BH3-only proteins. *Genes Dev.* **2005**, *19*, 1294–1305. [CrossRef] [PubMed]

142. Fletcher, J.I.; Meusburger, S.; Hawkins, C.J.; Riglar, D.T.; Lee, E.F.; Fairlie, W.D.; Huang, D.C.; Adams, J.M. Apoptosis is triggered when prosurvival Bcl-2 proteins cannot restrain Bax. *Proc. Natl. Acad. Sci. USA* **2008**, *105*, 18081–18087. [CrossRef] [PubMed]

143. Campbell, S.; Thibault, J.; Mehta, N.; Colman, P.M.; Barry, M.; Kvansakul, M. Structural insight into BH3 domain binding of vaccinia virus antiapoptotic F1L. *J. Virol.* **2014**, *88*, 8667–8677. [CrossRef] [PubMed]

144. Taylor, J.M.; Quilty, D.; Banadyga, L.; Barry, M. The vaccinia virus protein F1L interacts with Bim and inhibits activation of the pro-apoptotic protein Bax. *J. Biol. Chem.* **2006**, *281*, 39728–39739. [CrossRef] [PubMed]

145. Postigo, A.; Cross, J.R.; Downward, J.; Way, M. Interaction of F1L with the BH3 domain of Bak is responsible for inhibiting vaccinia-induced apoptosis. *Cell Death Differ.* **2006**, *13*, 1651–1662. [CrossRef] [PubMed]

146. Wasilenko, S.T.; Banadyga, L.; Bond, D.; Barry, M. The vaccinia virus F1L protein interacts with the proapoptotic protein Bak and inhibits Bak activation. *J. Virol.* **2005**, *79*, 14031–14043. [CrossRef] [PubMed]

147. Campbell, S.; Hazes, B.; Kvansakul, M.; Colman, P.; Barry, M. Vaccinia virus F1L interacts with Bak using highly divergent Bcl-2 homology domains and replaces the function of Mcl-1. *J. Biol. Chem.* **2010**, *285*, 4695–4708. [CrossRef] [PubMed]

148. Stewart, T.L.; Wasilenko, S.T.; Barry, M. Vaccinia virus F1L protein is a tail-anchored protein that functions at the mitochondria to inhibit apoptosis. *J. Virol.* **2005**, *79*, 1084–1098. [CrossRef] [PubMed]

149. Everett, H.; Barry, M.; Lee, S.F.; Sun, X.; Graham, K.; Stone, J.; Bleackley, R.C.; McFadden, G. M11L: A novel mitochondria-localized protein of myxoma virus that blocks apoptosis of infected leukocytes. *J. Exp. Med.* **2000**, *191*, 1487–1498. [CrossRef] [PubMed]

150. Wang, G.; Barrett, J.W.; Nazarian, S.H.; Everett, H.; Gao, X.; Bleackley, C.; Colwill, K.; Moran, M.F.; McFadden, G. Myxoma virus M11L prevents apoptosis through constitutive interaction with Bak. *J. Virol.* **2004**, *78*, 7097–7111. [CrossRef] [PubMed]

151. Douglas, A.E.; Corbett, K.D.; Berger, J.M.; McFadden, G.; Handel, T.M. Structure of M11L: A myxoma virus structural homolog of the apoptosis inhibitor, Bcl-2. *Protein Sci.* **2007**, *16*, 695–703. [CrossRef] [PubMed]

152. Westphal, D.; Ledgerwood, E.C.; Tyndall, J.D.; Hibma, M.H.; Ueda, N.; Fleming, S.B.; Mercer, A.A. The orf virus inhibitor of apoptosis functions in a Bcl-2-like manner, binding and neutralizing a set of BH3-only proteins and active Bax. *Apoptosis* **2009**, *14*, 1317–1330. [CrossRef] [PubMed]

153. Banadyga, L.; Veugelers, K.; Campbell, S.; Barry, M. The fowlpox virus BCL-2 homologue, FPV039, interacts with activated Bax and a discrete subset of BH3-only proteins to inhibit apoptosis. *J. Virol.* **2009**, *83*, 7085–7098. [CrossRef] [PubMed]

154. Neilan, J.G.; Lu, Z.; Afonso, C.L.; Kutish, G.F.; Sussman, M.D.; Rock, D.L. An African swine fever virus gene with similarity to the proto-oncogene Bcl-2 and the Epstein-Barr virus gene BHRF1. *J. Virol.* **1993**, *67*, 4391–4394. [PubMed]

155. Galindo, I.; Hernaez, B.; Diaz-Gil, G.; Escribano, J.M.; Alonso, C. A179L, a viral Bcl-2 homologue, targets the core Bcl-2 apoptotic machinery and its upstream BH3 activators with selective binding restrictions for Bid and Noxa. *Virology* **2008**, *375*, 561–572. [CrossRef] [PubMed]

156. Lin, P.W.; Huang, Y.J.; John, J.A.; Chang, Y.N.; Yuan, C.H.; Chen, W.Y.; Yeh, C.H.; Shen, S.T.; Lin, F.P.; Tsui, W.H.; et al. Iridovirus Bcl-2 protein inhibits apoptosis in the early stage of viral infection. *Apoptosis* **2008**, *13*, 165–176. [CrossRef] [PubMed]

157. Pattingre, S.; Tassa, A.; Qu, X.; Garuti, R.; Liang, X.H.; Mizushima, N.; Packer, M.; Schneider, M.D.; Levine, B. Bcl-2 antiapoptotic proteins inhibit Beclin 1-dependent autophagy. *Cell* **2005**, *122*, 927–939. [CrossRef] [PubMed]

158. Piya, S.; White, E.J.; Klein, S.R.; Jiang, H.; McDonnell, T.J.; Gomez-Manzano, C.; Fueyo, J. The E1B19K oncoprotein complexes with Beclin 1 to regulate autophagy in adenovirus-infected cells. *PLoS ONE* **2011**, *6*, e29467. [CrossRef] [PubMed]

159. Gerlic, M.; Faustin, B.; Postigo, A.; Yu, E.C.; Proell, M.; Gombosuren, N.; Krajewska, M.; Flynn, R.; Croft, M.; Way, M.; et al. Vaccinia virus F1L protein promotes virulence by inhibiting inflammasome activation. *Proc. Natl. Acad. Sci. USA* **2013**, *110*, 7808–7813. [CrossRef] [PubMed]

160. Yu, E.; Zhai, D.; Jin, C.; Gerlic, M.; Reed, J.C.; Liddington, R. Structural determinants of caspase-9 inhibition by the vaccinia virus protein, F1L. *J. Biol. Chem.* **2011**, *286*, 30748–30758. [CrossRef] [PubMed]

161. Zhai, D.; Yu, E.; Jin, C.; Welsh, K.; Shiau, C.W.; Chen, L.; Salvesen, G.S.; Liddington, R.; Reed, J.C. Vaccinia virus protein F1L is a caspase-9 inhibitor. *J. Biol. Chem.* **2010**, *285*, 5569–5580. [CrossRef] [PubMed]

162. Cooray, S.; Bahar, M.W.; Abrescia, N.G.; McVey, C.E.; Bartlett, N.W.; Chen, R.A.; Stuart, D.I.; Grimes, J.M.; Smith, G.L. Functional and structural studies of the vaccinia virus virulence factor N1 reveal a Bcl-2-like anti-apoptotic protein. *J. Gen. Virol.* **2007**, *88*, 1656–1666. [CrossRef] [PubMed]

163. De Motes, C.M.; Cooray, S.; Ren, H.; Almeida, G.M.F.; McGourty, K.; Bahar, M.W.; Stuart, D.I.; Grimes, J.M.; Graham, S.C.; Smith, G.L. Inhibition of Apoptosis and NF-κB Activation by Vaccinia Protein N1 Occur via Distinct Binding Surfaces and Make Different Contributions to Virulence. *PLoS Pathog.* **2011**, *7*, e1002430.

164. Fedosyuk, S.; Grishkovskaya, I.; de Almeida Ribeiro, E., Jr.; Skern, T. Characterization and structure of the vaccinia virus NF-κB antagonist A46. *J. Biol. Chem.* **2014**, *289*, 3749–3762. [CrossRef] [PubMed]

165. Kim, Y.; Lee, H.; Heo, L.; Seok, C.; Choe, J. Structure of vaccinia virus A46, an inhibitor of TLR4 signaling pathway, shows the conformation of VIPER motif. *Protein Sci.* **2014**, *23*, 906–914. [CrossRef] [PubMed]

166. Neidel, S.; Maluquer de Motes, C.; Mansur, D.S.; Strnadova, P.; Smith, G.L.; Graham, S.C. Vaccinia virus protein A49 is an unexpected member of the B-cell Lymphoma (Bcl)-2 protein family. *J. Biol. Chem.* **2015**, *290*, 5991–6002. [CrossRef] [PubMed]

167. Kalverda, A.P.; Thompson, G.S.; Vogel, A.; Schroder, M.; Bowie, A.G.; Khan, A.R.; Homans, S.W. Poxvirus K7 protein adopts a Bcl-2 fold: Biochemical mapping of its interactions with human DEAD box RNA helicase DDX3. *J. Mol. Biol.* **2009**, *385*, 843–853. [CrossRef] [PubMed]

168. Fedosyuk, S.; Bezerra, G.A.; Radakovics, K.; Smith, T.K.; Sammito, M.; Bobik, N.; Round, A.; Ten Eyck, L.F.; Djinovic-Carugo, K.; Uson, I.; et al. Vaccinia Virus Immunomodulator A46: A Lipid and Protein-Binding Scaffold for Sequestering Host TIR-Domain Proteins. *PLoS Pathog.* **2016**, *12*, e1006079. [CrossRef] [PubMed]

169. Oda, S.; Schroder, M.; Khan, A.R. Structural basis for targeting of human RNA helicase DDX3 by poxvirus protein K7. *Structure* **2009**, *17*, 1528–1537. [CrossRef] [PubMed]

170. Schroder, M.; Baran, M.; Bowie, A.G. Viral targeting of DEAD box protein 3 reveals its role in TBK1/IKKepsilon-mediated IRF activation. *EMBO J.* **2008**, *27*, 2147–2157. [CrossRef] [PubMed]

171. Merino, D.; Giam, M.; Hughes, P.D.; Siggs, O.M.; Heger, K.; O'Reilly, L.A.; Adams, J.M.; Strasser, A.; Lee, E.F.; Fairlie, W.D.; et al. The role of BH3-only protein Bim extends beyond inhibiting Bcl-2-like prosurvival proteins. *J. Cell Biol.* **2009**, *186*, 355–362. [CrossRef] [PubMed]

172. Poncet, D.; Larochette, N.; Pauleau, A.L.; Boya, P.; Jalil, A.A.; Cartron, P.F.; Vallette, F.; Schnebelen, C.; Bartle, L.M.; Skaletskaya, A.; et al. An anti-apoptotic viral protein that recruits Bax to mitochondria. *J. Biol. Chem.* **2004**, *279*, 22605–22614. [CrossRef] [PubMed]

173. Cross, J.R.; Postigo, A.; Blight, K.; Downward, J. Viral pro-survival proteins block separate stages in Bax activation but changes in mitochondrial ultrastructure still occur. *Cell Death Differ.* **2008**, *15*, 997–1008. [CrossRef] [PubMed]

174. Connolly, P.F.; Fearnhead, H.O. Viral hijacking of host caspases: An emerging category of pathogen-host interactions. *Cell Death Differ.* **2017**, *24*, 1401–1410. [CrossRef] [PubMed]

175. Collins-McMillen, D.; Kim, J.H.; Nogalski, M.T.; Stevenson, E.V.; Chan, G.C.; Caskey, J.R.; Cieply, S.J.; Yurochko, A.D. Human Cytomegalovirus Promotes Survival of Infected Monocytes via a Distinct Temporal Regulation of Cellular Bcl-2 Family Proteins. *J. Virol.* **2015**, *90*, 2356–2371. [CrossRef] [PubMed]

Article

Structural and Functional Insight into Canarypox Virus CNP058 Mediated Regulation of Apoptosis.

Mohd Ishtiaq Anasir, Amy A. Baxter ⓘ, Ivan K. H. Poon, Mark D. Hulett and Marc Kvansakul *

Department of Biochemistry and Genetics, La Trobe Institute for Molecular Science, La Trobe University, Melbourne, VIC 3086, Australia; 17818130@students.latrobe.edu.au (M.I.A.); A.Baxter@latrobe.edu.au (A.A.B.); i.poon@latrobe.edu.au (I.K.H.P.); m.hulett@latrobe.edu.au (M.D.H.)
* Correspondence: m.kvansakul@latrobe.edu.au; Tel.: +61-3-94792263

Academic Editor: Eric O. Freed
Received: 1 October 2017; Accepted: 18 October 2017; Published: 20 October 2017

Abstract: Programmed cell death or apoptosis is an important component of host defense systems against viral infection. The B-cell lymphoma 2 (Bcl-2) proteins family is the main arbiter of mitochondrially mediated apoptosis, and viruses have evolved sequence and structural mimics of Bcl-2 to subvert premature host cell apoptosis in response to viral infection. The sequencing of the canarypox virus genome identified a putative pro-survival Bcl-2 protein, CNP058. However, a role in apoptosis inhibition for CNP058 has not been identified to date. Here, we report that CNP058 is able to bind several host cell pro-death Bcl-2 proteins, including Bak and Bax, as well as several BH3 only-proteins including Bim, Bid, Bmf, Noxa, Puma, and Hrk with high to moderate affinities. We then defined the structural basis for CNP058 binding to pro-death Bcl-2 proteins by determining the crystal structure of CNP058 bound to Bim BH3. CNP058 adopts the conserved Bcl-2 like fold observed in cellular pro-survival Bcl-2 proteins, and utilizes the canonical ligand binding groove to bind Bim BH3. We then demonstrate that CNP058 is a potent inhibitor of ultraviolet (UV) induced apoptosis in a cell culture model. Our findings suggest that CNP058 is a potent inhibitor of apoptosis that is able to bind to BH3 domain peptides from a broad range of pro-death Bcl-2 proteins, and may play a key role in countering premature host apoptosis.

Keywords: poxvirus; avipoxvirus; apoptosis; X-ray crystallography; isothermal titration calorimetry; Bcl-2

1. Introduction

Apoptosis is a form of programmed cell death that can be triggered via external (extrinsic pathway or receptor mediated) or internal stimuli (intrinsic pathway or mitochondria mediated). Apoptosis plays a major role in the removal of damaged, unwanted or infected cells, impacting processes ranging from cellular homeostasis to the immune response against viral infection [1]. In order to thwart host apoptosis as part of an antiviral response, viruses have evolved to encode arrays of apoptosis regulatory proteins [1]. For instance, many viruses express decoy receptors to neutralize the Tumor Necrosis Factor receptor superfamily, the effectors of extrinsic apoptosis pathway [1]. In addition, the host intrinsic apoptosis pathway can be downregulated by viruses through the expression of B-cell lymphoma 2 (Bcl-2) mimics (vBcl-2), which nullify the activity of host cell apoptosis inducing Bcl-2 family members [2,3].

The Bcl-2 family proteins are the gatekeepers of the intrinsic apoptosis pathway, with the family members being characterized by the presence of at least one of the four conserved Bcl-2 homology (BH) domains. Bcl-2 proteins can be divided into two major classes: pro-death and pro-survival [4]. The pro-death Bcl-2 proteins are further sub-divided into two groups; (1) pro-death Bcl-2, such as

Bak and Bax that contain BH1-4 domains and (2) BH3-only proteins Bim, Bid, Puma, Bmf, Bik, Hrk, Noxa, and Bad that contain only the BH3 domain [4,5]. During cellular stress condition such as exposure to cytotoxic drugs, ultraviolet (UV) irradiation or viral infection, BH3-only proteins expression is up-regulated, resulting in the activation of Bak and Bax and/or neutralization of the pro-survival Bcl-2 proteins [6,7]. Upon their activation, Bak and Bax oligomerize to disrupt the outer mitochondrial membrane, facilitating the release of pro-apoptogenic factors such as cytochrome c, and ultimately triggering the caspase cascade and cell death [8]. The pro-survival Bcl-2 proteins include Bcl-2, Mcl-1, Bcl-x_L, A1, Bcl-w, and Bcl-b, contain all four BH domains, and are responsible for antagonizing the pro-death Bcl-2 proteins and BH3-only proteins in order to maintain mitochondrial membrane integrity [9]. Structural studies revealed that Bcl-2 proteins interact with each other via a BH3 domain-hydrophobic groove dependent manner [10].

Many viruses have been found to encode vBcl-2 proteins including adenovirus E1B19K [11], African swine fever virus A179L [12], Epstein-Barr virus BHRF1 [13], and Kaposi's Sarcoma herpes virus KSBcl-2 [14], which act either by directly binding and sequestering BH3 only proteins or by directly binding to Bak and Bax to prevent apoptosis initiation. Interestingly, poxviruses encode a series of highly diverse vBcl-2 proteins [14]. Sequence analysis showed that poxviral vBcl-2, such as Canarypox CNP058 and Fowlpox FPV039 harbor limited sequence identity with cellular Bcl-2, whilst others such as Vaccinia F1 and Myxoma M11L do not share any significant sequence identity [15]. Poxviral vBcl-2 proteins operate using several distinct mechanisms, for example, M11L inhibits apoptosis primarily by neutralizing Bak and Bax, whereas F1 targets Bim to inhibit apoptosis [16,17].

Canarypox CNP058 is a vBcl-2 protein discovered amongst certain avipoxviruses. In contrast to Fowlpox FPV039, which has been shown to adopt Bcl-2 fold [18] and is able to neutralize all BH3-only proteins, as well as Bak and Bax [18–20], little is known about CNP058. To understand the mechanism of apoptosis inhibition by CNP058, we measured the binding of CNP058 to peptides corresponding to BH3 domains of all pro-death Bcl-2 proteins, including Bak and Bax and all of the BH3-only proteins. We then determined the crystal structure of CNP058 bound to Bim BH3 domain, and demonstrate that CNP058 is able to potently inhibit UV light induced apoptosis in cell culture. Our findings demonstrate that CNP058 is a Bcl-2 protein that potently inhibits apoptosis, is able to interact with the BH3 domain of pro-death Bcl-2 proteins, and provide a platform to understand apoptosis inhibition by CNP058.

2. Materials and Methods

2.1. CNP058 Expression and Purification

Synthetic codon-optimized cDNA encoding for CNP058 lacking the C-terminal transmembrane domain (residues 1–143) (Genscript, Piscataway, NJ, USA) was cloned into the bacterial expression vector pGEX6P-1 (GE Healthcare, Chicago, IL, USA). The plasmid was transformed into *Escherichia coli* BL21 Star cells and grown in 2YT media supplemented with 1 mg/mL ampicillin. CNP058 was expressed using the auto-induction method for 24 h at 22 °C in a shaking incubator [21]. Bacterial cells were harvested by centrifugation at $3400 \times g$ (JLA 9.1000 rotor, Beckman Coulter Avanti J-E, Brea, CA, USA) for 15 min, resuspended in 100 mL lysis buffer (20 mM trisodium citrate pH 6.0, 200 mM NaCl), and lysed using a sonicator at 50 kHz for 4 cycles (15 s) with 30 s rest intervals in the presence of lysozyme (Sigma Aldrich, St. Louis, MO, USA) and DNAse I (deoxyribonuclease I from bovine pancreas, Sigma Aldrich). The resulting lysate was clarified by centrifugation at $31,000 \times g$ (JA 25.50 rotor, Beckman Coulter Avanti J-E) for 30 min. The supernatant was filtered with a 0.45 μM filter (Milipore, Burlington, MA, USA), loaded onto 5 mL of Glutathione Sepharose 4B resin (GE Healthcare) equilibrated with lysis buffer and subsequently washed with an additional 30 mL of lysis buffer. Human rhinovirus 3C protease (HRV3C protease) was added to the column and incubated overnight at 4 °C to liberate the target protein from the Glutathione-S-Transferase (GST) fusion tag. The cleaved target protein was eluted and concentrated to 5 mL prior to being subjected to size exclusion chromatography using a Superdex S75 16/600 column attached to an

ÄKTAxpress system (GE Healthcare) equilibrated in 20 mM trisodium citrate pH 6.0, 200 mM NaCl. The protein eluted as a single peak and displayed higher than 95% purity based on sodium dodecyl sulfate-polyacrylamide gel electrophoresis (SDS-PAGE) analysis.

2.2. Measurement of Interactions with BH3 Peptides

Affinities of CNP058 for different synthetic BH3 peptides (Mimotopes, Mulgrace, VIC, Australia) were measured using a MicroCal iTC200 system (GE Healthcare) at 25 °C. The measurements were performed in 20 mM trisodium citrate pH 6.0, 200 mM NaCl at a final protein concentration of 30 µM. BH3 domain peptides at a concentration of 300 µM were titrated into the protein sample using 19 injections of 2 µL per injection. All of the assays were performed in triplicates. Protein concentrations were measured using a UV spectrophotometer (Thermo Scientific, Scoresby, VIC, Australia) at a wavelength of 280 nm. BH3 peptides concentrations were calculated from the dry weight of peptide. The BH3 peptide sequences used in this study were: (1) *Gallus gallus* (GG) GG_Bak (Uniprot Id: Q5F404): 72-LGSTGSQVGRRLAIIGDDINKRYDAE-97; (2) *Homo sapiens* (HS) HS_Bax (Uniprot Id: Q07812): 50-VPQDASTKKLSECLKRI GDELDSNMELQ-77; (3) GG_Bid (Uniprot Id: Q8JGM8): 77-PEVNEAIVRTIAAQLA EIGDQLDKQIKAKVVNDL-110; (4) GG_Bmf (Uniprot Id: A9XRG9): 135-EARTEVQIARKLQCIADQFHRLHIQR-160; (5) GG_Bok (Uniprot Id: Q9I8I2): 67-VSAILLRLGDELEYIRPNVYRNIARQ-92; (6) GG_Noxa (RefSeq Id NP_001289026.1: ERDAVA ECALELRRIGDKADLQQKVL; (7) GG_Bik (Uniprot Id: E9JEC5): 43-ISSAIQVGHQLALIGD EFNRAYSRK-67; (8) HS_Bim (Uniprot Id: O43521-3): 51-DMRPEIWIAQELRRIGDEFNAYYARR-76; (9) HS_Puma (Uniprot Id: Q9BXH1-1): 130-EEQWAREIGAQLRRMADDLNAQ-YERR-155; (10) HS_Hrk (Uniprot Id: O00198-1): 26-RSSAAQLTA ARLKAIGDE-LHQRTMWR-51; (11) HS_Bad (Uniprot Id: Q92934-1): 103-NLWAAQRYGRELRRMSDEFVDSFKKG-128.

2.3. CNP058 Complex Crystallization and Data Collection

A complex of CNP058 with the Bim BH3 domain was prepared as previously described [22]. Briefly, CNP058-Bim BH3 complex was reconstituted by adding HS_Bim BH3 domain at a 1:1.25 molar ratio to CNP058. The reconstituted complex was used for crystallization trials using 96-well sitting drop trays (Swissci, Neuheim, Switzerland) with the vapour diffusion method at 20 °C. A total of 0.15 µL CNP058-Bim BH3 domain peptide complex was mixed with 0.15 µL of various crystallization conditions using a Gryphon nanodispenser robot (Art Robbins, Sunnyvale, CA, USA). Commercially available screening kits (Crystal Screen, PACT suite, JCSG-plus Screen and PEG/Ion Screen) were used as prepared by C3 for the initial crystallization screening, with hit optimization performed using 24-well hanging drop plates (EasyXtal DG-Tool, Qiagen, Hilden, Germany) of 1 + 1 µL protein:reservoir condition.

Crystals of CNP058 in complex with the Bim BH3 domain were obtained at 10 mg/mL in 0.2 M calcium chloride dihydrate, 0.1 M MES pH 6.0, 20% (*w/v*) PEG6000. This condition produced thick needle crystals of CNP058-Bim BH3 domain complex belonging to the C2 space group in the monoclinic crystal system. The final crystal contained one molecule of CNP058 bound to one molecule of Bim BH3 in the asymmetric unit and had a solvent content of 40.1%. The crystals were cryoprotected using 20% (*v/v*) ethylene glycol, and was flash cooled at 100 K using liquid nitrogen. All diffraction data were collected on the MX2 beamline at the Australian Synchrotron using an Eiger detector with an oscillation range of 0.1° per frame using a wavelength of 0.9537 Å. Diffraction data were integrated using XDS [23] and scaled using AIMLESS [24,25]. The structure of the CNP058-Bim BH3 domain complex was solved using molecular replacement with Phaser [26], with the structure of FPV039 (PDB: 5TZP) as a search model [18]. The final TFZ and LLG values were 8.9 and 102, respectively. The solution produced by Phaser was manually rebuilt over multiple cycles using Coot [27] and refined using PHENIX [28]. Data collection and refinement statistics details are summarized in Table 1. Coordinate files have been deposited in the Protein Data Bank under the accession code 5WOS. All of the images were generated using the PyMOL Molecular Graphics System, Version 1.8 Schrödinger,

LLC, New York, NY, USA). All of the software was accessed using the SBGrid suite [29,30]. All of the raw diffraction images were deposited on the SBGrid Data Bank [30] using accession numbers 5WOS.

Table 1. Crystallographic data collection and refinement statistics.

Data Collection and Refinement Statistics (Molecular Replacement)	
CNP058-Bim BH3 domain	
Data collection	
Space group	C121
No. of molecules in asymmetric unit	1 + 1
Cell dimensions	
a, b, c (Å)	73.79, 34.67, 71.73
$\alpha, \beta, \gamma,$ (°)	90.00, 114.92, 90.00
Wavelength (Å)	0.9537
Resolution (Å)	33.46–2.45 (2.55–2.45)
No. unique reflections	6021 (703)
R_{sym} or R_{merge}	0.097 (0.716)
$I/\sigma I$	6.1 (1.1)
$CC_{1/2}$	0.99 (0.45)
Wilson B-factor	40.9
Completeness (%)	97.2 (98.4)
Redundancy	2.7 (2.7)
Refinement	
Resolution (Å)	2.45
No. reflections	6017
R_{work}/R_{free}	0.2126/0.2453
No. atoms	
Protein	1340
Water	10
B-factors	
Protein	52.91
Water	56.11
R.m.s. deviations	
Bond lengths (Å)	0.003
Bond angles (°)	0.56
Ramachandran statistics (%)	
Favored	97.42
Allowed	2.58
Disallowed	0.00

Values in parentheses are for highest-resolution shell.

2.4. Cell Culture

Human epithelial cervical cancer (HeLa) cells were cultured in RPMI-1640 medium (Invitrogen, Carlsbad, CA, USA). All culture media were supplemented with 10% fetal bovine serum, 100 U/mL penicillin, and 100 μg/mL streptomycin (Invitrogen). Cell lines were cultured at 37 °C in a humidified atmosphere containing 5% CO_2 and detached from the flask with 0.25% trypsin and 0.5 μM EDTA (Invitrogen).

2.5. Transfection of HeLa Cells with GFP Constructs and Induction of Apoptosis

Full length CNP058 (residues 1–175) was cloned into the mammalian expression vector pcDNA3.1+N-eGFP (Genscript), vaccinia virus (WR) F1 in pEGFP-C3 was provided by Michele Barry (University of Alberta, Edmonton, AB, Canada) [31]. HeLa cells were grown to 60% confluency and transfected with plasmid constructs for green fluorescent protein (GFP)-tagged F1, GFP-tagged CNP058 or cytosolic GFP using X-tremeGENE 9 DNA Transfection Reagent (Roche, Basel, Switzerland) as per manufacturer's instructions. 24 h post transfection, cell supernatants were replaced with fresh growth media and cells were subjected to ultraviolet (UV) irradiation using a Stratalinker UV crosslinker (Stratagene, La Jolla, CA, USA), followed by 6 h incubation at 37 °C in a humidified atmosphere containing 5% CO_2.

2.6. Flow Cytometry

A two stain flow cytometry based apoptotic cell death assay, described previously in [32], was performed to analyze the ability of virus-encoded apoptosis inhibitors to inhibit apoptosis in HeLa cells subjected to UV irradiation [32]. Briefly, following apoptosis induction, trypsinized cells were stained with PE-conjugated annexin V stain (AV-PE) (BD Biosciences, San Jose, CA, USA) and TO-PRO-3 nucleic acid stain (Life Technologies, Carlsbad, CA, USA) as per the manufacturer's instructions and kept on ice until analysis by flow cytometry using the BD FACSCanto II Flow Cytometer and BD FACSDiva software v6.1.1 (BD Biosciences, St. Jose, CA, USA). Samples were subsequently analyzed with Flowjo software v8.8.6 (Tree Star, Ashland, OR, USA). A minimum of 9000 GFP-positive cell events were recorded per sample, with GFP-negative cells excluded from analysis and AV-PE/TO-PRO-3 staining used to gate apoptotic from non-apoptotic cell populations [32].

3. Results

To examine if CNP058 is able to interact with pro-death Bcl-2, truncated recombinant CNP058 encompassing the first 143 amino acids was expressed in *E. coli* and purified using a two step purification method of affinity chromatography followed by size exclusion chromatography. Since a number of vBcl-2 proteins have been shown to be dimers in solution, we examined size exclusion chromatograms to define the oligomeric state of CNP058 (Figure 1). Our results show that CNP058 is a monomer in solution.

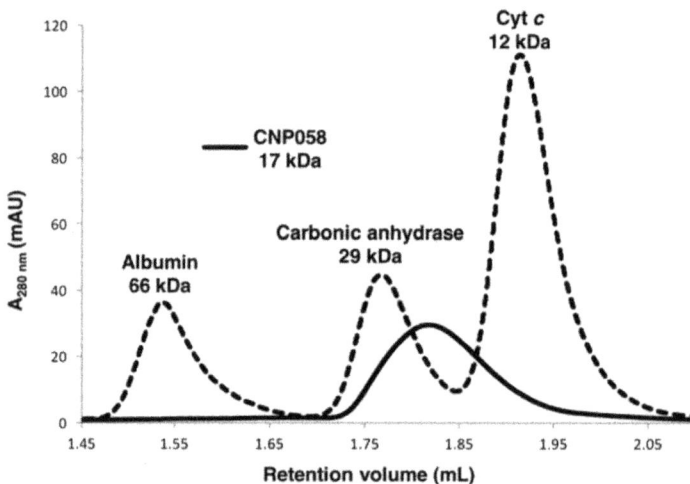

Figure 1. CNP058 is a monomer in solution. Size exclusion chromatography of CNP058 using a Superdex 200 Increase 3.2/300 column. The elution volume of the peak of interest (CNP058) is 1.77 mL (solid line). The molecular weight standards shown are albumin (66 kDa), carbonic anhydrase (29 kDa) and cytochrome *c* (Cyt *c*) (12 kDa) (all source from Sigma Aldrich), shown as dotted lines and labeled on top of the respective peaks. AU: absorbance units.

We then examined binding of CNP058 to peptides comprising the BH3 domain of all pro-death Bcl-2 proteins using isothermal titration calorimetry (ITC) (Figure 2). ITC showed that CNP058 engages

BH3 domains of Bid with high affinity (50 nM), whereas Bax, Bim, Hrk, Bmf interacted with 7–9-fold weaker affinity, and Bak, Puma and Noxa only displayed affinity in the micromolar region (Figure 2). In contrast, Bik, Bad, and Bok did not show any detectable affinity for CNP058.

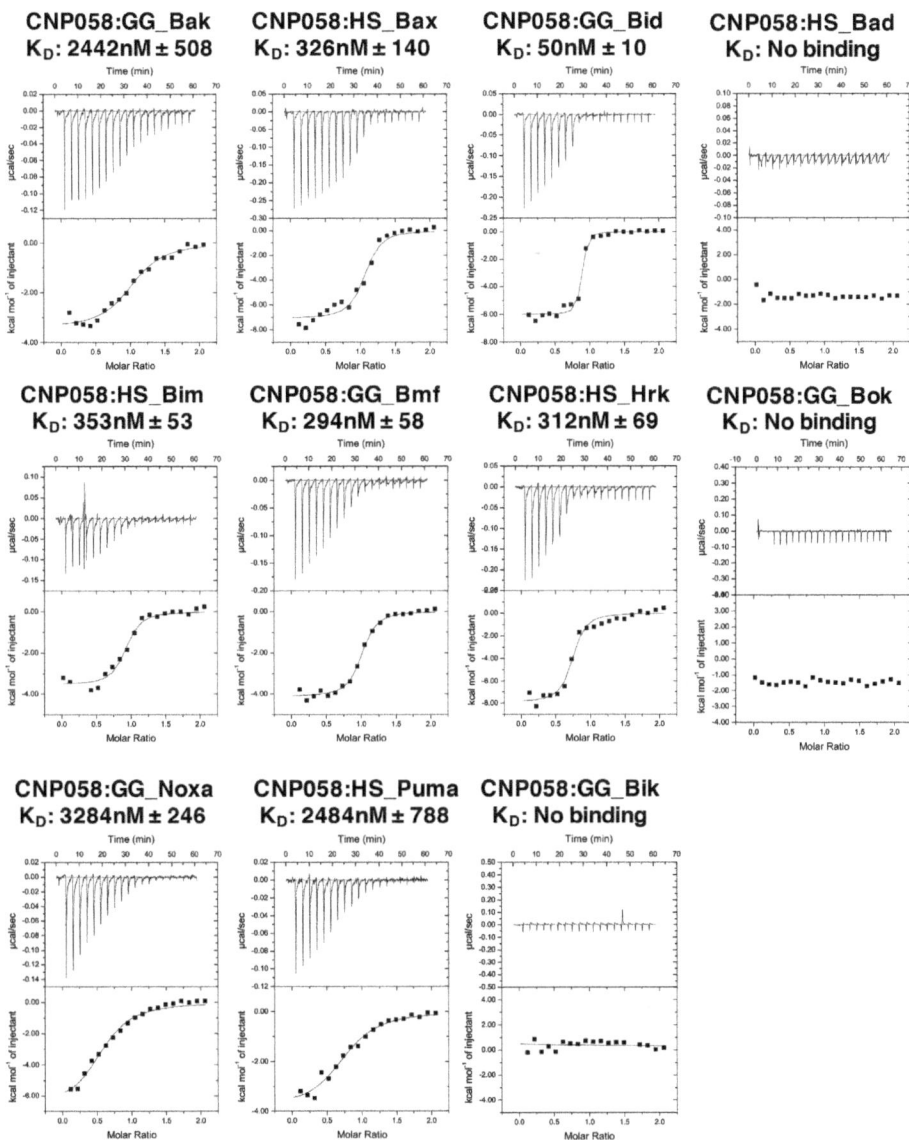

Figure 2. CNP058 interacts with Bak and Bax as well as all BH3-only proteins except Bad and Bik. Raw heats measured by isothermal titration calorimetry (ITC) for CNP058 interactions with Bcl-2 homology domain 3 (BH3) domain peptides of pro-death B-cell lymphoma 2 (Bcl-2) proteins from *Homo sapiens* (HS) and *Gallus gallus* (GG).

We then sought to understand the structural basis of CNP058 interaction with the BH3 domain of pro-death Bcl-2 by determining the structure of CNP058 in complex with the Bim BH3 domain.

The CNP058-Bim BH3 domain complex structure revealed that CNP058 adopts the conserved Bcl-2-like fold consisting of eight α-helices and possessing the conserved hydrophobic groove found in other Bcl-2 family members that is utilized to accommodate Bim BH3 domain (Figures 3, 4 and 5A, Table 1). The closest cellular Bcl-2 structural homolog as identified by a DALI search is Mcl-1 (root mean square deviation (r.m.s.d)) of 1.0 Å over 94 Cα atoms, Figure 5C) and the closest viral Bcl-2 homolog is FPV039 (r.m.s.d. of 1.2 Å over 114 Cα atoms, Figures 3 and 5B,D).

```
                         α1-helix                           α2-helix
                HHHHHHHHHHHHHHHHHHH                  HHHHHHHHHH
CNP058    1-MDPSVKKDEIYYTILNIIQNYFIEYCTGKNRNFHVEDE-----NTYIIVKNMCD-49
FPV039    1-MASSNMKDETYYIALNMIQNYIIEYNTNKPRKSFVIDS-----ISYDVLKAACK-49
Mcl-1   165-PPAEEEEDELYRQSLEIISRYLREQATGAKDTKPMGRSGATSRKALETLRRVGD-216
            .   :** *   *::*..*:  *   *.    .  :      :  ::   .
                --BH4-                               -----BH3-

                      α3-helix        α4-helix           α5-helix
            HHHH  HHHHHHHHHHHHHHH    HHHHHHHHHHHH      HHHHHHHHHHHH
CNP058   50-IILRDNIVEFRKD-----IDRCSDIENEIPEIVYDTIHDKITWGRVISIIAFGA-98
FPV039   50-SVIKTNYNEFDII-----ISRNIDF----NVIVTQVLEDKINWGRIITIIAFCA-94
Mcl-1   217-GVQRNHETAFQGMLRKLDIKNEDDVKSLSRVMIHVFSDGVTNWGRIVTLISFGA-272
            : : :      *       *..  *.      ::      ..  .***:::*:* *
            -----                                -------BH1--------

                         α6-helix        α7-helix α8-helix
            HHHHHHHH     HHHHHHHHHHHHH    HHHHH   HHHH
CNP058   99-YVTKVFKEKGRD-NVVDLMPDIITESLLSRCRSWLSDQNCWDGLKAYVYNNKK-150
FPV039   95-YYSKKVKQDTSPQYYDGIISEAITDAILSKYRSWFIDQDYWNGIRIYKNYSYI-147
Mcl-1   273-FVAKHLKTINQE-SCIEPLAESITDVLVRTKRDWLVKQRGWDGFVEFFHVEDL-322
            : :*  .*      ı ı **ı ıı   *.*ı .*  *ı*ı  :    .
                                         ----BH2-----

CNP058  151-FYYVTRYFRVAAFIITSLAVINLFL------175
FPV039  148-FNTASYCIFTASLIIASLAVFKICSFYM---175
Mcl-1   323-EGGI----RNVLLAFAGVAGVGAGLAYLIR-350
                 . : ::..:*  .
                -------------TM-------------
```

Figure 3. Structure based sequence alignment of CNP058 with FPV039 and cellular Bcl-2 protein Mcl-1. The α-helical secondary structure elements indicated (denoted as H and labeled as Helix 1–8) are based on the crystal structure of CNP058. Bcl-2 homology domains (BH domains) 1–4 are marked underneath the aligned sequences. Conserved residues are highlighted in yellow. "*" denotes conserved residues for all three proteins, ":" denotes highly similar residues between the three proteins, and "." denotes weakly similar residues between the three proteins. The NWGR or TWGR motifs are marked in bold. Uniprot accession codes: CNP058 = Q6VZT9, FPV039 = Q9J5G4 and Mcl-1 = Q07820; PDB accession codes: FPV039 = 5TZP and Mcl-1 = 2NL9.

The Bim BH3 domain is bound in the canonical hydrophobic ligand binding groove formed by α-helices 2–5 (Figure 6). Detailed inspection of the CNP058-Bim BH3 domain complex interface (Figure 6A) showed that Bim residues Ile58, Leu62, Ile65 and Phe69 extend into four hydrophobic pockets on the CNP058 binding groove. In addition, two salt bridges are found at the interface: Asp83[CNP058]-Arg63[Bim] and the highly conserved Arg89[CNP058]-Asp67[Bim], which is a hallmark of pro-survival Bcl-2:BH3 domain interactions. Notably, the highly conserved **NWGR** motif situated at the beginning of α5-helix of all Bcl-2 proteins is replaced with a **TWGR** motif in CNP058, however the Thr in the TWGR motif is not involved in any direct interactions with Bim.

C-terminus

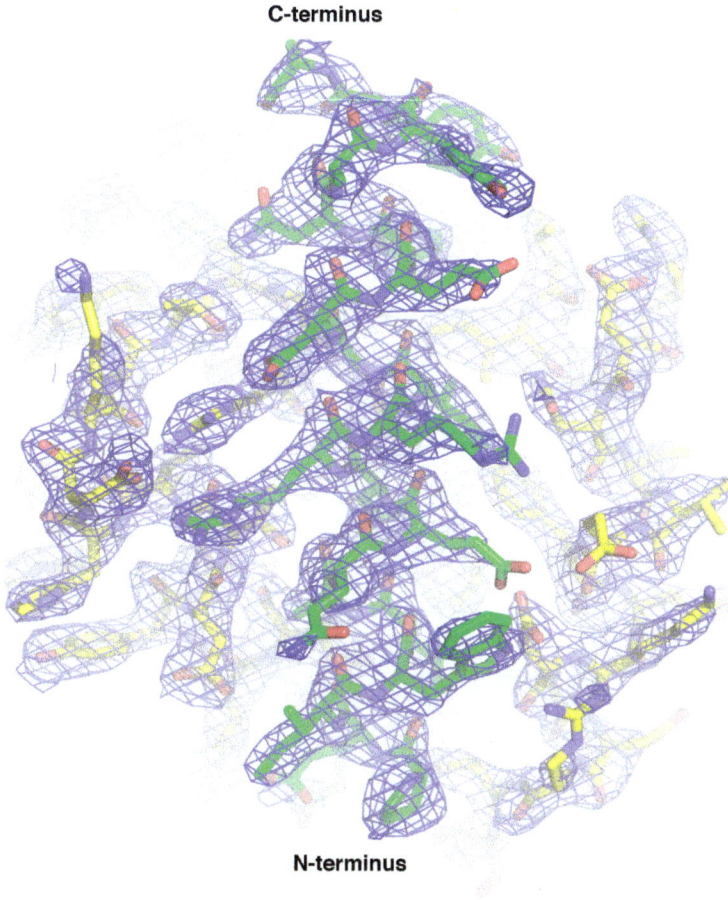

N-terminus

Figure 4. 2Fo-Fc electron density maps of CNP058:Bim BH3 domain complex. Electron density map encompassing the hydrophobic binding groove of CNP058 in complex with Bim BH3. CNP058 is shown as yellow sticks, whereas Bim BH3 is shown as green sticks. The electron density map is shown as a blue mesh contoured at 1 δ.

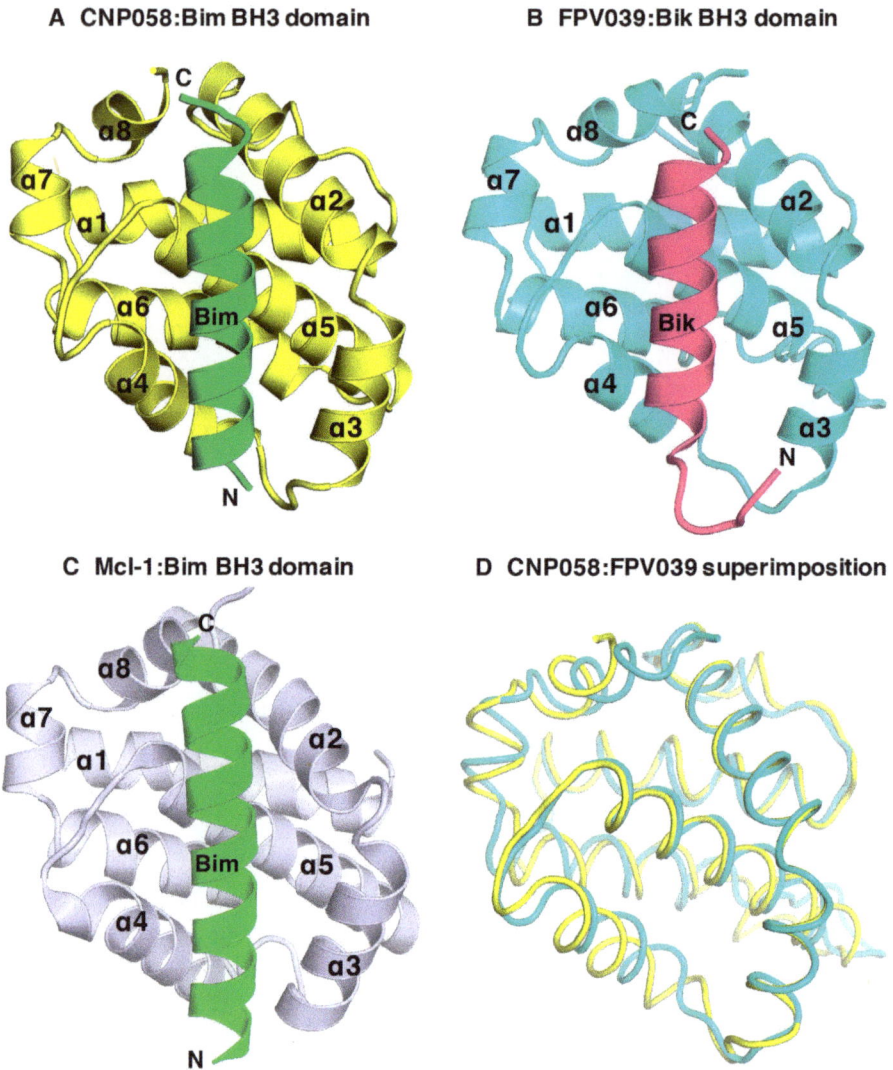

Figure 5. CNP058 interacts with pro-death Bcl-2 proteins in a BH3-domain hydrophobic groove dependent manner. (**A**) Cartoon representation of CNP058:Bim BH3 domain complex, (**B**) FPV039:Bik BH3 domain complex and (**C**) Mcl-1:Bim BH3 domain complex. The view is into the conserved hydrophobic ligand binding groove formed by α-helices 2–5. PDB ID: FPV039:Bik = 5TZP, Mcl-1:Bim = 2NL9. (**D**) CNP058:Bim complex (yellow) was superimposed onto FPV039:Bik complex (cyan). Bim and Bmf were removed from respective complex structures for clarity.

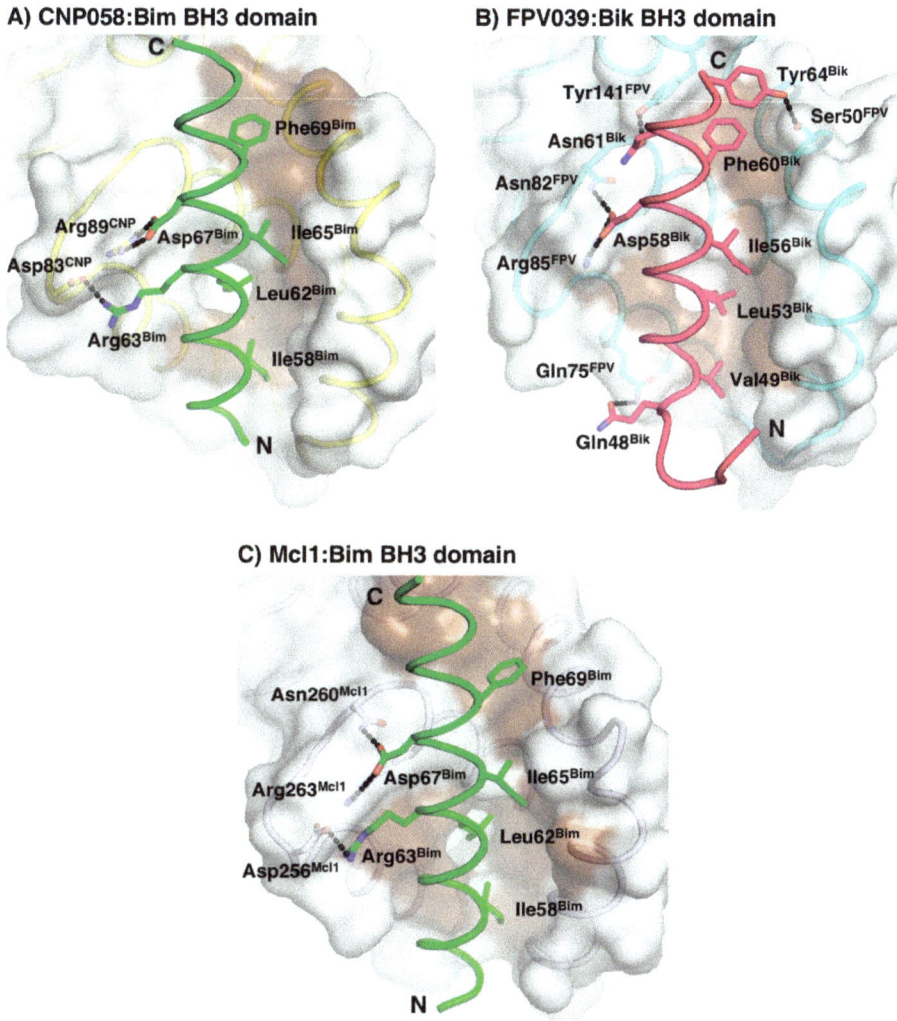

Figure 6. CNP058:Bim BH3 complex interface. Surface representation of (**A**) CNP058:Bim BH3 domain complex, (**B**) FPV039:Bik BH3 domain complex, and (**C**) Mcl-1:Bim BH3 domain complex. The view is into the conserved hydrophobic ligand binding groove formed by α-helices 2–5. The surfaces are shown in grey. Residues involved in interactions are shown as sticks and labeled. Hydrogen bonds and ionic interactions between CNP058 and Bim are denoted as black dotted lines. PDB ID: FPV039:Bik = 5TZP, Mcl-1:Bim = 2NL9.

To determine the functional relevance of CNP058's ability to engage peptides from pro-apoptotic Bcl-2 proteins, we then examined the capacity of CNP058 to inhibit apoptosis in HeLa cells triggered by UV irradiation (Figure 7). We observed that CNP058 acts as a potent inhibitor of UV induced apoptosis in HeLa cells when compared to the prototypical poxviral apoptosis inhibitor F1 from Vaccinia virus.

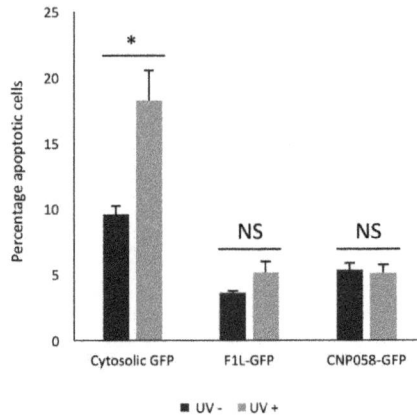

Figure 7. CNP058 is a potent inhibitor of UV induced apoptosis. The ability of HeLa cells expressing green fluorescent protein (GFP) tagged virus encoded apoptosis inhibitors CNP058 or F1 or cytosolic GFP to undergo apoptosis was determined 6 h after UV irradiation via flow cytometry. The percentage of apoptotic cells in total GFP-positive cell population was determined by AV-PE/TO-PRO-3 staining. Data are representative of three independent experiments, error bars represent the standard error of the mean (SEM), *n* = 3. Unpaired Student's two-tailed *t*-test was performed in Excel, * *p* < 0.05, NS: non-significant.

4. Discussion

Premature host cell apoptosis is a crucial mechanism as part of an innate response against viral infection to limit spread of viral infections. Consequently, many viruses including adenoviruses [11,33], herpesviruses [14] and poxviruses [18,34] evolved to encode vBcl-2 proteins that are important to subvert Bcl-2-mediated host cell apoptosis. Functional studies revealed that poxviruses encode a wide range of vBcl-2 proteins harboring diverse modes of action to interfere with apoptosis signaling [3,16,20,34]. For example, vaccinia virus F1 and deerpox virus DPV022 possess a very restricted binding profile where both only engage Bim with high affinity [34,35] (Table 3). In contrast, fowlpox virus FPV039, ORF virus ORFV125, and sheeppox virus SPPV14 target a wider range of cellular Bcl-2 proteins: FPV039 engages all BH3-only proteins and Bak and Bax [18,20], ORFV125 binds Bim, Bik, Hrk, Noxa, Puma, and Bax [36], whereas SPPV14 binds Bim, Bid, Bmf, Hrk, Puma, Bak and Bax [37].

Table 2. Affinities (in nM) of different pro-survival Bcl-2 proteins for peptides spanning the BH3 motif of endogenous pro-apoptotic Bcl-2 family members or Beclin-1 (measurements taken from: [16,18,34,35,37–48]).

	Poxviral Bcl-2							
Pro-Death	**SPPV14**	**M11L**	**MVA_F1**	**VAR_F1**	**DPV022**	**FPV039**	**CNP058**	**N1**
Bad	>2000	>1000	NB	NB	NB	653	NB	>1000
Bid	341	100	NB	3200	NB	2	50	152
Bik	>2000	>1000	NB	NB	NB	30	NB	n/a
Bim	26	5	250	NB	340	10	353	72
Bmf	67	100	NB	NB	NB	16	294	n/a
Hrk	63	>1000	NB	NB	NB	24	312	n/a
Noxa	>2000	>1000	NB	NB	NB	28	3284	n/a
Puma	65	>1000	NB	NB	NB	24	2484	n/a
Bak	46	50	4300	2640	6930	76	508	71
Bax	32	75	1850	960	4040	76	326	n/a
Beclin-1	n/a	n/a	n/a	n/a	NB	n/a	n/a	n/a

Table 3. *Cout.*

	Asfarviral Bcl-2		Herpesviral Bcl-2	
	A179L	BHRF1	Ks-Bcl-2	M11
Bad	258	>2000	>1000	NB
Bid	26	109	112	232
Bik	190	>2000	>1000	NB
Bim	6	18	29	131
Bmf	254	>2000	>1000	300
Hrk	1487	>1000	>1000	719
Noxa	1575	>2000	>1000	132
Puma	31	70	69	370
Bak	29	150	<50	76.3
Bax	26	1400	980	690
Beclin-1		n/a	n/a	40

	Human Bcl-2					Sponge Bcl-2
	Bcl-2	Bcl-w	Bcl-x$_L$	Mcl-1	A1	BHP2
Bad	16	30	5.3	>100,000	15,000	NB
Bid	6800	40	82	2100	1	NB
Bik	850	12	43	1700	58	NB
Bim	2.6	4.3	4.6	2.4	1	NB
Bmf	3	9.8	9.7	1100	180	NB
Hrk	320	49	3.7	370	46	3760
Noxa	>100,000	>100,000	>100,000	24	20	NB
Puma	3.3	5.1	6.3	5	1	NB
Bak	>1000	500	50	10	3	66
Bax	100	58	130	12	n/a	NB
Beclin-1	n/a	n/a	2300	n/a	n/a	n/a

CNP058 was identified as a hypothetical vBcl-2 in the canarypox genome that shares limited sequence identity with Mcl-1 [49] (Figure 3). Other vBcl-2 proteins identified in avipox genomes includes fowlpox FPV039, pigeonpox vBcl-2, penguinpox vBcl-2 and turkeypox vBcl-2 [18]. Phylogenetic analysis based on the sequences of the core 4b protein, DNA polymerase as well as vBcl-2 revealed that avipoxviruses are divided into two clades: fowlpox virus-like and canarypox virus-like [18,50,51]. Unlike FPV039, which is the prototypical vBcl-2 member of the fowlpox virus-like clade and has previously been shown to adopt Bcl-2-like fold and inhibits apoptosis by targeting all BH3-only proteins as well as Bak and Bax [18–20], not much is known about apoptosis regulation by CNP058, the prototypical vBcl-2 from the canarypox virus-like clade. To address this, we examined the ability of CNP058 to bind BH3 domains of all pro-death Bcl-2 proteins (Figure 2). Interestingly, CNP058 only displayed a high binding affinity towards Bid (50 nM), whilst Bim, Hrk, Bax, and Bmf were bound with moderate affinities of 353 nM, 312 nM, 326 nM and 294 nM, respectively. Furthermore, CNP058 bound the BH3 domains of Bak, Noxa, and Puma with low affinities of 2442 nM, 3284 nM, and 2484 nM, respectively. The binding profile for pro-apoptotic Bcl-2 proteins of CNP058 is in marked contrast to the one previously reported for FPV039, which demonstrated that it binds to both Noxa and Bad, an unusual feature not seen in cellular pro-survival Bcl-2 proteins, which only bind either Bad or Noxa [18]. Interestingly, despite sharing a high sequence similarity with FPV039 (38% identity, 58% similarity, Figure 3), CNP058 only displays binding to Noxa and not Bad, thus more resembling the binding profile of cellular pro-survival Bcl-2 proteins Mcl-1 and A1 [18]. The inability of CNP058 to interact with Bad BH3 domain suggests that Bad BH3 mimetics ABT-737 and ABT-263 would not be useful compounds to inhibit CNP058 in birds infected by canarypox virus, as these drugs are specific for pro-survival Bcl-2 proteins that interact with Bad BH3 [52]. Another interesting feature of the CNP058 pro-death Bcl-2 binding profile is the fact that it interacts with Noxa. To date only FPV039, A179L and ORFV125 have been identified as vBcl-2 proteins that interact with Noxa [18,36,39]. ORFV125 has been shown to interact with Noxa through immunoprecipitation experiment, however the affinity of the interaction has not yet been established, whereas FPV039 binds Noxa with tightly with an affinity of 28 nM, whilst the binding of A179L is of more modest affinity (1575 nM), as established using ITC [18,36,39]. Noxa has been shown to be upregulated in the presence of viral infection, double stranded RNA, and interferon, suggesting Noxa as a viral sensor to activate host intrinsic apoptosis [6,53]. The interaction of FPV039 and CNP058 with Noxa suggest that avipoxviruses may counter host intrinsic apoptosis triggered by Noxa during viral infection.

Despite being closely related to FPV039, a comparison of the binding profiles showed that CNP058 displays a distinct interaction profile with pro-death Bcl-2 proteins. CNP058 is similar to other vBcl-2, which only bind to a select subset of pro-death Bcl-2 proteins, whereas FPV039 and A179L are highly promiscuous and are able to engage all major pro-apoptotic Bcl-2 proteins [18,39]. In addition, the binding affinities of CNP058 towards pro-death Bcl-2 BH3 domains are significantly lower when compared to FPV039, with only the interaction of CNP058 with Bid in the low nanomolar range. In contrast, nearly all of the cellular pro-death BH3 domains interacted tightly with FPV039 with a high nanomolar affinity [18]. However, despite displaying substantially weaker affinities for peptides from all of the identified endogenous pro-apoptotic Bcl-2 members, we demonstrate that CNP058 is a potent inhibitor of UV irradiation induced apoptosis in transfected HeLa cells, and appears to be an equally potent inhibitor of apoptosis when compared to the prototypical poxviral pro-survival Bcl-2 protein F1 from vaccinia virus (Figure 7) [17,31,54–56]. However, when considering that our apoptosis assays were performed using cells transfected with CNP058 in the absence of infectious virus, the functional relevance for the comparable potency of apoptosis inhibition efficiency by F1 and CNP058 remain to be established.

At the structural level, CNP058 possesses longer α4, α5, and α6 helices in comparison to FPV039 (Figure 5D) [18]. These topological differences may contribute to the distinct binding profiles of the two proteins as α4 and α5 helices form core segments of the hydrophobic binding groove. Furthermore, the substitution of an Asn in the highly conserved **NWGR** for Thr may cause the difference in the binding profile between the two avipoxvirus vBcl-2s, since the threonine in **TWGR** motif of CNP058 is unable to form hydrogen bonds with bound BH3 domains, whereas the corresponding Asn in FPV039 is involved in hydrogen bonds in complexes with Bik and Bmf [18] (Figure 8). This suggests that the Asn in the **NWGR** motif is an important determinant for the interactions at the very least for avipox encoded pro-survival Bcl-2 proteins with the pro-death BH3 domains, a notion that is supported by similar findings for human Bcl-b [57], which binds only Bim and Bax amongst the cellular pro-apoptotic Bcl-2 proteins, in both cases with low half maximal inhibitory concentration (IC_{50}) values of 6.6 and 113 μM. Furthermore, the mouse Bcl-2 protein Boo features a SWSQ instead of the typical NWGR motif, and is unable to engage any pro-apoptotic Bcl-2 proteins [58], further underscoring the impact that substitutions in the NWGR motif have on pro-survival Bcl-2 proteins and their ability to bind to their pro-apoptotic cellular counterparts.

The overall structure of CNP058 is similar to other cellular Bcl-2 proteins, with an extended loop region connecting the α1 and α2 helices, a feature only seen in the two other vBcl-2 proteins, Bhrf1 and FPV039 [18,40]. However, CNP058 features a short loop region connecting helices α5–6, which is in marked contrast to the other avipoxviral Bcl-2 protein FPV039. Whereas, the α5–α6 loop in FPV039 was extended and comprised of 16 residues, the equivalent section in CNP058 features only a 3 residue loop, but significantly longer α5 and α6 helices. The impact of this difference on the ability of CNP058 to engage pro-apoptotic Bcl-2 proteins is not immediately obvious since this region is somewhat distant from the canonical ligand binding groove, and is thus not likely to be a significant determinant of pro-apoptotic Bcl-2 binding profiles.

A) CNP058:Bim BH3 domain

B) FPV039:Bik BH3 domain

C) Mcl1:Bim BH3 domain

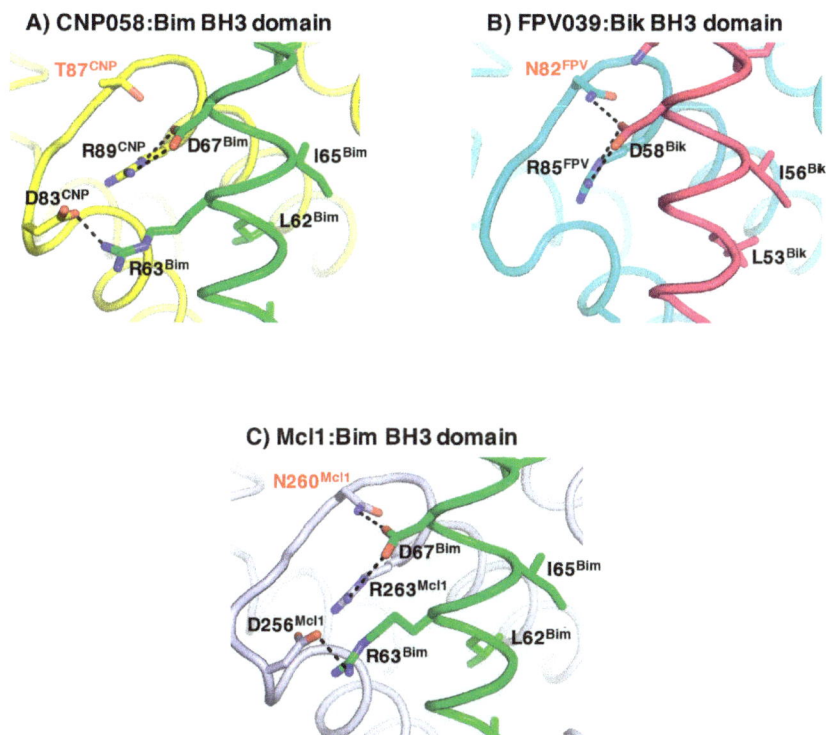

Figure 8. The role of the conserve NWGR motif. Detailed cartoon representation of the NWGR or equivalent motif in (**A**) CNP058:Bim BH3 domain complex, (**B**) FPV039:Bik BH3 domain complex and (**C**) Mcl-1:Bim BH3 domain complex. The view is into the conserved hydrophobic ligand binding groove formed by α-helices 2–5. Residues involved in interactions are shown as sticks and labelled. Hydrogen bonds and ionic interactions between pro-survival proteins and pro-apoptotic ligands are denoted as black dotted lines. The Asn residues in the NWGR motif in FPV039 and Mcl-1 as well as the Thr in the TWGR motif are labelled in red. PDB ID: FPV039:Bik = 5TZP, Mcl-1:Bim = 2NL9.

A comparison of the mode of ligand binding between the CNP058:Bim and FPV039:Bik complex reveals that binding of Bim to CNP058 buries a total of 1700 Å^2 of solvent accessible surface and an associated ΔG of interface formation and dissociation of −11.1 and 4.5 kcal/mol, respectively, whereas binding of Bik to FPV039 buries a total of 1450 Å^2 and an associated ΔG of interface formation and dissociation of −8.9 and 2.9 kcal/mol, respectively, suggesting that ligand binding to CNP058 is driven more by hydrophilic interactions as compared to FPV039. Furthermore, we examined the B-factor distribution along the key α3 and α4 helices that form the walls of the ligand binding groove in both CNP058 and FPV039 (Figure 9). Interestingly, in CNP058, significant B-factor differences are observed for the entire α3 helix and the lower part of helix α4 compared to the central helix α5, suggesting that both α3 and α4 are display considerably higher flexibility. In contrast, in the FPV039-Bik complex, only helix 4 displays higher B-factors when compared to the central helix α5, whereas helix α3 does not display significantly higher thermal motion.

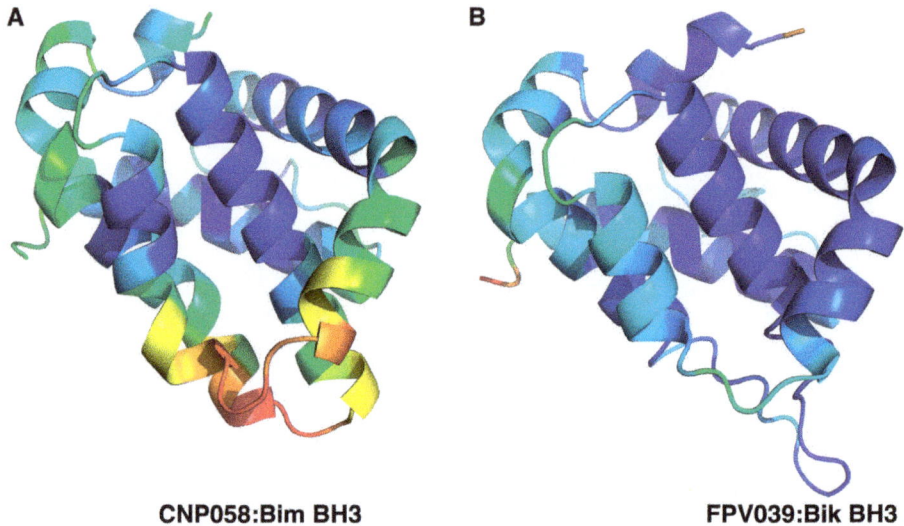

CNP058:Bim BH3　　　　　　　　　　　**FPV039:Bik BH3**

Figure 9. Cartoon representation of B-factor distribution of Cα atoms in (**A**) CNP058:Bim BH3 domain complex, (**B**) FPV039:Bik BH3 domain complex The view is into the conserved hydrophobic ligand binding groove formed by α-helices 2–5. BH3 domain ligands were removed for clarity. Colour are based on residues with low B-factor (blue) to high B-factor (red). PDB ID: FPV039:Bik = 5TZP.

Amongst the poxviruses, a number of sequence, structural, and functional homologs of Bcl-2 have been identified to date. These include proteins that displayed little or no over sequence identity with cellular Bcl-2 proteins such as myxoma virus M11L, deerpox virus DPV022, sheeppox virus SPPV14, variola virus F1, and vaccinia virus F1, N1, A49, A52, B14, and K7 [17,34,35,38,59–62]. These poxviral vBcl-2 proteins can be divided into three groups: (1) monomeric pro-survival vBcl-2 such as M11L, (2) dimeric vBcl-2 such as F1 and DPV022, and (3) A49, A52, B14, and K7, which all were identified as vBcl-2 proteins after their structures were determined experimentally, that do not interact with pro-death Bcl-2 and carry out other immunomodulatory functions such as nuclear factor kappa-light-chain-enhancer of activated B cells (NF-κB) pathway and interferon regulatory transcription factor 3 (IRF3) inhibitors [16,34,35,60–62]. Vaccinia virus N1 is unusual since it is dimeric, and functions as a dual inhibitory protein that modulates both intrinsic apoptosis and NF-κB signaling. Based on biochemical and structural studies, CNP058 and FPV039 are distinct from other poxvirus encoded vBcl-2 proteins since they both display readily identifiable sequence identity with cellular Bcl-2 proteins due to their obvious BH1 and BH2 motifs. Furthermore, at the structural level, they both closely resemble the cellular pro-survival protein Mcl-1. This suggests that FPV039 and CNP058 may be the closest homologs of cellular Bcl-2 proteins within the poxviruses.

In summary, our findings reveal that CNP058 adopts a Bcl-2-like fold similar to cellular pro-survival protein Mcl-1 and its avipoxvirus vBcl-2 counterpart FPV039. We showed that CNP058 binds to BH3 domains of pro-death proteins including Bak and Bax and most of the BH3-only proteins, and is able to protect HeLa cells against UV induced apoptosis. These findings establish CNP058 as a viral Bcl-2 protein with a broad pro-apoptotic Bcl-2 binding profile that is a potent inhibitor of apoptosis.

Acknowledgments: We thank staff at the MX beamlines at the Australian Synchrotron for help with X-ray data collection, and the CSIRO C3 Collaborative Crystallization Centre for assistance with crystallization and the Comprehensive Proteomics Platform at La Trobe University for core instrument support. This work was supported by the Australian Research Council (Fellowship FT130101349 to MK and Discovery Project DP170103790 to IHKP) and La Trobe University (Scholarship to MA).

Author Contributions: M.I.A.: Acquisition of data; Analysis and interpretation of data; Drafting and revising the article. A.A.B.: Acquisition of data; Analysis and interpretation of data; Drafting and revising the article. I.K.H.P.: Analysis and interpretation of data; Drafting and revising the article. M.D.H.: Analysis and interpretation of data; Drafting and revising the article. M.K.: Conception and design; Acquisition of data; Analysis and interpretation of data; Drafting and revising the article.

Conflicts of Interest: The authors declare no conflict of interest.

References

1. Benedict, C.A.; Norris, P.S.; Ware, C.F. To kill or be killed: Viral evasion of apoptosis. *Nat. Immunol.* **2002**, *3*, 1013–1018. [CrossRef] [PubMed]

2. Cuconati, A.; White, E. Viral homologs of bcl-2: Role of apoptosis in the regulation of virus infection. *Genes Dev.* **2002**, *16*, 2465–2478. [CrossRef] [PubMed]

3. Kvansakul, M.; Caria, S.; Hinds, M.G. The Bcl-2 family in host-virus interactions. *Viruses* **2017**, *9*, 290. [CrossRef] [PubMed]

4. Youle, R.J.; Strasser, A. TheBcl-2 protein family: Opposing activities that mediate cell death. *Nat. Rev. Mol. Cell Biol.* **2008**, *9*, 47–59. [CrossRef] [PubMed]

5. Kvansakul, M.; Hinds, M.G. The structural biology of BH3-only proteins. *Methods Enzymol.* **2014**, *544*, 49–74. [PubMed]

6. Eitz Ferrer, P.; Potthoff, S.; Kirschnek, S.; Gasteiger, G.; Kastenmuller, W.; Ludwig, H.; Paschen, S.A.; Villunger, A.; Sutter, G.; Drexler, I.; et al. Induction of Noxa-mediated apoptosis by modified vaccinia virus Ankara depends on viral recognition by cytosolic helicases, leading to IRF-3/IFN-β-dependent induction of pro-apoptotic Noxa. *PLoS Pathog.* **2011**, *7*, e1002083. [CrossRef] [PubMed]

7. Lomonosova, E.; Chinnadurai, G. BH3-only proteins in apoptosis and beyond: An overview. *Oncogene* **2008**, *27* (Suppl. 1), S2–S19. [CrossRef] [PubMed]

8. Westphal, D.; Kluck, R.M.; Dewson, G. Building blocks of the apoptotic pore: How Bax and Bak are activated and oligomerize during apoptosis. *Cell Death Differ.* **2014**, *21*, 196–205. [CrossRef] [PubMed]

9. Kelly, P.N.; Strasser, A. The role of Bcl-2 and its pro-survival relatives in tumourigenesis and cancer therapy. *Cell Death Differ.* **2011**, *18*, 1414–1424. [CrossRef] [PubMed]

10. Kvansakul, M.; Hinds, M.G. Structural biology of the Bcl-2 family and its mimicry by viral proteins. *Cell Death Dis.* **2013**, *4*, e909. [CrossRef] [PubMed]

11. White, E.; Cipriani, R. Specific disruption of intermediate filaments and the nuclear lamina by the 19-kDa product of the adenovirus *e1b* oncogene. *Proc. Natl. Acad. Sci. USA* **1989**, *86*, 9886–9890. [CrossRef] [PubMed]

12. Brun, A.; Rivas, C.; Esteban, M.; Escribano, J.M.; Alonso, C. African swine fever virus gene *a179l*, a viral homologue of Bcl-2, protects cells from programmed cell death. *Virology* **1996**, *225*, 227–230. [CrossRef] [PubMed]

13. Oudejans, J.J.; van den Brule, A.J.C.; Jiwa, N.M.; De Bruin, P.C.; Ossenkoppele, G.J.; van der Valk, P.; Walboomers, J.M.M.; Meijer, C.J.L.M. BHRF1, the Rpstein-Barr virus (EBV) homologue of the Bcl-2 protooncogene, is transcribed in EBV-associated B-cell lymphomas and in reactive lymphocytes. *Blood* **1995**, *86*, 1893–1902. [PubMed]

14. Nava, V.E.; Cheng, E.H.; Veliuona, M.; Zou, S.; Clem, R.J.; Mayer, M.L.; Hardwick, J.M. Herpesvirus saimiri encodes a functional homolog of the human Bcl-2 oncogene. *J. Virol.* **1997**, *71*, 4118–4122. [PubMed]

15. Taylor, J.M.; Barry, M. Near death experiences: Poxvirus regulation of apoptotic death. *Virology* **2006**, *344*, 139–150. [CrossRef] [PubMed]

16. Kvansakul, M.; van Delft, M.F.; Lee, E.F.; Gulbis, J.M.; Fairlie, W.D.; Huang, D.C.; Colman, P.M. A structural viral mimic of prosurvival Bcl-2: A pivotal role for sequestering proapoptotic Bax and Bak. *Mol. Cell* **2007**, *25*, 933–942. [CrossRef] [PubMed]

17. Campbell, S.; Thibault, J.; Mehta, N.; Colman, P.M.; Barry, M.; Kvansakul, M. Structural insight into BH3 domain binding of vaccinia virus antiapoptotic F1L. *J. Virol.* **2014**, *88*, 8667–8677. [CrossRef] [PubMed]

18. Anasir, M.I.; Caria, S.; Skinner, M.A.; Kvansakul, M. Structural basis of apoptosis inhibition by the fowlpox virus protein FPV039. *J. Biol. Chem.* **2017**, *292*, 9010–9021. [CrossRef] [PubMed]

19. Banadyga, L.; Gerig, J.; Stewart, T.; Barry, M. Fowlpox virus encodes a Bcl-2 homologue that protects cells from apoptotic death through interaction with the proapoptotic protein Bak. *J. Virol.* **2007**, *81*, 11032–11045. [CrossRef] [PubMed]

20. Banadyga, L.; Veugelers, K.; Campbell, S.; Barry, M. The fowlpox virus Bcl-2 homologue, FPV039, interacts with activated Bax and a discrete subset of BH3-only proteins to inhibit apoptosis. *J. Virol.* **2009**, *83*, 7085–7098. [CrossRef] [PubMed]

21. Studier, F.W. Protein production by auto-induction in high-density shaking cultures. *Protein Expr. Purif.* **2005**, *41*, 207–234. [CrossRef] [PubMed]

22. Kvansakul, M.; Czabotar, P.E. Preparing samples for crystallization of Bcl-2 family complexes. *Methods Mol. Biol.* **2016**, *1419*, 213–229. [PubMed]

23. Kabsch, W. XDS. *Acta Crystallogr. D Biol. Crystallogr.* **2010**, *66*, 125–132. [CrossRef] [PubMed]

24. Evans, P. Scaling and assessment of data quality. *Acta Crystallogr. D Biol. Crystallogr.* **2006**, *62*, 72–82. [CrossRef] [PubMed]

25. Winn, M.D.; Ballard, C.C.; Cowtan, K.D.; Dodson, E.J.; Emsley, P.; Evans, P.R.; Keegan, R.M.; Krissinel, E.B.; Leslie, A.G.; McCoy, A.; et al. Overview of the CCP4 suite and current developments. *Acta Crystallogr. D Biol. Crystallogr.* **2011**, *67*, 235–242. [CrossRef] [PubMed]

26. McCoy, A.J. Solving structures of protein complexes by molecular replacement with Phaser. *Acta Crystallogr. D Biol. Crystallogr.* **2007**, *63*, 32–41. [CrossRef] [PubMed]

27. Emsley, P.; Lohkamp, B.; Scott, W.G.; Cowtan, K. Features and development of Coot. *Acta Crystallogr. D Biol. Crystallogr.* **2010**, *66*, 486–501. [CrossRef] [PubMed]

28. Afonine, P.V.; Grosse-Kunstleve, R.W.; Echols, N.; Headd, J.J.; Moriarty, N.W.; Mustyakimov, M.; Terwilliger, T.C.; Urzhumtsev, A.; Zwart, P.H.; Adams, P.D. Towards automated crystallographic structure refinement with Phenix.Refine. *Acta Crystallogr. D Biol. Crystallogr.* **2012**, *68*, 352–367. [CrossRef] [PubMed]

29. Morin, A.; Eisenbraun, B.; Key, J.; Sanschagrin, P.C.; Timony, M.A.; Ottaviano, M.; Sliz, P. Collaboration gets the most out of software. *Elife* **2013**, *2*, e01456. [CrossRef] [PubMed]

30. Meyer, P.A.; Socias, S.; Key, J.; Ransey, E.; Tjon, E.C.; Buschiazzo, A.; Lei, M.; Botka, C.; Withrow, J.; Neau, D.; et al. Data publication with the structural biology data grid supports live analysis. *Nat. Commun.* **2016**, *7*, 10882. [CrossRef] [PubMed]

31. Campbell, S.; Hazes, B.; Kvansakul, M.; Colman, P.; Barry, M. Vaccinia virus F1L interacts with Bak using highly divergent Bcl-2 homology domains and replaces the function of Mcl-1. *J. Biol. Chem.* **2010**, *285*, 4695–4708. [CrossRef] [PubMed]

32. Jiang, L.; Tixeira, R.; Caruso, S.; Atkin-Smith, G.K.; Baxter, A.A.; Paone, S.; Hulett, M.D.; Poon, I.K. Monitoring the progression of cell death and the disassembly of dying cells by flow cytometry. *Nat. Protoc.* **2016**, *11*, 655–663. [CrossRef] [PubMed]

33. Han, J.; Sabbatini, P.; Perez, D.; Rao, L.; Modha, D.; White, E. The E1B 19K protein blocks apoptosis by interacting with and inhibiting the p53-inducible and death-promoting Bax protein. *Genes Dev.* **1995**, *10*, 461–477. [CrossRef]

34. Kvansakul, M.; Yang, H.; Fairlie, W.D.; Czabotar, P.E.; Fischer, S.F.; Perugini, M.A.; Huang, D.C.; Colman, P.M. Vaccinia virus anti-apoptotic F1L is a novel Bcl-2-like domain-swapped dimer that binds a highly selective subset of BH3-containing death ligands. *Cell Death Differ.* **2008**, *15*, 1564–1571. [CrossRef] [PubMed]

35. Burton, D.R.; Caria, S.; Marshall, B.; Barry, M.; Kvansakul, M. Structural basis of deerpox virus-mediated inhibition of apoptosis. *Acta Crystallogr. D Biol. Crystallogr.* **2015**, *71*, 1593–1603. [CrossRef] [PubMed]

36. Westphal, D.; Ledgerwood, E.C.; Tyndall, J.D.; Hibma, M.H.; Ueda, N.; Fleming, S.B.; Mercer, A.A. The orf virus inhibitor of apoptosis functions in a Bcl-2-like manner, binding and neutralizing a set of BH3-only proteins and active Bax. *Apoptosis* **2009**, *14*, 1317–1330. [CrossRef] [PubMed]

37. Okamoto, T.; Campbell, S.; Mehta, N.; Thibault, J.; Colman, P.M.; Barry, M.; Huang, D.C.; Kvansakul, M. Sheeppox virus SPPV14 encodes a Bcl-2-like cell death inhibitor that counters a distinct set of mammalian proapoptotic proteins. *J. Virol.* **2012**, *86*, 11501–11511. [CrossRef] [PubMed]

38. Marshall, B.; Puthalakath, H.; Caria, S.; Chugh, S.; Doerflinger, M.; Colman, P.M.; Kvansakul, M. Variola virus F1L is a Bcl-2-like protein that unlike its vaccinia virus counterpart inhibits apoptosis independent of bim. *Cell Death Dis.* **2015**, *6*, e1680. [CrossRef] [PubMed]

39. Banjara, S.; Caria, S.; Dixon, L.K.; Hinds, M.G.; Kvansakul, M. Structural insight into african swine fever virus A179l-mediated inhibition of apoptosis. *J. Virol.* **2017**, *91*. [CrossRef] [PubMed]

40. Kvansakul, M.; Wei, A.H.; Fletcher, J.I.; Willis, S.N.; Chen, L.; Roberts, A.W.; Huang, D.C.; Colman, P.M. Structural basis for apoptosis inhibition by Epstein-Barr virus BHRF1. *PLoS Pathog.* **2010**, *6*, e1001236. [CrossRef] [PubMed]

41. Flanagan, A.M.; Letai, A. BH3 domains define selective inhibitory interactions with BHRF-1 and KSHV Bcl-2. *Cell Death Differ.* **2008**, *15*, 580–588. [CrossRef] [PubMed]

42. Ku, B.; Woo, J.S.; Liang, C.; Lee, K.H.; Jung, J.U.; Oh, B.H. An insight into the mechanistic role of beclin 1 and its inhibition by prosurvival Bcl-2 family proteins. *Autophagy* **2008**, *4*, 519–520. [CrossRef] [PubMed]

43. Aoyagi, M.; Zhai, D.; Jin, C.; Aleshin, A.E.; Stec, B.; Reed, J.C.; Liddington, R.C. Vaccinia virus N1L protein resembles a B cell lymphoma-2 (Bcl-2) family protein. *Protein Sci.* **2007**, *16*, 118–124. [CrossRef] [PubMed]

44. Chen, L.; Willis, S.N.; Wei, A.; Smith, B.J.; Fletcher, J.I.; Hinds, M.G.; Colman, P.M.; Day, C.L.; Adams, J.M.; Huang, D.C. Differential targeting of prosurvival Bcl-2 proteins by their BH3-only ligands allows complementary apoptotic function. *Mol. Cell* **2005**, *17*, 393–403. [CrossRef] [PubMed]

45. Smits, C.; Czabotar, P.E.; Hinds, M.G.; Day, C.L. Structural plasticity underpins promiscuous binding of the prosurvival protein A1. *Structure* **2008**, *16*, 818–829. [CrossRef] [PubMed]

46. Willis, S.N.; Chen, L.; Dewson, G.; Wei, A.; Naik, E.; Fletcher, J.I.; Adams, J.M.; Huang, D.C. Proapoptotic bak is sequestered by Mcl-1 and Bcl-xL, but not Bcl-2, until displaced by BH3-only proteins. *Genes Dev.* **2005**, *19*, 1294–1305. [CrossRef] [PubMed]

47. Fletcher, J.I.; Meusburger, S.; Hawkins, C.J.; Riglar, D.T.; Lee, E.F.; Fairlie, W.D.; Huang, D.C.; Adams, J.M. Apoptosis is triggered when prosurvival Bcl-2 proteins cannot restrain Bax. *Proc. Natl. Acad. Sci. USA* **2008**, *105*, 18081–18087. [CrossRef] [PubMed]

48. Caria, S.; Hinds, M.G.; Kvansakul, M. Structural insight into an evolutionarily ancient programmed cell death regulator—The crystal structure of marine sponge BHP2 bound to LB-Bak-2. *Cell Death Dis.* **2017**, *8*, e2543. [CrossRef] [PubMed]

49. Tulman, E.R.; Afonso, C.L.; Lu, Z.; Zsak, L.; Kutish, G.F.; Rock, D.L. The genome of canarypox virus. *J. Virol.* **2003**, *78*, 353–366. [CrossRef]

50. Gyuranecz, M.; Foster, J.T.; Dan, A.; Ip, H.S.; Egstad, K.F.; Parker, P.G.; Higashiguchi, J.M.; Skinner, M.A.; Hofle, U.; Kreizinger, Z.; et al. Worldwide phylogenetic relationship of avian poxviruses. *J. Virol.* **2013**, *87*, 4938–4951. [CrossRef] [PubMed]

51. Jarmin, S.; Manvell, R.; Gough, R.E.; Laidlaw, S.M.; Skinner, M.A. Avipoxvirus phylogenetics: Identification of a PCR length polymorphism that discriminates between the two major clades. *J. Gen. Virol.* **2006**, *87*, 2191–2201. [CrossRef] [PubMed]

52. Kvansakul, M.; Hinds, M.G. The Bcl-2 family: Structures, interactions and targets for drug discovery. *Apoptosis* **2015**, *20*, 136–150. [CrossRef] [PubMed]

53. Sun, Y.; Leaman, D.W. Involvement of Noxa in cellular apoptotic responses to interferon, double-stranded RNA, and virus infection. *J. Biol. Chem.* **2005**, *280*, 15561–15568. [CrossRef] [PubMed]

54. Wasilenko, S.T.; Stewart, T.L.; Meyers, A.F.; Barry, M. Vaccinia virus encodes a previously uncharacterized mitochondrial-associated inhibitor of apoptosis. *Proc. Natl. Acad. Sci. USA* **2003**, *100*, 14345–14350. [CrossRef] [PubMed]

55. Caria, S.; Marshall, B.; Burton, R.L.; Campbell, S.; Pantaki-Eimany, D.; Hawkins, C.J.; Barry, M.; Kvansakul, M. The N terminus of the vaccinia virus protein F1L is an intrinsically unstructured region that is not involved in apoptosis regulation. *J. Biol. Chem.* **2016**, *291*, 14600–14608. [CrossRef] [PubMed]

56. Fischer, S.F.; Ludwig, H.; Holzapfel, J.; Kvansakul, M.; Chen, L.; Huang, D.C.; Sutter, G.; Knese, M.; Hacker, G. Modified Vaccinia virus ankara protein F1L is a novel BH3-domain-binding protein and acts together with the early viral protein E3l to block virus-associated apoptosis. *Cell Death Differ.* **2006**, *13*, 109–118. [CrossRef] [PubMed]

57. Rautureau, G.J.; Yabal, M.; Yang, H.; Huang, D.C.; Kvansakul, M.; Hinds, M.G. The restricted binding repertoire of Bcl-b leaves Bim as the universal BH3-only prosurvival Bcl-2 protein antagonist. *Cell Death Dis.* **2012**, *3*, e443. [CrossRef] [PubMed]

58. Rautureau, G.J.; Day, C.L.; Hinds, M.G. The structure of Boo/Diva reveals a divergent Bcl-2 protein. *Proteins* **2010**, *78*, 2181–2186. [CrossRef] [PubMed]

59. Douglas, A.E.; Corbett, K.D.; Berger, J.M.; McFadden, G.; Handel, T.M. Structure of M11L: A myxoma virus structural homolog of the apoptosis inhibitor, Bcl-2. *Protein Sci.* **2007**, *16*, 695–703. [CrossRef] [PubMed]

60. Neidel, S.; Maluquer de Motes, C.; Mansur, D.S.; Strnadova, P.; Smith, G.L.; Graham, S.C. Vaccinia virus protein A49 is an unexpected member of the B-cell lymphoma (Bcl)-2 protein family. *J. Biol. Chem.* **2015**, *290*, 5991–6002. [CrossRef] [PubMed]

61. Graham, S.C.; Bahar, M.W.; Cooray, S.; Chen, R.A.; Whalen, D.M.; Abrescia, N.G.; Alderton, D.; Owens, R.J.; Stuart, D.I.; Smith, G.L.; et al. Vaccinia virus proteins A52 and B14 share a Bcl-2-like fold but have evolved to inhibit NF-κB rather than apoptosis. *PLoS Pathog.* **2008**, *4*, e1000128. [CrossRef] [PubMed]

62. Oda, S.; Schroder, M.; Khan, A.R. Structural basis for targeting of human RNA helicase DDX3 by poxvirus protein K7. *Structure* **2009**, *17*, 1528–1537. [CrossRef] [PubMed]

viruses

MDPI

Article

Poxviruses Utilize Multiple Strategies to Inhibit Apoptosis

Daniel Brian Nichols *, William De Martini and Jessica Cottrell

Department of Biological Sciences, Seton Hall University, South Orange, NJ 07039, USA;
william.demartini@student.shu.edu (W.D.M.); jessica.cottrell@shu.edu (J.C.)
* Correspondence: daniel.nichols@shu.edu; Tel.: +1-973-761-9054

Received: 30 June 2017; Accepted: 2 August 2017; Published: 8 August 2017

Abstract: Cells have multiple means to induce apoptosis in response to viral infection. Poxviruses must prevent activation of cellular apoptosis to ensure successful replication. These viruses devote a substantial portion of their genome to immune evasion. Many of these immune evasion products expressed during infection antagonize cellular apoptotic pathways. Poxvirus products target multiple points in both the extrinsic and intrinsic apoptotic pathways, thereby mitigating apoptosis during infection. Interestingly, recent evidence indicates that poxviruses also hijack cellular means of eliminating apoptotic bodies as a means to spread cell to cell through a process called apoptotic mimicry. Poxviruses are the causative agent of many human and veterinary diseases. Further, there is substantial interest in developing these viruses as vectors for a variety of uses including vaccine delivery and as oncolytic viruses to treat certain human cancers. Therefore, an understanding of the molecular mechanisms through which poxviruses regulate the cellular apoptotic pathways remains a top research priority. In this review, we consider anti-apoptotic strategies of poxviruses focusing on three relevant poxvirus genera: *Orthopoxvirus*, *Molluscipoxvirus*, and *Leporipoxvirus*. All three genera express multiple products to inhibit both extrinsic and intrinsic apoptotic pathways with many of these products required for virulence.

Keywords: poxvirus; Vaccinia Virus; Molluscum Contagiosum Virus; Myxoma Virus; apoptosis; immune evasion; mitochondrial membrane permeabilization; protein kinase R; caspase; host defense

1. Introduction

1.1. Overview of Apoptotic Signaling Pathways

Apoptosis is a conserved process that can be triggered by both extrinsic and intrinsic stimuli. In both cases, activation leads to the accumulation of a class of cysteine proteases known as caspases that are found in most cell types [1,2]. Caspases are vital to apoptosis and are more abundant in organisms with higher complexity. However, caspases do have many non-apoptotic functions in processes such as immunity, learning, and cognition [3–5]. Caspases are classified by function. Initiator caspases enable the formation of protein platforms that regulate caspase activation while executioner caspases help with processes after mitochondrial outer membrane permeabilization (MOMP) [6]. MOMP is integral to the decision for a cell to commit to programmed cell death (PCD). Members of the B-cell lymphoma-2 (Bcl-2) family regulate MOMP and therefore determine whether a cell will undergo apoptosis. When apoptosis is initiated, the outer mitochondrial membrane releases intermembrane proteins such as cytochrome *c* into the cytosol. This event triggers the formation of the apoptosome, activation of executioner caspases, and proteolytic cleavage of numerous crucial cellular target proteins. Eventually, this results in the inactivation of DNase inhibitors, which allows the nuclear DNA to be fragmented [6]. This process is highly regulated and varies based on the main mechanisms of induction. The mechanisms of extrinsic, intrinsic, and double stranded RNA (dsRNA) induced apoptosis are discussed below (Figure 1).

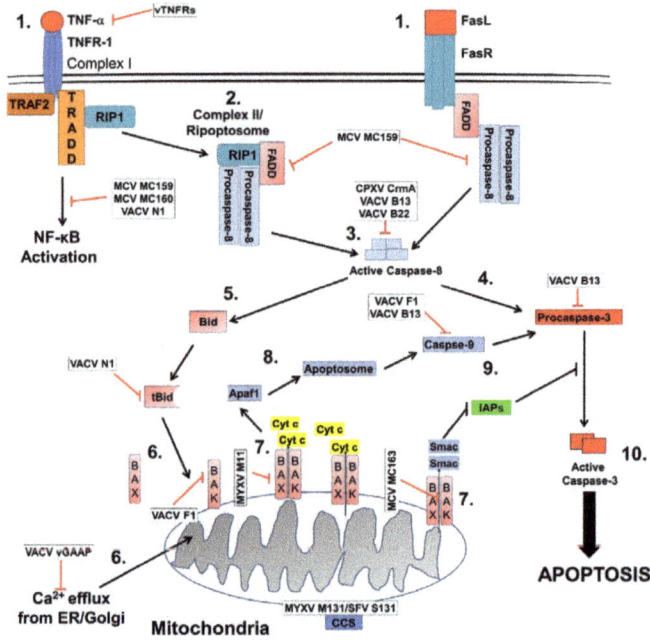

Figure 1. Overview of the extrinsic and intrinsic apoptotic pathways. (**1**) Tumor necrosis factor α (TNFα) or the Fas ligand (FasL) bind to the respective TNF-receptor (TNFR)-1 or FasR receptors. Fas-associated death domain protein (FADD) binds to the cytoplasmic region of FasR and forms a scaffold that recruits procaspase-8.For TNF, TNFR-associated death domain protein (TRADD) associates with the cytoplasmic death domain (DD) of the TNF-R1 and forms complex 1 which leads to nuclear factor κB (NF-κB) activation; (**2**) Alternatively, TNF can induce apoptosis when receptor-interacting protein 1 (RIP1) forms a cytoplasmic complex II consisting of RIP1, FADD, and procaspase-8; (**3**) Procaspase-8 oligomerization results in its autocleavage and activation where the initiator caspase-8 activates (**4**) caspase-3 or cleave additional substrates such as (**5**) BH3 interacting-doain death agonist (Bid) to truncated (t)Bid; (**6**) tBid activates Bcl-2 homologous antagonist killer (Bak)/Bcl-2-associated X protein (Bax) oligomers in the mitochondria. Alternatively, Bak/Bax can form pores in the mitochondria outer membrane in response to Ca^{2+} efflux from the endoplasmic reticulum (ER) or Golgi; (**7**) Bax/Bak pores result in mitochondria membrane permeabilization which leads to the subsequent release of cytochrome *c* and second mitochondria-derived activator of caspases/direct inhibitor of apoptosis protein with low pI) (Smac/DIABLO, referred to as "Smac" in the illustration) from the inner membrane space of the mitochondria to the cytosol; (**8**) Cytoplasmic cytochrome *c* binds Apaf1 leading to the formation of the apoptosome and the activation of initiator caspase-9; (**9**) caspase-9 in turn activates effector caspases such as caspase-3.Smac released form the mitochondria also binds inhibitor of apoptosis proteins (IAPs) which allows caspase-3 to become active and cleave target proteins; (**10**) Effector caspases in turn cleave target proteins resulting in the activation of apoptosis. Poxvirus proteins are indicated in the open boxes. Red lines indicate points in the pathway inhibited by viral proteins. Vaccinia virus (VACV) F1, Myxoma virus (MYXV) M11, MYXV M131, Shope Fibroma Virus (SFV) S131, and Molluscum Contagiosum Virus (MCV) MC163 localize to the mitochondria where these proteins antagonize mitochondria mediated responses in the intrinsic apoptotic pathway. MYXV M131/SFV S131 are depicted interacting with cellular copper chaperones for superoxide dismutase (CCS).

1.2. Extrinsic Apoptotic Pathway

Extrinsic apoptosis also known as death receptor mediated apoptosis typically involves activation of tumor necrosis factor (TNF) superfamily receptors [7]. Cytokines such as TNFα, Fas ligand (FasL), or TNF-related apoptosis-inducing ligand (TRAIL) associate with their respective receptor through an amino-terminal cysteine-rich domain (CRDs). These CRDs define their ligand specificity while a section of 60–70 amino acids known as the death domain (DD) is important for apoptosis induction [8,9]. Once a cytokine has bound to its cognate receptor, the recruitment of adaptor proteins such as Fas-associated death domain protein (FADD), TNF-receptor (TNFR)-associated death domain protein (TRADD), TNFR2-associated factor 2 (TRAF2) or receptor-interacting protein 1 (RIP1) can occur [10]. Next, these proteins assemble to form the death-inducing signaling complex (DISC). DISC provides the scaffold necessary to recruit and activate the initiator caspase, pro-caspase-8 through FADD's death effector domain (DED) [10]. Pro-caspase-8 is activated via proteolytic cleavage and the release of its active p18/p12 domain. The liberated caspase-8 activates downstream caspases-3, -6, and -7 which participate in the execution of the apoptotic process [8,11]. Regulation over complex II also known as a ripoptosome can occur in multiple ways. FLICE (FADD-like IL-1β-converting enzyme)/caspase-8 inhibitory protein (FLIP) recruitment to the DISC can interact with caspase-8 in the complex and inhibit PCD [12]. Second, the silencer of death domain protein (SODD) can act as an intracellular inhibitor of TNFR to prevent constitutive activation [13]. Finally, cellular inhibitor of apoptosis proteins (cIAPs) can control DISC formation. cIAPs can stimulate TNFα and help mediate the activation of nuclear factor κB (NF-κB) under certain conditions [14]. Poxviruses have multiple strategies to prevent the activation of these pathways and are discussed below.

1.3. Intrinsic Apoptotic Pathway

The intrinsic apoptotic pathway is regulated by Bcl-2 family members which work together to control the integrity of the outer mitochondrial membrane (OMM). During apoptosis, functionally redundant Bcl-2 family members such as Bcl-2-associated x protein (Bax) and Bcl-2 antagonist killer 1 (Bak) converge to direct mitochondrial membrane permeabilization (MMP) [15], which allows soluble mitochondrial proteins like cytochrome c to leak to the cytosol. MOMP also allows the release of the second mitochondria-derived activator of caspases/direct inhibitor of apoptosis protein with low isoelectric point (pI) (Smac/DIABLO) [6]. Smac/DIABLO inhibits cellular inhibitor of apoptosis proteins (cIAPs) thereby promoting apoptosis. Within the cytosol, cytochrome c and apoptotic protease factor-1 (APAF-1) engage to oligomerize into a caspase activation platform termed an apoptosome [16,17]. The apoptosome promotes the activation of initiator caspase-9, which in turn will activate executioner caspases-3 and -7. These caspases will begin a cascade of proteolytic cleavage on important cellular substrates, which will eventually lead to the nuclear DNA to be fragmented. It is important to note that the extrinsic and intrinsic pathways are not mutually exclusive with many aspects of the extrinsic pathway capable of inducing intrinsic apoptosis through the mitochondria. For example, active caspase-8 cleaves Bid into tBid. tBid in turn induces the formation of Bax/Bak oligomers within the OMM [6].

Bcl-2 protein function is conserved in vertebrates and limited data exists for function in invertebrates [18]. These proteins are globular, α-helical, and classified according to their anti-apoptotic or pro-apoptotic functions. All Bcl-2 proteins share conserved homology domains known as Bcl-2 homology (BH) regions [17]. Even close family relatives like Bcl-XL have BH regions. The anti-apoptotic family members include Bcl-2, Bcl-x, Mcl-1, and A1 and possess four conserved BH regions known as BH1-4. Pro-apoptotic family members are divided into two classes: "multidomain" which contain BH1-3 domains or "BH3-only" proteins. Bak, Bax, and Bcl-2 related ovarian killer (Bok) are mutlidomain family members while Bcl-2 like protein 11 (Bim), Bid, p53 upregulated modulator of apoptosis (PUMA), and Noxa are BH3-only family members. The regulation of OMM by Bcl2 family members is still unclear. However, direct activation, displacement, and embedding together are current models of the Bcl-2 family's role during apoptosis. Each model describes critical protein-protein and protein-membrane interactions of the Bcl-2 family during apoptosis and is extensively reviewed by Leber et al. [15]. In brief, BH3-only proteins, Bim and Bid are thought to act upstream of the multidomain proteins Bak and Bax to initiate apoptosis through direct binding interactions [19]. Bak and Bax then homo-oligomerize into proteolipid pores within the OMM [20]. The resulting pore formation releases intermembrane space proteins, which interact with various cellular proteins. These proteins activate proteases, caspases, and nucleases that will physically destroy the cell. The Bcl-2 family is conserved among vertebrates. Dynamic organelles such as mitochondria undergo fission and fusion regularly to yield an interconnected tubular mitochondrial network [16]. Cytokines and GTPases regulate these processes. Often MOMP coincides with fragmentation of the mitochondrial network suggesting the Bcl-2 proteins have important role in modulating the balance between fission and fusion [21]. The role of Bax and Bak in MOMP has been shown to be important but a precise mechanism is unclear [22–25]. After pro-apoptotic stimulation, mitochondrial fusion and fission machinery components are recruited to the scission sites and colocalize with Bax [26]. Studies showed that the process of MOMP and fission are linked [26–28]. However, several studies suggest that these processes are not interdependent on each other [29–33]. Cytochrome *c* cannot diffuse freely within the inner mitochondrial membrane (IMM). However, caspase dependent cristae rearrangements can occur during MOMP which allow cytochrome c to freely diffuse out of the IMM [34,35].

The mitochondria and the endoplasmic reticulum (ER) also interact during apoptosis.ER calcium Ca^{2+} levels can be influenced by the Bcl-2 family effecting storage and signaling [17]. Rong et al. demonstrated that ER-localized Bcl-2 and its BH4 domain can directly inhibit inositol 1,4,5-triphosphate (IP3) receptor on the ER [36]. The IP3 receptor is an IP3-gated Ca^{2+} channel that regulates many processes including cell proliferation and death [36]. In a mouse embryonic fibroblast model, multidomain proteins Bak and Bax were found to dysregulated Ca^{2+} uptake in the ER and mitochondria. Dysregulated Ca^{2+} levels inhibited apoptotic death, which was restored by introducing corrected calcium levels within the model [37]. BH3 domains like Bim and PUMA are also known to induce ER Ca^{2+} release; however, the mechanism by which this occurs is unknown [38]. Interestingly, DNA microarray analysis found PUMA was upregulated by ER stress. After a global RNA interference screen, PUMA was found to have a functional role in ER stress-mediated apoptosis [39,40]. Finally, dysregulated Ca^{2+} homeostasis can lead to unfolded protein response (UPR). UPR can compromise the ER functions of protein modification, folding, and secretion [17]. Bcl-2 family members such as Bak and Bax can help the ER recover from UPR by directly binding and regulating inositol requiring enzyme (IRE1) ER signaling.

The ER is the primarily recognized as the site of protein synthesis and folding of secreted, membrane-bound, and organelle targeted proteins. Proper ER function requires optimum levels of ATP and Ca^{2+} as well as an oxidizing environment conducive to disulphide-bond formation [41]. Cellular stress effecting energy levels, the redox state or Ca^{2+} concentration greatly reduce the folding capacity of the ER. As the folding capacity of the ER decreases, unfolded proteins aggregate and accumulate resulting in ER stress. Protein aggregation is toxic to cells and can lead to disease [42]. In response to ER stress, the unfolded protein response (UPR) can be activated by three ER transmembrane receptors: pancreatic ER kinase-like ER kinase (PERK), activating transcription factor 6 (ATF6), and IRE1. When unfolded proteins accumulate, GRP78, an ER chaperone, will dissociate from these receptors and trigger the unfolded protein response (UPR). The UPR response is typically a prosurvival response that functions to reduce the accumulation of unfolded proteins in the ER [43,44]. However, when protein accumulation becomes persistent it switches from a prosurvival signal to a pro-apoptotic signal. Signaling through PERK, ATF6, and IRE1 do not directly cause PCD but activate CCAAT-enhancer-binding protein (C/EBP) homologous protein (CHOP) and c-Jun N-terminal kinase (JNK) pathway which promote cell death [45,46]. CHOP has been found to interact with Bcl-2 proteins, which can promote or enhance apoptotic activation [47]. JNK can also regulate Bcl-2 by phosphorylation. Phosphorylated Bcl-2 is unable to sequester and inhibit pro-apoptotic BH3-only proteins and cannot control ER Ca^{2+} fluctuations [48]. ER stress induced JNK activation can target Bcl-2 proteins and lead to the activation of Bax and Bak proteins leading to the execution of apoptosis [49,50]. The caspases thought to be involved in ER stress include caspases 12, 3, 6, 7, 8, and 9. Currently, caspase 12 has been proposed to be the initiator caspase in ER stress-induced apoptosis but the evidence supporting this claim is still uncertain [51–53].

1.4. dsRNA Induced Apoptosis

Double stranded (ds)RNA has a well-established role in activating anti-viral responses through the upregulation of type I interferon (IFN-1) [54]. dsRNA has also been shown to play a role in caspase-8-dependent apoptosis [55,56]. The toll-like receptor 3 (TLR3) and the cytosolic protein kinase RNA-activated (PKR) are thought to trigger IFN-1 production and lead to PCD [55]. dsRNA released during a viral infection can stimulate the pattern-recognition receptor TLR3. When dsRNA binds to the TLR3, dimerization occurs, and the Toll/interleukin (IL)-1 receptor (TIR) cytoplasmic domain reorients, and enables the recruitment of the adapter molecule TIR domain-containing adapter inducing IFN_2 (TRIF).TRIF then recruits TNF receptor-associated factor (TRAF)-6 and the RIP-1 serine-threonine kinase that activates nuclear factor-κB (NF-κB) and pro-apoptotic gene expression. Alternatively or in combination with the above mechanism, TRIF can recruit TRAF-6, RIP-1, and TRAF3 and activate the IFN regulatory factor 3 (IRF-3) and the downstream IFN response [57]. TLR3 can induce apoptosis via caspase-8 even though it lacks a death domain [55,58–61]. Although TLR3's role in this signaling complex remains unclear, one study suggests that the C-terminal RIP homotypic interaction motif (RHIM) of RIP1 interacts with TRIF to trigger caspase-8 induced apoptosis [62]. Other data has shown that cIAPs negatively regulate TLR3-induced apoptosis [55,56,58,63]. Therefore, data-providing information about the molecular assembly of this TLR-3 inducing apoptosis complex is still needed.

Double stranded (ds)RNA can also induce IFN through the serine-threonine protein kinase PKR. The protein domains of PKR can be linked to the α subunit of eukaryotic translation initiation factor 2 (eIF-2α) family and the dsRNA-binding protein family [64]. When dsRNA induces IFN expression, PKR expression is stimulated. PKR inhibits protein synthesis through the phosphorylation of its cellular substrate eIF-2α which prevents the regeneration of the GTP in the ternary complex eIF2α-GTP-tRNA^{Met} [65]. Typically eIF-2α phosphorylation negatively regulates ribosome scanning and/or the direct attachment of internal ribosomal entry sites (IRES) [66] and leads to apoptosis. A number of viruses including poxviruses have evolved strategies to escape this control mechanism [67,68]. Data suggests that another cellular substrate of PKR, inhibitor of nuclear factor κ-B kinase subunit (IκB) α can participate in the regulation of this apoptotic pathway [69–71]. In many cells, heterodimers of NF-κB associate with IκBα proteins and render it inactive. However, when responding to activators, IκBα, can be phosphorylated. Once released from NF-κB, the IκBα's undergo proteolysis via the ubiquitin-proteasome pathway [72]. With the loss of these IkBs and the unmasking of its nuclear localization signal, NF-κB can translocate to the nucleus and upregulate the transcription of target genes [70]. TRAF proteins may help recruit the IκB kinase to exert control of the PKR induced apoptosis [73,74].

1.5. Poxvirus Background

Poxviruses have significant relevance to public health. For example, the Molluscum Contagiosum Virus (MCV) causes a common skin infection in humans resulting in persistent lesions that are difficult to control especially in immunocompromised individuals [75]. Emerging zoonotic diseases such as Monkeypox (MPXV) and Vaccinia Virus (VACV) infections continue to impact human health in infected areas [76,77]. Further, because the human population is no longer vaccinated, the intentional or unintentional release of Variola Virus (VARV), the causative agent of smallpox, remains a concern. Poxviruses also represent attractive vectors with many medically relevant uses ranging from vaccines to use as oncolytic viruses [78–80]. Therefore, understanding how these viruses modulate host immune responses, including cellular apoptotic pathways, remains a research priority.

Poxviruses are linear, dsDNA viruses that replicate exclusively in the cytoplasm of the host cell. These complex DNA viruses have large genomes of approximately 130–375 kbp that encode for numerous virus proteins [81]. Poxviruses have a remarkably complex virus life cycle. Upon entering a host cell, the virus transcribes the genome in a temporally regulated cascade of early, intermediate, and late genes. Early gene transcription occurs in the nucleocapsid with early transcription factors already bound to early promoters, and early messages are extruded through pores into the cytoplasm [81]. Given the complex nature of poxvirus replication, these viruses are exposed to an array of cellular proteins that detect and respond to infection. Many poxvirus early genes encode products with immune evasion functions [11,82,83]. In addition, some late transcription products with immunomodulatory function are packaged in viral particles called lateral bodies [84,85]. Presumably, these products are then delivered to the host cell and allow the virus to immediately begin subverting the host cell's innate immune responses. In all, about one third to one half of the poxvirus genome is devoted to immune evasion [82]. Many of these poxviral immune modulatory proteins are required for pathogenicity and to confer host range [84]. Given that the cell has multiple pathways to induce apoptosis and other cell death pathways (Reviewed in [11]) during a virus infection, it is not surprising that multiple poxvirus products are involved in antagonizing the cell's apoptotic response, thus allowing the virus to complete its life cycle and produce progeny virions (Table 1). While some of these viral proteins functions have been elucidated, the majority remain unknown. The study in how these viral products elicit their function in host cells has provided significant insights as to how viruses interact with host cells and how host cells sense and respond to virus infection. Here, we review anti-apoptotic strategies of several important poxviruses.

Table 1. Summary of poxvirus products with anti-apoptotic function.

Protein	Type of Protein	Virus	Function(s)
CrmA	Serpin	CPXV	Inhibits caspase(s) activity Reduces inflammation and promotes viral replication
B13 (SPI-2)	Serpin	VACV	Inhibits caspase(s) activity
B22 (SPI-1)	Serpin	VACV	Inhibits caspase(s) activity
SPI-3	Serpin	VACV	Inhibits caspase(s) activity. Inhibits cell fusion
SERP1	Serpin	MYXV	Inhibits caspase(s) activity Provides full virulence Reduces inflammation
SERP2	Serpin	MYXV	Inhibits caspase(s) activity Involved in lesion morphology Promotes myxomatosis
SERP3	Serpin	MYXV	Inhibits caspase(s) activity Involved in lesion morphology
CrmB	vTNFR	VACV	Mimics extracellular domain of TNFR1/2 Enhances virulence
CrmC	vTNFR	VACV	Mimics extracellular domain of TNFR1/2 Enhances virulence
CrmD	vTNFR	ECTV	Mimics extracellular domain of TNFR1/2 Possesses SECRET domain that binds to chemokines
CrmE	vTNFR	VACV	Mimics extracellular domain of TNFR1/2 Enhances virulence
M-T2	vTNFR	MYXV	Mimics extracellular domain of TNFR1/2.
			Secreted form inhibits TNF
			Intracellular form blocks virus induced lymphocyte apoptosis
T2	vTNFR	SFV	Mimics extracellular domain of TNFR1/2 Inhibits cellular TNF
A52	Bcl-2-like folds	VACV	Inhibits IL-1 induced NF-κB activation
B14	Bcl-2-like folds	VACV	Inhibits IL-1 induced NF-κB activation
A49	Bcl-2-like folds	VACV	Inhibits NF-κB activation through interactions with β-TrCP

Table 1. *Cont.*

Protein	Type of Protein	Virus	Function(s)
F1	Bcl-2-like folds	VACV	Inhibits staurosporine induced apoptosis Localizes to the mitochondria
N1	Bcl-2-like folds	VACV	Inhibits staurosporine induced apoptosis Interacts with Bad, Bax and Bid Inhibits NF-κB activation Localizes in cytosol
M11	Bcl-2-like folds	MYXV	Required for virulence Inhibits FasL and staurosporine induced apoptosis Interacts with Bak and Bax
DPV022	Bcl-2-like folds	DPV	Inhibits apoptosis induced by Bax and Bak Interacts with Bim, Bax, and Bak
SPPV14	Bcl-2-like folds	SPPV14	Inhibits intrinsic apoptosis by antagonizing Bak and Bax
6L	vGAAP	CMLV	Inhibits extrinsic and intrinsic apoptosis
			Forms ion channels reducing concentration of Ca^{2+} in golgi apparatus
M131	SOD Homolog	MYXV	Binds copper chaperones for superoxide dismutase (CCS)
			Cellular Cu-Zn SOD less active resulting in increased superoxide levels
			Protects cells from apoptosis
S131	SOD Homolog	SFV	Binds CCS
			Cellular Cu-Zn SOD less active resulting in increased superoxide levels
			Protects cells from apoptosis
MC163	SOD Homolog	MCV	Aids in virulence. Inhibits TNFα-induced apoptosis by preventing MMP. Localizes to the mitochondria Prevents staurosporine induced caspase 3 activation
MC066	Seleoncystein protein	MCV	Inhibits UV and hydrogen peroxide induced apoptosis
A45	SOD Homolog	VACV	Function currently unknown
E3	PKR antagonist	VACV	Inhibits PKR activation by sequestering dsRNA Binds to PKR Required for virulence
M029	E3 homolog	MYXV	Inhibits PKR activation Reduces/prevents cleavage of caspase-7 and PARP-1
SPV032	E3 homolog	SPV	Inhibits PKR activation Reduces/prevents cleavage of caspase-7 and PARP-1
D9/D10	Decapping enzymes	VACV	Inhibits PKR activation by reducing dsRNA accumulation
MC159	vFLIP	MCV	Inhibits TNFα and FasL induced apoptosis Interacts with FADD and procaspase-8 Prevent caspase 3 and caspase 8 activation Inhibits TNFα induced NFκB activation and MAVS-induced IRF-3 activation
MC160	vFLIP	MCV	Inhibits TNFα induced NFκB activation and MAVS-induced IRF-3 activation

Abbreviations: CPXV, Cowpox Virus; VACV, Vaccinia Virus; MYXV, Myxoma Virus; ECTV, Ectromelia Virus; SFV, Shope Fibroma Virus; MCV, Molluscum Contagiosum Virus; SPV. Swinepox Virus; CMLV, Camelpox Virus; DPV, Deerpox Virus; vTNFR, viral tumor necrosis factor receptor; vGAAP, viral Golgi anti-apoptotic protein; PKR, protein kinase R; vFLIP, viral FLICE inhibitory protein; SOD, superoxide dismutase; IL-1, interleukin-1; β-TrCP, β-transducing repeat containing protein; Bad, Bcl-2 associated death promoter; Bcl-associated X protein; Bid, BH3 interacting-domain death agonist; NF-κB, nuclear factor κ B; Bak, Bcl-2 homologous antagonist killer; Bim, Bcl-2-like protein 11; TNFα, tumor necrosis factor-α; MMP, mitochondrial membrane permeabilization; UV, ultraviolet; PARP-1, poly (ADP-ribose) polymerase 1; MAVS, mitochondrial antiviral-signaling protein; IRF-3, interferon regulatory transcription factor 3.

2. Poxvirus Inhibitors of the Extrinsic Pathway

2.1. TNF Receptor Orthologs

Decoy receptors are used by poxviruses to antagonize the host's ability to respond to infection. Through the inhibition of cytokines such as TNF binding to its receptor, poxviruses attenuate both inflammation and apoptosis. The TNF superfamily is composed of 19 members that bind to 20 cellular receptors of the TNF receptor superfamily [86]. TNF is first expressed as a transmembrane cytokine where it can be processed to a soluble cytokine by the metalloprotease TNFα-converting enzyme (TACE/ADAM17) [86–88]. The receptors for TNF bioactivity include TNFR1 and TNFR2. Whereas TNFR1 is expressed ubiquitously with conserved death-domain motifs TNFR2 is expression is restricted to immune and endothelial cells and lacks a death domain [89]. Many poxviruses employ molecular decoys, called viral TNF receptors (vTNFR), to mitigate the effects of TNFα. The contribution of vTNFRs to the pathogenesis of poxviruses has been examined in the literature. Extracellular TNF is captured by these secreted vTNFRs which lack transmembrane and signaling domains [90].

Five different vTNFRs have been described in orthopoxviruses: cytokine response modifier B (CrmB), CrmC, CrmD, CrmE and the viral homolog of CD30 (vCD30). vCD30 is encoded in ectromelia virus and binds the CD30L preventing its interaction with CD30 [91]. vCD30 is an inhibitor of T helper cell-mediated inflammation [91]. However, vCD30 is not a major virulence factor in the mousepox model [92]. The remaining vTNFRs bind cellular TNF. CrmE has been shown to bind TNF by crystallography [93]. The vTNFRs mimic the extracellular domain of cellular TNFR1/2 on the N-terminal region. The N-terminal region of these vTNFRs contain up to four conserved TNF binding cysteine rich domains (CRDs) [93]. VARV CrmB, cowpox virus (CPXV) CrmB and ectromeliavirus CrmD have an additional C-terminal extension, which has been named smallpox virus-encoded chemokine receptor (SECRET) domain. This domain has a high affinity for certain chemokines [94].

The differences between CrmB, CrmC, CrmD and CrmE are due to ligand affinity and their expression in orthopoxviruses [95]. Certain strains of VACV (USSR, Lister and Evans) express these vTNFRs via the *crmC* and *crmE* viral genes. The majority of the strains for VACV encode gene fragments related to vTNFR but do not encode a functional protein [82]. Investigation of vTNFRs by Reading et al. indicated that CrmB, CrmC and CrmE enhance the virulence of recombinant VACV [96].

The *crmB* gene is important in the pathogenicity of various poxviruses. CPXV lacking CrmB have increased lethal dose (LD)$_{50}$ in infected mice. In MPXV, the *crmB* gene is present in two copies and is the only vTNFR encoded in the genome [97]. In all viral species CrmB orthologs are expressed early in infection, while CPXV CrmC and CPXV CrmD are translated later in viral life cycle [98,99]. The only predicted gene to be active in VARV is the *crmB* gene [94,100]. A 2015 study by Pontejo et al. reported on the functional and binding properties of poxvirus vTNFRs [97]. CrmB from VARV is the most potent of the tested vTNFRs with a binding affinity for TNF stronger than the biopharmaceutical etancercept, a soluble form of the human TNF receptor 2 (hTNFR2) [97]. The binding affinity constant (K_d) of VARV CrmB to human TNF is 0.28 nM whereas the K_d of hTNFR2 is 0.3 nM. It was observed that CPXV CrmB possesses a higher binding affinity than hTNFR2 with mouse TNF with a K_d of 0.12 nM and 0.43 nM respectively [97].

Additionally, vTNFRs are present in *Leporipoxvirus* Shope Fibroma Virus (SFV). T2 was the first protein identified as a vTNFR.Similar to other vTNFRs, T2 sequesters TNF resulting in the inhibition of cellular TNF receptor activation and responses from downstream antiviral processes [101,102]. In the closely related MYXV, M-T2, described as the first "viroceptor", is an important virulence factor. Absence of M-T2 results in reduced pathogenicity in rabbit models. Two forms of M-T2 serve different functions. The secreted form of M-T2 binds and inhibits TNF while the intracellular version blocks virus induced lymphocyte apoptosis [103,104]. The anti-apoptotic function of intracellular M-T2 to inhibit TNFR1 induced cell death requires a highly conserved viral preligand assembly domain (vPLAD) located on the N-terminus [105]. These aforementioned interactions detail a fascinating mechanism by which poxviruses employ a protein decoy based defense mechanism in the continuing molecular arms race between the poxvirus and host's innate immune response.

2.2. Serine Protease Inhibitors (Serpins)

Poxviruses express several proteins to antagonize the function of caspases called serine protease inhibitors (serpins). Members of the serpin superfamily consist of a single polypeptide chain (370 to 390 amino acid residues) with a conserved domain of three βsheets and nine α helices [106]. The C-terminus portion of serpins possess a specific site called the reactive-site loop (RSL) which interacts with a serine or cysteine protease by acting as a substrate mimic. The RSL site is structurally located on a distorted α helix that extends from β-Sheet A [107]. The inhibitory function of serpins is executed through forming long lasting complexes with their target proteases with their substrate stable acyl-enzyme intermediates [108]. Specificity is primarily defined in serpins by the P1 residue within the RSL. For example, the serpin CrmA in CPXV has a P1 of aspartate which directs the protein to inhibit granzyme-B (serine protease) and caspases (thiol proteases) [109–111]. Specific mutations critical for serpin activity are related to human disease. When the wildtype methionine in P1 is mutated to an arginine in the serpin α-antitrypsin Pittsburgh elastase, the mutant gains the ability to inhibit trypsin like enzymes causing severe bleed disorder [107,112]. Serpins are present in numerous poxviruses including *Orthopoxviruses* CPXV and VACV and *Leporipoxvirus* MYXV. Several studies have shown their role in anti-inflammatory, anti-apoptotic and virulence processes. *Orthopoxviruses* and *Leporipoxviruses* encode for three serpins [113].

The first identified serpin was CrmA from CPXV. This serpin is also known as B13R in VACV. The CrmA inhibits caspase 1 (IL-β-converting enzyme, ICE) [110,114] whose activity produces mature proinflammatory cytokines such as IL-1β from proIL-1β. The proinflammatory cytokine IL-1β is important in controlling poxvirus infections [115,116]. Palumbo et al. reported that the deletion of CrmA from CPXV produces white inflammatory lesions in embryonated chicken eggs chorioallantoic membranes (CAMs) whereas typically wild type CPXV produce red non-inflammatory lesions. The CrmA deleted CPXV lesions exhibited lower amount of CPXV virus replication compared to wildtype [117,118].

In addition to being a viral inhibitor of inflammation, the CrmA protein has anti-apoptotic properties in culture cells [119,120]. CrmA inhibits apoptosis in swine cells infected with CPXV. CrmA inhibits the activation of multiple caspases, which are crucial initiator caspases in the extrinsic and intrinsic apoptotic pathway [2,121–123].

The VACV protein B13 (SPI-2) shares 92% amino acid with CPXV CrmA.Not surprisingly, B13 functions very similar to CPXV CrmA. Like CrmA, B13 inhibits multiple initiator caspases and can inhibit apoptosis induced by a variety of challenges including TNFα, FasL, staurosporine, and the DNA damaging agent doxorubicin (DOX) [2,11,119,120,124]. Recently, a study by Veyer et al. compared the anti-apoptotic activity of four different VACV proteins: B13, F1, N1 and viral Golgi anti-apoptotic protein (vGAAP) [125]. The authors utilized recombinant VACV strain (vv811), which lacks 55 genes including those coding for several VACV anti-apoptotic proteins. When expressed in strain vv811, B13 was the most potent inhibitor of both the extrinsic and intrinsic apoptotic pathways [125].

Vaccinia virus B22R (SPI-1) gene encodes for a similar protein to SPI-2/(CrmA). B22R is 44% identical to B13Rwith a different reactive center [126,127]. Both serpins SPI-1 and SPI-2 are expressed early in the viral infection process and remain inside the host cell [126–128]. In a study by Shisler et al. the importance of SP1-1/B22 in VACV was examined by means of a mutated VACV lacking the SPI-1/B22R gene. Due to this gene depletion, viral replication in A549 cells was lowered by almost two logs in one-step growth curve. Not surprising, there was a reduction in virus particles as well intermediate and late mRNA, viral late protein and cleave proteins. A549 cells lacking the SPI-1 gene were found to be sensitive to TNF induced apoptosis [129]. The SPI-3 protein, known for inhibiting cell fusion [130] is yet another VACV serpin but is not required for virulence in VACV and CPXV in intranasal inoculated mice [131,132].

The *Leporipoxvirus* MYXV genome encodes for two intact serpins SERP1 and SERP2 and a truncated SERP3 [133]. The SERP1 serpin in MYXV is required in vivo for full virulence with mutations in both genes of SERP1 causing significant attenuation. SERP1 is a late virally expressed protein and reduces inflammation following MYXV infection [134]. Like SPI-3, SERP1 possesses an arginine residue at the P1 in the RSL and possesses a similar proteinase inhibitory profile which suggests they have similar functional in vivo [135,136]. Wang et al. investigated the phenotypic effects of swapping SPI-3 and SERP1 from their native virus genome, MYXV and CPXV respectively while not modifying the wildtype promoters in order to maintain their viral temporal expression. Despite their similarity, these two serpins are not interchangeable between MYXV and CPXV [135].

In keeping with the theme of switching serpins between different poxviruses, Nathaniel et al. studied the effect of swapping CPXV CrmA with MYXV SERP2. The serpin SERP2 possesses arginine residue at the P1 in the RSL similarly as CrmA but only shares a 35% amino acid identity but does possess similar functionality for inhibition of caspase 1 and granzyme-B. MYXV lacking the SERP2 serpin are attenuate in rabbits [137–139]. Despite these functional similarities, CrmA and SERP2 are not fully interchangeable. SERP2 does not inhibit inflammation, but restores viral load in CPXV infected CAMs. CrmA restores partial MYXV virulence however lesion morphology was not fully recovered [140].

SERP-3 was characterized 2001 by Guerin et al. as the third serpin in MYXV. The SERP-3 serpin has a significant amount of deletions compared to other viral serpins and does not share much amino acid identity to SERP-1 (19%) or SERP-2 (31%). SERP-3 contains several conserved motifs found commonly in serpins. *Serp3* transcripts are detectable at 8 h post infection and as late 16h using reverse transcriptase PCR in MYXV infected RK13 cells [141]. Rabbits inoculated with MYXV without the *serp3* gene produced small, thin and less congested lesion as compared to wildtype. Eight days post infection the rabbits displayed symptoms of respiratory and conjunctival bacterial infection rather than the severe symptoms found with myxomatosis [141]. Based on the aforementioned findings, serpin activity is at its full efficacy in its native virus despite the functional similarities two serpins originating from different poxviruses. The mode of action of these serpins needs to be elucidated.

2.3. The Molluscum Contagiosum Virus Death Effector Domain Containing Proteins MC159 and MC160

Relative to other poxviruses, MCV utilizes unique strategies to antagonize the extrinsic apoptotic pathway [83]. Many of the apoptotic modulators present in VACV are absent in the MCV genome [142,143]. MC lesions are characterized by increased hyperplasia and hypertrophy [75]. Two types of MC lesions have been described, inflamed (I-MC) and non-inflamed MC (NI-MC) [144]. Of interest, NI-MC lesions appear to have limited apoptotic responses. A study conducted by Vermi et al. found caspase-3 to be in the inactive in NI-MC lesions. However, abundant apoptotic cell death is present at the site of I-MC lesions [144]. Such observations may highlight a struggle between host innate immune responses and the ability of MCV to subvert host cell immune responses including apoptosis. Given the lack of apoptotic responses at NI-MC lesions and overall persistence of MCV infections, it is expected that MCV produce several viral proteins that dampen host cell apoptotic responses. However, due to the lack of a cell culture system or available animal model to study MCV processes, identification of MCV proteins that regulate cellular apoptosis has thus far relied on ectopic expression or use of surrogate viruses that express MCV proteins [83].

The Molluscum Contagiosum Virus MCV encodes two viral proteins, MC159 and MC160, each possessing two tandem death effector domains (DEDs) [143]. DEDs are involved in protein-protein interaction and are found in a variety of pro- and anti-apoptotic signaling molecules. MC159 and MC160 are predicted to be expressed from early gene promoters during an MCV infection [143]. Both MC159 and MC160 belong to a family of proteins collectively referred to as viral FLICE-like inhibitory proteins (vFLIPs) [83]. This family of viral proteins regulates several host pathways involved in innate immune response including pathways that lead to the activation of apoptosis, NF-κB, interferon, and necroptosis [83,145–152]. In addition to the MCV proteins, vFLIPs are also expressed by several gamma herpesviruses including human herpes virus 8 and equine herpes virus [12,153,154].

Death effector domainsare found in several pro-apoptotic host proteins including FADD and procaspase-8. FADD possesses a single DED while procaspase-8 contains two tandem DEDs, similar in arrangement to the DEDs of MC159 and MC160 [83,155–157]. During activation of the extrinsic pathway by inducers such as FasL, FADD and procaspase-8 assemble through DED interactions at the receptor to form the DISC [158]. Upon interaction with FADD, procaspase-8 in turn forms oligomeric filaments consisting of many molecules of procaspase-8 [159–162]. The procaspase-8 self-association induces its autocleavage into the active effector caspase-8, which in turn cleaves pro-apoptotic products leading to activation of the extrinsic apoptotic pathway [163]. The cellular FLICE-inhibitory protein (cFLIP) also contains two tandem death effector domains and is capable of regulating apoptosis upon association with procaspase-8 [164,165].

Of the two MCV DED-containing proteins, the MC159 protein has been the best characterized in terms of its role during apoptotic signaling. The MC159 protein is comprised of 241 amino acids with the two tandem DEDs located at the N-terminus [83]. The crystal structure of MC159 was one of the first DED-containing proteins to be solved. Structural analysis revealed that the MC159 DEDs tightly associate in a dumbbell conformation via hydrophobic interactions between the two DEDs [166, 167]. Both MC159 DEDs contain a conserved RxDL motif (69–72 in DED1 and 166–169 in DED2). The RxDL motif is conserved in several DED-containing proteins including FADD, procaspase-8, and the homologous MC160 protein. The arginine and aspartate residues present in the RxDL motif interact with upstream glutamates in their respective DEDs (E24 inDED1 and E111 in DED2) to form a network of hydrogen bonds collectively referred to as the charge triad [166,167]. This interaction provides a key element in DED-folding and orients the side chains of adjacent amino acid residues involved in protein-protein interactions. Alanine substitutions at any of these key residues disrupt MC159 function [166,168].

MC159 expression inhibits apoptosis induced by TNFα and FasL [12,153,155,168]. Through DED interactions, MC159 associates with both FADD and procaspase-8 [153,155]. By interacting with both FADD and procaspase-8 DEDs, MC159 expression blocks the formation of death effector filaments and caps these filaments thereby blocking caspase-8 activation [162,169]. Several MC159 point mutations have been identified that lose the ability to inhibit apoptosis [168]. Many of these loss of function mutations were the result of replacing charged amino acids in or nearby the charge triad of DED1 with alanine substitutions (R69A, D71A; E18A, E19A, D21A). These MC159 mutants no longer inhibit the formation of DED-filaments in cells stimulated with Fas [168]. Further, a study by Yang et al., found that several of these MC159 mutants that could no longer inhibit apoptosis correlated with a loss of the ability to form a ternary complex with Fas/FADD [166]. Fu et al. recently published a study using cryoelectron microscopy to determine the filament structure of oligomerized caspase-8 DEDs [162]. MC159 interacts with filamentous FADD and caspase-8 to prevent further caspase-8 oligomerization. Thus, by MC159 binding and capping caspase-8 oligomers, caspase-8 activation is prevented [162]. Interestingly, the capping mechanism of MC159 is unique when compared to cFLIP. cFLIP binds to the DISC dependent on the recruitment of FADD and procaspase-8.Fu et al. present a model where cFLIP assembles with caspase-8 during oligomerization [162]. Thus, the presence of cFLIP in the oligomers likely reduces the activation of caspase-8, thereby preventing apoptosis [162].

Of note, the C-terminal portion of MC159 also possesses three TRAF3 binding motifs (PxQxS/T), which mediate the MC159-TRAF3 interaction [152]. MC159 recruits TRAF2 and TRAF3 into DISC complexes. Further, MC159 mutants lacking TRAF-binding motifs only partially protect Jurkat cells from Fas-induced cell death. Therefore, MC159 utilizes both TRAF-dependent and TRAF-independent mechanisms to prevent Fas-induced cell death [152]. However, the TRAF-dependent mechanisms are not currently well understood.

A significant interest in the field is to determine how MC159 contributes to pathogenicity during an MCV infection. Unfortunately, the lack of a suitable system to study MC159 in the context of an MCV infection is not available. Several studies have used surrogate viruses to express MC159. Shisler and Moss utilized a recombinant VACV with a *crmA* deletion expressing MC159 [155]. Relative to the parental VACV, the recombinant MC159 virus prevents Fas-induced activation of caspase-3 and caspase-8, thus blocking apoptosis [155]. Given that VACV and MCV likely trigger similar innate immune responses, expressing MC159 in the context of VACV is expected to mimic the mechanism by which MC159 would function in the context of an MCV infection. More recently, Huttman et al. utilized murine cytomegalovirus (MCMV) as a surrogate virus to express MC159 [170]. In their study, the authors replaced the MCMV *M36* gene, an MCMV gene that codes for a caspase-8 inhibitor, with MC159. The ΔM36::MC159 virus inhibits TNF-induced apoptosis whereas the ΔM36 MCMV virus does not [170]. Therefore, in the context of surrogate viral infections, MC159 functions similar to what has been reported in ectopic expression studies, at least in terms of apoptotic signaling. However, Huttmann et al. do stress the need for a cell culture system to study MC159, and other MCV immune evasion molecules, in the context of an MCV infection [170].

Like MC159, the MC160 protein contains two tandem death effector domains. However, the 371 amino acid MC160 protein possesses a much longer C-terminus than MC159 [83]. The DEDs of MC160 are 45% and 33% similar to the corresponding DEDs of the MC159 protein [155]. Using the available structure of MC159 protein as a template [166], we recently reported on homology modeling of the MC160 protein [171]. Based on structural alignments between the MC159 and the MC160 DEDs, homology modeling predicts that key hydrogen bonding interactions, such as those present in the charge triads of DED1 and DED2 of MC159, are also present in MC160 [171]. Therefore, the molecular modeling predicts that the overall structures of the MC159 and MC160 DEDs are similar. It should be noted that this prediction will need to be verified with structural studies. Like MC159, MC160 also binds to both FADD and procaspase-8 through MC160 DEDs [147,155]. Unlike MC159, MC160 expression does not appear to affect the extrinsic apoptotic pathways when transfected or expressed in a surrogate VACV [155]. However, there is a single report of anti-apoptotic activity associated with MC160 expression in HEK 293 cells challenged with either Fas or TNF [154]. When MC160 was expressed in a surrogate VACV, MC160 expression failed to prevent the cleavage of caspase-8, -3, or poly[ADP-ribose] polymerase 1(PARP-1) [155]. MC160 also contains caspase cleavage sites and can be a target of caspase-mediated cleavage [155]. Co-expression with MC159 can inhibit caspase-mediated cleavage of MC160 [155]. Interesting parallels can be drawn when comparing MC159/MC160 to cellular cFLIP. Several isoforms of cFLIP exists in cells including the shorter cFLIP$_S$ and the long cFLIP$_L$. cFLIP$_S$ inhibits caspase activation, presumably by associating with procaspase-8 and preventing caspase autoprocessing when cFLIP$_S$ inserts into DED oligomers [165]. The longer cFLIP$_L$ forms heterodimers with procaspase-8 and promotes caspase-8 activation [172]. Like MC160, cFLIP$_L$ is cleaved by procaspase-8 [172]. However, in the case of MC160, independent expression of MC160 does not activate caspase-8 [155]. Interestingly, the caspase-8/cFLIP$_L$ heterodimer is only partially active capable of cleaving selected substrates [173]. Therefore, cFLIP$_L$ may affect caspase-8 activity in a manner dependent on the concentration of cFLIP$_L$, with high concentrations of cFLIPL inhibiting caspase-8 and low concentrations activating [162]. Whether MC160 alters caspase-8 activity in a similar manner remains to be determined. Interestingly, some of the amino acid residues present in MC159 implicated in FADD binding are altered in MC160 relative to MC159 [166]. For example, the MC159 mutant E18A, E19A, and D21A lost the ability to form a ternary complex with Fas/FADD and therefore no longer inhibited apoptosis. Based on structural alignments, equivalent residues in the MC160 DED1 are A16, E17, and D19. Therefore, it is tempting to speculate that MC160 may not bind FADD as strongly as MC159.Whether this variation, and others, throughout the MC160 amino acid sequence accounts for its inability to inhibit apoptosis is not known.

In the context of MCV infection, both MC159 and MC160 are predicted to be expressed concurrently. Therefore, MC160 may play a role in regulating apoptotic responses that is dependent on the co-expression of MC159. The majority of studies focusing on MC159 or MC160 independently express these two viral proteins. Shisler and Moss co-transfected MC159 and MC160 expression vectors, but did not observe enhanced anti-apoptotic effects with cells challenged with Fas antibodies and expressing both MC159 and MC160 relative to cells expressing MC159 alone. Aside from apoptosis, additional functions have been described for both MC159 and MC160.Both MC159 and MC160 inhibit TNFα-induced NF-κB activation and MAVS-mediated induction of IRF-3 and subsequent activation of type I interferons [83,145–147,149,150]. MC159 has also been reported to inhibit PKR-induced NF-κB activation and PKR-induced apoptosis [174].

3. Poxvirus Inhibitors of the Intrinsic Pathway

3.1. Poxvirus Proteins with Bcl-2-Like Folds

Several poxviruses express proteins that adopt a Bcl-2-like fold. VACV has at least eleven proteins with either confirmed or predicted Bcl-2-like structure [11]. Interestingly, these poxvirus proteins share little sequence similarities at the amino acid level to Bcl-2 family member proteins. Despite the pro- and anti-apoptotic roles of cellular Bcl-2 family members, the majority of these poxvirus proteins that share these structural similarities do not actually inhibit apoptosis. Instead, many of these poxvirus proteins have evolved to inhibit cellular innate immune signaling networks [11,82]. For example, Graham et al. demonstrated that VACV proteins A52 and B14 function to inhibit NF-κB activation, despite the presence of the Bcl-2-like fold [175]. Expression of both A52 and B14 in HEK 293 cells dampen NF-κB activation induced by IL-1α. B14 expression, but not A52, prevents activation of NF-κB in response to TNFα as well [175]. Neither A52 nor B14 inhibit apoptosis. The lack of apoptotic inhibition is attributed to missing hydrophobic BH3-peptide binding grooves that are absent in both A52 and B14 [175]. VACV A49 also adopts the Bcl-2 fold, but does not possess the surface groove required to bind BH3 proteins [176]. A49 inhibits NF-κB activation by binding β-transducing repeat containing protein (β-TrCP) as a means to block the ubiquitination of IκBα [177].

Wasilenko et al. originally identified the F1 protein as a novel VACV inhibitor of apoptosis [178]. VACV mutant vv811 expressing F1 is protected from apoptosis induced by staurosporine in Jurkat cells. F1 localizes to the mitochondria via a C-terminal hydrophobic domain where F1 prevents the loss of the mitochondrial membrane potential and subsequent release of cytochrome *c* [178]. Mutant viruses lacking F1L induce apoptosis mediated by Bak/Bax [179,180]. The crystal structure of F1 revealed the characteristic Bcl-2-like fold with affinity for pro-apoptotic proteins with BH3 domains [181]. Several studies have shown F1 interacts with Bim, Bak, and Bax to prevent oligomerization of Bak and Bax and subsequent release of cytochrome *c* into the mitochondria [179,181]. More recently, Campbell et al. reported that F1 achieves its anti-apoptotic function through sequestering Bim [182]. The F1L mutant (A115W) which retains the ability to bind Bak, but not Bim$_L$, could not protect cells from mitochondrial mediated apoptosis [182]. In addition, the F1 protein works synergistically with the Vaccinia Growth Factor to counteract infection-induced cell death via a pathway involving Bad [183]. The N-terminus of F1 has also been shown to bind and directly inhibit caspase-9 indicating that F1 may inhibit apoptosis at both the mitochondria and at the level of caspase-9 [184]. In contrast, a recent study by Caria et al. reported that the N-terminal region of F1 is not involved in apoptosis as deletion of the N-terminus does not affect inhibition of apoptosis during a viral infection [185]. However, the F1 N terminus does function as an inhibitor of NLR family pyrin domain containing 1(NLRP1) inflammasome activity during infection [186]. VARV also encodes a homolog of VACV F1. However, unlike VACV F1L, which can inhibit both Bak and Bax-mediated apoptosis, VARV F1 could only block Bax-mediated apoptosis [187].

Unlike the F1 protein, VACV N1 is not located in the mitochondria, but rather is a cytoplasmic dimeric protein [188]. Interestingly, targeting N1 to the mitochondria results in a loss of function suggesting that cytosolic localization of N1 is critical for its anti-apoptotic function [189]. Despite the lack of any Bcl-2 homology at the sequence level, N1 adopts a Bcl-2-like fold with structural similarity to cellular Bcl-X$_L$ [190,191]. Cooray et al. showed that N1 expression protects cells from staurosporine induced apoptosis through interaction with pro-apoptotic proteins Bad, Bax, and Bid [190]. However, Postigo and Way found that N1 overexpression did not protect against staurosporine or Bax overexpression induced apoptosis [192]. In addition, when expressed in vv811, N1 did not protect cells against VACV-induced apoptosis compared to either B13 or F1 [125]. Also, infection with the N1L knock out in VACV Western Reserve (WR) does not induce cell death [125,192]. Further, N1L and F1L did not genetically interact, as would be expected from the complementarity of their interacting proteins [192]. Though N1 expression may inhibit the intrinsic apoptotic pathway under certain conditions, the primary function of N1 is thought to be inhibition of NF-κB activation [11].

N1 interacts with both TRAF family member-associated NF-κB activator (TANK) and IKKγ to block NF-κB activation induced by a variety of signals including multiple toll-like receptors, TNFα and IL-1 [193]. Inhibition of NF-κB by the N1 protein is distinct from its ability to block apoptosis. De Motes et al. created separate N1 mutants with mutations in the Bcl-2- like surface groove and mutants in the dimer interface [189]. Mutations in the Bcl-2 groove resulted in N1 mutant proteins that lost the ability to protect cells from apoptosis, whereas mutations in the dimer interface resulted in mutants that could longer inhibit NF-κB activation. Interestingly, only N1 mutants that lost the ability to inhibit NF-κB activation resulted in attenuated VACV infection whereas N1 mutant proteins that could no longer inhibit apoptosis (but still blocked NF-κB) retained virulence in a mouse model [189].

Aside from the orthopoxviruses, several other poxviruses express proteins that adopt a Bcl-2 fold. One of the best characterized proteins is the MYXV M11.M11 is 166 amino acids in length with a C-terminal transmembrane domain that allows M11 to insert into the outer membrane of the mitochondria [194]. The M11L gene is required for virulence. Mutant MYXV with an M11 deletion results in attenuated disease in rabbits with rabbits making a full recovery [195]. M11 inhibits apoptosis induced by multiple challenges including staurosporine and FasL [196–198]. The anti-apoptotic activity of M11 was originally linked to M11 association with the peripheral benzodiazepine receptor, a component of the mitochondria permeability transition pore (MPTP) [196]. M11 expression prevents mitochondria membrane permeability and thereby prevents downstream apoptotic effects. However, it was observed that M11 provides protects against mitochondria membrane permeability even in PBR-deficient cells suggesting that M11 uses multiple mechanisms to inhibit apoptosis [196]. Despite lacking sequence homologies to Bcl-2, the crystal structure of M11 identified M11 is a structural mimic of pro-apoptotic Bcl-2 family member proteins [199,200]. M11 interacts with both Bak and Bax and through BH3 domains thereby sequestering these proteins and preventing Bak and Bax oligomerization [197,198,200]. In the case of Bax, M11 does not prevent Bax from translocating to the mitochondria in response to apoptotic signals. Instead, M11 blocks Bax activation at the mitochondria by blocking a Bax conformational change [197].

Since the characterization of N1, F1, and M11, additional poxvirus proteins have been identified with predicted Bcl-2-like folds that inhibit apoptosis. However, identification of such proteins has been difficult as these poxvirus proteins lack sequence homologies to cellular Bcl-2 family members. To circumvent this challenge, Okamoto et al. used the sequence of M11L as the query in a BLASTP search [201]. This approach led to the identification of six additional poxvirus proteins (Deerpox virus DPV022, Swinepox virus SPV12, Shope fibroma virus gp011L, Deer poxvirus DPV022, Lumpy skin disease virus LD17, and Sheeppox virus SPPV14) with similarities to M11 and Bcl-2 family proteins. Of the proteins identified, DPV022, LD17, and SPPV14 significantly prevented cell death in response to ectoposide [201]. Further experiments confirmed that SPPV14 inhibits the intrinsic apoptotic pathway by antagonizing Bak and Bax mediated apoptosis thus preventing the release of cytochrome *c* from the mitochondria to the cytoplasm. Interestingly, SPPV14 can functionally replace F1L and inhibit VACV induced apoptosis when expressed in a VACV with an F1L deletion [201]. Structural analysis of DPV022 identified the Bcl-2 fold [202]. DPV022 inhibits apoptosis through interactions with Bim, Bax, and Bak [202,203].

3.2. vGAAP

Camelpoxvirus (CMLV) gene *6L* encodes a hydrophobic protein of 237-amino acids with multiple transmembrane domains [204,205]. This protein was named vGAAP. CMLV vGAAP is expressed early during infection and localizes to the golgi apparatus [204]. To study the virulence of this protein, VACV genomes were screened for equivalent genes. VACV vGAAP is expressed in three strains of VACV (Lister, USSR, and Evans) [204]. The VACV vGAAP expression pattern is identical to that of CMLV and localizes predominantly to the Golgi. Interestingly, homologs of vGAAP are found eukaryotic cells. Both human hGAAP and vGAAP inhibit apoptosis and seem to have overlapping function as vGAAP can complement the loss of hGAAP [204]. hGAAP is 73% identical to vGAAP [204].

When transiently expressed in cells VACV vGAAP inhibits both the intrinsic and extrinsic apoptotic pathways induced by a multitude of challenges including staurosporine, TNFα/cycloheximide (CHX), Fas antibodies, doxorubicin, cisplatin, C2 ceramide, and apoptosis induced by the overexpression of Bax [204]. Interestingly, VACV virus mutants with vGAAP deleted show increased signs of disease and increased viral titers when compared to wild-type and vGAAP revertant virusesin a mouse model [204]. Both vGAAP and hGAAP form ion channels resulting in a passive leak of Ca^{2+} thereby reducing the concentration of Ca^{2+} in the Golgi apparatus [206,207]. Presumably, this leak of Ca^{2+} from intracellular stores affects apoptotic pathways mediated by the release of Ca^{2+}. When expressed in cells, vGAAP forms oligomers [206,207]. However, mutant vGAAP proteins that lose the ability to oligomerize retain both anti-apoptotic function and the ability to modulate Ca^{2+} content [207]. vGAAP is the first poxvirus protein identified that forms an ion channel [11,206].

3.3. Poxvirus Superoxide Dismutase Homologs

Many poxviruses encode for Cu-Zn superoxide dismutase (SOD) homologs [143,208–211]. The majority ofthese SOD homologs do not possess enzymatic activity as they are missing critical regions necessary for SOD enzymatic function. One exception is the *Amsactamoorei* entomopoxvirusAMV255 open reading frame which does possess superoxide dismutase activity, though this product is not required for replication [212]. The best studied are the leporipoxvirus SOD homologs present in MYXV and SFV. Neither the M131 nor the S131 proteins possess SOD enzymatic activity [209]. However, leporipoxvirus SOD expression does inhibit the activity of cellular SODS [209]. Both MYXV and SFV Cu-Zn SODs function as decoy proteins by binding copper chaperones for superoxide dismutase (CCS) [211]. Presumably, by leporipoxvirus SODs binding CCS, the levels of active cellular SODs aredecreased. As a result, cellular Cu-Zn SOD is less active and the levels of superoxide increases during infection [210,211]. MYXV M131 expression protects Jurkat cells from apoptosis triggered by both Fas and staurosporine [210]. Conversely, mutant MYXV viruses lacking M131 lose the ability to protect cells from apoptosis [210]. The SFV SOD also contributes to virulence as mutant SFV lacking S131 produce an attenuated infection relative to the parental strain with significantly smaller tumors [210]. Thus, by inhibiting apoptosis, the leporipoxvirus SODs may at least partially be responsible for tumorigenesis observed during infection.

The Molluscum Contagiosum Virus also encodes a Cu/Zn SOD homolog. The MCV SOD homolog, MC163, was originally reported to have homology to sweet potato SOD [143]. Like the SFV proteins, the MC163 protein is predicted to be inactive as it lacks amino acids critical for SOD enzymatic activity. We recently reported that MC163 contains a mitochondrial localization sequence in the N-terminal region [213]. As predicted, MC163 co-localized with pMTurquoise2-mito, a cyan fluorescent protein engineered with a mitochondrial localization sequence (MLS) sequence [213]. Using truncated MC163 mutant proteins, we confirmed that the N-terminal region of MC163 was required for its mitochondria localization. MC163 is the second MCV protein reported to localize to the mitochondria. MC007 was reported to target to the cell's mitochondria as a means to sequester host retinoblastoma (Rb), which may contribute to dysregulation of the cell cycle during an MCV infection [214]. MC163 expression inhibits apoptosis induced by several challenges including TNF/cycloheximide (CHX) and staurosporine in HeLa cells. Using the 5,5',6,6'-tetrachloro-1,1',3,3'-tetraethylbenzimidazolylcarbocyanine iodide (JC-1) mitochondrial membrane potential dye, we demonstrated that MC163 expression prevents mitochondrial membrane permeabilization, an important precursor in the activation of the cell's intrinsic apoptotic responses [213]. Whether MC163 inhibits apoptosis in a manner similar to the leporipoxvirus SODs by binding CCS as a means to increase reactive oxygen species or functions via a novel mechanism to inhibit apoptosis is the subject of an ongoing investigation.

Though the lack of a tissue culture model has prevented us from studying MC163 during the context of an MCV infection, it is likely that MCV expresses both MC159 and MC163 as a means to further dampen apoptotic signaling events. It is worth noting that MC159 expression could not inhibit staurosporine-induced apoptosis [152], while MC163 expression prevented caspase-3 activation induced by staurosporine [213]. Therefore, MCV may produce MC159 to prevent activation of the extrinsic pathway, while MC163 expression blocks intrinsic apoptosis. In addition to MC159 and MC163, MCV also encodes a third known inhibitor of apoptosis.MC66 was identified as a poxvirus selenocysteine protein with homology to cellular glutathione peroxidases [215]. MC066 expression inhibits apoptosis induced by both UV radiation and hydrogen peroxide. Therefore, during an MCV infection, the anti-apoptotic actions of MC066. MC159, and MC163 along with the ability of MC007 to bind and sequester cellular Rb at the mitochondria likely contribute to the persistence of MCV neoplasms.

The VACV A45R open reading frame has also been identified as a Cu-Zn SOD homolog, conserved in multiple orthopoxviruses [208]. However, whether A45 affects cellular apoptotic pathways is at this time not known. A45 lacks any apparent SOD activity. Interestingly, the 13.5 kDa A45 protein is incorporated into VACV virions [208]. However, the function of A45 in the context of a VACV infection is not well understood. Deletion of the *A45R* gene had no apparent deleterious effect on either VACV replication or virulence in murine or rabbit models [208]. It is possible that the numerous anti-apoptotic proteins produced during a VACV infection could mask the loss of A45. Alternatively, VACV A45 may function differently than what has been reported for MCV and leporipoxvirus SODS.

4. Inhibition of dsRNA-Induced Apoptosis

Double stranded RNA is an important pathogen associated molecular pattern (PAMP) produced in cells during viral infection. Though poxviruses possess linear, dsDNA genomes, numerous studies have demonstrated that dsRNA accumulates in the cell during a poxvirus infection [216–220]. Poxviruses do not terminate transcripts efficiently [221]. Therefore, overlapping transcripts of RNA are synthesized from the virus' RNA polymerase as the enzyme transcribes genes oriented in opposite directions [81,218,220,222–224]. This event occurs most frequently during intermediate and late gene transcription. It should be noted, that some dsRNA products can also be detected as a result of overlapping RNAs produced from early gene transcripts [220].

Poxvirus dsRNA induces several innate immune responses including apoptosis [225]. Host cells have several cytoplasmic proteins capable of sensing and responding to viral dsRNA. In the case of poxviruses, one of the best characterized innate immune responses involves activation of protein kinase R. PKR is a serine/threonine kinase that becomes active upon binding dsRNA (Figure 2). Once activated, PKR mediates a variety of antiviral effects including phosphorylation of eIF2α to prevent initiation of protein translation [226]. Activation of PKR also induces the activation of NF-κB, up regulation of type I IFNs, and induction of apoptosis [226,227]. Upon activation, PKR can induce apoptosis through several pathways including one involving FADD and procaspase-8 [227]. Cellular RIG-1 and MDA5 also sense cytoplasmic dsRNA [228–233]. These proteins subsequently signal through MAVS complexes culminating in the activation of NF-κB and IRF-3 resulting in the production of type I IFNs [228,234]. In addition to activating IFN, MAVS signaling complexes induce apoptosis [235–238]. Interestingly, Ferrer et al. reported that Modified Vaccinia Ankara (MVA) and MVAΔFlL viruses induce apoptosis through a pathway at least in part dependent on RIG-I, MDA-5 and MAVS indicating that MAVS signaling towards apoptosis may play a role in poxvirus infected cells [238]. PKR, RIG-1, and MDA5 are all capable of sensing poxvirus dsRNA and inducing antiviral responses [225,238–243].

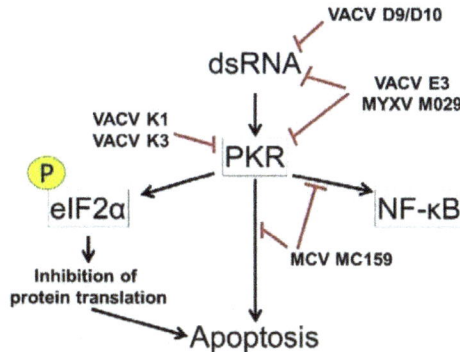

Figure 2. Viral double stranded RNA (dsRNA) activates protein kinase R (PKR). Upon binding dsRNA, PKR becomes activated and elicits several antiviral responses. Active PKR phosphorylates eukaryotic initiation factor 2α (eIF2α) resulting in inhibition of protein translation. PKR can also mediate apoptosis and NF-κB activation. Several poxvirus proteins inhibit PKR activation using a variety of mechanisms. The red lines indicate points in the pathway targeted by several poxvirus immune evasion molecules.

4.1. VACV E3

Given the importance of these host pattern recognition receptors (PRRs) in recognizing viral dsRNA to induce an antiviral state, it is not surprising that poxviruses have evolved strategies to counter these host innate immune responses. For example, VACV produces several proteins to mitigate the effects of PKR. The best studied of these proteins is VACV E3. *E3L* is an early gene that encodes for 19 kDa and 25 kDa isoforms due to an alternative AUG initiation site [11,244]. The N-terminal region of the E3 protein contains a Z-DNA binding domain while the C-terminus contains the dsRNA binding domain [245–247]. VACV *E3L* is one of several known vaccinia host range genes. Deletion of E3 results in viruses that are incapable of growing in several cell lines including HeLa and Vero, but retain the ability to replicate in BHK21 and CEF cells [225,248–253]. While the C-terminal region of E3 is required for replication in HeLa cells, the N-terminal region of E3 is dispensable for replication in most cell lines [248,252]. Though the C-terminal region is required for most of the described functions in cell culture, both the N- and C-terminal regions of E3 are required for pathogenicity in mice models [246,254,255].

The primary role of VACV E3 has been attributed to inhibition of PKR-mediated effects including eIF2α phosphorylation and apoptosis [11,225,245,256,257]. E3L deficient VACV induce PKR activation and apoptosis in HeLa cell lines [225,239]. The importance of PKR in responding to poxvirus infection is highlighted by the observation that replication of E3L deficient viruses is restored when PKR is knocked down in HeLa lines stably expressing small interfering (si)RNA specific for PKR [242]. The most accepted mechanism as to how E3 elicits its effects is that E3 binds viral dsRNA through the C-terminal domain, thus sequestering dsRNA from cellular PRRs [11]. Liu and Moss recently demonstrated that E3 colocalizes with viral factories and dsRNA during VACV infection in A549 cells [258]. Therefore, the E3 protein prevents or reduces the cell's capabilities to identify viral dsRNA. However, E3 may have additional functions aside from binding dsRNA. Dueck et al. utilized alanine scanning of the dsRNA binding domain to identify E3 mutants with reduced biological function [259]. Of interest, one particular mutant D103A retained the ability to bind polyI:C, a synthetic dsRNA mimic, but could no longer inhibit PKR or block apoptosis as assessed by cleavage of the caspase substrate PARP-1 [259]. In addition to binding dsRNA, E3 physically associates with PKR [260,261]. In light of these observations, the E3 protein may use several mechanisms including binding viral dsRNA as well as novel uncharacterized mechanisms to prevent the activation of PKR. The E3 protein is capable of dampening numerous antiviral pathways mediated by PKR and other cellular PRRs including PKR-induced eIF2α phosphorylation, activation of apoptosis, and activation of NF-κB and type I IFNs [11,262–264].

4.2. Poxvirus E3 Homologs

E3 is highly conserved in most genera of poxviruses with avipoxviruses and MCV being the notable exceptions. However, the role of these poxvirus E3 homologs play during viral infection has remained largely unexplored. Myskiw et al. expressed E3 homologs from Nigeria goat and sheeppox virus, yaba monkey tumor virus, swinepox, and MYXV in recombinant VACV with E3L deleted [265]. All of the tested E3 homologs bind dsRNA as determined by the ability to immune precipitate with poly I:C. Of the E3 homologs tested, MYXVM029 and swinepox SPV032 displayed the strongest affinity for dsRNA. Interestingly, only MYXV M029 and swinepox SPV032 expressing VACV viruses prevented the activation of PKR and prevented or reduced the cleavage of apoptotic markers caspase-7 and PARP-1 [265]. In addition, when expressed individually in VACV M029 and SPV032 compensate for the loss of E3 and restore growth in the nonpermissive HeLa cell line. However, none of the E3 orthologs restore VACV pathogenicity in mice due to the deletion of E3L [265]. Recently, it has been reported that not all poxviruses produce the same amount of dsRNA during infection. For example, the MPXV produces significantly less dsRNA than VACV, and therefore, may not be as subject to dsRNA mediated responses as VACV [266]. Therefore, E3 homologs from other poxviruses may be able to bind dsRNA with weaker affinity than that of VACV E3, but still contribute to virulence during infection of these virus' natural hosts. It is also possible that the E3 homologs do contribute to virulence when expressed by their respective viruses using mechanisms similar or different from those employed by VACV E3. Rahman et al. reported that MYXV deleted for M029 is attenuated in susceptible European rabbits and produces no signs of myxomatosis [267]. MYXV *M029L* is also a host range gene with deletion of M029L resulting in mutant viruses that either lose their ability to replicate or grow to lower titers in multiple cell lines including RK13, BSC40, and NIH3T3 [267]. Unlike E3, MYXV M029 lacks the N-terminal region required for binding Z DNA [268]. Similar to observations reported regarding E3L, deletion of M029L from MYXV results in a virus that upregulates activates PKR [267]. When E3L is stably expressed in the RK13 cell lines, M029L deleted viruses grow as efficiently as wild-type, further highlighting that M029 and E3 proteins have similar functions [267]. Similar to observations with E3 deletions, knocking down PKR restores replication in M029 deleted viruses [267]. M029 also immunopreciptates with PKR in an association that is dependent on the presence of dsRNA [267].

4.3. Poxvirus Decapping Enzymes

In addition to E3 and E3 homologs, poxviruses encode multiple proteins to inhibit the effects of dsRNA-induced PKR activation. VACV alone has at least four proteins (K1, C7, E3, and K3) that block inhibit PKR activation by various mechanisms [82,269]. Recently, the VACV decapping enzymes D9 and D10 were reported to prevent activation of PKR as well [217,270]. D9 and D10 prevent the accumulation of dsRNA by decapping viral mRNAs. These decapped viral mRNAs are then subject to degradation by the host exonuclease Xrn1 [217]. Transfecting cells with siRNA specific to Xrn1 results in accumulation of viral dsRNA and subsequent activation of both PKR and the cellular RNaseL pathways during VACV infection [217]. Further, siRNA targeting Xrn1 significantly reduces viral titers. Interestingly, E3 was well expressed even when Xrn1 was depleted. Therefore, the data presented by Burgess and Mohr suggest that E3 can become overwhelmed by excess amounts of dsRNA that accumulate in the absence of Xrn1 [217]. However, VACV with an E3L deletioninduce PKR and eIF2α phosphorylation to higher amounts and at earlier times than what is observed in viruses lacking functional decapping enzymes [258]. This observation correlated with much less viral protein production in vΔE3L viruses than what was observed in vD9muD10mu viruses. Further, D9 and D10 are not detected in viruses with E3 deleted [258]. Therefore, both E3 and the decapping enzymes are necessary to prevent dsRNA induced anti-viral pathways.

Liu et al. further demonstrated that catalytic site mutations in D9 and D10 result in a mutant VACV with reduced replication kinetics in BS-C-1 and HeLa cells [270]. The vD9muD10mu VACV activate PKR and RNaseL in BS-C-1 cells due to increased levels of dsRNA produced during infection. Interestingly, the E3 present in the vD9muD10mu VACV is not sufficient to block PKR activation in the absence of active D9 and D10 [217,270]. Presumably, this increase in levels of dsRNA could also trigger apoptotic pathways through PKR, thereby preventing virus infection. Utilizing clustered regularly interspaced short palindromic repeats-CRISPR associated protein 9 (CRISPR-Cas9) to create double knockouts of PKR and RNase L as well as triple knockouts of Xrn1, PKR, and RNase L, Liu and Moss demonstrated that in the absence of PKR and RNase L, protein synthesis is restored in vD9muD10mu as well as vΔE3L viruses [258]. However, while vΔE3L viral titers were restored in double and triple knockout cells to that of the control virus expressing E3 and decapping enzymes in control cells, the vD9muD10mu titers remained low in double and triple knockout A549 cells suggesting additional inhibitory mechanisms may be in play [258]. By expressing both E3 and the decapping enzymes, VACV ensures that the levels of dsRNA are kept at concentrations below the threshold necessary to trigger the cells innate immune response to dsRNA [217,258,270].

4.4. MCV MC159 Inhibits PKR-Induced Apoptosis

Responses to dsRNA in MCV have not been well characterized. The replication and transcription processes of MCV and VACV are expected to be similar, and therefore, MCV likely produces dsRNA during infection. However, due to the lack of a suitable cell culture model to study MCV, MCV dsRNA production during an MCV infection has never been demonstrated. In addition, the activation state of PKR during an MCV infection has yet to be determined. PKR is well expressed in keratinocytes and capable of sensing dsRNA [271]. Therefore, it seems reasonable to speculate that MCV must counteract the effects of PKR to cause a persistent infection. MCV lacks an E3 homolog to bind any MCV dsRNA that may be produced during infection [142,143]. A single study by Gil et al. reported that the MC159 protein blocks PKR-induced activation of apoptosis in HeLa cells when MC159 and PKR are each expressed from a recombinant VACV [174]. MC159 expression also blocks PKR-induced NF-κB activation. However, MC159 could not prevent the phosphorylation of eIF2α and could not overcome the additional antiviral effects of PKR [174]. Therefore, how MCV inhibits the antiviral effects of PKR remain largely unknown. Like VACV, MCV also encodes decapping enzymes (MC098 and MC099) [142,143]. Whether these MCV products have a similar function as the aforementioned VACV enzymes remains to be determined.

5. Apoptotic Mimicry

During apoptosis, the inner leaflet of the cell membrane becomes reversed exposing phosphatidylserine (PS) on the outside of apoptotic bodies. The now exposed PS serves as a signal for phagocytes to clear the apoptotic bodies. This process induces a strong anti-inflammatory signal. Therefore, apoptotic cells, unlike necrotic cells, do not induce an inflammatory response [272].

Vaccinia virus mature virions (MV) contain PS in the viral envelopes [273–275]. The PS present in the VACV envelope is likely derived from the PS-rich ER luminal leaflet [272,276–279]. Mercer and Helenius reported that VACV utilizes macropinocytosis as a means to enter cells via a process dependent on the presence of PS [280]. Laliberte and Moss further demonstrated that purified VACV MVs extracted with NP-40 detergent could be reconstituted with PS resulting in restoration of infectivity [281]. As this method of viral entrance is similar to the uptake of apoptotic bodies, the term apoptotic mimicry is used to refer to virus entry that use similar methods dependent on PS to gain entry into the cells [272]. Indeed, VACV is one of several enveloped viruses to utilize this mechanism to gain entry into host cells (Reviewed in [272]). Essentially, VACV mimics an apoptotic body to trick the cell into ingesting VACV. Blebbistatin, an inhibitor of micropinocytosis, prevents entry of VACV WR, thus confirming a role for macropinocytosis in poxvirus entry [280,281]. Both forms of VACV virions, MV and extracellular virions (EV), use macropinocytosisas a means to gain entry into host cells [85,280,282–285]. The soluble serum protein Gas6 servers as a bridge interacting with both PS in the VACV envelope and the cellular TAM receptor Axl [286]. Gas6 significantly enhances the entrance of EV in HMEC cells.Further, anti-Axl antibodies reduce the ability of VACV to infect HMECs. Conversely, HEK 293T cells overexpressing Axl enhance VACV EV infectivity [286].

Given that apoptotic clearing is associated with dampening inflammatory responses, it is possible that VACV and other poxviruses utilize apoptotic mimicry as a means to subvert immune response [272]. For example, transcription of toll-like receptor and cytokine suppressor molecules SOCS1 and SOCS3 (suppressor of cytokine signaling 1 and 3) is enhanced upon PS-Gas6 binding to TAM receptors [85,287]. In general, virus interaction with TAM receptors is thought to play a role in suppression of innate immune responses. Bhattacharyya et al. used pseudotyped human immunodeficiency virus 1 (HIV-1)-derived viruses to demonstrate minimal upregulation of type I IFN mRNAs and upregulation of SOCS1 mRNA in wild-type bone-marrow-derived dendritic cells (BMDCs) [288]. However, in TAM knockout cells, infection resulted in a significant upregulation of type I IFN mRNAs. Therefore, virus-induced activation of TAM receptors represents a means to prevent type I IFN expression [288]. During apoptotic clearance, transforming growth factor-β (TGF-β) and IL-10 are produced to prevent inflammatory responses [85,272,289,290]. Interestingly, VACV infection also induces anti-inflammatory cytokines [85,272,291,292]. Therefore, VACV and other poxviruses may utilize apoptotic mimicry as a general means to suppress host immune responses by taking advantage of cellular mechanisms to clear apoptotic bodies. In the case of MCV, NI-MC lesions are almost completely undetected by immune cells [144]. Therefore, it is tempting to speculate that VACV, MCV, and other poxviruses might use apoptotic mimicry as a means to remain undetected by the immune system [85,272]. If apoptotic mimicry does play a role in poxvirus immune suppression, this mechanism could delay poxvirus detection, thereby allowing the virus to spread and cause persistent infections.

6. Conclusions

Multiple poxvirus proteins from several genera have been described to date that inhibit apoptosis. The study of the molecular mechanisms by which these proteins elicit their anti-apoptotic function has led to a better understanding of how these viruses have adapted to survive in host cells. In addition, through the study of poxvirus anti-apoptotic proteins, researchers have gained a better understanding of how host cells detect and respond to virus infection. This knowledge is fundamentally important as poxviruses continue to be evaluated for use as vaccine vectors and oncolytic viruses. Given the large size of poxvirus genomes, it is important to note that additional viral products that antagonize cell death may be present and have yet to be discovered.

Author Contributions: All authors made substantial contributions in writing this manuscript.

Conflicts of Interest: The authors have no conflicts of interest to declare.

References

1. Gartner, A.; Milstein, S.; Ahmed, S.; Hodgkin, J.; Hengartner, M.O. A conserved checkpoint pathway mediates DNA damage-induced apoptosis and cell cycle arrest in *C. elegans*. *Mol. Cell* **2000**, *5*, 435–443. [CrossRef]
2. Garcia-Calvo, M.; Peterson, E.P.; Leiting, B.; Ruel, R.; Nicholson, D.W.; Thornberry, N.A. Inhibition of human caspases by peptide-based and macromolecular inhibitors. *J. Biol. Chem.* **1998**, *273*, 32608–32613. [CrossRef] [PubMed]
3. Fuchs, Y.; Steller, H. Programmed cell death in animal development and disease. *Cell* **2011**, *147*, 742–758. [CrossRef] [PubMed]
4. Thornberry, N.A.; Lazebnik, Y. Caspases: Enemies within. *Science* **1998**, *281*, 1312–1316. [CrossRef] [PubMed]
5. Nakajima, Y.I.; Kuranaga, E. Caspase-dependent non-apoptotic processes in development. *Cell Death Differ.* **2017**, *28*, 1422–1430. [CrossRef] [PubMed]

6. Kiraz, Y.; Adan, A.; Kartal Yandim, M.; Baran, Y. Major apoptotic mechanisms and genes involved in apoptosis. *Tumour Biol.* **2016**, *37*, 8471–8486. [CrossRef] [PubMed]
7. Micheau, O.; Tschopp, J. Induction of TNF receptor I-mediated apoptosis via two sequential signaling complexes. *Cell* **2003**, *114*, 181–190. [CrossRef]
8. Ashkenazi, A.; Dixit, V.M. Death receptors: Signaling and modulation. *Science* **1998**, *281*, 1305–1308. [CrossRef] [PubMed]
9. Bodmer, J.L.; Schneider, P.; Tschopp, J. The molecular architecture of the TNF superfamily. *Trends Biochem. Sci.* **2002**, *27*, 19–26. [CrossRef]
10. Scaffidi, C.; Kirchhoff, S.; Krammer, P.H.; Peter, M.E. Apoptosis signaling in lymphocytes. *Curr. Opin. Immunol.* **1999**, *11*, 277–285. [CrossRef]
11. Veyer, D.L.; Carrara, G.; Maluquer de Motes, C.; Smith, G.L. Vaccinia virus evasion of regulated cell death. *Immunol. Lett.* **2017**, *186*, 68–80. [CrossRef] [PubMed]
12. Irmler, M.; Thome, M.; Hahne, M.; Schneider, P.; Hofmann, K.; Steiner, V.; Bodmer, J.L.; Schroter, M.; Burns, K.; Mattmann, C.; et al. Inhibition of death receptor signals by cellular FLIP. *Nature* **1997**, *388*, 190–195. [PubMed]
13. Takada, H.; Chen, N.J.; Mirtsos, C.; Suzuki, S.; Suzuki, N.; Wakeham, A.; Mak, T.W.; Yeh, W.C. Role of SODD in regulation of tumor necrosis factor responses. *Mol. Cell. Biol.* **2003**, *23*, 4026–4033. [CrossRef] [PubMed]
14. Bertrand, M.J.; Milutinovic, S.; Dickson, K.M.; Ho, W.C.; Boudreault, A.; Durkin, J.; Gillard, J.W.; Jaquith, J.B.; Morris, S.J.; Barker, P.A. cIAP1 and cIAP2 facilitate cancer cell survival by functioning as E3 ligases that promote RIP1 ubiquitination. *Mol. Cell* **2008**, *30*, 689–700. [CrossRef] [PubMed]
15. Leber, B.; Lin, J.; Andrews, D.W. Embedded together: The life and death consequences of interaction of the Bcl-2 family with membranes. *Apoptosis* **2007**, *12*, 897–911. [CrossRef] [PubMed]
16. Autret, A.; Martin, S.J. Emerging role for members of the Bcl-2 family in mitochondrial morphogenesis. *Mol. Cell* **2009**, *36*, 355–363. [CrossRef] [PubMed]
17. Chipuk, J.E.; Moldoveanu, T.; Llambi, F.; Parsons, M.J.; Green, D.R. The Bcl-2 family reunion. *Mol. Cell* **2010**, *37*, 299–310. [CrossRef] [PubMed]
18. Hardwick, J.M.; Soane, L. Multiple functions of BCL-2 family proteins. *Cold Spring Harb. Perspect. Biol.* **2013**, *5*, a008722. [CrossRef] [PubMed]
19. Kuwana, T.; Bouchier-Hayes, L.; Chipuk, J.E.; Bonzon, C.; Sullivan, B.A.; Green, D.R.; Newmeyer, D.D. BH3 domains of BH3-only proteins differentially regulate Bax-mediated mitochondrial membrane permeabilization both directly and indirectly. *Mol. Cell* **2005**, *17*, 525–535. [CrossRef] [PubMed]
20. Kuwana, T.; Mackey, M.R.; Perkins, G.; Ellisman, M.H.; Latterich, M.; Schneiter, R.; Green, D.R.; Newmeyer, D.D. Bid, Bax, and lipids cooperate to form supramolecular openings in the outer mitochondrial membrane. *Cell* **2002**, *111*, 331–342. [CrossRef]
21. Karbowski, M.; Norris, K.L.; Cleland, M.M.; Jeong, S.Y.; Youle, R.J. Role of Bax and Bak in mitochondrial morphogenesis. *Nature* **2006**, *443*, 658–662. [CrossRef] [PubMed]
22. Brooks, C.; Dong, Z. Regulation of mitochondrial morphological dynamics during apoptosis by Bcl-2 family proteins: A key in Bak? *Cell Cycle* **2007**, *6*, 3043–3047. [CrossRef] [PubMed]
23. Brooks, C.; Wei, Q.; Feng, L.; Dong, G.; Tao, Y.; Mei, L.; Xie, Z.J.; Dong, Z. Bak regulates mitochondrial morphology and pathology during apoptosis by interacting with mitofusins. *Proc. Natl. Acad. Sci. USA* **2007**, *104*, 11649–11654. [CrossRef] [PubMed]
24. Delivani, P.; Adrain, C.; Taylor, R.C.; Duriez, P.J.; Martin, S.J. Role for CED-9 and Egl-1 as regulators of mitochondrial fission and fusion dynamics. *Mol. Cell* **2006**, *21*, 761–773. [CrossRef] [PubMed]
25. Sheridan, C.; Delivani, P.; Cullen, S.P.; Martin, S.J. Bax- or Bak-induced mitochondrial fission can be uncoupled from cytochrome *c* release. *Mol. Cell* **2008**, *31*, 570–585. [CrossRef] [PubMed]
26. Frank, S.; Gaume, B.; Bergmann-Leitner, E.S.; Leitner, W.W.; Robert, E.G.; Catez, F.; Smith, C.L.; Youle, R.J. The role of dynamin-related protein 1, a mediator of mitochondrial fission, in apoptosis. *Dev. Cell* **2001**, *1*, 515–525. [CrossRef]
27. Cassidy-Stone, A.; Chipuk, J.E.; Ingerman, E.; Song, C.; Yoo, C.; Kuwana, T.; Kurth, M.J.; Shaw, J.T.; Hinshaw, J.E.; Green, D.R.; et al. Chemical inhibition of the mitochondrial division dynamin reveals its role in Bax/Bak-dependent mitochondrial outer membrane permeabilization. *Dev. Cell* **2008**, *14*, 193–204. [CrossRef] [PubMed]

28. Gandre-Babbe, S.; van der Bliek, A.M. The novel tail-anchored membrane protein Mff controls mitochondrial and peroxisomal fission in mammalian cells. *Mol. Biol. Cell* **2008**, *19*, 2402–2412. [CrossRef] [PubMed]

29. Arnoult, D.; Rismanchi, N.; Grodet, A.; Roberts, R.G.; Seeburg, D.P.; Estaquier, J.; Sheng, M.; Blackstone, C. Bax/Bak-dependent release of DDP/TIMM8a promotes Drp1-mediated mitochondrial fission and mitoptosis during programmed cell death. *Curr. Biol.* **2005**, *15*, 2112–2118. [CrossRef] [PubMed]

30. Estaquier, J.; Arnoult, D. Inhibiting Drp1-mediated mitochondrial fission selectively prevents the release of cytochrome *c* during apoptosis. *Cell Death Differ.* **2007**, *14*, 1086–1094. [CrossRef] [PubMed]

31. Ishihara, N.; Nomura, M.; Jofuku, A.; Kato, H.; Suzuki, S.O.; Masuda, K.; Otera, H.; Nakanishi, Y.; Nonaka, I.; Goto, Y.; et al. Mitochondrial fission factor Drp1 is essential for embryonic development and synapse formation in mice. *Nat. Cell Biol.* **2009**, *11*, 958–966. [CrossRef] [PubMed]

32. Parone, P.A.; James, D.I.; Da Cruz, S.; Mattenberger, Y.; Donze, O.; Barja, F.; Martinou, J.C. Inhibiting the mitochondrial fission machinery does not prevent Bax/Bak-dependent apoptosis. *Mol. Cell. Biol.* **2006**, *26*, 7397–7408. [CrossRef] [PubMed]

33. Parone, P.A.; Martinou, J.C. Mitochondrial fission and apoptosis: An ongoing trial. *Biochim. Biophys. Acta* **2006**, *1763*, 522–530. [CrossRef] [PubMed]

34. Scorrano, L.; Ashiya, M.; Buttle, K.; Weiler, S.; Oakes, S.A.; Mannella, C.A.; Korsmeyer, S.J. A distinct pathway remodels mitochondrial cristae and mobilizes cytochrome *c* during apoptosis. *Dev. Cell* **2002**, *2*, 55–67. [CrossRef]

35. Sun, L.L.; Sun, L.R.; Wang, G.Y. Mitochondrial membrane potential at HL-60 cell apoptosis induced by cytarabine. *Zhongguo Shi Yan Xue Ye Xue Za Zhi* **2007**, *15*, 1196–1199. (In Chinese) [PubMed]

36. Rong, Y.P.; Bultynck, G.; Aromolaran, A.S.; Zhong, F.; Parys, J.B.; De Smedt, H.; Mignery, G.A.; Roderick, H.L.; Bootman, M.D.; Distelhorst, C.W. The BH4 domain of Bcl-2 inhibits ER calcium release and apoptosis by binding the regulatory and coupling domain of the IP3 receptor. *Proc. Natl. Acad. Sci. USA* **2009**, *106*, 14397–14402. [CrossRef] [PubMed]

37. Scorrano, L.; Oakes, S.A.; Opferman, J.T.; Cheng, E.H.; Sorcinelli, M.D.; Pozzan, T.; Korsmeyer, S.J. BAX and BAK regulation of endoplasmic reticulum Ca^{2+}: A control point for apoptosis. *Science* **2003**, *300*, 135–139. [CrossRef] [PubMed]

38. Rodriguez, D.; Rojas-Rivera, D.; Hetz, C. Integrating stress signals at the endoplasmic reticulum: The BCL-2 protein family rheostat. *Biochim. Biophys. Acta* **2011**, *1813*, 564–574. [CrossRef] [PubMed]

39. Reimertz, C.; Kogel, D.; Rami, A.; Chittenden, T.; Prehn, J.H. Gene expression during ER stress-induced apoptosis in neurons: Induction of the BH3-only protein Bbc3/PUMA and activation of the mitochondrial apoptosis pathway. *J. Cell Biol.* **2003**, *162*, 587–597. [CrossRef] [PubMed]

40. Futami, T.; Miyagishi, M.; Taira, K. Identification of a network involved in thapsigargin-induced apoptosis using a library of small interfering RNA expression vectors. *J. Biol. Chem.* **2005**, *280*, 826–831. [CrossRef] [PubMed]

41. Gaut, J.R.; Hendershot, L.M. The modification and assembly of proteins in the endoplasmic reticulum. *Curr. Opin. Cell Biol.* **1993**, *5*, 589–595. [CrossRef]

42. Kaufman, R.J. Orchestrating the unfolded protein response in health and disease. *J. Clin. Investig.* **2002**, *110*, 1389–1398. [CrossRef] [PubMed]

43. Schroder, M.; Kaufman, R.J. The mammalian unfolded protein response. *Annu. Rev. Biochem.* **2005**, *74*, 739–789. [CrossRef] [PubMed]

44. Schroder, M.; Kaufman, R.J. ER stress and the unfolded protein response. *Mutat. Res.* **2005**, *569*, 29–63. [CrossRef] [PubMed]

45. Wang, X.Z.; Kuroda, M.; Sok, J.; Batchvarova, N.; Kimmel, R.; Chung, P.; Zinszner, H.; Ron, D. Identification of novel stress-induced genes downstream of chop. *EMBO J.* **1998**, *17*, 3619–3630. [CrossRef] [PubMed]

46. Zinszner, H.; Kuroda, M.; Wang, X.; Batchvarova, N.; Lightfoot, R.T.; Remotti, H.; Stevens, J.L.; Ron, D. CHOP is implicated in programmed cell death in response to impaired function of the endoplasmic reticulum. *Genes Dev.* **1998**, *12*, 982–995. [CrossRef] [PubMed]

47. Matsumoto, M.; Minami, M.; Takeda, K.; Sakao, Y.; Akira, S. Ectopic expression of CHOP (GADD153) induces apoptosis in M1 myeloblastic leukemia cells. *FEBS Lett.* **1996**, *395*, 143–147. [CrossRef]

48. Szegezdi, E.; Logue, S.E.; Gorman, A.M.; Samali, A. Mediators of endoplasmic reticulum stress-induced apoptosis. *EMBO Rep.* **2006**, *7*, 880–885. [CrossRef] [PubMed]

49. Lei, K.; Davis, R.J. JNK phosphorylation of Bim-related members of the Bcl2 family induces Bax-dependent apoptosis. *Proc. Natl. Acad. Sci. USA* **2003**, *100*, 2432–2437. [CrossRef] [PubMed]

50. Morishima, N.; Nakanishi, K.; Tsuchiya, K.; Shibata, T.; Seiwa, E. Translocation of Bim to the endoplasmic reticulum (ER) mediates ER stress signaling for activation of caspase-12 during ER stress-induced apoptosis. *J. Biol. Chem.* **2004**, *279*, 50375–50381. [CrossRef] [PubMed]

51. Nakagawa, T.; Zhu, H.; Morishima, N.; Li, E.; Xu, J.; Yankner, B.A.; Yuan, J. Caspase-12 mediates endoplasmic-reticulum-specific apoptosis and cytotoxicity by amyloid-β. *Nature* **2000**, *403*, 98–103. [CrossRef] [PubMed]

52. Saleh, M.; Mathison, J.C.; Wolinski, M.K.; Bensinger, S.J.; Fitzgerald, P.; Droin, N.; Ulevitch, R.J.; Green, D.R.; Nicholson, D.W. Enhanced bacterial clearance and sepsis resistance in caspase-12-deficient mice. *Nature* **2006**, *440*, 1064–1068. [CrossRef] [PubMed]

53. Szegezdi, E.; Fitzgerald, U.; Samali, A. Caspase-12 and ER-stress-mediated apoptosis: The story so far. *Ann. N. Y. Acad. Sci.* **2003**, *1010*, 186–194. [CrossRef] [PubMed]

54. Levy, H.B.; Law, L.W.; Rabson, A.S. Inhibition of tumor growth by polyinosinic-polycytidylic acid. *Proc. Natl. Acad. Sci. USA* **1969**, *62*, 357–361. [CrossRef] [PubMed]

55. Estornes, Y.; Toscano, F.; Virard, F.; Jacquemin, G.; Pierrot, A.; Vanbervliet, B.; Bonnin, M.; Lalaoui, N.; Mercier-Gouy, P.; Pacheco, Y.; et al. dsRNA induces apoptosis through an atypical death complex associating TLR3 to caspase-8. *Cell Death Differ.* **2012**, *19*, 1482–1494. [CrossRef] [PubMed]

56. Feoktistova, M.; Geserick, P.; Kellert, B.; Dimitrova, D.P.; Langlais, C.; Hupe, M.; Cain, K.; MacFarlane, M.; Hacker, G.; Leverkus, M. cIAPs block Ripoptosome formation, a RIP1/caspase-8 containing intracellular cell death complex differentially regulated by cFLIP isoforms. *Mol. Cell* **2011**, *43*, 449–463. [CrossRef] [PubMed]

57. Takeuchi, O.; Akira, S. Innate immunity to virus infection. *Immunol. Rev.* **2009**, *227*, 75–86. [CrossRef] [PubMed]

58. Friboulet, L.; Pioche-Durieu, C.; Rodriguez, S.; Valent, A.; Souquere, S.; Ripoche, H.; Khabir, A.; Tsao, S.W.; Bosq, J.; Lo, K.W.; et al. Recurrent overexpression of c-IAP2 in EBV-associated nasopharyngeal carcinomas: Critical role in resistance to Toll-like receptor 3-mediated apoptosis. *Neoplasia* **2008**, *10*, 1183–1194. [CrossRef] [PubMed]

59. Funami, K.; Sasai, M.; Ohba, Y.; Oshiumi, H.; Seya, T.; Matsumoto, M. Spatiotemporal mobilization of Toll/IL-1 receptor domain-containing adaptor molecule-1 in response to dsRNA. *J. Immunol.* **2007**, *179*, 6867–6872. [CrossRef] [PubMed]

60. Salaun, B.; Coste, I.; Rissoan, M.C.; Lebecque, S.J.; Renno, T. TLR3 can directly trigger apoptosis in human cancer cells. *J. Immunol.* **2006**, *176*, 4894–4901. [CrossRef] [PubMed]

61. Salaun, B.; Lebecque, S.; Matikainen, S.; Rimoldi, D.; Romero, P. Toll-like receptor 3 expressed by melanoma cells as a target for therapy? *Clin. Cancer Res.* **2007**, *13*, 4565–4574. [CrossRef] [PubMed]

62. Meylan, E.; Burns, K.; Hofmann, K.; Blancheteau, V.; Martinon, F.; Kelliher, M.; Tschopp, J. RIP1 is an essential mediator of Toll-like receptor 3-induced NF-κB activation. *Nat. Immunol.* **2004**, *5*, 503–507. [CrossRef] [PubMed]

63. Tenev, T.; Bianchi, K.; Darding, M.; Broemer, M.; Langlais, C.; Wallberg, F.; Zachariou, A.; Lopez, J.; MacFarlane, M.; Cain, K.; et al. The Ripoptosome, a signaling platform that assembles in response to genotoxic stress and loss of IAPs. *Mol. Cell* **2011**, *43*, 432–448. [CrossRef] [PubMed]

64. Raven, J.F.; Koromilas, A.E. PERK and PKR: Old kinases learn new tricks. *Cell Cycle* **2008**, *7*, 1146–1150. [CrossRef] [PubMed]

65. Farrell, P.J.; Balkow, K.; Hunt, T.; Jackson, R.J.; Trachsel, H. Phosphorylation of initiation factor eIF-2 and the control of reticulocyte protein synthesis. *Cell* **1977**, *11*, 187–200. [CrossRef]

66. Sonenberg, N.; Hinnebusch, A.G. Regulation of translation initiation in eukaryotes: Mechanisms and biological targets. *Cell* **2009**, *136*, 731–745. [CrossRef] [PubMed]

67. Dabo, S.; Meurs, E.F. dsRNA-dependent protein kinase PKR and its role in stress, signaling and HCV infection. *Viruses* **2012**, *4*, 2598–2635. [CrossRef] [PubMed]

68. Dar, A.C.; Sicheri, F. X-ray crystal structure and functional analysis of vaccinia virus K3L reveals molecular determinants for PKR subversion and substrate recognition. *Mol. Cell* **2002**, *10*, 295–305. [CrossRef]

69. Baldwin, A.S., Jr. The NF-κB and IκB proteins: New discoveries and insights. *Annu. Rev. Immunol.* **1996**, *14*, 649–683. [CrossRef] [PubMed]

70. Gil, J.; Alcami, J.; Esteban, M. Induction of apoptosis by double-stranded-RNA-dependent protein kinase (PKR) involves the α subunit of eukaryotic translation initiation factor 2 and NF-κB. *Mol. Cell. Biol.* **1999**, *19*, 4653–4663. [CrossRef] [PubMed]

71. Kumar, A.; Haque, J.; Lacoste, J.; Hiscott, J.; Williams, B.R. Double-stranded RNA-dependent protein kinase activates transcription factor NF-κB by phosphorylating IκB. *Proc. Natl. Acad. Sci. USA* **1994**, *91*, 6288–6292. [CrossRef] [PubMed]

72. Roff, M.; Thompson, J.; Rodriguez, M.S.; Jacque, J.M.; Baleux, F.; Arenzana-Seisdedos, F.; Hay, R.T. Role of IκBα ubiquitination in signal-induced activation of NFκB in vivo. *J. Biol. Chem.* **1996**, *271*, 7844–7850. [CrossRef] [PubMed]

73. Bonnet, M.C.; Daurat, C.; Ottone, C.; Meurs, E.F. The N-terminus of PKR is responsible for the activation of the NF-κB signaling pathway by interacting with the IKK complex. *Cell. Signal.* **2006**, *18*, 1865–1875. [CrossRef] [PubMed]

74. Gil, J.; Garcia, M.A.; Gomez-Puertas, P.; Guerra, S.; Rullas, J.; Nakano, H.; Alcami, J.; Esteban, M. TRAF family proteins link PKR with NF-κB activation. *Mol. Cell. Biol.* **2004**, *24*, 4502–4512. [CrossRef] [PubMed]

75. Chen, X.; Anstey, A.V.; Bugert, J.J. *Molluscum contagiosum* virus infection. *Lancet Infect. Dis.* **2013**, *13*, 877–888. [CrossRef]

76. Abrahao, J.S.; Campos, R.K.; de Souza Trindade, G.; da Fonseca, F.G.; Ferreira, P.C.P.; Kroon, E.G. Outbreak of severe zoonotic vaccinia virus infection, Southeastern Brazil. *Emerg. Infect. Dis.* **2015**, *21*, 695–698. [CrossRef] [PubMed]

77. McCollum, A.M.; Damon, I.K. Human monkeypox. *Clin. Infect. Dis.* **2014**, *58*, 260–267. [CrossRef] [PubMed]

78. Sharp, D.W.; Lattime, E.C. Recombinant Poxvirus and the Tumor Microenvironment: Oncolysis, Immune Regulation and Immunization. *Biomedicines* **2016**, *4*, 19. [CrossRef] [PubMed]

79. Chan, W.M.; Rahman, M.M.; McFadden, G. Oncolytic myxoma virus: The path to clinic. *Vaccine* **2013**, *31*, 4252–4258. [CrossRef] [PubMed]

80. Sanchez-Sampedro, L.; Perdiguero, B.; Mejias-Perez, E.; Garcia-Arriaza, J.; Di Pilato, M.; Esteban, M. The evolution of poxvirus vaccines. *Viruses* **2015**, *7*, 1726–1803. [CrossRef] [PubMed]

81. Moss, B. *Poxviridae*: The viruses and their replication. In *Fields*; Lippincott Williams and Willkins: Philadelphia, PA, USA, 2013.

82. Smith, G.L.; Benfield, C.T.; Maluquer de Motes, C.; Mazzon, M.; Ember, S.W.; Ferguson, B.J.; Sumner, R.P. Vaccinia virus immune evasion: Mechanisms, virulence and immunogenicity. *J. Gen. Virol.* **2013**, *94*, 2367–2392. [CrossRef] [PubMed]

83. Shisler, J.L. Immune evasion strategies of molluscum contagiosum virus. *Adv. Virus Res.* **2015**, *92*, 201–252. [PubMed]

84. Schmidt, F.I.; Bleck, C.K.; Reh, L.; Novy, K.; Wollscheid, B.; Helenius, A.; Stahlberg, H.; Mercer, J. Vaccinia virus entry is followed by core activation and proteasome-mediated release of the immunomodulatory effector VH1 from lateral bodies. *Cell Rep.* **2013**, *4*, 464–476. [CrossRef] [PubMed]

85. Bidgood, S.R.; Mercer, J. Cloak and Dagger: Alternative Immune Evasion and Modulation Strategies of Poxviruses. *Viruses* **2015**, *7*, 4800–4825. [CrossRef] [PubMed]

86. Aggarwal, B.B. Signalling pathways of the TNF superfamily: A double-edged sword. *Nat. Rev. Immunol.* **2003**, *3*, 745–756. [CrossRef] [PubMed]

87. Black, R.A.; Rauch, C.T.; Kozlosky, C.J.; Peschon, J.J.; Slack, J.L.; Wolfson, M.F.; Castner, B.J.; Stocking, K.L.; Reddy, P.; Srinivasan, S.; et al. A metalloproteinase disintegrin that releases tumour-necrosis factor-α from cells. *Nature* **1997**, *385*, 729–733. [CrossRef] [PubMed]

88. Issuree, P.D.; Maretzky, T.; McIlwain, D.R.; Monette, S.; Qing, X.; Lang, P.A.; Swendeman, S.L.; Park-Min, K.H.; Binder, N.; Kalliolias, G.D.; et al. iRHOM2 is a critical pathogenic mediator of inflammatory arthritis. *J. Clin. Investig.* **2013**, *123*, 928–932. [CrossRef] [PubMed]

89. Locksley, R.M.; Killeen, N.; Lenardo, M.J. The TNF and TNF receptor superfamilies: Integrating mammalian biology. *Cell* **2001**, *104*, 487–501. [CrossRef]

90. Alcami, A. Viral mimicry of cytokines, chemokines and their receptors. *Nat. Rev. Immunol.* **2003**, *3*, 36–50. [CrossRef] [PubMed]

91. Saraiva, M.; Smith, P.; Fallon, P.G.; Alcami, A. Inhibition of type 1 cytokine-mediated inflammation by a soluble CD30 homologue encoded by ectromelia (mousepox) virus. *J. Exp. Med.* **2002**, *196*, 829–839. [CrossRef] [PubMed]

92. Alejo, A.; Saraiva, M.; Ruiz-Arguello, M.B.; Viejo-Borbolla, A.; de Marco, M.F.; Salguero, F.J.; Alcami, A. A method for the generation of ectromelia virus (ECTV) recombinants: In vivo analysis of ECTV vCD30 deletion mutants. *PLoS ONE* **2009**, *4*, e5175. [CrossRef] [PubMed]

93. Graham, S.C.; Bahar, M.W.; Abrescia, N.G.; Smith, G.L.; Stuart, D.I.; Grimes, J.M. Structure of CrmE, a virus-encoded tumour necrosis factor receptor. *J. Mol. Biol.* **2007**, *372*, 660–671. [CrossRef] [PubMed]

94. Alejo, A.; Ruiz-Arguello, M.B.; Ho, Y.; Smith, V.P.; Saraiva, M.; Alcami, A. A chemokine-binding domain in the tumor necrosis factor receptor from variola (smallpox) virus. *Proc. Natl. Acad. Sci. USA* **2006**, *103*, 5995–6000. [CrossRef] [PubMed]

95. Rahman, M.M.; Barrett, J.W.; Brouckaert, P.; McFadden, G. Variation in ligand binding specificities of a novel class of poxvirus-encoded tumor necrosis factor-binding protein. *J. Biol. Chem.* **2006**, *281*, 22517–22526. [CrossRef] [PubMed]

96. Reading, P.C.; Khanna, A.; Smith, G.L. Vaccinia virus CrmE encodes a soluble and cell surface tumor necrosis factor receptor that contributes to virus virulence. *Virology* **2002**, *292*, 285–298. [CrossRef] [PubMed]

97. Pontejo, S.M.; Alejo, A.; Alcami, A. Comparative Biochemical and Functional Analysis of Viral and Human Secreted Tumor Necrosis Factor (TNF) Decoy Receptors. *J. Biol. Chem.* **2015**, *290*, 15973–15984. [CrossRef] [PubMed]

98. Hu, F.Q.; Smith, C.A.; Pickup, D.J. Cowpox virus contains two copies of an early gene encoding a soluble secreted form of the type II TNF receptor. *Virology* **1994**, *204*, 343–356. [CrossRef] [PubMed]

99. Smith, C.A.; Hu, F.Q.; Smith, T.D.; Richards, C.L.; Smolak, P.; Goodwin, R.G.; Pickup, D.J. Cowpox virus genome encodes a second soluble homologue of cellular TNF receptors, distinct from CrmB, that binds TNF but not LTα. *Virology* **1996**, *223*, 132–147. [CrossRef] [PubMed]

100. Saraiva, M.; Alcami, A. CrmE, a novel soluble tumor necrosis factor receptor encoded by poxviruses. *J. Virol.* **2001**, *75*, 226–233. [CrossRef] [PubMed]

101. Schreiber, M.; McFadden, G. The myxoma virus TNF-receptor homologue (T2) inhibits tumor necrosis factor-α in a species-specific fashion. *Virology* **1994**, *204*, 692–705. [CrossRef] [PubMed]

102. Schreiber, M.; Rajarathnam, K.; McFadden, G. Myxoma virus T2 protein, a tumor necrosis factor (TNF) receptor homolog, is secreted as a monomer and dimer that each bind rabbit TNFα, but the dimer is a more potent TNF inhibitor. *J. Biol. Chem.* **1996**, *271*, 13333–13341. [CrossRef] [PubMed]

103. Sedger, L.; McFadden, G. M-T2: A poxvirus TNF receptor homologue with dual activities. *Immunol. Cell Biol.* **1996**, *74*, 538–545. [CrossRef] [PubMed]

104. Xu, X.; Nash, P.; McFadden, G. Myxoma virus expresses a TNF receptor homolog with two distinct functions. *Virus Genes* **2000**, *21*, 97–109. [CrossRef] [PubMed]

105. Sedger, L.M.; Osvath, S.R.; Xu, X.M.; Li, G.; Chan, F.K.; Barrett, J.W.; McFadden, G. Poxvirus tumor necrosis factor receptor (TNFR)-like T2 proteins contain a conserved preligand assembly domain that inhibits cellular TNFR1-induced cell death. *J. Virol.* **2006**, *80*, 9300–9309. [CrossRef] [PubMed]

106. Huber, R.; Carrell, R.W. Implications of the three-dimensional structure of α 1-antitrypsin for structure and function of serpins. *Biochemistry* **1989**, *28*, 8951–8966. [CrossRef] [PubMed]

107. Gettins, P.; Patston, P.A.; Schapira, M. The role of conformational change in serpin structure and function. *Bioessays* **1993**, *15*, 461–467. [CrossRef] [PubMed]

108. Lawrence, D.A.; Olson, S.T.; Palaniappan, S.; Ginsburg, D. Serpin reactive center loop mobility is required for inhibitor function but not for enzyme recognition. *J. Biol. Chem.* **1994**, *269*, 27657–27662. [PubMed]

109. Quan, L.T.; Caputo, A.; Bleackley, R.C.; Pickup, D.J.; Salvesen, G.S. Granzyme B is inhibited by the cowpox virus serpin cytokine response modifier A. *J. Biol. Chem.* **1995**, *270*, 10377–10379. [CrossRef] [PubMed]

110. Komiyama, T.; Ray, C.A.; Pickup, D.J.; Howard, A.D.; Thornberry, N.A.; Peterson, E.P.; Salvesen, G. Inhibition of interleukin-1β converting enzyme by the cowpox virus serpin CrmA. An example of cross-class inhibition. *J. Biol. Chem.* **1994**, *269*, 19331–19337. [PubMed]

111. Owen, M.C.; Brennan, S.O.; Lewis, J.H.; Carrell, R.W. Mutation of antitrypsin to antithrombin. α1-antitrypsin Pittsburgh (358 Met →Arg), a fatal bleeding disorder. *N. Engl. J. Med.* **1983**, *309*, 694–698. [CrossRef] [PubMed]

112. Law, R.H.; Zhang, Q.; McGowan, S.; Buckle, A.M.; Silverman, G.A.; Wong, W.; Rosado, C.J.; Langendorf, C.G.; Pike, R.N.; Bird, P.I.; et al. An overview of the serpin superfamily. *Genome Biol.* **2006**, *7*, 216. [CrossRef] [PubMed]

113. Silverman, G.A.; Bird, P.I.; Carrell, R.W.; Church, F.C.; Coughlin, P.B.; Gettins, P.G.; Irving, J.A.; Lomas, D.A.; Luke, C.J.; Moyer, R.W.; et al. The serpins are an expanding superfamily of structurally similar but functionally diverse proteins. Evolution, mechanism of inhibition, novel functions, and a revised nomenclature. *J. Biol. Chem.* **2001**, *276*, 33293–33296. [CrossRef] [PubMed]

114. Best, S.M. Viral subversion of apoptotic enzymes: Escape from death row. *Annu. Rev. Microbiol.* **2008**, *62*, 171–192. [CrossRef] [PubMed]

115. Spriggs, M.K.; Hruby, D.E.; Maliszewski, C.R.; Pickup, D.J.; Sims, J.E.; Buller, R.M.; VanSlyke, J. Vaccinia and cowpox viruses encode a novel secreted interleukin-1-binding protein. *Cell* **1992**, *71*, 145–152. [CrossRef]

116. Adamson, B.; Norman, T.M.; Jost, M.; Cho, M.Y.; Nunez, J.K.; Chen, Y.; Villalta, J.E.; Gilbert, L.A.; Horlbeck, M.A.; Hein, M.Y.; et al. A Multiplexed single-cell CRISPR screening platform enables systematic dissection of the unfolded protein response. *Cell* **2016**, *167*, 1867–1882. [CrossRef] [PubMed]

117. Palumbo, G.J.; Buller, R.M.; Glasgow, W.C. Multigenic evasion of inflammation by poxviruses. *J. Virol.* **1994**, *68*, 1737–1749. [PubMed]

118. Palumbo, G.J.; Pickup, D.J.; Fredrickson, T.N.; McIntyre, L.J.; Buller, R.M. Inhibition of an inflammatory response is mediated by a 38-kDa protein of cowpox virus. *Virology* **1989**, *172*, 262–273. [CrossRef]

119. Dobbelstein, M.; Shenk, T. Protection against apoptosis by the vaccinia virus SPI-2 (B13R) gene product. *J. Virol.* **1996**, *70*, 6479–6485. [PubMed]

120. Kettle, S.; Alcami, A.; Khanna, A.; Ehret, R.; Jassoy, C.; Smith, G.L. Vaccinia virus serpin B13R (SPI-2) inhibits interleukin-1beta-converting enzyme and protects virus-infected cells from TNF- and Fas-mediated apoptosis, but does not prevent IL-1β-induced fever. *J. Gen. Virol.* **1997**, *78*, 677–685. [CrossRef] [PubMed]

121. Ray, C.A.; Pickup, D.J. The mode of death of pig kidney cells infected with cowpox virus is governed by the expression of the *crmA* gene. *Virology* **1996**, *217*, 384–391. [CrossRef] [PubMed]

122. Zhou, Q.; Snipas, S.; Orth, K.; Muzio, M.; Dixit, V.M.; Salvesen, G.S. Target protease specificity of the viral serpin CrmA. Analysis of five caspases. *J. Biol. Chem.* **1997**, *272*, 7797–7800. [CrossRef] [PubMed]

123. Ekert, P.G.; Silke, J.; Vaux, D.L. Caspase inhibitors. *Cell Death Differ.* **1999**, *6*, 1081–1086. [CrossRef] [PubMed]

124. Ray, C.A.; Black, R.A.; Kronheim, S.R.; Greenstreet, T.A.; Sleath, P.R.; Salvesen, G.S.; Pickup, D.J. Viral inhibition of inflammation: Cowpox virus encodes an inhibitor of the interleukin-1β converting enzyme. *Cell* **1992**, *69*, 597–604. [CrossRef]

125. Veyer, D.L.; Maluquer de Motes, C.; Sumner, R.P.; Ludwig, L.; Johnson, B.F.; Smith, G.L. Analysis of the anti-apoptotic activity of four vaccinia virus proteins demonstrates that B13 is the most potent inhibitor in isolation and during viral infection. *J. Gen. Virol.* **2014**, *95*, 2757–2768. [CrossRef] [PubMed]

126. Kotwal, G.J.; Moss, B. Vaccinia virus encodes two proteins that are structurally related to members of the plasma serine protease inhibitor superfamily. *J. Virol.* **1989**, *63*, 600–606. [PubMed]

127. Smith, G.L.; Howard, S.T.; Chan, Y.S. Vaccinia virus encodes a family of genes with homology to serine proteinase inhibitors. *J. Gen. Virol.* **1989**, *70*, 2333–2343. [CrossRef] [PubMed]

128. Kettle, S.; Blake, N.W.; Law, K.M.; Smith, G.L. Vaccinia virus serpins B13R (SPI-2) and B22R (SPI-1) encode M_r 38.5 and 40K, intracellular polypeptides that do not affect virus virulence in a murine intranasal model. *Virology* **1995**, *206*, 136–147. [CrossRef]

129. Shisler, J.L.; Isaacs, S.N.; Moss, B. Vaccinia virus serpin-1 deletion mutant exhibits a host range defect characterized by low levels of intermediate and late mRNAs. *Virology* **1999**, *262*, 298–311. [CrossRef] [PubMed]

130. Turner, P.C.; Moyer, R.W. Orthopoxvirus fusion inhibitor glycoprotein SPI-3 (open reading frame K2L) contains motifs characteristic of serine proteinase inhibitors that are not required for control of cell fusion. *J. Virol.* **1995**, *69*, 5978–5987. [PubMed]

131. Law, K.M.; Smith, G.L. A vaccinia serine protease inhibitor which prevents virus-induced cell fusion. *J. Gen. Virol.* **1992**, *73*, 549–557. [CrossRef] [PubMed]

132. Thompson, J.P.; Turner, P.C.; Ali, A.N.; Crenshaw, B.C.; Moyer, R.W. The effects of serpin gene mutations on the distinctive pathobiology of cowpox and rabbitpox virus following intranasal inoculation of Balb/c mice. *Virology* **1993**, *197*, 328–338. [CrossRef] [PubMed]

133. Cameron, C.; Hota-Mitchell, S.; Chen, L.; Barrett, J.; Cao, J.X.; Macaulay, C.; Willer, D.; Evans, D.; McFadden, G. The complete DNA sequence of myxoma virus. *Virology* **1999**, *264*, 298–318. [CrossRef] [PubMed]

134. Macen, J.L.; Upton, C.; Nation, N.; McFadden, G. SERP1, a serine proteinase inhibitor encoded by myxoma virus, is a secreted glycoprotein that interferes with inflammation. *Virology* **1993**, *195*, 348–363. [CrossRef] [PubMed]

135. Wang, Y.X.; Turner, P.C.; Ness, T.L.; Moon, K.B.; Schoeb, T.R.; Moyer, R.W. The cowpox virus SPI-3 and myxoma virus SERP1 serpins are not functionally interchangeable despite their similar proteinase inhibition profiles in vitro. *Virology* **2000**, *272*, 281–292. [CrossRef] [PubMed]

136. Turner, P.C.; Baquero, M.T.; Yuan, S.; Thoennes, S.R.; Moyer, R.W. The cowpox virus serpin SPI-3 complexes with and inhibits urokinase-type and tissue-type plasminogen activators and plasmin. *Virology* **2000**, *272*, 267–280. [CrossRef] [PubMed]

137. Petit, F.; Bertagnoli, S.; Gelfi, J.; Fassy, F.; Boucraut-Baralon, C.; Milon, A. Characterization of a myxoma virus-encoded serpin-like protein with activity against interleukin-1β-converting enzyme. *J. Virol.* **1996**, *70*, 5860–5866. [PubMed]

138. Messud-Petit, F.; Gelfi, J.; Delverdier, M.; Amardeilh, M.F.; Py, R.; Sutter, G.; Bertagnoli, S. Serp2, an inhibitor of the interleukin-1β-converting enzyme, is critical in the pathobiology of myxoma virus. *J. Virol.* **1998**, *72*, 7830–7839. [PubMed]

139. Turner, P.C.; Sancho, M.C.; Thoennes, S.R.; Caputo, A.; Bleackley, R.C.; Moyer, R.W. Myxoma virus Serp2 is a weak inhibitor of granzyme B and interleukin-1β converting enzyme in vitro and unlike CrmA cannot block apoptosis in cowpox virus-infected cells. *J. Virol.* **1999**, *73*, 6394–6404. [PubMed]

140. Nathaniel, R.; MacNeill, A.L.; Wang, Y.X.; Turner, P.C.; Moyer, R.W. *Cowpox virus* CrmA, *Myxoma virus* SERP2 and baculovirus P35 are not functionally interchangeable caspase inhibitors in poxvirus infections. *J. Gen. Virol.* **2004**, *85 Pt 5*, 1267–1278. [CrossRef] [PubMed]

141. Guerin, J.L.; Gelfi, J.; Camus, C.; Delverdier, M.; Whisstock, J.C.; Amardeihl, M.F.; Py, R.; Bertagnoli, S.; Messud-Petit, F. Characterization and functional analysis of Serp3: A novel myxoma virus-encoded serpin involved in virulence. *J. Gen. Virol.* **2001**, *82*, 1407–1417. [CrossRef] [PubMed]

142. Senkevich, T.G.; Bugert, J.J.; Sisler, J.R.; Koonin, E.V.; Darai, G.; Moss, B. Genome sequence of a human tumorigenic poxvirus: Prediction of specific host response-evasion genes. *Science* **1996**, *273*, 813–816. [CrossRef] [PubMed]

143. Senkevich, T.G.; Koonin, E.V.; Bugert, J.J.; Darai, G.; Moss, B. The genome of molluscum contagiosum virus: Analysis and comparison with other poxviruses. *Virology* **1997**, *233*, 19–42. [CrossRef] [PubMed]

144. Vermi, W.; Fisogni, S.; Salogni, L.; Scharer, L.; Kutzner, H.; Sozzani, S.; Lonardi, S.; Rossini, C.; Calzavara-Pinton, P.; LeBoit, P.E.; et al. Spontaneous regression of highly immunogenic *Molluscum contagiosum* virus (MCV)-induced skin lesions is associated with plasmacytoid dendritic cells and IFN-DC infiltration. *J. Investig. Dermatol.* **2011**, *131*, 426–434. [CrossRef] [PubMed]

145. Murao, L.E.; Shisler, J.L. The MCV MC159 protein inhibits late, but not early, events of TNF-α-induced NF-κB activation. *Virology* **2005**, *340*, 255–264. [CrossRef] [PubMed]

146. Nichols, D.B.; Shisler, J.L. The MC160 protein expressed by the dermatotropic poxvirus molluscum contagiosum virus prevents tumor necrosis factor α-induced NF-κB activation via inhibition of Iκ kinase complex formation. *J. Virol.* **2006**, *80*, 578–586. [CrossRef] [PubMed]

147. Nichols, D.B.; Shisler, J.L. Poxvirus MC160 protein utilizes multiple mechanisms to inhibit NF-κB activation mediated via components of the tumor necrosis factor receptor 1 signal transduction pathway. *J. Virol.* **2009**, *83*, 3162–3174. [CrossRef] [PubMed]

148. Randall, C.M.; Shisler, J.L. Molluscum Contagiosum Virus: Persistence Pays Off. *Future Virol.* **2013**, *8*, 561–573. [CrossRef]

149. Randall, C.M.; Biswas, S.; Selen, C.V.; Shisler, J.L. Inhibition of interferon gene activation by death-effector domain-containing proteins from the molluscum contagiosum virus. *Proc. Natl. Acad. Sci. USA* **2014**, *111*, E265–E272. [CrossRef] [PubMed]

150. Randall, C.M.; Jokela, J.A.; Shisler, J.L. The MC159 protein from the molluscum contagiosum poxvirus inhibits NF-κB activation by interacting with the IκB kinase complex. *J. Immunol.* **2012**, *188*, 2371–2379. [CrossRef] [PubMed]

151. Chan, F.K.; Shisler, J.; Bixby, J.G.; Felices, M.; Zheng, L.; Appel, M.; Orenstein, J.; Moss, B.; Lenardo, M.J. A role for tumor necrosis factor receptor-2 and receptor-interacting protein in programmed necrosis and antiviral responses. *J. Biol. Chem.* **2003**, *278*, 51613–51621. [CrossRef] [PubMed]

152. Thurau, M.; Everett, H.; Tapernoux, M.; Tschopp, J.; Thome, M. The TRAF3-binding site of human molluscipox virus FLIP molecule MC159 is critical for its capacity to inhibit Fas-induced apoptosis. *Cell Death Differ.* **2006**, *13*, 1577–1585. [CrossRef] [PubMed]

153. Bertin, J.; Armstrong, R.C.; Ottilie, S.; Martin, D.A.; Wang, Y.; Banks, S.; Wang, G.H.; Senkevich, T.G.; Alnemri, E.S.; Moss, B.; et al. Death effector domain-containing herpesvirus and poxvirus proteins inhibit both Fas- and TNFR1-induced apoptosis. *Proc. Natl. Acad. Sci. USA* **1997**, *94*, 1172–1176. [CrossRef] [PubMed]

154. Hu, S.; Vincenz, C.; Buller, M.; Dixit, V.M. A novel family of viral death effector domain-containing molecules that inhibit both CD-95- and tumor necrosis factor receptor-1-induced apoptosis. *J. Biol. Chem.* **1997**, *272*, 9621–9624. [CrossRef] [PubMed]

155. Shisler, J.L.; Moss, B. Molluscum contagiosum virus inhibitors of apoptosis: The MC159 v-FLIP protein blocks Fas-induced activation of procaspases and degradation of the related MC160 protein. *Virology* **2001**, *282*, 14–25. [CrossRef] [PubMed]

156. Boldin, M.P.; Goncharov, T.M.; Goltsev, Y.V.; Wallach, D. Involvement of MACH, a novel MORT1/FADD-interacting protease, in Fas/APO-1- and TNF receptor-induced cell death. *Cell* **1996**, *85*, 803–815. [CrossRef]

157. Muzio, M.; Chinnaiyan, A.M.; Kischkel, F.C.; O'Rourke, K.; Shevchenko, A.; Ni, J.; Scaffidi, C.; Bretz, J.D.; Zhang, M.; Gentz, R.; et al. FLICE, a novel FADD-homologous ICE/CED-3-like protease, is recruited to the CD95 (Fas/APO-1) death-inducing signaling complex. *Cell* **1996**, *85*, 817–827. [CrossRef]

158. Taylor, R.C.; Cullen, S.P.; Martin, S.J. Apoptosis: Controlled demolition at the cellular level. *Nat.Rev. Mol.Cell Biol.* **2008**, *9*, 231–241. [CrossRef] [PubMed]

159. Esposito, D.; Sankar, A.; Morgner, N.; Robinson, C.V.; Rittinger, K.; Driscoll, P.C. Solution NMR investigation of the CD95/FADD homotypic death domain complex suggests lack of engagement of the CD95 C terminus. *Structure* **2010**, *18*, 1378–1390. [CrossRef] [PubMed]

160. Jang, T.H.; Zheng, C.; Li, J.; Richards, C.; Hsiao, Y.S.; Walz, T.; Wu, H.; Park, H.H. Structural study of the RIPoptosome core reveals a helical assembly for kinase recruitment. *Biochemistry* **2014**, *53*, 5424–5431. [CrossRef] [PubMed]

161. Wang, L.; Yang, J.K.; Kabaleeswaran, V.; Rice, A.J.; Cruz, A.C.; Park, A.Y.; Yin, Q.; Damko, E.; Jang, S.B.; Raunser, S.; et al. The Fas-FADD death domain complex structure reveals the basis of DISC assembly and disease mutations. *Nat. Struct. Mol. Biol.* **2010**, *17*, 1324–1329. [CrossRef] [PubMed]

162. Fu, T.M.; Li, Y.; Lu, A.; Li, Z.; Vajjhala, P.R.; Cruz, A.C.; Srivastava, D.B.; DiMaio, F.; Penczek, P.A.; Siegel, R.M.; et al. Cryo-EM Structure of caspase-8 tandem DED filament reveals assembly and regulation mechanisms of the death-inducing signaling complex. *Mol. Cell* **2016**, *64*, 236–250. [CrossRef] [PubMed]

163. Boatright, K.M.; Renatus, M.; Scott, F.L.; Sperandio, S.; Shin, H.; Pedersen, I.M.; Ricci, J.E.; Edris, W.A.; Sutherlin, D.P.; Green, D.R.; et al. A unified model for apical caspase activation. *Mol. Cell* **2003**, *11*, 529–541. [CrossRef]

164. Tschopp, J.; Irmler, M.; Thome, M. Inhibition of Fas death signals by FLIPs. *Curr. Opin. Immunol.* **1998**, *10*, 552–558. [CrossRef]

165. Hughes, M.A.; Powley, I.R.; Jukes-Jones, R.; Horn, S.; Feoktistova, M.; Fairall, L.; Schwabe, J.W.; Leverkus, M.; Cain, K.; MacFarlane, M. Co-operative and hierarchical binding of c-FLIP and caspase-8: A unified model defines how c-FLIP isoforms differentially control cell fate. *Mol. Cell* **2016**, *61*, 834–849. [CrossRef] [PubMed]

166. Yang, J.K.; Wang, L.; Zheng, L.; Wan, F.; Ahmed, M.; Lenardo, M.J.; Wu, H. Crystal structure of MC159 reveals molecular mechanism of DISC assembly and FLIP inhibition. *Mol. Cell* **2005**, *20*, 939–949. [CrossRef] [PubMed]

167. Li, F.Y.; Jeffrey, P.D.; Yu, J.W.; Shi, Y. Crystal structure of a viral FLIP: Insights into FLIP-mediated inhibition of death receptor signaling. *J. Biol. Chem.* **2006**, *281*, 2960–2968. [CrossRef] [PubMed]

168. Garvey, T.L.; Bertin, J.; Siegel, R.M.; Wang, G.H.; Lenardo, M.J.; Cohen, J.I. Binding of FADD and caspase-8 to molluscum contagiosum virus MC159 v-FLIP is not sufficient for its antiapoptotic function. *J. Virol.* **2002**, *76*, 697–706. [CrossRef] [PubMed]

169. Siegel, R.M.; Martin, D.A.; Zheng, L.; Ng, S.Y.; Bertin, J.; Cohen, J.; Lenardo, M.J. Death-effector filaments: Novel cytoplasmic structures that recruit caspases and trigger apoptosis. *J. Cell Biol.* **1998**, *141*, 1243–1253. [CrossRef] [PubMed]

170. Huttmann, J.; Krause, E.; Schommartz, T.; Brune, W. Functional comparison of molluscum contagiosum virus vFLIP MC159 with murine cytomegalovirus M36/vICA and M45/vIRA proteins. *J. Virol.* **2015**, *90*, 2895–2905. [CrossRef] [PubMed]

171. Beaury, M.; Velagapudi, U.K.; Weber, S.; Soto, C.; Talele, T.T.; Nichols, D.B. The molluscum contagiosum virus death effector domain containing protein MC160 RxDL motifs are not required for its known viral immune evasion functions. *Virus Genes* **2017**, *53*, 522–531. [CrossRef] [PubMed]

172. Yu, J.W.; Jeffrey, P.D.; Shi, Y. Mechanism of procaspase-8 activation by c-FLIP$_L$. *Proc. Natl. Acad. Sci. USA* **2009**, *106*, 8169–8174. [CrossRef] [PubMed]

173. Pop, C.; Oberst, A.; Drag, M.; Van Raam, B.J.; Riedl, S.J.; Green, D.R.; Salvesen, G.S. FLIP$_L$ induces caspase 8 activity in the absence of interdomain caspase 8 cleavage and alters substrate specificity. *Biochem. J.* **2011**, *433*, 447–457. [CrossRef] [PubMed]

174. Gil, J.; Rullas, J.; Alcami, J.; Esteban, M. MC159L protein from the poxvirus molluscum contagiosum virus inhibits NF-κB activation and apoptosis induced by PKR. *J. Gen. Virol.* **2001**, *82*, 3027–3034. [CrossRef] [PubMed]

175. Graham, S.C.; Bahar, M.W.; Cooray, S.; Chen, R.A.; Whalen, D.M.; Abrescia, N.G.; Alderton, D.; Owens, R.J.; Stuart, D.I.; Smith, G.L.; et al. Vaccinia virus proteins A52 and B14 Share a Bcl-2-like fold but have evolved to inhibit NF-κB rather than apoptosis. *PLoS Pathog.* **2008**, *4*, e1000128. [CrossRef] [PubMed]

176. Neidel, S.; Maluquer de Motes, C.; Mansur, D. S.; Strnadova, P.; Smith, G.L.; Graham, S.C. Vaccinia virus protein A49 is an unexpected member of the B-cell Lymphoma (Bcl)-2 protein family. *J. Biol. Chem.* **2015**, *290*, 5991–6002. [CrossRef] [PubMed]

177. Mansur, D.S.; Maluquer de Motes, C.; Unterholzner, L.; Sumner, R.P.; Ferguson, B.J.; Ren, H.; Strnadova, P.; Bowie, A.G.; Smith, G.L. Poxvirus targeting of E3 ligase beta-TrCP by molecular mimicry: a mechanism to inhibit NF-κB activation and promote immune evasion and virulence. *PLoS Pathog.* **2013**, *9*, e1003183. [CrossRef] [PubMed]

178. Wasilenko, S.T.; Stewart, T.L.; Meyers, A.F.; Barry, M. Vaccinia virus encodes a previously uncharacterized mitochondrial-associated inhibitor of apoptosis. *Proc. Natl. Acad. Sci. USA* **2003**, *100*, 14345–14350. [CrossRef] [PubMed]

179. Fischer, S.F.; Ludwig, H.; Holzapfel, J.; Kvansakul, M.; Chen, L.; Huang, D.C.; Sutter, G.; Knese, M.; Hacker, G. Modified vaccinia virus Ankara protein F1L is a novel BH3-domain-binding protein and acts together with the early viral protein E3L to block virus-associated apoptosis. *Cell Death Differ.* **2006**, *13*, 109–118. [CrossRef] [PubMed]

180. Postigo, A.; Cross, J.R.; Downward, J.; Way, M. Interaction of F1L with the BH3 domain of Bak is responsible for inhibiting vaccinia-induced apoptosis. *Cell Death Differ.* **2006**, *13*, 1651–1662. [CrossRef] [PubMed]

181. Kvansakul, M.; Yang, H.; Fairlie, W.D.; Czabotar, P.E.; Fischer, S.F.; Perugini, M.A.; Huang, D.C.; Colman, P.M. Vaccinia virus anti-apoptotic F1L is a novel Bcl-2-like domain-swapped dimer that binds a highly selective subset of BH3-containing death ligands. *Cell Death Differ.* **2008**, *15*, 1564–1571. [CrossRef] [PubMed]

182. Campbell, S.; Thibault, J.; Mehta, N.; Colman, P.M.; Barry, M.; Kvansakul, M. Structural insight into BH3 domain binding of vaccinia virus antiapoptotic F1L. *J. Virol.* **2014**, *88*, 8667–8677. [CrossRef] [PubMed]

183. Postigo, A.; Martin, M.C.; Dodding, M.P.; Way, M. Vaccinia-induced epidermal growth factor receptor-MEK signalling and the anti-apoptotic protein F1L synergize to suppress cell death during infection. *Cell Microbiol.* **2009**, *11*, 1208–1218. [CrossRef] [PubMed]

184. Zhai, D.; Yu, E.; Jin, C.; Welsh, K.; Shiau, C.W.; Chen, L.; Salvesen, G.S.; Liddington, R.; Reed, J.C. Vaccinia virus protein F1L is a caspase-9 inhibitor. *J. Biol. Chem.* **2010**, *285*, 5569–5580. [CrossRef] [PubMed]

185. Caria, S.; Marshall, B.; Burton, R.L.; Campbell, S.; Pantaki-Eimany, D.; Hawkins, C.J.; Barry, M.; Kvansakul, M. The N Terminus of the vaccinia virus protein F1L is an intrinsically unstructured region that is not involved in apoptosis regulation. *J. Biol. Chem.* **2016**, *291*, 14600–14608. [CrossRef] [PubMed]

186. Gerlic, M.; Faustin, B.; Postigo, A.; Yu, E.C.; Proell, M.; Gombosuren, N.; Krajewska, M.; Flynn, R.; Croft, M.; Way, M.; et al. Vaccinia virus F1L protein promotes virulence by inhibiting inflammasome activation. *Proc. Natl. Acad. Sci. USA* **2013**, *110*, 7808–7813. [CrossRef] [PubMed]

187. Marshall, B.; Puthalakath, H.; Caria, S.; Chugh, S.; Doerflinger, M.; Colman, P.M.; Kvansakul, M. Variola virus F1L is a Bcl-2-like protein that unlike its vaccinia virus counterpart inhibits apoptosis independent of Bim. *Cell Death Dis.* **2015**, *6*, e1680. [CrossRef] [PubMed]

188. Bartlett, N.; Symons, J.A.; Tscharke, D.C.; Smith, G.L. The vaccinia virus N1L protein is an intracellular homodimer that promotes virulence. *J. Gen. Virol.* **2002**, *83*, 1965–1976. [CrossRef] [PubMed]

189. De Motes, C.M.; Cooray, S.; Ren, H.; Almeida, G.M.; McGourty, K.; Bahar, M.W.; Stuart, D.I.; Grimes, J.M.; Graham, S.C.; Smith, G.L. Inhibition of apoptosis and NF-κB activation by vaccinia protein N1 occur via distinct binding surfaces and make different contributions to virulence. *PLoS Pathog.* **2011**, *7*, e1002430.

190. Cooray, S.; Bahar, M.W.; Abrescia, N.G.; McVey, C.E.; Bartlett, N.W.; Chen, R.A.; Stuart, D.I.; Grimes, J.M.; Smith, G.L. Functional and structural studies of the vaccinia virus virulence factor N1 reveal a Bcl-2-like anti-apoptotic protein. *J. Gen. Virol.* **2007**, *88*, 1656–1666. [CrossRef] [PubMed]

191. Aoyagi, M.; Zhai, D.; Jin, C.; Aleshin, A.E.; Stec, B.; Reed, J.C.; Liddington, R.C. Vaccinia virus N1L protein resembles a B cell lymphoma-2 (Bcl-2) family protein. *Protein Sci.* **2007**, *16*, 118–124. [CrossRef] [PubMed]

192. Postigo, A.; Way, M. The vaccinia virus-encoded Bcl-2 homologues do not act as direct Bax inhibitors. *J. Virol.* **2012**, *86*, 203–213. [CrossRef] [PubMed]

193. DiPerna, G.; Stack, J.; Bowie, A.G.; Boyd, A.; Kotwal, G.; Zhang, Z.; Arvikar, S.; Latz, E.; Fitzgerald, K.A.; Marshall, W.L. Poxvirus protein N1L targets the I-κB kinase complex, inhibits signaling to NF-κB by the tumor necrosis factor superfamily of receptors, and inhibits NF-κB and IRF3 signaling by toll-like receptors. *J. Biol. Chem.* **2004**, *279*, 36570–36578. [CrossRef] [PubMed]

194. Everett, H.; Barry, M.; Lee, S.F.; Sun, X.; Graham, K.; Stone, J.; Bleackley, R.C.; McFadden, G. M11L: A novel mitochondria-localized protein of myxoma virus that blocks apoptosis of infected leukocytes. *J. Exp. Med.* **2000**, *191*, 1487–1498. [CrossRef] [PubMed]

195. Graham, K.A.; Opgenorth, A.; Upton, C.; McFadden, G. Myxoma virus M11L ORF encodes a protein for which cell surface localization is critical in manifestation of viral virulence. *Virology* **1992**, *191*, 112–124. [CrossRef]

196. Everett, H.; Barry, M.; Sun, X.; Lee, S.F.; Frantz, C.; Berthiaume, L.G.; McFadden, G.; Bleackley, R.C. The myxoma poxvirus protein, M11L, prevents apoptosis by direct interaction with the mitochondrial permeability transition pore. *J. Exp. Med.* **2002**, *196*, 1127–1139. [CrossRef] [PubMed]

197. Su, J.; Wang, G.; Barrett, J.W.; Irvine, T.S.; Gao, X.; McFadden, G. Myxoma virus M11L blocks apoptosis through inhibition of conformational activation of Bax at the mitochondria. *J. Virol.* **2006**, *80*, 1140–1151. [CrossRef] [PubMed]

198. Wang, G.; Barrett, J.W.; Nazarian, S.H.; Everett, H.; Gao, X.; Bleackley, C.; Colwill, K.; Moran, M.F.; McFadden, G. Myxoma virus M11L prevents apoptosis through constitutive interaction with Bak. *J. Virol.* **2004**, *78*, 7097–7111. [CrossRef] [PubMed]

199. Douglas, A.E.; Corbett, K.D.; Berger, J.M.; McFadden, G.; Handel, T.M. Structure of M11L: A myxoma virus structural homolog of the apoptosis inhibitor, Bcl-2. *Protein Sci.* **2007**, *16*, 695–703. [CrossRef] [PubMed]

200. Kvansakul, M.; van Delft, M.F.; Lee, E.F.; Gulbis, J.M.; Fairlie, W.D.; Huang, D.C.; Colman, P.M. A structural viral mimic of prosurvival Bcl-2: A pivotal role for sequestering proapoptotic Bax and Bak. *Mol. Cell* **2007**, *25*, 933–942. [CrossRef] [PubMed]

201. Okamoto, T.; Campbell, S.; Mehta, N.; Thibault, J.; Colman, P.M.; Barry, M.; Huang, D.C.; Kvansakul, M. Sheeppox virus SPPV14 encodes a Bcl-2-like cell death inhibitor that counters a distinct set of mammalian proapoptotic proteins. *J. Virol.* **2012**, *86*, 11501–11511. [CrossRef] [PubMed]

202. Burton, D.R.; Caria, S.; Marshall, B.; Barry, M.; Kvansakul, M. Structural basis of Deerpox virus-mediated inhibition of apoptosis. *Acta Crystallogr. D Biol. Crystallogr.* **2015**, *71*, 1593–1603. [CrossRef] [PubMed]

203. Banadyga, L.; Lam, S.C.; Okamoto, T.; Kvansakul, M.; Huang, D.C.; Barry, M. Deerpox virus encodes an inhibitor of apoptosis that regulates Bak and Bax. *J. Virol.* **2011**, *85*, 1922–1934. [CrossRef] [PubMed]

204. Gubser, C.; Bergamaschi, D.; Hollinshead, M.; Lu, X.; van Kuppeveld, F.J.; Smith, G.L. A new inhibitor of apoptosis from vaccinia virus and eukaryotes. *PLoS Pathog.* **2007**, *3*, e17. [CrossRef] [PubMed]

205. Carrara, G.; Saraiva, N.; Gubser, C.; Johnson, B.F.; Smith, G.L. Six-transmembrane topology for Golgi anti-apoptotic protein (GAAP) and Bax inhibitor 1 (BI-1) provides model for the transmembrane Bax inhibitor-containing motif (TMBIM) family. *J. Biol. Chem.* **2012**, *287*, 15896–15905. [CrossRef] [PubMed]

206. Carrara, G.; Saraiva, N.; Parsons, M.; Byrne, B.; Prole, D.L.; Taylor, C.W.; Smith, G.L. Golgi anti-apoptotic proteins are highly conserved ion channels that affect apoptosis and cell migration. *J. Biol. Chem.* **2015**, *290*, 11785–11801. [CrossRef] [PubMed]

207. Saraiva, N.; Prole, D.L.; Carrara, G.; de Motes, C.M.; Johnson, B.F.; Byrne, B.; Taylor, C.W.; Smith, G.L. Human and viral Golgi anti-apoptotic proteins (GAAPs) oligomerize via different mechanisms and monomeric GAAP inhibits apoptosis and modulates calcium. *J. Biol. Chem.* **2013**, *288*, 13057–13067. [CrossRef] [PubMed]

208. Almazan, F.; Tscharke, D.C.; Smith, G.L. The vaccinia virus superoxide dismutase-like protein (A45R) is a virion component that is nonessential for virus replication. *J. Virol.* **2001**, *75*, 7018–7029. [CrossRef] [PubMed]

209. Cao, J.X.; Teoh, M.L.; Moon, M.; McFadden, G.; Evans, D.H. Leporipoxvirus Cu-Zn superoxide dismutase homologs inhibit cellular superoxide dismutase, but are not essential for virus replication or virulence. *Virology* **2002**, *296*, 125–135. [CrossRef] [PubMed]

210. Teoh, M.L.; Turner, P.V.; Evans, D.H. Tumorigenic poxviruses up-regulate intracellular superoxide to inhibit apoptosis and promote cell proliferation. *J. Virol.* **2005**, *79*, 5799–5811. [CrossRef] [PubMed]

211. Teoh, M.L.; Walasek, P.J.; Evans, D.H. Leporipoxvirus Cu,Zn-superoxide dismutase (SOD) homologs are catalytically inert decoy proteins that bind copper chaperone for SOD. *J. Biol. Chem.* **2003**, *278*, 33175–33184. [CrossRef] [PubMed]

212. Becker, M.N.; Greenleaf, W.B.; Ostrov, D.A.; Moyer, R.W. *Amsacta moorei* entomopoxvirus expresses an active superoxide dismutase. *J. Virol.* **2004**, *78*, 10265–10275. [CrossRef] [PubMed]

213. Coutu, J.; Ryerson, M.R.; Bugert, J.; Brian Nichols, D. The Molluscum Contagiosum Virus protein MC163 localizes to the mitochondria and dampens mitochondrial mediated apoptotic responses. *Virology* **2017**, *505*, 91–101. [CrossRef] [PubMed]

214. Mohr, S.; Grandemange, S.; Massimi, P.; Darai, G.; Banks, L.; Martinou, J.C.; Zeier, M.; Muranyi, W. Targeting the retinoblastoma protein by MC007L, gene product of the molluscum contagiosum virus: Detection of a novel virus-cell interaction by a member of the poxviruses. *J. Virol.* **2008**, *82*, 10625–10633. [CrossRef] [PubMed]

215. Shisler, J.L.; Senkevich, T.G.; Berry, M.J.; Moss, B. Ultraviolet-induced cell death blocked by a selenoprotein from a human dermatotropic poxvirus. *Science* **1998**, *279*, 102–105. [CrossRef] [PubMed]

216. Duesberg, P.H.; Colby, C. On the biosynthesis and structure of double-stranded RNA in vaccinia virus-infected cells. *Proc. Natl. Acad. Sci. USA* **1969**, *64*, 396–403. [CrossRef] [PubMed]

217. Burgess, H.M.; Mohr, I. Cellular 5′-3′ mRNA exonuclease Xrn1 controls double-stranded RNA accumulation and anti-viral responses. *Cell Host Microbe* **2015**, *17*, 332–344. [CrossRef] [PubMed]

218. Jacobs, B.L.; Langland, J.O. When two strands are better than one: The mediators and modulators of the cellular responses to double-stranded RNA. *Virology* **1996**, *219*, 339–349. [CrossRef] [PubMed]

219. Weber, F.; Wagner, V.; Rasmussen, S.B.; Hartmann, R.; Paludan, S.R. Double-stranded RNA is produced by positive-strand RNA viruses and DNA viruses but not in detectable amounts by negative-strand RNA viruses. *J. Virol.* **2006**, *80*, 5059–5064. [CrossRef] [PubMed]

220. Willis, K.L.; Langland, J.O.; Shisler, J.L. Viral double-stranded RNAs from vaccinia virus early or intermediate gene transcripts possess PKR activating function, resulting in NF-κB activation, when the K1 protein is absent or mutated. *J. Biol. Chem.* **2011**, *286*, 7765–7778. [CrossRef] [PubMed]

221. Broyles, S.S. Vaccinia virus transcription. *J. Gen. Virol.* **2003**, *84*, 2293–2303. [CrossRef] [PubMed]

222. Colby, C.; Duesberg, P.H. Double-stranded RNA in vaccinia virus infected cells. *Nature* **1969**, *222*, 940–944. [CrossRef] [PubMed]

223. Colby, C.; Jurale, C.; Kates, J.R. Mechanism of synthesis of vaccinia virus double-stranded ribonucleic acid in vivo and in vitro. *J. Virol.* **1971**, *7*, 71–76. [PubMed]

224. Boone, R.F.; Parr, R.P.; Moss, B. Intermolecular duplexes formed from polyadenylylated vaccinia virus RNA. *J. Virol.* **1979**, *30*, 365–374. [PubMed]

225. Kibler, K.V.; Shors, T.; Perkins, K.B.; Zeman, C.C.; Banaszak, M.P.; Biesterfeldt, J.; Langland, J.O.; Jacobs, B.L. Double-stranded RNA is a trigger for apoptosis in vaccinia virus-infected cells. *J. Virol.* **1997**, *71*, 1992–2003. [PubMed]

226. Munir, M.; Berg, M. The multiple faces of proteinkinase R in antiviral defense. *Virulence* **2013**, *4*, 85–89. [CrossRef] [PubMed]

227. Kang, R.; Tang, D. PKR-dependent inflammatory signals. *Sci. Signal.* **2012**, *5*, 47. [CrossRef] [PubMed]

228. Seth, R.B.; Sun, L.; Ea, C.K.; Chen, Z.J. Identification and characterization of MAVS, a mitochondrial antiviral signaling protein that activates NF-κB and IRF 3. *Cell* **2005**, *122*, 669–682. [CrossRef] [PubMed]

229. Kawai, T.; Takahashi, K.; Sato, S.; Coban, C.; Kumar, H.; Kato, H.; Ishii, K.J.; Takeuchi, O.; Akira, S. IPS-1, an adaptor triggering RIG-I- and Mda5-mediated type I interferon induction. *Nat. Immunol.* **2005**, *6*, 981–988. [CrossRef] [PubMed]

230. Kato, H.; Takeuchi, O.; Sato, S.; Yoneyama, M.; Yamamoto, M.; Matsui, K.; Uematsu, S.; Jung, A.; Kawai, T.; Ishii, K.J.; et al. Differential roles of MDA5 and RIG-I helicases in the recognition of RNA viruses. *Nature* **2006**, *441*, 101–105. [CrossRef] [PubMed]

231. Yoneyama, M.; Kikuchi, M.; Natsukawa, T.; Shinobu, N.; Imaizumi, T.; Miyagishi, M.; Taira, K.; Akira, S.; Fujita, T. The RNA helicase RIG-I has an essential function in double-stranded RNA-induced innate antiviral responses. *Nat. Immunol.* **2004**, *5*, 730–737. [CrossRef] [PubMed]

232. Hornung, V.; Ellegast, J.; Kim, S.; Brzozka, K.; Jung, A.; Kato, H.; Poeck, H.; Akira, S.; Conzelmann, K.K.; Schlee, M.; et al. 5'-Triphosphate RNA is the ligand for RIG-I. *Science* **2006**, *314*, 994–997. [CrossRef] [PubMed]

233. Pichlmair, A.; Schulz, O.; Tan, C.P.; Naslund, T.I.; Liljestrom, P.; Weber, F.; Reis e Sousa, C. RIG-I-mediated antiviral responses to single-stranded RNA bearing 5'-phosphates. *Science* **2006**, *314*, 997–1001. [CrossRef] [PubMed]

234. Fitzgerald, K.A.; McWhirter, S.M.; Faia, K.L.; Rowe, D.C.; Latz, E.; Golenbock, D.T.; Coyle, A.J.; Liao, S.M.; Maniatis, T. IKKε and TBK1 are essential components of the IRF3 signaling pathway. *Nat. Immunol.* **2003**, *4*, 491–496. [CrossRef] [PubMed]

235. Heylbroeck, C.; Balachandran, S.; Servant, M.J.; DeLuca, C.; Barber, G.N.; Lin, R.; Hiscott, J. The IRF-3 transcription factor mediates Sendai virus-induced apoptosis. *J. Virol.* **2000**, *74*, 3781–3792. [CrossRef] [PubMed]

236. Huang, Y.; Liu, H.; Li, S.; Tang, Y.; Wei, B.; Yu, H.; Wang, C. MAVS-MKK7-JNK2 defines a novel apoptotic signaling pathway during viral infection. *PLoS Pathog.* **2014**, *10*, e1004020. [CrossRef] [PubMed]

237. Lei, Y.; Moore, C.B.; Liesman, R.M.; O'Connor, B.P.; Bergstralh, D.T.; Chen, Z.J.; Pickles, R.J.; Ting, J.P. MAVS-mediated apoptosis and its inhibition by viral proteins. *PLoS ONE* **2009**, *4*, e5466. [CrossRef] [PubMed]

238. Eitz Ferrer, P.; Potthoff, S.; Kirschnek, S.; Gasteiger, G.; Kastenmuller, W.; Ludwig, H.; Paschen, S.A.; Villunger, A.; Sutter, G.; Drexler, I.; et al. Induction of Noxa-mediated apoptosis by modified vaccinia virus Ankara depends on viral recognition by cytosolic helicases, leading to IRF-3/IFN-β-dependent induction of pro-apoptotic Noxa. *PLoS Pathog.* **2011**, *7*, e1002083. [CrossRef] [PubMed]

239. Lee, S.B.; Esteban, M. The interferon-induced double-stranded RNA-activated human p68 protein kinase inhibits the replication of vaccinia virus. *Virology* **1993**, *193*, 1037–1041. [CrossRef] [PubMed]

240. Chang, H.W.; Jacobs, B.L. Identification of a conserved motif that is necessary for binding of the vaccinia virus E3L gene products to double-stranded RNA. *Virology* **1993**, *194*, 537–547. [CrossRef] [PubMed]

241. Rivas, C.; Gil, J.; Melkova, Z.; Esteban, M.; Diaz-Guerra, M. Vaccinia virus E3L protein is an inhibitor of the interferon (IFN)-induced 2-5A synthetase enzyme. *Virology* **1998**, *243*, 406–414. [CrossRef] [PubMed]

242. Zhang, P.; Jacobs, B.L.; Samuel, C.E. Loss of protein kinase PKR expression in human HeLa cells complements the vaccinia virus E3L deletion mutant phenotype by restoration of viral protein synthesis. *J. Virol.* **2008**, *82*, 840–848. [CrossRef] [PubMed]

243. Myskiw, C.; Arsenio, J.; Booy, E.P.; Hammett, C.; Deschambault, Y.; Gibson, S.B.; Cao, J. RNA species generated in vaccinia virus infected cells activate cell type-specific MDA5 or RIG-I dependent interferon gene transcription and PKR dependent apoptosis. *Virology* **2011**, *413*, 183–193. [CrossRef] [PubMed]

244. Yuwen, H.; Cox, J.H.; Yewdell, J.W.; Bennink, J.R.; Moss, B. Nuclear localization of a double-stranded RNA-binding protein encoded by the vaccinia virus E3L gene. *Virology* **1993**, *195*, 732–744. [CrossRef] [PubMed]

245. Chang, H.W.; Watson, J.C.; Jacobs, B.L. The E3L gene of vaccinia virus encodes an inhibitor of the interferon-induced, double-stranded RNA-dependent protein kinase. *Proc. Natl. Acad. Sci. USA* **1992**, *89*, 4825–4829. [CrossRef] [PubMed]

246. Kim, Y.G.; Muralinath, M.; Brandt, T.; Pearcy, M.; Hauns, K.; Lowenhaupt, K.; Jacobs, B.L.; Rich, A. A role for Z-DNA binding in vaccinia virus pathogenesis. *Proc. Natl. Acad. Sci. USA* **2003**, *100*, 6974–6979. [CrossRef] [PubMed]

247. Herbert, A.; Alfken, J.; Kim, Y.G.; Mian, I.S.; Nishikura, K.; Rich, A. A Z-DNA binding domain present in the human editing enzyme, double-stranded RNA adenosine deaminase. *Proc. Natl. Acad. Sci. USA* **1997**, *94*, 8421–8426. [CrossRef] [PubMed]

248. Chang, H.W.; Uribe, L.H.; Jacobs, B.L. Rescue of vaccinia virus lacking the E3L gene by mutants of E3L. *J. Virol.* **1995**, *69*, 6605–6608. [PubMed]

249. Langland, J.O.; Jacobs, B.L. The role of the PKR-inhibitory genes, E3L and K3L, in determining vaccinia virus host range. *Virology* **2002**, *299*, 133–141. [CrossRef] [PubMed]

250. Beattie, E.; Denzler, K.L.; Tartaglia, J.; Perkus, M.E.; Paoletti, E.; Jacobs, B.L. Reversal of the interferon-sensitive phenotype of a vaccinia virus lacking E3L by expression of the reovirus S4 gene. *J. Virol.* **1995**, *69*, 499–505. [PubMed]

251. Beattie, E.; Paoletti, E.; Tartaglia, J. Distinct patterns of IFN sensitivity observed in cells infected with vaccinia K3L- and E3L-mutant viruses. *Virology* **1995**, *210*, 254–263. [CrossRef] [PubMed]

252. Shors, T.; Kibler, K.V.; Perkins, K.B.; Seidler-Wulff, R.; Banaszak, M.P.; Jacobs, B.L. Complementation of vaccinia virus deleted of the E3L gene by mutants of E3L. *Virology* **1997**, *239*, 269–276. [CrossRef] [PubMed]

253. Lee, S.B.; Esteban, M. The interferon-induced double-stranded RNA-activated protein kinase induces apoptosis. *Virology* **1994**, *199*, 491–496. [CrossRef] [PubMed]

254. Brandt, T.A.; Jacobs, B.L. Both carboxy- and amino-terminal domains of the vaccinia virus interferon resistance gene, E3L, are required for pathogenesis in a mouse model. *J. Virol.* **2001**, *75*, 850–856. [CrossRef] [PubMed]

255. Brandt, T.; Heck, M.C.; Vijaysri, S.; Jentarra, G.M.; Cameron, J.M.; Jacobs, B.L. The N-terminal domain of the vaccinia virus E3L-protein is required for neurovirulence, but not induction of a protective immune response. *Virology* **2005**, *333*, 263–270. [CrossRef] [PubMed]

256. Garcia, M.A.; Guerra, S.; Gil, J.; Jimenez, V.; Esteban, M. Anti-apoptotic and oncogenic properties of the dsRNA-binding protein of vaccinia virus, E3L. *Oncogene* **2002**, *21*, 8379–8387. [CrossRef] [PubMed]

257. Marchal, J.A.; Lopez, G.J.; Peran, M.; Comino, A.; Delgado, J.R.; Garcia-Garcia, J.A.; Conde, V.; Aranda, F.M.; Rivas, C.; Esteban, M.; et al. The impact of PKR activation: From neurodegeneration to cancer. *FASEB J.* **2014**, *28*, 1965–1974. [CrossRef] [PubMed]

258. Liu, R.; Moss, B. Opposing Roles of double-stranded RNA effector pathways and viral defense proteins revealed with CRISPR-Cas9 knockout cell lines and vaccinia virus mutants. *J. Virol.* **2016**, *90*, 7864–7879. [CrossRef] [PubMed]

259. Dueck, K.J.; Hu, Y.S.; Chen, P.; Deschambault, Y.; Lee, J.; Varga, J.; Cao, J. Mutational analysis of vaccinia virus E3 protein: The biological functions do not correlate with its biochemical capacity to bind double-stranded RNA. *J. Virol.* **2015**, *89*, 5382–5394. [CrossRef] [PubMed]

260. Sharp, T.V.; Moonan, F.; Romashko, A.; Joshi, B.; Barber, G.N.; Jagus, R. The vaccinia virus E3L gene product interacts with both the regulatory and the substrate binding regions of PKR: Implications for PKR autoregulation. *Virology* **1998**, *250*, 302–315. [CrossRef] [PubMed]

261. Romano, P.R.; Zhang, F.; Tan, S.L.; Garcia-Barrio, M.T.; Katze, M.G.; Dever, T.E.; Hinnebusch, A.G. Inhibition of double-stranded RNA-dependent protein kinase PKR by vaccinia virus E3: Role of complex formation and the E3 N-terminal domain. *Mol. Cell. Biol.* **1998**, *18*, 7304–7316. [CrossRef] [PubMed]

262. Marq, J.B.; Hausmann, S.; Luban, J.; Kolakofsky, D.; Garcin, D. The double-stranded RNA binding domain of the vaccinia virus E3L protein inhibits both RNA- and DNA-induced activation of interferon β. *J. Biol. Chem.* **2009**, *284*, 25471–25478. [CrossRef] [PubMed]

263. Myskiw, C.; Arsenio, J.; van Bruggen, R.; Deschambault, Y.; Cao, J. Vaccinia virus E3 suppresses expression of diverse cytokines through inhibition of the PKR, NF-κB, and IRF3 pathways. *J. Virol.* **2009**, *83*, 6757–6768. [CrossRef] [PubMed]

264. Valentine, R.; Smith, G.L. Inhibition of the RNA polymerase III-mediated dsDNA-sensing pathway of innate immunity by vaccinia virus protein E3. *J. Gen. Virol.* **2010**, *91*, 2221–2229. [CrossRef] [PubMed]

265. Myskiw, C.; Arsenio, J.; Hammett, C.; van Bruggen, R.; Deschambault, Y.; Beausoleil, N.; Babiuk, S.; Cao, J. Comparative analysis of poxvirus orthologues of the vaccinia virus E3 protein: Modulation of protein kinase R activity, cytokine responses, and virus pathogenicity. *J. Virol.* **2011**, *85*, 12280–12291. [CrossRef] [PubMed]

266. Arndt, W.D.; White, S.D.; Johnson, B.P.; Huynh, T.; Liao, J.; Harrington, H.; Cotsmire, S.; Kibler, K.V.; Langland, J.; Jacobs, B.L. Monkeypox virus induces the synthesis of less dsRNA than vaccinia virus, and is more resistant to the anti-poxvirus drug, IBT, than vaccinia virus. *Virology* **2016**, *497*, 125–135. [CrossRef] [PubMed]

267. Rahman, M.M.; Liu, J.; Chan, W.M.; Rothenburg, S.; McFadden, G. Myxoma virus protein M029 is a dual function immunomodulator that inhibits PKR and also conscripts RHA/DHX9 to promote expanded host tropism and viral replication. *PLoS Pathog.* **2013**, *9*, e1003465. [CrossRef] [PubMed]

268. White, S.D.; Jacobs, B.L. The amino terminus of the vaccinia virus E3 protein is necessary to inhibit the interferon response. *J. Virol.* **2012**, *86*, 5895–5904. [CrossRef] [PubMed]

269. Backes, S.; Sperling, K.M.; Zwilling, J.; Gasteiger, G.; Ludwig, H.; Kremmer, E.; Schwantes, A.; Staib, C.; Sutter, G. Viral host-range factor C7 or K1 is essential for modified vaccinia virus Ankara late gene expression in human and murine cells, irrespective of their capacity to inhibit protein kinase R-mediated phosphorylation of eukaryotic translation initiation factor 2α. *J. Gen. Virol.* **2010**, *91*, 470–482. [PubMed]

270. Liu, S.W.; Katsafanas, G.C.; Liu, R.; Wyatt, L.S.; Moss, B. Poxvirus decapping enzymes enhance virulence by preventing the accumulation of dsRNA and the induction of innate antiviral responses. *Cell Host Microbe* **2015**, *17*, 320–331. [CrossRef] [PubMed]

271. Kalali, B.N.; Kollisch, G.; Mages, J.; Muller, T.; Bauer, S.; Wagner, H.; Ring, J.; Lang, R.; Mempel, M.; Ollert, M. Double-stranded RNA induces an antiviral defense status in epidermal keratinocytes through TLR3-, PKR-, and MDA5/RIG-I-mediated differential signaling. *J. Immunol.* **2008**, *181*, 2694–2704. [CrossRef] [PubMed]

272. Amara, A.; Mercer, J. Viral apoptotic mimicry. *Nat. Rev. Microbiol.* **2015**, *13*, 461–469. [CrossRef] [PubMed]

273. Oie, M. Reversible inactivation and reactivation of vaccinia virus by manipulation of viral lipid composition. *Virology* **1985**, *142*, 299–306. [CrossRef]

274. Sodeik, B.; Doms, R.W.; Ericsson, M.; Hiller, G.; Machamer, C.E.; van't Hof, W.; van Meer, G.; Moss, B.; Griffiths, G. Assembly of vaccinia virus: Role of the intermediate compartment between the endoplasmic reticulum and the Golgi stacks. *J. Cell Biol.* **1993**, *121*, 521–541. [CrossRef] [PubMed]

275. Cluett, E.B.; Machamer, C.E. The envelope of vaccinia virus reveals an unusual phospholipid in Golgi complex membranes. *J. Cell Sci.* **1996**, *109*, 2121–2131. [PubMed]

276. Kay, J.G.; Koivusalo, M.; Ma, X.; Wohland, T.; Grinstein, S. Phosphatidylserine dynamics in cellular membranes. *Mol. Biol. Cell* **2012**, *23*, 2198–2212. [CrossRef] [PubMed]

277. Leventis, P.A.; Grinstein, S. The distribution and function of phosphatidylserine in cellular membranes. *Annu. Rev. Biophys.* **2010**, *39*, 407–427. [CrossRef] [PubMed]

278. Chlanda, P.; Carbajal, M.A.; Cyrklaff, M.; Griffiths, G.; Krijnse-Locker, J. Membrane rupture generates single open membrane sheets during vaccinia virus assembly. *Cell Host Microbe* **2009**, *6*, 81–90. [CrossRef] [PubMed]

279. Maruri-Avidal, L.; Weisberg, A.S.; Moss, B. Direct formation of vaccinia virus membranes from the endoplasmic reticulum in the absence of the newly characterized L2-interacting protein A30.5. *J. Virol.* **2013**, *87*, 12313–12326. [CrossRef] [PubMed]

280. Mercer, J.; Helenius, A. Vaccinia virus uses macropinocytosis and apoptotic mimicry to enter host cells. *Science* **2008**, *320*, 531–535. [CrossRef] [PubMed]

281. Laliberte, J.P.; Moss, B. Appraising the apoptotic mimicry model and the role of phospholipids for poxvirus entry. *Proc. Natl. Acad. Sci. USA* **2009**, *106*, 17517–17521. [CrossRef] [PubMed]

282. Schmidt, F.I.; Bleck, C.K.; Helenius, A.; Mercer, J. Vaccinia extracellular virions enter cells by macropinocytosis and acid-activated membrane rupture. *EMBO J.* **2011**, *30*, 3647–3661. [CrossRef] [PubMed]

283. Schmidt, F.I.; Bleck, C.K.; Mercer, J. Poxvirus host cell entry. *Curr. Opin Virol.* **2012**, *2*, 20–27. [CrossRef] [PubMed]

284. Huang, C.Y.; Lu, T.Y.; Bair, C.H.; Chang, Y.S.; Jwo, J.K.; Chang, W. A novel cellular protein, VPEF, facilitates vaccinia virus penetration into HeLa cells through fluid phase endocytosis. *J. Virol.* **2008**, *82*, 7988–7999. [CrossRef] [PubMed]

285. Sandgren, K.J.; Wilkinson, J.; Miranda-Saksena, M.; McInerney, G.M.; Byth-Wilson, K.; Robinson, P.J.; Cunningham, A.L. A differential role for macropinocytosis in mediating entry of the two forms of vaccinia virus into dendritic cells. *PLoS Pathog.* **2010**, *6*, e1000866. [CrossRef] [PubMed]

286. Morizono, K.; Xie, Y.; Olafsen, T.; Lee, B.; Dasgupta, A.; Wu, A.M.; Chen, I.S. The soluble serum protein Gas6 bridges virion envelope phosphatidylserine to the TAM receptor tyrosine kinase Axl to mediate viral entry. *Cell Host Microbe* **2011**, *9*, 286–298. [CrossRef] [PubMed]

287. Rothlin, C.V.; Ghosh, S.; Zuniga, E.I.; Oldstone, M.B.; Lemke, G. TAM receptors are pleiotropic inhibitors of the innate immune response. *Cell* **2007**, *131*, 1124–1136. [CrossRef] [PubMed]

288. Bhattacharyya, S.; Zagorska, A.; Lew, E.D.; Shrestha, B.; Rothlin, C.V.; Naughton, J.; Diamond, M.S.; Lemke, G.; Young, J.A. Enveloped viruses disable innate immune responses in dendritic cells by direct activation of TAM receptors. *Cell Host Microbe* **2013**, *14*, 136–147. [CrossRef] [PubMed]

289. Fadok, V.A.; Bratton, D.L.; Konowal, A.; Freed, P.W.; Westcott, J.Y.; Henson, P.M. Macrophages that have ingested apoptotic cells in vitro inhibit proinflammatory cytokine production through autocrine/paracrine mechanisms involving TGF-β, PGE2, and PAF. *J. Clin. Investig.* **1998**, *101*, 890–898. [CrossRef] [PubMed]

290. Voll, R.E.; Herrmann, M.; Roth, E.A.; Stach, C.; Kalden, J.R.; Girkontaite, I. Immunosuppressive effects of apoptotic cells. *Nature* **1997**, *390*, 350–351. [CrossRef] [PubMed]

291. Liu, L.; Xu, Z.; Fuhlbrigge, R.C.; Pena-Cruz, V.; Lieberman, J.; Kupper, T.S. Vaccinia virus induces strong immunoregulatory cytokine production in healthy human epidermal keratinocytes: A novel strategy for immune evasion. *J. Virol.* **2005**, *79*, 7363–7370. [CrossRef] [PubMed]

292. Hayasaka, D.; Ennis, F.A.; Terajima, M. Pathogeneses of respiratory infections with virulent and attenuated vaccinia viruses. *Virol. J.* **2007**, *4*, 22. [CrossRef] [PubMed]

Review

EBV and Apoptosis: The Viral Master Regulator of Cell Fate?

Leah Fitzsimmons [1] (iD) **and Gemma L. Kelly** [2,3,*]

[1] Institute of Cancer and Genomic Sciences and Centre for Human Virology, College of Medical and Dental Sciences, University of Birmingham, Edgbaston, Birmingham B15 2TT, UK; l.fitzsimmons@bham.ac.uk

[2] Molecular Genetics of Cancer Division, The Walter and Eliza Hall Institute for Medical Research, Parkville, Melbourne, VIC 3052, Australia

[3] Department of Medical Biology, The University of Melbourne, Parkville, Melbourne, VIC 3052, Australia

* Correspondence: gkelly@wehi.edu.au; Tel.: +61-3-9345-2497

Received: 10 October 2017; Accepted: 9 November 2017; Published: 13 November 2017

Abstract: Epstein–Barr virus (EBV) was first discovered in cells from a patient with Burkitt lymphoma (BL), and is now known to be a contributory factor in 1–2% of all cancers, for which there are as yet, no EBV-targeted therapies available. Like other herpesviruses, EBV adopts a persistent latent infection in vivo and only rarely reactivates into replicative lytic cycle. Although latency is associated with restricted patterns of gene expression, genes are never expressed in isolation; always in groups. Here, we discuss (1) the ways in which the latent genes of EBV are known to modulate cell death, (2) how these mechanisms relate to growth transformation and lymphomagenesis, and (3) how EBV genes cooperate to coordinately regulate key cell death pathways in BL and lymphoblastoid cell lines (LCLs). Since manipulation of the cell death machinery is critical in EBV pathogenesis, understanding the mechanisms that underpin EBV regulation of apoptosis therefore provides opportunities for novel therapeutic interventions.

Keywords: EBV; apoptosis; genetic cooperation; latency; virus cancers; p53; BCL-2 family; growth transformation

---

1. Introduction

All viruses possess methods to negotiate and subvert the cell death pathways of their hosts, but for viruses such as influenza that lead to acute infections, the battle for host cell survival is ultimately lost when the host clears the virus. Persistent viruses however, such as those of the herpesvirus family that are carried by the host for life, must avoid elimination by immune cells, whilst continuing to disseminate within the host and avoid being lost through normal cell turnover. Human herpesvirus 4, also known as Epstein–Barr virus (EBV), is extremely efficient at establishing a persistent life-long infection in human B cells. Most primary infections occur asymptomatically in early childhood and by adulthood the majority (95% world-wide) of the adult population are infected with EBV [1]. Somewhat paradoxically, given the ubiquitous and asymptomatic nature of infection, EBV is the archetypal human tumour virus. EBV was first associated with an unusual and aggressive form of childhood lymphoma in 1964 [2,3], and is now known to contribute to over 200,000 new cancer diagnoses each year [4]. These unusual characteristics are, in part, due to the many and various ways EBV has evolved to exhibit exquisite control over cell death. Here, we review the mechanisms by which EBV genes co-operate to regulate cell death and how this may contribute to lymphomagenesis by focussing on evidence from the study of in vitro transformation of B cells and Burkitt lymphoma, the cancer in which EBV was discovered.

The existence of EBV was first suspected by Denis Parsons Burkitt, a surgeon working in post-war Uganda. In 1958, he published the first detailed clinical study of a strange lymphoma affecting the jaws and abdomens of children across sub-Saharan Africa [5]. He was struck by both the prevalence and the poor prognosis of the disease: the tumours accounted for more cases of cancer in childhood than all other malignancies combined and furthermore, were often rapidly fatal [6]. Burkitt lymphoma (BL), as it came to be known, also had an unusual geographical distribution which overlapped almost perfectly with that of arthropod borne infectious diseases such as Yellow Fever and Rift Valley Fever [7–10]. A chance meeting with the then animal virologist, Anthony Epstein, led to the search for a possible viral cause of BL and in 1964 the first micrographs showing the unmistakable icosahedral structures of a previously undescribed herpesvirus were published [2]. It did however take several more years until Epstein–Barr virus (named after Epstein and his PhD student, Yvonne Barr) was convincingly shown to be an aetiological agent in BL and other cancers (reviewed in detail elsewhere [11,12]).

2. 'Transforming' Cell Death: Latent Genes

The oncogenic potential of EBV was first demonstrated experimentally in 1967, when it was observed that co-culturing lethally-irradiated, EBV-producing BL cells with primary human lymphocytes led to the outgrowth of permanently proliferating lymphoblastoid cell lines (LCLs) [13,14]. This process of growth transformation, by which EBV is able to immortalise B lymphocytes that would otherwise senesce and die in vitro, provided an important model for the study of EBV infection and has therefore been extensively researched. Clearly, the activation of resting B cells into cell cycle is imperative for growth transformation, but inhibition of cell death is equally essential. Although the virus encodes around 100 open reading frames (ORFs), the vast majority of EBV genes have replicative, immune suppressive or structural functions and as such are only expressed during the viral lytic cycle [1]. EBV establishes a largely latent infection in infected cells (the virus is lytic in around 1% of cells at any given time), where it is maintained as an episome by virtue of its ability to 'piggy back' onto the host replication machinery during mitosis [15,16]. During these latent infections, only a small subset of EBV-encoded gene products is expressed [17,18]. In established LCLs (immortalised B cells) EBV displays a Latency III pattern of infection. Latency III is the most extensive form of latent infection, involving the expression of ten EBV-encoded proteins and a variety of non-coding RNAs. These are the Epstein–Barr Nuclear Antigens (EBNA-1, EBNA-2, EBNAs-3A, -3B, -3C and EBNA-LP), Latent Membrane Proteins (LMP1, LMP2A and LMP2B), and the viral BCL-2 homologue, BHRF1; as well as two non-coding RNAs (EBER1 and EBER2), and two families of microRNAs encoded within the *BamHI* A rightward transcripts (BARTs) and the BHRF1 locus (BHRF1 miRNAs), respectively (Figure 1) [18–24]. These EBV latent gene products are expressed at different time points post-infection of B cells, finally leading to growth transformation.

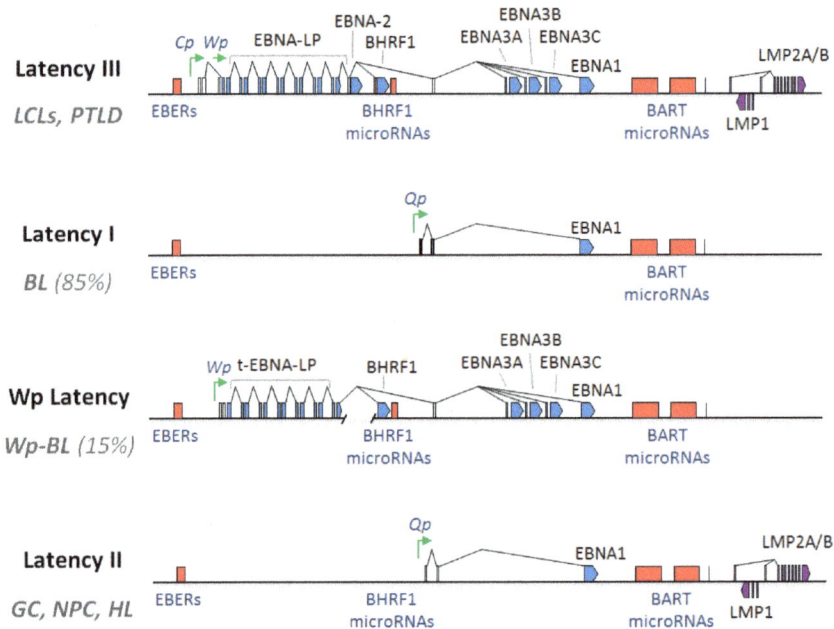

Figure 1. Patterns of latent gene expression found in Epstein–Barr virus (EBV)-associated malignancies and growth transformed B cell lines. Schematic showing: the Latency III EBV gene expression programme, as found in B cells transformed in vitro into lymphoblastoid cell lines (LCLs); Latency I EBV gene expression as found in the majority (85%) of EBV-positive Burkitt lymphomas (BL); Wp-restricted latency (Wp Latency), as found in a minority (15%) of EBV-positive BLs (termed Wp-BL); and Latency II EBV gene expression, which is found in EBV-positive Hodgkin lymphoma (HL) as well as the EBV-associated epithelial malignancies, nasopharyngeal carcinoma (NPC) and gastric carcinoma (GC). Latent proteins (EBNA1, EBNA2, EBNA3A, EBNA3B, EBNA3C, EBNA-LP, BHRF1, LMP1 and LMP2A/B) are shown in blue. Non-coding RNAs (EBERs, miR-BHRF1s and miR-BARTs) are shown in red, and selected latent promoters (Cp, Wp and Qp) are shown in green. Connecting lines denote splicing patterns, whilst blocks indicate exons. In Wp-BL, EBNA-LP is truncated due to a genomic deletion and is therefore denoted as t-EBNA-LP.

2.1. Dynamics of Early Infection

Upon infection of resting B cells, EBV gene expression, driven by host cell RNA polymerase II, begins almost immediately; the Wp promoter that drives early latent gene expression reaches maximal activity around 8–12 h post-infection (PI). These long and differentially spliced Wp-transcripts preferentially encode EBNA-LP, EBNA-2 and BHRF1 [25,26]. The nuclear antigens (EBNAs-LP and -2) then transactivate the Cp and LMP promoters [27–29], leading to the expression of EBNA1, EBNA3A, -3B and -3C and LMP1, 2A and 2B, respectively, which reach peak expression at 2–3 days PI [25,30]. Importantly however, there is a delay between maximal expression of latent transcripts and the proteins they encode. The EBNA2, EBNA-LP and BHRF1 proteins reach levels comparable to those in established LCLs at around 72 h [25,31], whereas LMP1 protein is low or undetectable until 5 days PI. [19,32]. Expression of EBV non-coding RNAs is similarly delayed: they are not detected at appreciable levels until several days after infection (Figure 2). Many of these EBV genes are reported to have roles in cell proliferation and/or survival.

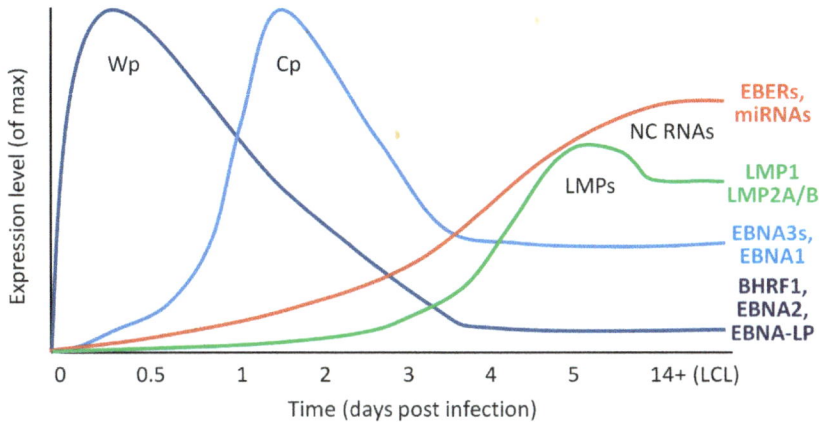

Figure 2. Temporal patterns of latent gene expression during growth transformation of primary resting B cells. Schematic showing the general transcription patterns of different classes of latent EBV genes during in vitro growth transformation of primary, resting B cells. Wp-derived transcripts preferentially give rise to BHRF1, EBNA2 and EBNA-LP in order to kick start cells into cycle, though they also encode EBNA-3A, -3B and -3C (EBNA3s) and EBNA1. Cp can encode all EBNAs and BHRF1. NC RNAs include EBER1, EBER2, miR-BARTs and miR-BHRF1s. Data are cumulative estimations based on transcriptional data published by Tierney et al. [33], Shannon-Lowe et al. [30], and Amoroso et al. [24].

2.2. EBNA-2 and EBNA-LP

EBNA-2 and EBNA-LP are the first proteins to be expressed following infection of B cells. EBNA2 is a functional mimic of cellular Notch [34–36] and is responsible for kick-starting cell cycle activation through its RBP-Jκ-mediated pleiotropic effects on chromatin organisation and gene regulation [37–40]. Therefore, it is not surprising that EBNA2 expression is essential for B cell transformation [41]. EBNA2 can also inhibit intrinsic cell death through interactions with, and upregulation of, cellular proteins. EBNA-2 can directly bind and inhibit the orphan nuclear receptor Nur77 [42,43] which is reported to bind and modulate the function of several pro-survival BCL-2 family members [44]. Additionally, EBNA-2 expression was shown to upregulate the pro-survival BCL-2 family protein, BFL-1/A1, at the mRNA level via binding to RBP-Jκ/CBF1 [45] and co-ordinately downregulate the BCL-2 family death inducer, BIK [46]. More recently, EBNA-2 has also been shown to contribute to the activation of *MYC*, that can both increase proliferation and sensitise cells to apoptosis, through long-range interactions [47]. EBNA-LP, the transcriptional coactivator of EBNA2 [48], is essential for efficient B cell transformation [49–51], but has so far had few survival functions attributed to it in the context of LCLs. Interestingly, however, it has been reported that EBNA-LP (also called EBNA-5), can bind Fte-1/S3a, which is able to contribute to cell survival by interacting with PARP [52]. Another study found that EBNA-LP could interact with p14^{ARF} in a yeast 2-hybrid system and colocalised with p14^{ARF} and p53 transcripts in LCLs [53]. EBNA-LP has also been shown to interact with BCL-2 in the presence of HAX-1 in pull down experiments using glutathione S-transferase fusion proteins in the primate kidney cell line, COS-7 [54]. Therefore, the possible survival functions of EBNA-LP during transformation warrant further investigation.

2.3. EBNA-3A, -3B and -3C

The EBNA-3s are a family of three large proteins (3A, 3B, 3C), which likely arose by gene duplication, that predominantly function as regulators of host cell and virus transcription. Like EBNA-2, the EBNA-3 proteins do not bind DNA directly, but instead transactivate or repress gene expression via interactions with transcription factors, such as RBP-Jκ, for which all four EBNAs compete (reviewed in [55]). Although they share less than 30% amino acid identity, EBNA-3A, -3B and -3C exhibit structural similarity [56,57] and display some overlap in terms of some of the loci and processes that they regulate. Interestingly, only EBNA-3C is essential for B cell transformation; although B cells infected with viruses lacking EBNA-3A display a growth impairment and readily undergo apoptosis [58–60]. In contrast, EBNA-3B is dispensable for B cell transformation and LCLs generated with an EBNA-3B knock-out (KO) virus exhibit comparable resistance to apoptosis as those transformed with wild-type (wt) EBV [61,62]. Extensive analyses of cells infected with EBNA-3 KO or conditional viruses that encode estrogen-inducible EBNA-3 proteins revealed that EBNA-3A and -3C are able to co-operatively downregulate the apoptosis inducing, BH3-only protein BIM [63–65] as well as the tumour suppressors p16^{INK4a} and p14^{ARF} [58,59,66–69] through epigenetic silencing. EBNA-3C can also reportedly interact with p53 as well as binding and stabilising the p53 regulators, ING4, ING5, MDM2 and Gemin3 [70–72]. Although the downregulation of BIM and p14^{ARF} by EBNA-3A and -3C occurs through epigenetic silencing at their transcriptional start sites (TSS) [59,63,64,66,68], the EBNA3 proteins regulate many other genes by long range interactions up to 50 kb away from a TSS [69,73]. It is estimated that collectively, the EBNA-3 proteins can bind more than 7000 sites on the cellular genome; therefore, it is likely that other cell survival genes regulated by the EBNA3s will be identified in the future.

2.4. EBNA-1

The primary function of EBNA-1 protein is to tether EBV episomes onto host cell chromatin to ensure the viral DNA is replicated during cell mitosis. Therefore, EBNA1 expression is essential for the maintenance of latent infection [74–76]. There is evidence that EBNA-1 may also affect the survival of EBV-infected cells, but interestingly the literature in this regard is conflicting, with reports that EBNA1 is both pro- and anti-apoptotic in function [77]. EBNA-1 has been shown to cause genomic instability by triggering reactive oxygen species production and the DNA damage response (DDR) [78,79]. Conversely however, EBNA1 may stabilise p53 to counteract DDR by binding to the p53 regulator, USP7 [80] and further block downstream caspase activation by upregulation of Survivin [81]. EBNA-1 reportedly binds and regulates the promoters of many other cellular genes but the functional consequences and implications of these interactions for cell survival are not yet fully elucidated [82–85].

2.5. LMP1

LMP1 promotes cell growth and survival and is essential for B cell transformation [86,87]. Functionally, it is a CD40 homologue [88,89] that constitutively mimics cellular TNF signalling through its two cytoplasmic signalling domains, CTAR1 and CTAR2 (sometimes referred to as TES1 and 2) [90–92]. CTAR1 binds TRAF and consequently stimulates the non-canonical NF-κB cascade via p100/RelB [93,94], whilst CTAR2 activates both the canonical pathway and JNK signalling through binding to TRADD, IRF7 and BS69 and downstream activation of IKKβ [95–97]. Consequently, LMP1 upregulates NF-κB responsive BCL-2 pro-survival proteins, including BCL-2, MCL-1 and BFL-1/A1 [98–102] and the cyto-protective, JNK-regulated chemokines, CCL3 and CCL4 [103]. Additionally, a recent genome-wide CRISPR/Cas9 loss-of-function screen in BLs and LCLs in vitro identified another previously identified NF-κB/LMP1 transcriptional target, cFLIP [104], which can suppress the extrinsic apoptotic and necroptosis pathways, as critically important for the survival of LCLs [105]. Conversely, LMP1 has also been shown to be capable of inducing Fas-mediated apoptosis in B cells with coincident cleavage of Caspase 8 and BID [106,107]. Interestingly, this pro-death activity, which is independent of the CTARs and instead maps to the transmembrane region is masked when LMP1 is expressed at moderate levels and is only evident where LMP1 is over-expressed [98].

2.6. LMP2A and LMP2B

LMP2A and 2B are expressed from a series of complex, overlapping transcripts, each driven by its own unique promoter that, like the LMP1p, are transactivated by EBNA-2 during transformation [108–110].

LMP2A, like LMP1, is a constitutive mimic of cellular receptor signalling. Whereas LMP1 is a CD40 homologue, LMP2A emulates B cell receptor (BCR) cross-linking, therefore during infection of B cells, the LMPs are able to recapitulate the signals essential for normal B cell development and function [1]. In vitro, LMP2A is dispensable for the initial transformation of B cells [111–114], but has been shown to provide critical ongoing cell survival signals, including inhibition of TGF-β-associated apoptosis through activation of PI3K/Akt/mTOR [105,115,116]. Furthermore, LMP2A can rescue "crippled" B cells; that is germinal centre B cells with Ig mutations that consequently lack a functional BCR, and so would otherwise rapidly undergo programmed cell death [117,118]. LMP2A function is dependent on recruitment of and interaction with Src family tyrosine kinases and LMP2A must be phosphorylated at residues Y74 and Y85 by Syk in order to activate pro-survival PI3K/Akt/mTOR signalling [119–121]. LMP2B shares most of its exons with LMP2A, but lacks the N-terminal Syk domain [110,122]. Much less is known about the function of LMP2B compared to LMP2A, but it has been shown to functionally oppose LMP2A and instead may potentiate LMP1 signalling in order to fine tune signalling in infected B cells [123].

2.7. BHRF1 and BALF1

A number of viruses, including EBV, encode homologues of cellular pro-survival BCL-2 family proteins in order to evade apoptosis triggered as part of the anti-viral response. These viral BCL-2 proteins (vBCL-2s) vary in the degree of similarity they display to their cellular counterparts in terms of sequence, structure and function (reviewed in [124]). Although most vBCL-2 proteins are regarded as regulators of the intrinsic apoptosis pathway, some also regulate additional cellular processes [125–128].

EBV encodes two BCL-2 homologues, called BHRF1 and BALF1. These viral proteins have long been known to be expressed at high levels during lytic cycle, presumably to keep host cells alive to ensure efficient virus replication. They have more recently been shown to also be expressed during the early stages of primary B cell infection in vitro: BHRF1 and BALF1 cDNAs can be detected by day 1 PI [33,129]. Recombinant viruses lacking either BHRF1 or BALF1 are able to transform resting B cells in vitro, but are slightly impaired. However, the loss of both BHRF1 and BALF1 completely abrogates the ability of the virus to transform B cells, suggesting redundancy in their function in this regard [129]. Whilst BHRF1 protein has been shown to also be constitutively expressed in established latent LCLs [19], no antibody to BALF1 exists and therefore the expression of BALF1 protein in LCLs has not been examined. Therefore, although BALF1 transcripts can be detected at low levels in LCLs [33], it is not known whether these arise from a small number of lytically infected cells or, like BHRF1, are expressed during latency.

Both BHRF1 and BALF1 contain all four of the functionally important BCL-2 homology domains, BH1-4 [130,131]. BHRF1 is relatively well-characterised: its BH3 binding groove exhibits 25% amino acid sequence similarity to cellular BCL-2 [132] and structurally most closely resembles BCL-x_L. Like cellular BCL-2 pro-survival proteins, BHRF1 protein has a hydrophobic groove formed by α-helices 2–5 of its 7-helix bundle [133], into which BH3 ligands can bind. BHRF1 is potently anti-apoptotic in vitro and can confer protection to different cell types from a wide variety of death-inducing stimuli [19,131,133–142]. Furthermore, ectopic expression of BHRF1 from a retrovirus vector in *Eμ-Myc* mouse lymphoma cells in vivo can confer protection from DNA damaging chemotherapeutics [133].

The mechanism by which BHRF1 confers apoptosis protection is through binding and sequestration of cellular pro-apoptotic proteins. BHRF1 can bind to the cellular BH3-only proteins BIM, PUMA and BID and the apoptosis effector (or executioner) protein, BAK [124] (see Figure 3). One report attributes all apoptosis protection to the ability of BHRF1 to bind BIM [134,135], however work from our laboratory has shown that BHRF1 can also protect in the absence of BIM (Fitzsimmons et al. manuscript in preparation). By comparison to BHRF1, BALF1 is much less well characterised. BALF1 was identified as a BCL-2 homologue by predicted structural similarity to known vBCL-2s, and though it has been shown to modulate apoptosis, it remains controversial as to whether it promotes or inhibits cell death [130,143].

a

	Cellular Pro-survival				vBCL2
	BCL-2	BCL-x$_L$	MCL-1	A1	BHRF1
BIM	2.6	4.6	2.4	1	18
PUMA	3.3	6.3	5	1	70
BID	2000+	82	2000+	1	109
BAD	16	5.3	2000+	2000+	2000+
NOXA	2000+	2000+	24	20	2000+
BAK	2000+	50	10	3	150
BAX	100	130	12	n/a	1400

(Pro-death, rows BIM–BAX)

b

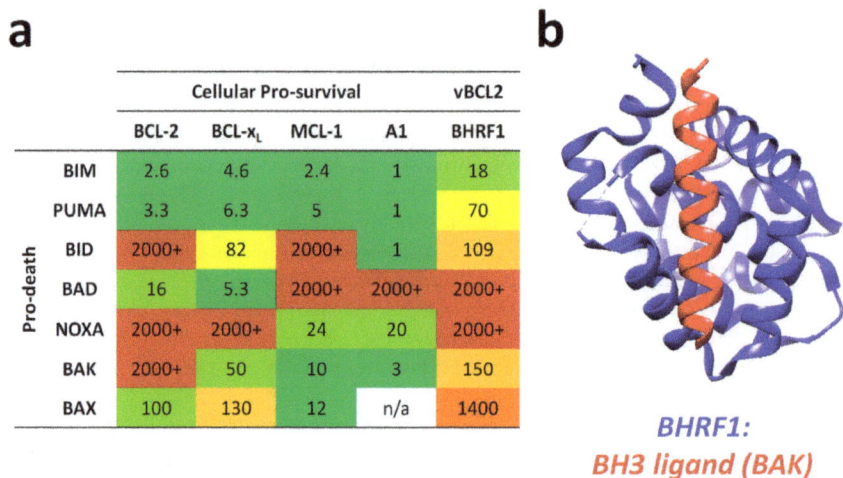

BHRF1:
BH3 ligand (BAK)

Figure 3. Binding specificities and affinities of BCL-2 family members and EBV BHRF1. (**a**) Interactions between pro-survival and pro-death BCL-2 family members. Reported as K$_d$ in nM as determined by surface plasmon resonance (BCL-2, BCL-x$_L$ and MCL-1) or isothermal calorimetry. Sources were: BCL-2 and BCL-x$_L$ [144–146], MCL-1 [144], A1 [147] and BHRF1 [124,133]. Colour coding was applied as follows: green 1–10 nM, pale green 11–50 nM, yellow 51–100 nM, pale orange 101–1000 nM, orange 1001–2000 nM, red 2000–100,000 nM. (**b**) Ribbon structure representation of BHRF1 (blue) bound to the BH3 domain of BAK (red). This graphic was prepared using the UCSF Chimera software package (developed by the Resource for Biocomputing, Visualization, and Informatics at the University of California, San Francisco [148], supported by NIGMS P41-GM103311) using pdb accession code 2XPX [133].

2.8. Non-Coding RNAs

A number of viral non-coding RNAs are expressed during transformation. Some of these transcripts are reported to regulate cell death during transformation and latency; however, their functions are less well understood than those of the latent proteins. The Epstein–Barr virus encoded RNAs (EBERs) are two highly transcribed small (~170 nt) nuclear RNAs (EBER1 and EBER2) [23,149] whose function(s) remain somewhat enigmatic. The EBERs are abundantly expressed in established LCLs but they are not detectable until around 72 h after infection of primary B cells [150]. EBV also encodes around 40 mature microRNAs (miRs/miRNAs), derived from 25 precursor (pre-miRNA) transcripts [151–154]. MicroRNAs selectively bind to and inhibit mRNAs by causing transcript instability, degradation, or impaired translation. Similar to cellular miRNAs, EBV encoded pre-miRNAs are processed by Drosher and Dicer into mature miRNAs and incorporated into the RISC complex, which then binds the target transcript to which the miRNA confers specificity by virtue of its 6 nt 'seed sequence' (reviewed in [155,156]). The EBV miRNAs can be divided into two families, according to their position within the viral genome. The BHRF1 miRNAs reside either side of the BHRF1 ORF, whereas the BART family of miRNAs are derived from the *Bam*HI A rightward transcripts (BARTs). Despite abundant BHRF1 and BART primary transcript expression at early time points during transformation, processing of their derivative miRNAs is delayed, peaking at around 72–120 h PI [24].

2.8.1. EBERs

The EBERs are ubiquitously expressed from their own viral promoters in all forms of latent infection. There have been conflicting reports regarding the ability of EBER-KO recombinant viruses to transform B cells, although it should be noted that no previous studies have found the EBERs to be essential for transformation. Interestingly, different groups have used different strains of EBV in their experiments, suggesting that the role of the EBERs in transformation may be context dependent and influenced by the other co-expressed viral proteins. Swaminathan et al. used the transformation incompetent P3HR1 virus strain, which harbours a large genomic deletion spanning the EBNA2 locus, and recombined it with a large fragment of the prototypic B95.8 EBV that was either wt for or lacked the EBER locus (lacking both EBER1 and EBER2). The resulting recombinant wt or EBER-KO EBVs were both transformation competent [157] and were found to give rise to LCLs that grew at the same rate as wt EBV-derived LCLs [158]. However, an alternative EBER-KO virus made in the Akata virus strain was found to be less than 50% as efficient at transforming B cells than wt virus and the resulting LCLs exhibited a growth impairment in limiting dilution assays compared to wt comparators [159]. When expression of the individual EBERs was restored in this background, it was found that an EBER2 knock-in (KI) virus behaved like wt Akata virus, but the EBER1-KI (lacking EBER2) virus was as impaired as the double EBER-KO, suggesting that only EBER2 is important for transformation and that EBERs play distinct, non-overlapping roles in this process [160]. Recently however, it has been shown that EBER-KO viruses (both double and single KOs) made in the B95.8 background could transform B cells with equal efficiency to wt virus [161]. It may be relevant that this study used adult PBMCs as a source of B cells, whereas the Akata-derived virus studies were carried out on cord blood-derived cells. Interestingly, the B95.8 study identified two apoptosis-related genes as EBER targets by microarray analyses. Deletion of EBER1 led to a 6-fold downregulation of GAS2, whilst EBER2-KO cells showed around a 40-fold reduction in SASH1 [161]. Both GAS2 and SASH1 are reported to be downstream targets of Caspase 3 and have been shown to contribute to apoptosis induction [162,163]. Somewhat counterintuitively, this suggests that EBERs may prime LCLs for cell death, although further mechanistic studies have yet to be published.

2.8.2. BHRF1 microRNAs

The BHRF1 family of miRNAs consists of one miRNA encoded upstream (miR-BHRF1-1) and two miRNAs encoded downstream (miR-BHRF1-2 and miR-BHRF1-3) of the BHRF1 ORF. During viral transformation of B cells, the BHRF1 miRNAs are processed from the polycistronic Wp and Cp-driven latent transcripts that encode the EBNA proteins and latent BHRF1 protein [164–166]. EBV genomes deleted for individual miR-BHRF1s show a mild impairment in transformation compared to wt controls and the contribution of these miRNAs was found to be cumulative. Viruses lacking miR-BHRF1-2 (Δ2) and -3 (Δ3) were the most transformation deficient, whereas viruses deleted for miR-BHRF1-1 (Δ1) transformed B cells almost as efficiently as wt EBV [167]. Accordingly, when all three BHRF1 miRNAs were deleted, the resulting Δ123 viruses showed a marked transformation defect, which was attributed to increased apoptotic cell death early after infection. Additionally, the resultant LCLs exhibited reduced S-phase entry and proliferated less well than controls [168,169]. In humanised NSG mice, animals infected with the Δ123 virus displayed lower viral loads at early time points after infection compared to mice infected with wt virus, but there was no difference in tumorigenic potential between Δ123 and wt EBV [170]. Interestingly, it was shown that ΔmiR-BHRF1 LCLs exhibit lower expression levels of BHRF1 transcripts and protein as well as other latent transcripts [31,168], indicating that at least some of the effect of miR-BHRF1 deletions were indirect. Indeed, whilst EBV lacking BHRF1 protein transformed around 50% as efficiently as wt controls, viruses lacking both BHRF1 protein and miRNAs were incapable of transforming B cells [31,129]. These results suggest that the BHRF1 miRNAs and BHRF1 protein function cooperatively to control cell cycle entry and apoptosis during primary infection.

Other than regulating the expression of the pro-survival, BHRF1 protein, it is also somewhat unclear how the BHRF1 miRNAs might further regulate apoptosis. This is, in part, because it can be difficult to experimentally confirm bona fide miRNA targets and because viral miRNAs sometimes display unique functional features compared to their cellular counterparts, such as tolerating 'bulges' in their mRNA seed region binding sites [165,171] (and reviewed in [172]). What is known is that EBV miRNAs bind and may contribute to the regulation of hundreds of viral and cellular transcripts, many of which have been ascribed roles in cell survival. As well as being primarily responsible for the regulation of BHRF1 protein, miR-BHRF1-2 has also been found by luciferase reporter assays to inhibit expression of the apoptosis-related genes, BACH1 and KDM4B [165], whilst miR-BHRF1-3 has been found to downregulate PTEN protein, thereby reducing cell cycle progression and, through the PI3K/Akt pathway, apoptosis [31]. Interestingly, although PTEN and BACH1 are both pro-apoptotic [173,174], KDM4B (also called JMJD2B) is a gene induced by p53 that can repress the induction of select p53 targets including p21, PIG3 and PUMA and thereby suppress cell cycle arrest and cell death [175]. This suggests that BHRF1 miRNAs might function to modulate the cell death response.

2.8.3. BART microRNAs

The *BamHI* A region, which is transcriptionally active in several EBV-associated malignancies, has long been known to give rise to a complex variety of transcripts driven from two promoters (P1 and P2) [176–179]. It took until 2004 to discover that introns from these transcripts give rise to microRNAs [20]. It had previously been proposed that these transcripts encoded proteins [180–182], but whilst they could be transcribed and investigated in vitro, there was little evidence for the existence of *BamHI* A-derived proteins in EBV-positive cell lines [183,184]. The BART miRNAs are reported to regulate a number of cell survival and growth-related transcripts. However, the majority of studies linking BART miRNAs to survival were carried out in nasopharyngeal carcinoma (NPC) where they are highly expressed. Although the BART miRNAs are expressed at lower levels in LCLs compared to epithelial cells, they still number thousands of copies per cell in total, and are consistently detected in all types of latency [166]. Interestingly, the expression of individual BART miRNAs exhibits hierarchy, and the less abundant miRNAs are only present at tens of copies per cell in LCLs [24,165].

The prototypic EBV strain, B95.8, from which many of the widely used recombinant EBVs are derived, harbours a large (12 kbp) deletion compared to other EBV isolates [185,186], and consequently lacks most of the BART microRNAs. The readiness of the B95.8 strain and its recombinant derivatives to transform resting B cells suggests that the BART miRNAs are dispensable for transformation. Accordingly, Seto et al. found that a recombinant EBV lacking both BART and BHRF1 miRNAs was equally as efficient in transformation assays as a virus deficient in only BHRF1 miRNAs [169]. However, another study found that two B95.8 derivative viruses that had been 'repaired' for the BART miRNA sequences using different strategies both caused more efficient outgrowth of B cells PI. Interestingly, there was greater variation between the two different miR-BART repaired recombinant virus strains than between the B95.8 strain and the less efficient repaired strain [187].

In order to identify targets of EBV miRNAs, Dolken et al. carried out immunoprecipitation of RISC complexes (containing miRNAs and their targets) in two EBV-infected BL cell lines and then identified the target mRNA using microarray analyses. They showed that miR-BART3 and miR-BART16 downregulated IPO7 and TOM22 transcripts, respectively [188], both of which encode transport proteins and are reported to have pro-apoptotic functions [189,190]. Another group used a similar technique to investigate the cellular targets of EBV-encoded miRNA in Latency III BL cells (expressing all of the miR-BARTs), but first crosslinked the RISC complexes and RNA species together by UV-irradiation [191]. In this study, Riley et al. identified 132 apoptosis-associated genes as miR-BART targets, notably: the previously reported PTEN, IPO7 and TOM22, in addition to the EBV oncogenes, LMP1 and BHRF1; multiple pro-death BCL-2 family members, PUMA, BAK, BID, PMAIP1 (NOXA), BIM, and pro-survival MCL1; as well as CASP2, CASP3 and MYC. Only a small subset of these targets was validated in luciferase reporter assays [192]. Finally, a third group used an alternative method to crosslink miRNAs and their targets together in B95.8-derived LCLs (expressing only five BART miRNAs), followed by RNA sequencing to identify the targets. They also reported a significant enrichment among their hits for genes that can directly or indirectly influence cell survival including MCL1; the apoptosome component, APAF1; and the miRNA processing factor, DICER [165].

The use of RNA sequencing technologies has also led to the identification of a large number of novel EBV transcripts in latently infected cells [193], some of which are likely to turn out to be bona fide, functional ncRNAs [194]. Therefore, it is possible that the repertoire of EBV-encoded ncRNAs that regulate cell survival may expand in the near future.

3. Lytic Cycle Genes and Transformation

Although EBV is usually found to be latent in infected cells both in vivo and in vitro, the virus must periodically enter into the lytic cycle to generate infectious viral progeny to infect new cells within the host and thereby guarantee persistence and dissemination into new hosts. In vivo, the lytic cycle is thought to occur in terminally differentiated plasma cells and epithelial cells (reviewed in [195]). In vitro, replication occurs sporadically in a small population of cells in latent LCLs and tumour cell lines. Although the frequency and rate of lytic cycle activation can vary between cell lines and types, usually around 1–5% of cells are found to express the lytic cycle marker gene, BZLF1, in LCL cultures at any given time [196]. The lytic cycle is initiated by expression of BZLF1 and BRLF1, the 'immediate early' lytic genes, and then proceeds in two further phases of virus gene expression. BZLF1 binds and activates promoters containing Z-response elements (ZREs) [197,198], and seems to preferentially bind ZREs that are highly methylated [199,200]. BRLF1 can enhance transcription directly, by binding DNA at GC-rich promoter sequences to activate transcription [201], or indirectly, where DNA-binding is not necessary [202,203]. These transcription factors stimulate the expression of a second wave of lytic-associated viral genes, known as the early lytic genes. Early genes include those required for viral DNA synthesis, inhibition of apoptosis and immune evasion. Once the viral DNA has replicated, the late lytic genes are expressed, many of which encode structural or packaging elements of the virus.

The lytic cycle is also closely linked with cell death since the thousands of EBV particles that are produced eventually cause any cell undergoing lytic replication to lyse, releasing infectious virus [204,205]. Conversely, EBV must also ensure that cell death is not triggered too early after lytic cycle activation to allow time for progeny to be synthesised and packaged efficiently. It is unsurprising then, that EBV lytic genes have been ascribed both pro- and anti-apoptotic functions. For example, BZLF1 indirectly induces cell death via inhibition of the NF-κB family protein, p65 [206] and downregulation of CD74 [207]. Conversely, the viral BCL-2 protein, BHRF1, has been shown to directly block BZLF1 toxicity, which has the dual consequences of blocking apoptosis and contributing to the evasion of NK cell recognition and killing of BZLF1-expressing EBV-positive B cells [208]. Another group found that induction of EBV-positive cell lines into the lytic cycle also induced cell death in cell cultures. Interestingly however, the lytic BZLF1-positive cells were resistant to death compared to the cells that remained latent, further suggesting that EBV expresses potent anti-apoptotic genes upon lytic reactivation. Furthermore, cells that were treated with acyclovir (ACV) to specifically block late lytic cycle EBV gene expression exhibited an increased propensity to undergo apoptosis, suggesting that both early and late lytic genes contribute to cell survival during virus replication [209]. The BH3-only BCL-2 family protein, BIM, has also been shown to be downregulated in a two-step fashion during lytic cycle, suggesting that it is a common target for both early and late EBV lytic genes [210].

In primary B cells, a number of lytic-cycle-associated EBV genes have been shown to be transcribed early during transformation (1–4 days PI.) including BZLF1, BRLF1, BCRF1 (viral IL-10), BALF1, BHRF1 and BMRF1 in the absence of any production of infectious virions [129,199,211,212]. Conversely, inhibition of apoptosis by miR-BART20-5p has been shown to indirectly inhibit the lytic cycle, suggesting that latent apoptosis regulatory genes may also play a role in EBV replication [213]. There also remain a number of EBV genes that are assigned as 'lytic' about which almost nothing is known. LF3, for example, is classified as a lytic gene since it is robustly transcriptionally upregulated during the lytic cycle [214]. However, RNA-sequencing and quantitative PCR studies showed that LF3 is also highly transcribed in largely latent cell lines [33,193]. Together, these studies demonstrate the interplay between apoptosis and the EBV lytic cycle and raise the possibilities that (1) the definition between latency and lytic cycle may be less well defined than current dogma dictates, (2) some lytic genes may regulate cell death and (3) poorly described EBV genes may contribute to cell death regulation during latency as well as lytic cycle. Further work is required to better understand the role of lytic cycle genes in the regulation of cell death during transformation and in malignancy in order to determine whether these processes might be useful therapeutic targets for EBV-associated diseases.

4. Cooperative Cell Death Inhibition by EBV

4.1. Counteracting the DNA Damage Response

A number of studies have provided insight into the interplay between latent proteins and host cell survival pathways through analyses of early infection (the first 7–10 days PI.) during the establishment of latency (Figure 2). EBV infected B cells begin to proliferate rapidly at around 4 days PI. [32,215–217], when the Wp transcript-encoded EBNA and BHRF1 proteins are maximally expressed [19,25,30,31]. Coincident with entry into cell cycle, activated lymphoblasts initiate the DNA damage response, including robust upregulation of p53 and related pro-apoptotic genes [216,218–220]. Many oncogenic viruses (including SV40, HPV and adenovirus) must directly inactivate p53 in order to transform cells (reviewed [221]), yet in the case of EBV, p53 signalling remains intact in established LCLs [218]. However, DNA damage signalling and p53-induced apoptosis must be countered in order for EBV to transform resting B cells [222]. Furthermore, CRISPR screens have revealed that the p53 inhibitors, MDM2 and MDM4, remain critical for the ongoing survival of established LCLs [105]. Despite markers of DNA damage including γH2AX and chromosomal abnormalities appearing within the first seven days following infection [216,223], few cells are found to undergo apoptosis at these early time points [32,215,217]. It is hypothesised that EBNA3C is largely responsible for preventing p53-driven apoptosis during early time points PI. Since it has been reported to bind and/or stabilise several p53 inhibitors including Gemin3, Aurora B kinase, ING4, ING5 and MDM2 [70,72,224–226] and also to repress the p53 activator p14^{ARF} [66]. Interestingly, inhibition of p53-mediated apoptosis also remains essential long-term as a CRISPR screen has recently shown that MDM2 and MDM4 are critical for the survival of established LCLs [105].

Following activation of p53, cell death proceeds through the BCL-2 family/intrinsic pathway to apoptosis, via p53-induced upregulation of the BH3-only proteins PUMA and NOXA. These BH3-only proteins initiate cell death by binding to the BCL-2 pro-survival proteins, thereby releasing the executioners of apoptosis BAX and BAK. These proteins form pores in the mitochondrial outer membrane (MOMP) which commits the cell to apoptosis. BAK and BAX can also be directly activated by the BH3-only proteins, BIM, tBID and PUMA, though only PUMA is upregulated by p53 activation (reviewed in [227]) (see Figure 4). Many human viruses have evolved strategies to interfere with the intrinsic apoptotic pathway (reviewed in [228]). EBV itself encodes two multi-domain BCL-2 homologues, BHRF1 and BALF1, as discussed earlier. BHRF1 may contribute to the blockade of p53-dependent DDR signalling as it is able to bind and inhibit pro-apoptotic PUMA, [133] and can efficiently inhibit DNA-damage-induced cell death in the absence of any other EBV genes (Fitzsimmons et al. in preparation). LMP1 has been shown to block apoptosis through upregulation of cellular BCL-2 pro-survival proteins [98,99,101,102]. As discussed throughout this review, EBNA1, EBNA2, the EBNA3s, and both families of EBV miRNAs, also regulate a variety of different BCL-2 family members and associated intrinsic apoptosis regulatory elements. Consistent with this complex regulation by multiple EBV genes, a recent study showed that EBV-infected lymphoblasts and LCLs do not rely on any single BCL-2 family member for survival. Instead, the sensitivity of EBV-infected cells to BH3-only proteins changes throughout the process of growth transformation [32].

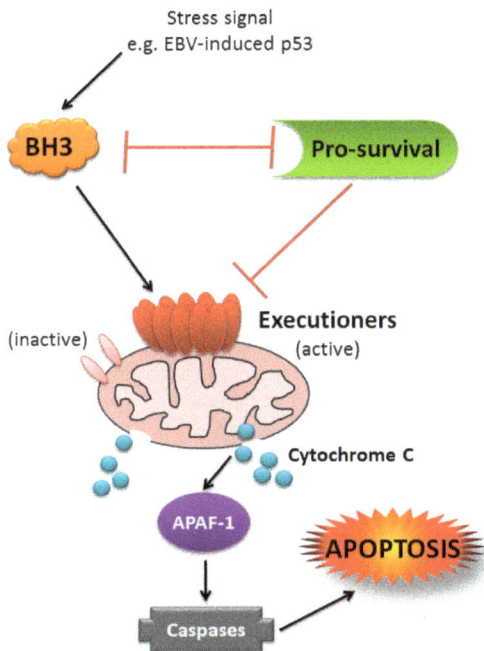

Figure 4. Model of BCL-2 family-mediated intrinsic apoptosis. Schematic of interactions among different classes of BCL-2 family members in the intrinsic apoptosis pathway. Pro-survival, BCL-2-like proteins, including BHRF1 (green), inhibit both classes of pro-death proteins, though the affinities and specificities for these binding partners vary. BH3-only, pro-death proteins (orange) inhibit pro-survival BCL-2s through reciprocal binding and some can directly activate executioner pro-death BCL-2 homologues (only BIM, PUMA and t-BID are able to perform direct activation). The executioner BCL-2s, BAK and BAX exist as inactive monomers until activated by BH3-only proteins or disinhibited by pro-survival BCL-2s. Upon activation, BAK and/or BAX multimerise to form pores in the mitochondrial outer membrane, causing the release of cytochrome C (and other pro-apoptotic factors), which then activate APAF-1, leading to the assembly of the apoptosome, caspase cleavage and consequent cell destruction.

4.2. Cooperation in Transformation

The in vitro growth transformation of B cells by EBV requires that the virus efficiently activates the cell cycle such that the infected cells will proliferate continually, but also necessitates robust and comprehensive inhibition of cell death [229]. Studies of transformed cells and on the process of transformation have been instrumental in revealing how the latent genes of EBV operate a multi-faceted, fail-safe system to regulate key cell fate signalling pathways. For example, multiple BCL-2 family members are known to be regulated by latent EBV genes and many are targeted by more than one viral gene product. Despite this belt-and-braces approach to overcoming cell death, EBV-driven transformation is not efficient: only a fraction of infected cells survive to become continuously proliferating LCLs [230,231]. This may be because EBV-mediated rewiring of the B cell proliferation machinery concomitantly triggers tumour suppressor, cell death and anti-viral pathways, meaning that LCLs remain constantly 'primed' for cell death. However, more likely it demonstrates how evolution has selected for a carefully orchestrated balance between the agonistic and antagonistic functions of the various multifunctional latent genes that allow EBV to respond dynamically to different stimuli and trigger or block cell death efficiently, depending on circumstance. Detailed understanding of the combinatorial and/or redundant effects of EBV genes in apoptosis regulation in a given disease setting is imperative to inform therapeutic treatments. To this end, the recent development of new methods such as CRISPR/Cas9 to quickly and precisely genetically modify the EBV genome within infected cells [232] will be central for dissecting the cooperation between EBV genes in cell death regulation.

4.3. Cooperative Inhibition of Apoptosis in Malignancy

EBV is known to contribute to various forms of malignancy of lymphoid and epithelial cell origin, together accounting for around 200,000 new cancer diagnoses annually [4]. Although the role of the virus is not fully understood in these EBV-associated cancers, there is a large body of evidence to suggest that the virus contributes to the inhibition of cell death in many cases. Whilst the transformation of resting B cells into continuously proliferating LCLs is a useful model for EBV infection, the pattern of latent gene expression seen in LCLs (Latency III) is rarely seen in EBV-positive tumours. If fact, it is usually only detected in patients who are immunocompromised, for example in EBV-driven post-transplant lymphoproliferative disease (PTLD) or in HIV-positive individuals [1]. In these cases, where the immunological imperative for restricted latency is relieved, the cooperative contribution of the Latency III genes is likely to largely reflect those seen in LCLs. Accordingly, although relatively few mechanistic studies exist on EBV-mediated apoptosis inhibition in PTLD, there are reports that key genes implicated in cell death regulation in LCLs, such as Syk, IL10 and BIM, are also deregulated in this malignancy [233–235]. In all other known EBV-associated malignancies, EBV usually adopts more restricted and less immunogenic latencies that vary among different EBV-associated cancers (Figure 1). This means that the most abundant or functionally important EBV targets for therapy may differ among different EBV-associated malignancies. Another important consideration in this regard is that there is evidence that EBV genes work together to cooperatively modulate cell survival pathways in diverse EBV-associated cancers, even in BLs, which exhibit the most restricted form of EBV latency (Latency I). Whilst this raises the possibility that targeting a single EBV protein in malignancy may be ineffective, it does suggest that therapies which disrupt viral hijacking of these critical pathways might be of clinical benefit across multiple EBV malignancies. Here, we review in detail the role of EBV genes in the cooperative inhibition of apoptosis in BL.

5. Burkitt Lymphoma (BL)

The frequency and geographical incidence of BL in malarial endemic regions of sub-Saharan Africa is consistent with an infectious aetiology and this led to the identification of BL as the first cancer to be associated with EBV infection. Accordingly, EBV was found to reside in more than 95% of BLs in high incidence or endemic BL (eBL) areas. BL also occurs worldwide at a lower incidence, and in these cases, known as sporadic BLs (spBL), EBV is found in 15–85% of tumours, varying by geographical region [236–239].

5.1. c-MYC

Common to all cases of BL is the hallmark reciprocal translocation between the *c-MYC* oncogene and a constitutively active immunoglobulin (Ig) gene promoter/enhancer [240–242]. This unchecked overexpression of the c-MYC protein reprograms cells for maximum proliferative capacity [243–245]. Conversely however, *c-MYC* also sensitises cells to apoptosis under conditions of stress [246]. Therefore, in this tumour setting, constitutive c-MYC expression circumvents the requirement for EBV to drive proliferation, but it does place an imperative on EBV to block cellular apoptosis during oncogenesis. One well-characterised pro-apoptotic pathway activated by *c-MYC* is the ARF-MDM2-p53-axis. Accordingly, around 30% of BL biopsies [247] and 70% of BL cell lines have mutated p53 sequences [248–250]. Furthermore, in BLs where p53 has a wild type sequence, p14^{ARF} or MDM2 is frequently altered in their expression in order to circumvent p53-dependent cell death [251,252]. Additionally, deregulation of the downstream BCL-2 family pathway accelerates MYC-induced lymphomagenesis in the *Eμ-Myc* mouse model of BL, which carries a *c-MYC*; IgM heavy chain enhancer transgene. Blocking the intrinsic apoptotic pathway through genetic deletion of pro-apoptotic BCL-2 family genes or constitutive overexpression of BCL-2 pro-survival proteins can all accelerate lymphoma onset in this mouse model [253–259]. Furthermore, silencing of the pro-apoptotic BH3-only genes PUMA and BIM has been described in human BL tumours [260,261].

Interestingly, *c-MYC* itself is frequently mutated in BL, and this is hypothesised to reflect a need to avoid apoptosis activation whilst retaining or enhancing proliferative capability [262–265]. Indeed, one *c-MYC* mutant that commonly occurs in BL has been reported to be incapable of inducing BIM and consequently cannot trigger cells into apoptosis [266]. Outside of the p53 and BCL-2 pathways, mutations in other pro-survival signalling pathways may also play a role in BL. Recent genome-wide studies have identified the transcription factor E2A (TCF-3), ID3, and cyclin D3 as frequently mutated in BL biopsies and cell lines [262,265,267–269]. Interestingly, these targets all contribute to a third survival axis, which is ultimately co-ordinated by PI3K—hence several different diversion mechanisms may contribute to apoptosis escape in BL (reviewed in [270]).

5.2. The Contribution of EBV

In EBV-positive BL, the virus is present as multiple viral episomes in every tumour cell and is maintained indefinitely, suggesting a selective advantage associated with virus retention. Treating BL cells with a dominant negative derivative of the viral episome maintenance protein, EBNA1 (dnEBNA1), enforces loss of viral genomes and induces apoptosis, suggesting that EBV is essential for the continued survival of these BL cells [271,272]. Incidences of EBV-negative BL do occur, albeit with lower frequency, and interestingly analyses have revealed EBV-positive and -negative BLs to be genetically distinct; differing in the number of somatic mutations, frequency of chromosomal copy number changes and the precise *c-MYC* translocations [262,273–277]. This suggests that EBV-positive and -negative BLs arise via slightly different pathogenic routes and furthermore, that the role of EBV in BL needs to be examined in the context of virus positive lymphomas. Given that subversion of apoptosis is an essential step in *MYC*-driven lymphomagenesis [278], it has long been suggested that EBV may provide protection against intrinsic apoptosis in the established tumour.

5.3. Restricted EBV Latency in BL

5.3.1. Latency I

In 85% of eBLs, the virus expresses only EBNA1, the EBER transcripts and the miR-BARTs, collectively termed Latency I (Figure 1). EBNA1, EBERs and miR-BARTs are found to be expressed in all EBV-associated malignancies, but when expressed in the absence of the other latent proteins, this Latency I pattern is the most restricted form of viral gene expression found in EBV-associated cancers. Thus, the Latency I gene products constitute the minimum ongoing contribution of EBV to cancer. The nature of this ongoing viral contribution to BL has been much discussed, but remains somewhat controversial. In 1994, a report that spontaneous loss of EBV from a sporadic BL cell line called Akata-BL increased the sensitivity of the cells to apoptosis, impaired their growth and rendered them non-tumorigenic in a mouse model, provided compelling evidence that the virus remains critical for the survival of BL cells. Furthermore, it demonstrated for the first time a system in which isogenic cells that contained or had 'lost' the virus could be directly compared [279]. This group later showed that the 'EBV-loss' phenotype could be reversed by restoring the virus genome—a key result which directly implicated the virus as a protective agent in BL [280]. A different group used hydroxyurea to eliminate the viral genome from two BL cell lines; Mutu-BL and Akata-BL, but found that only Akata-BL-derived EBV-loss cells showed an increase in apoptosis sensitivity [281]. An alternative strategy was used by the Sugden group, who found that enforced loss of EBV genomes using dnEBNA1 consistently induces apoptosis in BL cells [271,272,282–284]. Although this approach convincingly demonstrates the complete dependency of BL cells on EBV for their ongoing survival, the toxic nature of the phenotype makes it difficult to isolate EBV-loss clones for comparison with EBV-positive cells, as was previously done.

Recently we reported a large-scale study investigating the rates of spontaneous EBV loss from BL cell lines and found that, whilst the phenomenon is rare (loss clones comprised just 3% of those screened), it was consistently associated with an increase in apoptosis sensitivity. Importantly, and consistent with other laboratories, we showed that reinfection of EBV-loss BL cells with recombinant EBV and the establishment of a Latency I infection could reverse the apoptosis-sensitive phenotype [280,285,286].

The mechanism of apoptosis protection by EBV in Latency I BL appears to be cooperative and to exhibit redundancy. Whilst EBERs are reported to promote survival by upregulating IL10 in some clonal EBV-loss variants of Akata-BL [287], we and others have found that neither EBNA1 nor EBERs alone can restore apoptosis protection to EBV-loss BLs [285,286,288]. Interestingly however, the Sample group, who first showed that EBERs alone could not protect BL cells, did find that BL cells expressing both EBNA1 and EBERs were tumorigenic in vivo, although to a lesser extent than BL cells infected with wt EBV [288]. Additionally, whilst we have found that expression of miR-BARTs alone could not confer any protection to EBV-loss BL cells [285], miR-BARTs, when expressed in combination with EBNA1, have been shown to reduce dnEBNA1-induced cell death [187]. Furthermore, when EBV-loss cells were reinfected with a recombinant, Latency I-restricted EBV that lacks many of the miR-BARTs, they exhibited a moderate, but significant reduction in apoptosis resistance [285]. As mentioned previously (see Section 2.6), LMP2A can regulate important cell death pathways. Accordingly, ectopic expression of LMP2A has been shown to protect some BL cell lines from apoptosis [121,289], and can enhance tumorigenicity in a mouse model [290,291]. However, whilst LMP2 transcripts are detectable in BL tumour material [292–294], absolute quantitation has shown that the abundance of these transcripts in Latency I cells is low compared to those found in other forms of latency [33,193]. Additionally, LMP2A protein expression is a consistent feature of LCLs and Latency II malignancies, however is rarely detected in BL cells [33,292]. Therefore, it has been extensively demonstrated that optimal cell death inhibition in Latency I BL can only be conferred by the presence and cooperation of all of the Latency I-associated EBV genes. Although the precise mechanisms by which each Latency I gene product inhibits cell death in BL has not been fully elucidated, the collective presence of all Latency I products can block the induction of both BIM and PUMA following apoptotic challenge [285]. BIM and PUMA translation is likely inhibited by miR-BARTs as their 3'-UTRs contain multiple predicted binding sites for a variety of BART miRNAs [165,192,295,296], though miR-BARTs alone are insufficient to appreciably downregulate BIM and/or PUMA in BL cells [285]. EBERs meanwhile, may contribute to the suppression of BIM and PUMA induction via activation of AKT/PI3K signalling [297], which has been shown to regulate BIM and PUMA expression at both transcriptional and post-translational level [298]. PUMA is also subject to regulation by the TGF-β pathway [299,300], which EBNA1 has been shown to modulate through downregulation of SMAD2 [82,301]. Caspase 3 expression is also reduced in both EBV-positive and miR-BART expressing Latency I BL cells compared to levels in dnEBNA1-treated control cells [187]. Therefore, it is possible that there is a third, downstream pro-apoptotic target for Latency I EBV genes. The importance of BIM and PUMA in BL cell survival has also been demonstrated by the finding that their expression is often selected against by epigenetic silencing [64,260], though further work is required to establish whether these changes occur more commonly in EBV-negative versus EBV-positive tumours. In summary, the Latency I EBV genes (EBNA1, EBERs and miR-BARTs), collectively suppress induction of pro-apoptotic BIM and PUMA via cooperative targeting of their transcription and translation in order to overcome a c-MYC-induced sensitivity to apoptosis.

5.3.2. Wp-Restricted Latency

In the remaining ~15% of eBLs, EBV adopts a more extensive pattern of gene expression involving EBNA-3A, -3B and -3C, a truncated form of EBNA-LP and the BHRF1 protein, in addition to EBNA1, EBERs and miR-BARTs [19,302,303]. More recently, we have also found that two of the three BHRF1 miRNAs are also expressed in these tumours [24]. This pattern of gene expression, termed Wp-restricted Latency (Wp-BL) is imposed by a large deletion in the virus genome and is associated with marked resistance to cell death [302]. The exact size of the deletion varies between tumour samples, but always encompasses the coding region of the EBNA2 gene, placing BHRF1 adjacent to the promoter-encoding Wp repeat region [303]. This promoter, which is silent in Latency I BL, drives high levels of BHRF1 expression and also transcribes the EBNA proteins [19]. EBNA-3A and -3C are known to cooperatively downregulate BIM and p16^{INK4a} through epigenetic silencing leading to reduced sensitivity to cell death [59,64,65,304]. However, it has been shown that short-term expression of EBNA3s, either individually or in combination, cannot protect BL cells to the same extent as Wp-restricted latency [19]. Wp-BLs also express a variant of EBNA-LP that is truncated (t-EBNA-LP) compared to the form expressed in Latency III as a result of the characteristic EBV genomic deletion. Interestingly, this t-EBNA-LP has also been shown to contribute to cell survival via inhibition of protein phosphatase 2A (PP2A), a function apparently specific to the truncated form and not shared by the full length EBNA-LP [305]. BHRF1, by virtue of the fact that it is a viral BCL-2 homologue, can efficiently inhibit cell death even in the absence of other EBV genes via binding to and inhibition of cellular pro-apoptotic proteins BIM, PUMA, BID and BAK [19,133]. The propensity of BHRF1 to bind particular pro-death cellular proteins may vary under different conditions of cell stress (unpublished data from our lab). Therefore, EBV likely exhibits potent cell death inhibition in Wp-BL through a cooperative mechanism involving the EBNA3, t-EBNA-LP and BHRF1 proteins and possibly miR-BHRF1s, though the role of the miRNAs may be an indirect effect on regulation of BHRF1 itself [31].

6. Future Perspectives

Cooperation among EBV genes to inhibit cell death is clearly important for both growth transformation and the pathogenesis of BL; however, relatively little is known about how EBV genes cooperate outside of these models. There is evidence in the literature to suggest that EBV targets common pathways in different disease settings. For example, EBV is now known to be an etiologic factor in around 9% of gastric carcinomas (GC). As in other EBV-associated cancers, the virus is present as a latent infection in these tumours, although there are conflicting reports as to whether it establishes Latency I or II infection (the latter involves expression of LMP1 and LMP2 proteins in addition to the Latency I genes) [306]. Like BL, EBV-positive and -negative cases of GC are genetically distinct. EBV-positive GCs can be distinguished by marked hypermethylation of genes, including universal silencing of p16^{INK4A}, and frequent characteristic mutations in PIK3CA, ARID1A and BCOR genes [306,307]. The silencing of p16^{INK4A} and the 80% incidence of mutations in the PI3K family gene, PIK3CA, suggest that the same key signalling pathways identified in BL and LCLs are also important in GC pathogenesis. Similarly, in the setting of NPC, where EBV establishes a Latency II infection, the *INK4A* locus (encoding p16^{INK4A} and p14^{ARF}) is commonly mutated or silenced. Therefore, possible cooperation between EBV genes to regulate key survival signalling pathways, in particular BCL-2 family signalling, PI3K signalling and the ARF-MDM2-p53-axis, in EBV-associated cancers warrants further investigation.

Whilst it is clear that the known latent genes of EBV make a considerable contribution to the apoptosis phenotype of EBV-associated cancers, the roles of other EBV-encoded genes in this regard require further attention. Lytic-cycle-associated transcripts can be readily detected in EBV-positive cancers [33,193,267,306,308] and though these are thought to originate from the small percentage of cells within the population that spontaneously undergo lytic replication, more detailed single cell analyses are required to confirm whether any of these transcripts (and where relevant, the proteins they encode) are detectable in otherwise latently infected tumour cells. For example, the expression of BHRF1 protein was initially thought to be restricted to the lytic cycle [131], however it is now known to be expressed in latently infected Wp-BL and LCLs [19]. Of note in this regard, BHRF1 transcripts have also been reported in 15–20% of otherwise latent EBV-positive GC biopsies, whilst BARF1 transcripts were detected in 40–100% of samples (6/13 and 9/9 biopsies, respectively) [309,310]. Furthermore, BALF1 has also been detected in 80% of NPC biopsies (13/16) [311]. This analysis is further complicated by the findings that BARF1 can be secreted from EBV infected cells [312] and certain EBV gene products, including BART miRNAs and LMP proteins, can be released from infected cells in exosomes, and thereby influence the survival of neighbouring cells [313] (EBV exosomes reviewed in [314]). Evidence is also emerging that novel splice variants of LMP2A (and possibly other latent genes) may be expressed during the lytic cycle or in rarely infected cell types, such as NK or T cells, raising the possibility for additional or modified apoptosis-regulating functionality by latent genes in these settings [315,316]. New techniques to genetically modify EBV genomes in situ, such as CRISPR/Cas9 [232,317], will be critical to dissect the contribution(s) of these lesser characterised EBV genes.

Whilst the evidence that EBV can regulate apoptosis is compelling, future investigations may also uncover mechanisms through which EBV can modulate additional forms of cell death. For example, the LMP1 protein has been reported to modulate autophagy by both NF-κB-dependent and -independent mechanisms [318,319] and autophagy has recently been reported to play an important role in cell survival during growth transformation of B cells [217]. Furthermore, a number of recent studies have identified viral modulators of emerging cell death processes, such as necroptosis and pyroptosis in viruses including cytomegalovirus, vaccinia virus and hepatitis C virus [320–322]. Whilst little is known about EBV in this regard, these findings further open up the field of viral modulation of cell death. Therefore, whilst the study of growth transformation and BL have had a profound impact on our understanding of EBV manipulation of cell survival, future work taking into account the possibility of cooperation among the full spectrum of EBV genes in regulating both well-characterised and emerging cell death pathways will be key to the development of new therapeutic approaches.

Acknowledgments: This work was supported by funding from: Cancer Research UK, Programme Grant C5575/A15032; Cancer Council Victoria, grant-in-aid #1086157 awarded to G.L.K.; the National Health and Medical Research Council, Project Grant #1086291 awarded to G.L.K. and a bequest from the Estate of Anthony Redstone.

Author Contributions: L.F. wrote the manuscript and it was critically edited by G.L.K.

Conflicts of Interest: The authors declare no conflict of interest.

References

1. Rickinson, A.A.E.K. *Fields Virology*, 4th ed.; Lippincott, Williams and Wilkins: Philadelphia, PA, USA, 2001; Volume 2, p. 3063.
2. Epstein, M.A.; Achong, B.G.; Barr, Y.M. Virus Particles in Cultured Lymphoblasts from Burkitt's Lymphoma. *Lancet* **1964**, *1*, 702–703. [CrossRef]
3. Epstein, M.A.; Barr, Y.M.; Achong, B.G. A Second Virus-Carrying Tissue Culture Strain (Eb2) of Lymphoblasts from Burkitt's Lymphoma. *Pathol. Biol. (Paris)* **1964**, *12*, 1233–1234.
4. Cohen, J.I.; Fauci, A.S.; Varmus, H.; Nabel, G.J. Epstein-Barr virus: An important vaccine target for cancer prevention. *Sci. Transl. Med.* **2011**, *3*, 107fs7. [CrossRef] [PubMed]
5. Burkitt, D. A sarcoma involving the jaws in African children. *Br. J. Surg.* **1958**, *46*, 218–223. [CrossRef] [PubMed]
6. Burkitt, D.P.; Wright, D.H. *Burkitt's Lymphoma*; Churchill Livingstone: London, UK, 1970.
7. Burkitt, D. A Children's Cancer with Geographical Limitations. *Cancer Prog.* **1963**, *92*, 102–113. [PubMed]
8. Burkitt, D.P. Observations on the geography of malignant lymphoma. *East Afr. Med. J.* **1961**, *38*, 511–514. [PubMed]
9. Haddow, A.J. An Improved Map for the Study of Burkitt's Lymphoma Syndrome in Africa. *East Afr. Med. J.* **1963**, *40*, 429–432. [PubMed]
10. Haddow, A.J. Age Incidence in Burkitt's Lymphoma Syndrome. *East Afr. Med. J.* **1964**, *41*, 1–6. [PubMed]
11. Kelly, G.L.; Rickinson, A.B. Burkitt lymphoma: Revisiting the pathogenesis of a virus-associated malignancy. *Hematol. Am. Soc. Hematol. Educ. Progr.* **2007**, *2007*, 277–284. [CrossRef] [PubMed]
12. Crawford, D.H.; Rickinson, A.; Johannessen, I.O. *Cancer Virus: The Story of Epstein-Barr Virus*; OUP Oxford: Oxford, UK, 2014.
13. Pope, J.H.; Horne, M.K.; Scott, W. Transformation of foetal human keukocytes in vitro by filtrates of a human leukaemic cell line containing herpes-like virus. *Int. J. Cancer* **1968**, *3*, 857–866. [CrossRef] [PubMed]

14. Henle, W.; Diehl, V.; Kohn, G.; Zur Hausen, H.; Henle, G. Herpes-type virus and chromosome marker in normal leukocytes after growth with irradiated Burkitt cells. *Science* **1967**, *157*, 1064–1065. [CrossRef] [PubMed]

15. Lindahl, T.; Adams, A.; Bjursell, G.; Bornkamm, G.W.; Kaschka-Dierich, C.; Jehn, U. Covalently closed circular duplex DNA of Epstein-Barr virus in a human lymphoid cell line. *J. Mol. Biol.* **1976**, *102*, 511–530. [CrossRef]

16. Hammerschmidt, W.; Sugden, B. Replication of Epstein-Barr viral DNA. *Cold Spring Harb. Perspect. Biol.* **2013**, *5*, a013029. [CrossRef] [PubMed]

17. Rowe, D.T.; Rowe, M.; Evan, G.I.; Wallace, L.E.; Farrell, P.J.; Rickinson, A.B. Restricted expression of EBV latent genes and T-lymphocyte-detected membrane antigen in Burkitt's lymphoma cells. *EMBO J.* **1986**, *5*, 2599–2607. [PubMed]

18. Rowe, M.; Rowe, D.T.; Gregory, C.D.; Young, L.S.; Farrell, P.J.; Rupani, H.; Rickinson, A.B. Differences in B cell growth phenotype reflect novel patterns of Epstein-Barr virus latent gene expression in Burkitt's lymphoma cells. *EMBO J.* **1987**, *6*, 2743–2751. [PubMed]

19. Kelly, G.L.; Long, H.M.; Stylianou, J.; Thomas, W.A.; Leese, A.; Bell, A.I.; Bornkamm, G.W.; Mautner, J.; Rickinson, A.B.; Rowe, M. An Epstein-Barr virus anti-apoptotic protein constitutively expressed in transformed cells and implicated in burkitt lymphomagenesis: The Wp/BHRF1 link. *PLoS Pathog.* **2009**, *5*, e1000341. [CrossRef] [PubMed]

20. Pfeffer, S.; Zavolan, M.; Grasser, F.A.; Chien, M.; Russo, J.J.; Ju, J.; John, B.; Enright, A.J.; Marks, D.; Sander, C.; et al. Identification of virus-encoded microRNAs. *Science* **2004**, *304*, 734–736. [CrossRef] [PubMed]

21. Xing, L.; Kieff, E. Epstein-Barr virus BHRF1 micro- and stable RNAs during latency III and after induction of replication. *J. Virol.* **2007**, *81*, 9967–9975. [CrossRef] [PubMed]

22. Cai, X.; Schafer, A.; Lu, S.; Bilello, J.P.; Desrosiers, R.C.; Edwards, R.; Raab-Traub, N.; Cullen, B.R. Epstein-Barr virus microRNAs are evolutionarily conserved and differentially expressed. *PLoS Pathog.* **2006**, *2*, e23. [CrossRef] [PubMed]

23. Rosa, M.D.; Gottlieb, E.; Lerner, M.R.; Steitz, J.A. Striking similarities are exhibited by two small Epstein-Barr virus-encoded ribonucleic acids and the adenovirus-associated ribonucleic acids VAI and VAII. *Mol. Cell. Biol.* **1981**, *1*, 785–796. [CrossRef] [PubMed]

24. Amoroso, R.; Fitzsimmons, L.; Thomas, W.A.; Kelly, G.L.; Rowe, M.; Bell, A.I. Quantitative studies of Epstein-Barr virus-encoded microRNAs provide novel insights into their regulation. *J. Virol.* **2011**, *85*, 996–1010. [CrossRef] [PubMed]

25. Alfieri, C.; Birkenbach, M.; Kieff, E. Early events in Epstein-Barr virus infection of human B lymphocytes. *Virology* **1991**, *181*, 595–608. [CrossRef]

26. Austin, P.J.; Flemington, E.; Yandava, C.N.; Strominger, J.L.; Speck, S.H. Complex transcription of the Epstein-Barr virus BamHI fragment H rightward open reading frame 1 (BHRF1) in latently and lytically infected B lymphocytes. *Proc. Natl. Acad. Sci. USA* **1988**, *85*, 3678–3682. [CrossRef] [PubMed]

27. Abbot, S.D.; Rowe, M.; Cadwallader, K.; Ricksten, A.; Gordon, J.; Wang, F.; Rymo, L.; Rickinson, A.B. Epstein-Barr virus nuclear antigen 2 induces expression of the virus-encoded latent membrane protein. *J. Virol.* **1990**, *64*, 2126–2134. [PubMed]

28. Zimber-Strobl, U.; Kremmer, E.; Grasser, F.; Marschall, G.; Laux, G.; Bornkamm, G.W. The Epstein-Barr virus nuclear antigen 2 interacts with an EBNA2 responsive cis-element of the terminal protein 1 gene promoter. *EMBO J.* **1993**, *12*, 167–175. [PubMed]

29. Schlager, S.; Speck, S.H.; Woisetschlager, M. Transcription of the Epstein-Barr virus nuclear antigen 1 (EBNA1) gene occurs before induction of the BCR2 (Cp) EBNA gene promoter during the initial stages of infection in B cells. *J. Virol.* **1996**, *70*, 3561–3570. [PubMed]

30. Shannon-Lowe, C.; Baldwin, G.; Feederle, R.; Bell, A.; Rickinson, A.; Delecluse, H.J. Epstein-Barr virus-induced B-cell transformation: Quantitating events from virus binding to cell outgrowth. *J. Gen. Virol.* **2005**, *86*, 3009–3019. [CrossRef] [PubMed]

31. Bernhardt, K.; Haar, J.; Tsai, M.H.; Poirey, R.; Feederle, R.; Delecluse, H.J. A Viral microRNA Cluster Regulates the Expression of PTEN, p27 and of a bcl-2 Homolog. *PLoS Pathog.* **2016**, *12*, e1005405. [CrossRef] [PubMed]

32. Price, A.M.; Tourigny, J.P.; Forte, E.; Salinas, R.E.; Dave, S.S.; Luftig, M.A. Analysis of Epstein-Barr virus-regulated host gene expression changes through primary B-cell outgrowth reveals delayed kinetics of latent membrane protein 1-mediated NF-κB activation. *J. Virol.* **2012**, *86*, 11096–11106. [CrossRef] [PubMed]

33. Tierney, R.J.; Shannon-Lowe, C.D.; Fitzsimmons, L.; Bell, A.I.; Rowe, M. Unexpected patterns of Epstein-Barr virus transcription revealed by a High throughput PCR array for absolute quantification of viral mRNA. *Virology* **2015**, *474*, 117–130. [CrossRef] [PubMed]

34. Hofelmayr, H.; Strobl, L.J.; Marschall, G.; Bornkamm, G.W.; Zimber-Strobl, U. Activated Notch1 can transiently substitute for EBNA2 in the maintenance of proliferation of LMP1-expressing immortalized B cells. *J. Virol.* **2001**, *75*, 2033–2040. [CrossRef] [PubMed]

35. Strobl, L.J.; Hofelmayr, H.; Marschall, G.; Brielmeier, M.; Bornkamm, G.W.; Zimber-Strobl, U. Activated Notch1 modulates gene expression in B cells similarly to Epstein-Barr viral nuclear antigen 2. *J. Virol.* **2000**, *74*, 1727–1735. [CrossRef] [PubMed]

36. Sakai, T.; Taniguchi, Y.; Tamura, K.; Minoguchi, S.; Fukuhara, T.; Strobl, L.J.; Zimber-Strobl, U.; Bornkamm, G.W.; Honjo, T. Functional replacement of the intracellular region of the Notch1 receptor by Epstein-Barr virus nuclear antigen 2. *J. Virol.* **1998**, *72*, 6034–6039. [PubMed]

37. Grossman, S.R.; Johannsen, E.; Tong, X.; Yalamanchili, R.; Kieff, E. The Epstein-Barr virus nuclear antigen 2 transactivator is directed to response elements by the J kappa recombination signal binding protein. *Proc. Natl. Acad. Sci. USA* **1994**, *91*, 7568–7572. [CrossRef] [PubMed]

38. Yalamanchili, R.; Tong, X.; Grossman, S.; Johannsen, E.; Mosialos, G.; Kieff, E. Genetic and biochemical evidence that EBNA 2 interaction with a 63-kDa cellular GTG-binding protein is essential for B lymphocyte growth transformation by EBV. *Virology* **1994**, *204*, 634–641. [CrossRef] [PubMed]

39. Zimber-Strobl, U.; Strobl, L.J.; Meitinger, C.; Hinrichs, R.; Sakai, T.; Furukawa, T.; Honjo, T.; Bornkamm, G.W. Epstein-Barr virus nuclear antigen 2 exerts its transactivating function through interaction with recombination signal binding protein RBP-J kappa, the homologue of Drosophila Suppressor of Hairless. *EMBO J.* **1994**, *13*, 4973–4982. [PubMed]

40. Henkel, T.; Ling, P.D.; Hayward, S.D.; Peterson, M.G. Mediation of Epstein-Barr virus EBNA2 transactivation by recombination signal-binding protein J kappa. *Science* **1994**, *265*, 92–95. [CrossRef] [PubMed]

41. Tierney, R.; Nagra, J.; Hutchings, I.; Shannon-Lowe, C.; Altmann, M.; Hammerschmidt, W.; Rickinson, A.; Bell, A. Epstein-Barr virus exploits BSAP/Pax5 to achieve the B-cell specificity of its growth-transforming program. *J. Virol.* **2007**, *81*, 10092–10100. [CrossRef] [PubMed]

42. Lee, J.M.; Lee, K.H.; Weidner, M.; Osborne, B.A.; Hayward, S.D. Epstein-Barr virus EBNA2 blocks Nur77-mediated apoptosis. *Proc. Natl. Acad. Sci. USA* **2002**, *99*, 11878–11883. [CrossRef] [PubMed]

43. Lee, J.M.; Lee, K.H.; Farrell, C.J.; Ling, P.D.; Kempkes, B.; Park, J.H.; Hayward, S.D. EBNA2 is required for protection of latently Epstein-Barr virus-infected B cells against specific apoptotic stimuli. *J. Virol.* **2004**, *78*, 12694–12697. [CrossRef] [PubMed]

44. Godoi, P.H.; Wilkie-Grantham, R.P.; Hishiki, A.; Sano, R.; Matsuzawa, Y.; Yanagi, H.; Munte, C.E.; Chen, Y.; Yao, Y.; Marassi, F.M.; et al. Orphan Nuclear Receptor NR4A1 Binds a Novel Protein Interaction Site on Anti-apoptotic B Cell Lymphoma Gene 2 Family Proteins. *J. Biol. Chem.* **2016**, *291*, 14072–14084. [CrossRef] [PubMed]

45. Pegman, P.M.; Smith, S.M.; D'Souza, B.N.; Loughran, S.T.; Maier, S.; Kempkes, B.; Cahill, P.A.; Simmons, M.J.; Gelinas, C.; Walls, D. Epstein-Barr virus nuclear antigen 2 trans-activates the cellular antiapoptotic bfl-1 gene by a CBF1/RBPJ kappa-dependent pathway. *J. Virol.* **2006**, *80*, 8133–8144. [CrossRef] [PubMed]

46. Campion, E.M.; Hakimjavadi, R.; Loughran, S.T.; Phelan, S.; Smith, S.M.; D'Souza, B.N.; Tierney, R.J.; Bell, A.I.; Cahill, P.A.; Walls, D. Repression of the proapoptotic cellular BIK/NBK gene by Epstein-Barr virus antagonizes transforming growth factor beta1-induced B-cell apoptosis. *J. Virol.* **2014**, *88*, 5001–5013. [CrossRef] [PubMed]

47. Wood, C.D.; Veenstra, H.; Khasnis, S.; Gunnell, A.; Webb, H.M.; Shannon-Lowe, C.; Andrews, S.; Osborne, C.S.; West, M.J. MYC activation and BCL2L11 silencing by a tumour virus through the large-scale reconfiguration of enhancer-promoter hubs. *eLife* **2016**, *5*. [CrossRef] [PubMed]

48. Sinclair, A.J.; Palmero, I.; Peters, G.; Farrell, P.J. EBNA-2 and EBNA-LP cooperate to cause G0 to G1 transition during immortalization of resting human B lymphocytes by Epstein-Barr virus. *EMBO J.* **1994**, *13*, 3321–3328. [PubMed]

49. Mannick, J.B.; Cohen, J.I.; Birkenbach, M.; Marchini, A.; Kieff, E. The Epstein-Barr virus nuclear protein encoded by the leader of the EBNA RNAs is important in B-lymphocyte transformation. *J. Virol.* **1991**, *65*, 6826–6837. [PubMed]

50. Hammerschmidt, W.; Sugden, B. Genetic analysis of immortalizing functions of Epstein-Barr virus in human B lymphocytes. *Nature* **1989**, *340*, 393–397. [CrossRef] [PubMed]

51. Tierney, R.J.; Kao, K.Y.; Nagra, J.K.; Rickinson, A.B. Epstein-Barr virus BamHI W repeat number limits EBNA2/EBNA-LP coexpression in newly infected B cells and the efficiency of B-cell transformation: A rationale for the multiple W repeats in wild-type virus strains. *J. Virol.* **2011**, *85*, 12362–12375. [CrossRef] [PubMed]

52. Kashuba, E.; Yurchenko, M.; Szirak, K.; Stahl, J.; Klein, G.; Szekely, L. Epstein-Barr virus-encoded EBNA-5 binds to Epstein-Barr virus-induced Fte1/S3a protein. *Exp. Cell Res.* **2005**, *303*, 47–55. [CrossRef] [PubMed]

53. Kashuba, E.; Mattsson, K.; Pokrovskaja, K.; Kiss, C.; Protopopova, M.; Ehlin-Henriksson, B.; Klein, G.; Szekely, L. EBV-encoded EBNA-5 associates with P14ARF in extranucleolar inclusions and prolongs the survival of P14ARF-expressing cells. *Int. J. Cancer* **2003**, *105*, 644–653. [CrossRef] [PubMed]

54. Matsuda, G.; Nakajima, K.; Kawaguchi, Y.; Yamanashi, Y.; Hirai, K. Epstein-Barr virus (EBV) nuclear antigen leader protein (EBNA-LP) forms complexes with a cellular anti-apoptosis protein Bcl-2 or its EBV counterpart BHRF1 through HS1-associated protein X-1. *Microbiol. Immunol.* **2003**, *47*, 91–99. [CrossRef] [PubMed]

55. Allday, M.J.; Bazot, Q.; White, R.E. The EBNA3 Family: Two Oncoproteins and a Tumour Suppressor that Are Central to the Biology of EBV in B Cells. *Curr. Top. Microbiol. Immunol.* **2015**, *391*, 61–117. [PubMed]

56. Le Roux, A.; Kerdiles, B.; Walls, D.; Dedieu, J.F.; Perricaudet, M. The Epstein-Barr virus determined nuclear antigens EBNA-3A, -3B, and -3C repress EBNA-2-mediated transactivation of the viral terminal protein 1 gene promoter. *Virology* **1994**, *205*, 596–602. [CrossRef] [PubMed]

57. Yenamandra, S.P.; Sompallae, R.; Klein, G.; Kashuba, E. Comparative analysis of the Epstein-Barr virus encoded nuclear proteins of EBNA-3 family. *Comput. Biol. Med.* **2009**, *39*, 1036–1042. [CrossRef] [PubMed]

58. Hertle, M.L.; Popp, C.; Petermann, S.; Maier, S.; Kremmer, E.; Lang, R.; Mages, J.; Kempkes, B. Differential gene expression patterns of EBV infected EBNA-3A positive and negative human B lymphocytes. *PLoS Pathog.* **2009**, *5*, e1000506. [CrossRef] [PubMed]

59. Skalska, L.; White, R.E.; Franz, M.; Ruhmann, M.; Allday, M.J. Epigenetic repression of p16(INK4A) by latent Epstein-Barr virus requires the interaction of EBNA3A and EBNA3C with CtBP. *PLoS Pathog.* **2010**, *6*, e1000951. [CrossRef] [PubMed]

60. Tomkinson, B.; Robertson, E.; Kieff, E. Epstein-Barr virus nuclear proteins EBNA-3A and EBNA-3C are essential for B-lymphocyte growth transformation. *J. Virol.* **1993**, *67*, 2014–2025. [PubMed]

61. Tomkinson, B.; Kieff, E. Use of second-site homologous recombination to demonstrate that Epstein-Barr virus nuclear protein 3B is not important for lymphocyte infection or growth transformation in vitro. *J. Virol.* **1992**, *66*, 2893–2903. [PubMed]

62. Chen, A.; Divisconte, M.; Jiang, X.; Quink, C.; Wang, F. Epstein-Barr virus with the latent infection nuclear antigen 3B completely deleted is still competent for B-cell growth transformation in vitro. *J. Virol.* **2005**, *79*, 4506–4509. [CrossRef] [PubMed]

63. Paschos, K.; Parker, G.A.; Watanatanasup, E.; White, R.E.; Allday, M.J. BIM promoter directly targeted by EBNA3C in polycomb-mediated repression by EBV. *Nucleic Acids Res.* **2012**, *40*, 7233–7246. [CrossRef] [PubMed]

64. Paschos, K.; Smith, P.; Anderton, E.; Middeldorp, J.M.; White, R.E.; Allday, M.J. Epstein-barr virus latency in B cells leads to epigenetic repression and CpG methylation of the tumour suppressor gene Bim. *PLoS Pathog.* **2009**, *5*, e1000492. [CrossRef] [PubMed]

65. Anderton, E.; Yee, J.; Smith, P.; Crook, T.; White, R.E.; Allday, M.J. Two Epstein-Barr virus (EBV) oncoproteins cooperate to repress expression of the proapoptotic tumour-suppressor Bim: Clues to the pathogenesis of Burkitt's lymphoma. *Oncogene* **2008**, *27*, 421–433. [CrossRef] [PubMed]

66. Maruo, S.; Zhao, B.; Johannsen, E.; Kieff, E.; Zou, J.; Takada, K. Epstein-Barr virus nuclear antigens 3C and 3A maintain lymphoblastoid cell growth by repressing p16INK4A and p14ARF expression. *Proc. Natl. Acad. Sci. USA* **2011**, *108*, 1919–1924. [CrossRef] [PubMed]

67. Maruo, S.; Wu, Y.; Ishikawa, S.; Kanda, T.; Iwakiri, D.; Takada, K. Epstein-Barr virus nuclear protein EBNA3C is required for cell cycle progression and growth maintenance of lymphoblastoid cells. *Proc. Natl. Acad. Sci. USA* **2006**, *103*, 19500–19505. [CrossRef] [PubMed]

68. Skalska, L.; White, R.E.; Parker, G.A.; Turro, E.; Sinclair, A.J.; Paschos, K.; Allday, M.J. Induction of p16(INK4a) is the major barrier to proliferation when Epstein-Barr virus (EBV) transforms primary B cells into lymphoblastoid cell lines. *PLoS Pathog.* **2013**, *9*, e1003187. [CrossRef]

69. Jiang, S.; Willox, B.; Zhou, H.; Holthaus, A.M.; Wang, A.; Shi, T.T.; Maruo, S.; Kharchenko, P.V.; Johannsen, E.C.; Kieff, E.; et al. Epstein-Barr virus nuclear antigen 3C binds to BATF/IRF4 or SPI1/IRF4 composite sites and recruits Sin3A to repress CDKN2A. *Proc. Natl. Acad. Sci. USA* **2014**, *111*, 421–426. [CrossRef] [PubMed]

70. Cai, Q.; Guo, Y.; Xiao, B.; Banerjee, S.; Saha, A.; Lu, J.; Glisovic, T.; Robertson, E.S. Epstein-Barr virus nuclear antigen 3C stabilizes Gemin3 to block p53-mediated apoptosis. *PLoS Pathog.* **2011**, *7*, e1002418. [CrossRef] [PubMed]

71. Yi, F.; Saha, A.; Murakami, M.; Kumar, P.; Knight, J.S.; Cai, Q.; Choudhuri, T.; Robertson, E.S. Epstein-Barr virus nuclear antigen 3C targets p53 and modulates its transcriptional and apoptotic activities. *Virology* **2009**, *388*, 236–247. [CrossRef] [PubMed]

72. Saha, A.; Murakami, M.; Kumar, P.; Bajaj, B.; Sims, K.; Robertson, E.S. Epstein-Barr virus nuclear antigen 3C augments Mdm2-mediated p53 ubiquitination and degradation by deubiquitinating Mdm2. *J. Virol.* **2009**, *83*, 4652–4669. [CrossRef] [PubMed]

73. McClellan, M.J.; Wood, C.D.; Ojeniyi, O.; Cooper, T.J.; Kanhere, A.; Arvey, A.; Webb, H.M.; Palermo, R.D.; Harth-Hertle, M.L.; Kempkes, B.; et al. Modulation of enhancer looping and differential gene targeting by Epstein-Barr virus transcription factors directs cellular reprogramming. *PLoS Pathog.* **2013**, *9*, e1003636. [CrossRef] [PubMed]

74. Sears, J.; Ujihara, M.; Wong, S.; Ott, C.; Middeldorp, J.; Aiyar, A. The amino terminus of Epstein-Barr Virus (EBV) nuclear antigen 1 contains AT hooks that facilitate the replication and partitioning of latent EBV genomes by tethering them to cellular chromosomes. *J. Virol.* **2004**, *78*, 11487–11505. [CrossRef] [PubMed]

75. Mackey, D.; Sugden, B. Applications of oriP plasmids and their mode of replication. *Methods Enzymol.* **1999**, *306*, 308–328. [PubMed]

76. Leight, E.R.; Sugden, B. EBNA-1: A protein pivotal to latent infection by Epstein-Barr virus. *Rev. Med. Virol.* **2000**, *10*, 83–100. [CrossRef]

77. Frappier, L. Contributions of Epstein-Barr nuclear antigen 1 (EBNA1) to cell immortalization and survival. *Viruses* **2012**, *4*, 1537–1547. [CrossRef] [PubMed]

78. Gruhne, B.; Sompallae, R.; Masucci, M.G. Three Epstein-Barr virus latency proteins independently promote genomic instability by inducing DNA damage, inhibiting DNA repair and inactivating cell cycle checkpoints. *Oncogene* **2009**, *28*, 3997–4008. [CrossRef] [PubMed]

79. Gruhne, B.; Sompallae, R.; Marescotti, D.; Kamranvar, S.A.; Gastaldello, S.; Masucci, M.G. The Epstein-Barr virus nuclear antigen-1 promotes genomic instability via induction of reactive oxygen species. *Proc. Natl. Acad. Sci. USA* **2009**, *106*, 2313–2318. [CrossRef] [PubMed]

80. Saridakis, V.; Sheng, Y.; Sarkari, F.; Holowaty, M.N.; Shire, K.; Nguyen, T.; Zhang, R.G.; Liao, J.; Lee, W.; Edwards, A.M.; et al. Structure of the p53 binding domain of HAUSP/USP7 bound to Epstein-Barr nuclear antigen 1 implications for EBV-mediated immortalization. *Mol. Cell* **2005**, *18*, 25–36. [CrossRef] [PubMed]

81. Lu, J.; Murakami, M.; Verma, S.C.; Cai, Q.L.; Haldar, S.; Kaul, R.; Wasik, M.A.; Middeldorp, J.; Robertson, E.S. Epstein-Barr Virus nuclear antigen 1 (EBNA1) confers resistance to apoptosis in EBV-positive B-lymphoma cells through up-regulation of survivin. *Virology* **2011**, *410*, 64–75. [CrossRef] [PubMed]

82. Wood, V.H.; O'Neil, J.D.; Wei, W.; Stewart, S.E.; Dawson, C.W.; Young, L.S. Epstein-Barr virus-encoded EBNA1 regulates cellular gene transcription and modulates the STAT1 and TGFbeta signaling pathways. *Oncogene* **2007**, *26*, 4135–4147. [CrossRef] [PubMed]

83. Canaan, A.; Haviv, I.; Urban, A.E.; Schulz, V.P.; Hartman, S.; Zhang, Z.; Palejev, D.; Deisseroth, A.B.; Lacy, J.; Snyder, M.; et al. EBNA1 regulates cellular gene expression by binding cellular promoters. *Proc. Natl. Acad. Sci. USA* **2009**, *106*, 22421–22426. [CrossRef] [PubMed]

84. Dresang, L.R.; Vereide, D.T.; Sugden, B. Identifying sites bound by Epstein-Barr virus nuclear antigen 1 (EBNA1) in the human genome: Defining a position-weighted matrix to predict sites bound by EBNA1 in viral genomes. *J. Virol.* **2009**, *83*, 2930–2940. [CrossRef] [PubMed]

85. Lu, F.; Wikramasinghe, P.; Norseen, J.; Tsai, K.; Wang, P.; Showe, L.; Davuluri, R.V.; Lieberman, P.M. Genome-wide analysis of host-chromosome binding sites for Epstein-Barr Virus Nuclear Antigen 1 (EBNA1). *Virol. J.* **2010**, *7*, 262. [CrossRef] [PubMed]

86. Kaye, K.M.; Izumi, K.M.; Kieff, E. Epstein-Barr virus latent membrane protein 1 is essential for B-lymphocyte growth transformation. *Proc. Natl. Acad. Sci. USA* **1993**, *90*, 9150–9154. [CrossRef] [PubMed]

87. Dirmeier, U.; Neuhierl, B.; Kilger, E.; Reisbach, G.; Sandberg, M.L.; Hammerschmidt, W. Latent membrane protein 1 is critical for efficient growth transformation of human B cells by epstein-barr virus. *Cancer Res.* **2003**, *63*, 2982–2989. [PubMed]

88. Rastelli, J.; Homig-Holzel, C.; Seagal, J.; Muller, W.; Hermann, A.C.; Rajewsky, K.; Zimber-Strobl, U. LMP1 signaling can replace CD40 signaling in B cells in vivo and has unique features of inducing class-switch recombination to IgG1. *Blood* **2008**, *111*, 1448–1455. [CrossRef] [PubMed]

89. Uchida, J.; Yasui, T.; Takaoka-Shichijo, Y.; Muraoka, M.; Kulwichit, W.; Raab-Traub, N.; Kikutani, H. Mimicry of CD40 signals by Epstein-Barr virus LMP1 in B lymphocyte responses. *Science* **1999**, *286*, 300–303. [CrossRef] [PubMed]

90. Izumi, K.M.; Kieff, E.D. The Epstein-Barr virus oncogene product latent membrane protein 1 engages the tumor necrosis factor receptor-associated death domain protein to mediate B lymphocyte growth transformation and activate NF-kappaB. *Proc. Natl. Acad. Sci. USA* **1997**, *94*, 12592–12597. [CrossRef] [PubMed]

91. Izumi, K.M.; Kaye, K.M.; Kieff, E.D. The Epstein-Barr virus LMP1 amino acid sequence that engages tumor necrosis factor receptor associated factors is critical for primary B lymphocyte growth transformation. *Proc. Natl. Acad. Sci. USA* **1997**, *94*, 1447–1452. [CrossRef] [PubMed]

92. Mosialos, G.; Birkenbach, M.; Yalamanchili, R.; VanArsdale, T.; Ware, C.; Kieff, E. The Epstein-Barr virus transforming protein LMP1 engages signaling proteins for the tumor necrosis factor receptor family. *Cell* **1995**, *80*, 389–399. [CrossRef]

93. Luftig, M.; Prinarakis, E.; Yasui, T.; Tsichritzis, T.; Cahir-McFarland, E.; Inoue, J.; Nakano, H.; Mak, T.W.; Yeh, W.C.; Li, X.; et al. Epstein-Barr virus latent membrane protein 1 activation of NF-kappaB through IRAK1 and TRAF6. *Proc. Natl. Acad. Sci. USA* **2003**, *100*, 15595–15600. [CrossRef] [PubMed]

94. Edwards, R.H.; Marquitz, A.R.; Raab-Traub, N. Changes in expression induced by Epstein-Barr Virus LMP1-CTAR1: Potential role of bcl3. *MBio* **2015**, *6*. [CrossRef] [PubMed]

95. Gewurz, B.E.; Mar, J.C.; Padi, M.; Zhao, B.; Shinners, N.P.; Takasaki, K.; Bedoya, E.; Zou, J.Y.; Cahir-McFarland, E.; Quackenbush, J.; et al. Canonical NF-κB activation is essential for Epstein-Barr virus latent membrane protein 1 TES2/CTAR2 gene regulation. *J. Virol.* **2011**, *85*, 6764–6773. [CrossRef] [PubMed]

96. Ersing, I.; Bernhardt, K.; Gewurz, B.E. NF-kappaB and IRF7 pathway activation by Epstein-Barr virus Latent Membrane Protein 1. *Viruses* **2013**, *5*, 1587–1606. [CrossRef] [PubMed]

97. Kung, C.P.; Raab-Traub, N. Epstein-Barr virus latent membrane protein 1 modulates distinctive NF-κB pathways through C-terminus-activating region 1 to regulate epidermal growth factor receptor expression. *J. Virol.* **2010**, *84*, 6605–6614. [CrossRef] [PubMed]

98. Pratt, Z.L.; Zhang, J.; Sugden, B. The latent membrane protein 1 (LMP1) oncogene of Epstein-Barr virus can simultaneously induce and inhibit apoptosis in B cells. *J. Virol.* **2012**, *86*, 4380–4393. [CrossRef] [PubMed]

99. D'Souza, B.; Rowe, M.; Walls, D. The bfl-1 gene is transcriptionally upregulated by the Epstein-Barr virus LMP1, and its expression promotes the survival of a Burkitt's lymphoma cell line. *J. Virol.* **2000**, *74*, 6652–6658. [CrossRef] [PubMed]

100. D'Souza, B.N.; Edelstein, L.C.; Pegman, P.M.; Smith, S.M.; Loughran, S.T.; Clarke, A.; Mehl, A.; Rowe, M.; Gelinas, C.; Walls, D. Nuclear factor κB -dependent activation of the antiapoptotic bfl-1 gene by the Epstein-Barr virus latent membrane protein 1 and activated CD40 receptor. *J. Virol.* **2004**, *78*, 1800–1816. [CrossRef] [PubMed]

101. Henderson, S.; Rowe, M.; Gregory, C.; Croom-Carter, D.; Wang, F.; Longnecker, R.; Kieff, E.; Rickinson, A. Induction of bcl-2 expression by Epstein-Barr virus latent membrane protein 1 protects infected B cells from programmed cell death. *Cell* **1991**, *65*, 1107–1115. [CrossRef]

102. Wang, S.; Rowe, M.; Lundgren, E. Expression of the Epstein Barr virus transforming protein LMP1 causes a rapid and transient stimulation of the Bcl-2 homologue Mcl-1 levels in B-cell lines. *Cancer Res.* **1996**, *56*, 4610–4613. [PubMed]

103. Tsai, S.C.; Lin, S.J.; Lin, C.J.; Chou, Y.C.; Lin, J.H.; Yeh, T.H.; Chen, M.R.; Huang, L.M.; Lu, M.Y.; Huang, Y.C.; et al. Autocrine CCL3 and CCL4 induced by the oncoprotein LMP1 promote Epstein-Barr virus-triggered B cell proliferation. *J. Virol.* **2013**, *87*, 9041–9052. [CrossRef] [PubMed]

104. Zhao, B.; Barrera, L.A.; Ersing, I.; Willox, B.; Schmidt, S.C.; Greenfeld, H.; Zhou, H.; Mollo, S.B.; Shi, T.T.; Takasaki, K.; et al. The NF-κB genomic landscape in lymphoblastoid B cells. *Cell Rep.* **2014**, *8*, 1595–1606. [CrossRef] [PubMed]

105. Ma, Y.; Walsh, M.J.; Bernhardt, K.; Ashbaugh, C.W.; Trudeau, S.J.; Ashbaugh, I.Y.; Jiang, S.; Jiang, C.; Zhao, B.; Root, D.E.; et al. CRISPR/Cas9 Screens Reveal Epstein-Barr Virus-Transformed B Cell Host Dependency Factors. *Cell Host Microbe* **2017**, *21*, 580–591. [CrossRef] [PubMed]

106. Le Clorennec, C.; Ouk, T.S.; Youlyouz-Marfak, I.; Panteix, S.; Martin, C.C.; Rastelli, J.; Adriaenssens, E.; Zimber-Strobl, U.; Coll, J.; Feuillard, J.; et al. Molecular basis of cytotoxicity of Epstein-Barr virus (EBV) latent membrane protein 1 (LMP1) in EBV latency III B cells: LMP1 induces type II ligand-independent autoactivation of CD95/Fas with caspase 8-mediated apoptosis. *J. Virol.* **2008**, *82*, 6721–6733. [CrossRef] [PubMed]

107. Le Clorennec, C.; Youlyouz-Marfak, I.; Adriaenssens, E.; Coll, J.; Bornkamm, G.W.; Feuillard, J. EBV latency III immortalization program sensitizes B cells to induction of CD95-mediated apoptosis via LMP1: Role of NF-κB, STAT1, and p53. *Blood* **2006**, *107*, 2070–2078. [CrossRef] [PubMed]

108. Laux, G.; Dugrillon, F.; Eckert, C.; Adam, B.; Zimber-Strobl, U.; Bornkamm, G.W. Identification and characterization of an Epstein-Barr virus nuclear antigen 2-responsive cis element in the bidirectional promoter region of latent membrane protein and terminal protein 2 genes. *J. Virol.* **1994**, *68*, 6947–6958. [PubMed]

109. Laux, G.; Perricaudet, M.; Farrell, P.J. A spliced Epstein-Barr virus gene expressed in immortalized lymphocytes is created by circularization of the linear viral genome. *EMBO J.* **1988**, *7*, 769–774. [PubMed]

110. Sample, J.; Liebowitz, D.; Kieff, E. Two related Epstein-Barr virus membrane proteins are encoded by separate genes. *J. Virol.* **1989**, *63*, 933–937. [PubMed]

111. Kim, O.J.; Yates, J.L. Mutants of Epstein-Barr virus with a selective marker disrupting the TP gene transform B cells and replicate normally in culture. *J. Virol.* **1993**, *67*, 7634–7640. [PubMed]

112. Longnecker, R.; Miller, C.L.; Tomkinson, B.; Miao, X.Q.; Kieff, E. Deletion of DNA encoding the first five transmembrane domains of Epstein-Barr virus latent membrane proteins 2A and 2B. *J. Virol.* **1993**, *67*, 5068–5074. [PubMed]

113. Longnecker, R.; Miller, C.L.; Miao, X.Q.; Tomkinson, B.; Kieff, E. The last seven transmembrane and carboxy-terminal cytoplasmic domains of Epstein-Barr virus latent membrane protein 2 (LMP2) are dispensable for lymphocyte infection and growth transformation in vitro. *J. Virol.* **1993**, *67*, 2006–2013. [PubMed]

114. Speck, P.; Kline, K.A.; Cheresh, P.; Longnecker, R. Epstein-Barr virus lacking latent membrane protein 2 immortalizes B cells with efficiency indistinguishable from that of wild-type virus. *J. Gen. Virol.* **1999**, *80*, 2193–2203. [CrossRef] [PubMed]

115. Fukuda, M.; Longnecker, R. Epstein-Barr virus latent membrane protein 2A mediates transformation through constitutive activation of the Ras/PI3-K/Akt Pathway. *J. Virol.* **2007**, *81*, 9299–9306. [CrossRef] [PubMed]

116. Cen, O.; Longnecker, R. Latent Membrane Protein 2 (LMP2). *Curr. Top. Microbiol. Immunol.* **2015**, *391*, 151–180. [PubMed]

117. Mancao, C.; Altmann, M.; Jungnickel, B.; Hammerschmidt, W. Rescue of "crippled" germinal center B cells from apoptosis by Epstein-Barr virus. *Blood* **2005**, *106*, 4339–4344. [CrossRef] [PubMed]

118. Mancao, C.; Hammerschmidt, W. Epstein-Barr virus latent membrane protein 2A is a B-cell receptor mimic and essential for B-cell survival. *Blood* **2007**, *110*, 3715–3721. [CrossRef] [PubMed]

119. Merchant, M.; Longnecker, R. LMP2A survival and developmental signals are transmitted through Btk-dependent and Btk-independent pathways. *Virology* **2001**, *291*, 46–54. [CrossRef] [PubMed]

120. Merchant, M.; Swart, R.; Katzman, R.B.; Ikeda, M.; Ikeda, A.; Longnecker, R.; Dykstra, M.L.; Pierce, S.K. The effects of the Epstein-Barr virus latent membrane protein 2A on B cell function. *Int. Rev. Immunol.* **2001**, *20*, 805–835. [CrossRef] [PubMed]

121. Swart, R.; Ruf, I.K.; Sample, J.; Longnecker, R. Latent membrane protein 2A-mediated effects on the phosphatidylinositol 3-Kinase/Akt pathway. *J. Virol.* **2000**, *74*, 10838–10845. [CrossRef] [PubMed]

122. Laux, G.; Economou, A.; Farrell, P.J. The terminal protein gene 2 of Epstein-Barr virus is transcribed from a bidirectional latent promoter region. *J. Gen. Virol.* **1989**, *70*, 3079–3084. [CrossRef] [PubMed]

123. Rovedo, M.; Longnecker, R. Epstein-Barr virus latent membrane protein 2A preferentially signals through the Src family kinase Lyn. *J. Virol.* **2008**, *82*, 8520–8528. [CrossRef] [PubMed]

124. Kvansakul, M.; Hinds, M.G. Structural biology of the Bcl-2 family and its mimicry by viral proteins. *Cell Death Dis.* **2013**, *4*, e909. [CrossRef] [PubMed]

125. Cooray, S.; Bahar, M.W.; Abrescia, N.G.; McVey, C.E.; Bartlett, N.W.; Chen, R.A.; Stuart, D.I.; Grimes, J.M.; Smith, G.L. Functional and structural studies of the vaccinia virus virulence factor N1 reveal a Bcl-2-like anti-apoptotic protein. *J. Gen. Virol.* **2007**, *88*, 1656–1666. [CrossRef] [PubMed]

126. Aoyagi, M.; Zhai, D.; Jin, C.; Aleshin, A.E.; Stec, B.; Reed, J.C.; Liddington, R.C. Vaccinia virus N1L protein resembles a B cell lymphoma-2 (Bcl-2) family protein. *Protein Sci.* **2007**, *16*, 118–124. [CrossRef] [PubMed]

127. DiPerna, G.; Stack, J.; Bowie, A.G.; Boyd, A.; Kotwal, G.; Zhang, Z.; Arvikar, S.; Latz, E.; Fitzgerald, K.A.; Marshall, W.L. Poxvirus protein N1L targets the I-κB kinase complex, inhibits signaling to NF-κB by the tumor necrosis factor superfamily of receptors, and inhibits NF-κB and IRF3 signaling by toll-like receptors. *J. Biol. Chem.* **2004**, *279*, 36570–36578. [CrossRef] [PubMed]

128. Graham, S.C.; Bahar, M.W.; Cooray, S.; Chen, R.A.; Whalen, D.M.; Abrescia, N.G.; Alderton, D.; Owens, R.J.; Stuart, D.I.; Smith, G.L.; et al. Vaccinia virus proteins A52 and B14 Share a Bcl-2-like fold but have evolved to inhibit NF-κB rather than apoptosis. *PLoS Pathog.* **2008**, *4*, e1000128. [CrossRef] [PubMed]

129. Altmann, M.; Hammerschmidt, W. Epstein-Barr virus provides a new paradigm: A requirement for the immediate inhibition of apoptosis. *PLoS Biol.* **2005**, *3*, e404. [CrossRef] [PubMed]

130. Marshall, W.L.; Yim, C.; Gustafson, E.; Graf, T.; Sage, D.R.; Hanify, K.; Williams, L.; Fingeroth, J.; Finberg, R.W. Epstein-Barr virus encodes a novel homolog of the bcl-2 oncogene that inhibits apoptosis and associates with Bax and Bak. *J. Virol.* **1999**, *73*, 5181–5185. [PubMed]

131. Henderson, S.; Huen, D.; Rowe, M.; Dawson, C.; Johnson, G.; Rickinson, A. Epstein-Barr virus-coded BHRF1 protein, a viral homologue of Bcl-2, protects human B cells from programmed cell death. *Proc. Natl. Acad. Sci. USA* **1993**, *90*, 8479–8483. [CrossRef] [PubMed]

132. Foight, G.W.; Keating, A.E. Locating Herpesvirus Bcl-2 Homologs in the Specificity Landscape of Anti-Apoptotic Bcl-2 Proteins. *J. Mol. Biol.* **2015**, *427*, 2468–2490. [CrossRef] [PubMed]

133. Kvansakul, M.; Wei, A.H.; Fletcher, J.I.; Willis, S.N.; Chen, L.; Roberts, A.W.; Huang, D.C.; Colman, P.M. Structural basis for apoptosis inhibition by Epstein-Barr virus BHRF1. *PLoS Pathog.* **2010**, *6*, e1001236. [CrossRef] [PubMed]

134. Flanagan, A.M.; Letai, A. BH3 domains define selective inhibitory interactions with BHRF-1 and KSHV BCL-2. *Cell Death Differ.* **2008**, *15*, 580–588. [CrossRef] [PubMed]

135. Desbien, A.L.; Kappler, J.W.; Marrack, P. The Epstein-Barr virus Bcl-2 homolog, BHRF1, blocks apoptosis by binding to a limited amount of Bim. *Proc. Natl. Acad. Sci. USA* **2009**, *106*, 5663–5668. [CrossRef] [PubMed]

136. Fanidi, A.; Hancock, D.C.; Littlewood, T.D. Suppression of c-Myc-induced apoptosis by the Epstein-Barr virus gene product BHRF1. *J. Virol.* **1998**, *72*, 8392–8395. [PubMed]

137. Foghsgaard, L.; Jaattela, M. The ability of BHRF1 to inhibit apoptosis is dependent on stimulus and cell type. *J. Virol.* **1997**, *71*, 7509–7517. [PubMed]

138. Kawanishi, M.; Tada-Oikawa, S.; Kawanishi, S. Epstein-Barr virus BHRF1 functions downstream of Bid cleavage and upstream of mitochondrial dysfunction to inhibit TRAIL-induced apoptosis in BJAB cells. *Biochem. Biophys. Res. Commun.* **2002**, *297*, 682–687. [CrossRef]

139. McCarthy, N.J.; Hazlewood, S.A.; Huen, D.S.; Rickinson, A.B.; Williams, G.T. The Epstein-Barr virus gene BHRF1, a homologue of the cellular oncogene Bcl-2, inhibits apoptosis induced by gamma radiation and chemotherapeutic drugs. *Adv. Exp. Med. Biol.* **1996**, *406*, 83–97. [PubMed]

140. Watanabe, A.; Maruo, S.; Ito, T.; Ito, M.; Katsumura, K.R.; Takada, K. Epstein-Barr virus-encoded Bcl-2 homologue functions as a survival factor in Wp-restricted Burkitt lymphoma cell line P3HR-1. *J. Virol.* **2010**, *84*, 2893–2901. [CrossRef] [PubMed]

141. Yee, J.; White, R.E.; Anderton, E.; Allday, M.J. Latent Epstein-Barr Virus Can Inhibit Apoptosis in B Cells by Blocking the Induction of NOXA Expression. *PLoS ONE* **2011**, *6*, e28506. [CrossRef] [PubMed]

142. Zhao, E.G.; Song, Q.; Cross, S.; Misko, I.; Lees-Miller, S.P.; Lavin, M.F. Resistance to etoposide-induced apoptosis in a Burkitt's lymphoma cell line. *Int. J. Cancer* **1998**, *77*, 755–762. [CrossRef]

143. Bellows, D.S.; Howell, M.; Pearson, C.; Hazlewood, S.A.; Hardwick, J.M. Epstein-Barr virus BALF1 is a BCL-2-like antagonist of the herpesvirus antiapoptotic BCL-2 proteins. *J. Virol.* **2002**, *76*, 2469–2479. [CrossRef] [PubMed]

144. Chen, L.; Willis, S.N.; Wei, A.; Smith, B.J.; Fletcher, J.I.; Hinds, M.G.; Colman, P.M.; Day, C.L.; Adams, J.M.; Huang, D.C. Differential targeting of prosurvival Bcl-2 proteins by their BH3-only ligands allows complementary apoptotic function. *Mol. Cell* **2005**, *17*, 393–403. [CrossRef] [PubMed]

145. Willis, S.N.; Chen, L.; Dewson, G.; Wei, A.; Naik, E.; Fletcher, J.I.; Adams, J.M.; Huang, D.C. Proapoptotic Bak is sequestered by Mcl-1 and Bcl-xL, but not Bcl-2, until displaced by BH3-only proteins. *Genes. Dev.* **2005**, *19*, 1294–1305. [CrossRef] [PubMed]

146. Fletcher, J.I.; Meusburger, S.; Hawkins, C.J.; Riglar, D.T.; Lee, E.F.; Fairlie, W.D.; Huang, D.C.; Adams, J.M. Apoptosis is triggered when prosurvival Bcl-2 proteins cannot restrain Bax. *Proc. Natl. Acad. Sci. USA* **2008**, *105*, 18081–18087. [CrossRef] [PubMed]

147. Smits, C.; Czabotar, P.E.; Hinds, M.G.; Day, C.L. Structural plasticity underpins promiscuous binding of the prosurvival protein A1. *Structure* **2008**, *16*, 818–829. [CrossRef] [PubMed]

148. Pettersen, E.F.; Goddard, T.D.; Huang, C.C.; Couch, G.S.; Greenblatt, D.M.; Meng, E.C.; Ferrin, T.E. UCSF Chimera—A visualization system for exploratory research and analysis. *J. Comput. Chem.* **2004**, *25*, 1605–1612. [CrossRef] [PubMed]

149. Lerner, M.R.; Andrews, N.C.; Miller, G.; Steitz, J.A. Two small RNAs encoded by Epstein-Barr virus and complexed with protein are precipitated by antibodies from patients with systemic lupus erythematosus. *Proc. Natl. Acad. Sci. USA* **1981**, *78*, 805–809. [CrossRef] [PubMed]

150. Shannon-Lowe, C.; Adland, E.; Bell, A.I.; Delecluse, H.J.; Rickinson, A.B.; Rowe, M. Features distinguishing Epstein-Barr virus infections of epithelial cells and B cells: Viral genome expression, genome maintenance, and genome amplification. *J. Virol.* **2009**, *83*, 7749–7760. [CrossRef] [PubMed]

151. Zhu, J.Y.; Pfuhl, T.; Motsch, N.; Barth, S.; Nicholls, J.; Grasser, F.; Meister, G. Identification of novel Epstein-Barr virus microRNA genes from nasopharyngeal carcinomas. *J. Virol.* **2009**, *83*, 3333–3341. [CrossRef] [PubMed]

152. Chen, S.J.; Chen, G.H.; Chen, Y.H.; Liu, C.Y.; Chang, K.P.; Chang, Y.S.; Chen, H.C. Characterization of Epstein-Barr virus miRNAome in nasopharyngeal carcinoma by deep sequencing. *PLoS ONE* **2010**, *5*. [CrossRef] [PubMed]

153. Grundhoff, A.; Sullivan, C.S.; Ganem, D. A combined computational and microarray-based approach identifies novel microRNAs encoded by human gamma-herpesviruses. *RNA* **2006**, *12*, 733–750. [CrossRef] [PubMed]

154. Edwards, R.H.; Marquitz, A.R.; Raab-Traub, N. Epstein-Barr virus BART microRNAs are produced from a large intron prior to splicing. *J. Virol.* **2008**, *82*, 9094–9106. [CrossRef] [PubMed]

155. Bartel, D.P. MicroRNAs: Genomics, biogenesis, mechanism, and function. *Cell* **2004**, *116*, 281–297. [CrossRef]

156. Ambros, V. The functions of animal microRNAs. *Nature* **2004**, *431*, 350–355. [CrossRef] [PubMed]

157. Swaminathan, S.; Tomkinson, B.; Kieff, E. Recombinant Epstein-Barr virus with small RNA (EBER) genes deleted transforms lymphocytes and replicates in vitro. *Proc. Natl. Acad. Sci. USA* **1991**, *88*, 1546–1550. [CrossRef] [PubMed]

158. Swaminathan, S.; Huneycutt, B.S.; Reiss, C.S.; Kieff, E. Epstein-Barr virus-encoded small RNAs (EBERs) do not modulate interferon effects in infected lymphocytes. *J. Virol.* **1992**, *66*, 5133–5136. [PubMed]

159. Yajima, M.; Kanda, T.; Takada, K. Critical role of Epstein-Barr Virus (EBV)-encoded RNA in efficient EBV-induced B-lymphocyte growth transformation. *J. Virol.* **2005**, *79*, 4298–4307. [CrossRef] [PubMed]

160. Wu, Y.; Maruo, S.; Yajima, M.; Kanda, T.; Takada, K. Epstein-Barr virus (EBV)-encoded RNA 2 (EBER2) but not EBER1 plays a critical role in EBV-induced B-cell growth transformation. *J. Virol.* **2007**, *81*, 11236–11245. [CrossRef] [PubMed]

161. Gregorovic, G.; Bosshard, R.; Karstegl, C.E.; White, R.E.; Pattle, S.; Chiang, A.K.; Dittrich-Breiholz, O.; Kracht, M.; Russ, R.; Farrell, P.J. Cellular gene expression that correlates with EBER expression in Epstein-Barr Virus-infected lymphoblastoid cell lines. *J. Virol.* **2011**, *85*, 3535–3545. [CrossRef] [PubMed]

162. Benetti, R.; Del Sal, G.; Monte, M.; Paroni, G.; Brancolini, C.; Schneider, C. The death substrate Gas2 binds m-calpain and increases susceptibility to p53-dependent apoptosis. *EMBO J.* **2001**, *20*, 2702–2714. [CrossRef] [PubMed]

163. Burgess, J.T.; Bolderson, E.; Adams, M.N.; Baird, A.M.; Zhang, S.D.; Gately, K.A.; Umezawa, K.; O'Byrne, K.J.; Richard, D.J. Activation and cleavage of SASH1 by caspase-3 mediates an apoptotic response. *Cell Death Dis.* **2016**, *7*, e2469. [CrossRef] [PubMed]

164. Xia, T.; O'Hara, A.; Araujo, I.; Barreto, J.; Carvalho, E.; Sapucaia, J.B.; Ramos, J.C.; Luz, E.; Pedroso, C.; Manrique, M.; et al. EBV microRNAs in primary lymphomas and targeting of CXCL-11 by ebv-mir-BHRF1-3. *Cancer Res.* **2008**, *68*, 1436–1442. [CrossRef] [PubMed]

165. Skalsky, R.L.; Corcoran, D.L.; Gottwein, E.; Frank, C.L.; Kang, D.; Hafner, M.; Nusbaum, J.D.; Feederle, R.; Delecluse, H.J.; Luftig, M.A.; et al. The viral and cellular microRNA targetome in lymphoblastoid cell lines. *PLoS Pathog.* **2012**, *8*, e1002484. [CrossRef] [PubMed]

166. Pratt, Z.L.; Kuzembayeva, M.; Sengupta, S.; Sugden, B. The microRNAs of Epstein-Barr Virus are expressed at dramatically differing levels among cell lines. *Virology* **2009**, *386*, 387–397. [CrossRef] [PubMed]

167. Feederle, R.; Haar, J.; Bernhardt, K.; Linnstaedt, S.D.; Bannert, H.; Lips, H.; Cullen, B.R.; Delecluse, H.J. The members of an Epstein-Barr virus microRNA cluster cooperate to transform B lymphocytes. *J. Virol.* **2011**, *85*, 9801–9810. [CrossRef] [PubMed]

168. Feederle, R.; Linnstaedt, S.D.; Bannert, H.; Lips, H.; Bencun, M.; Cullen, B.R.; Delecluse, H.J. A viral microRNA cluster strongly potentiates the transforming properties of a human herpesvirus. *PLoS Pathog.* **2011**, *7*, e1001294. [CrossRef] [PubMed]

169. Seto, E.; Moosmann, A.; Gromminger, S.; Walz, N.; Grundhoff, A.; Hammerschmidt, W. Micro RNAs of Epstein-Barr virus promote cell cycle progression and prevent apoptosis of primary human B cells. *PLoS Pathog.* **2010**, *6*. [CrossRef] [PubMed]

170. Wahl, A.; Linnstaedt, S.D.; Esoda, C.; Krisko, J.F.; Martinez-Torres, F.; Delecluse, H.J.; Cullen, B.R.; Garcia, J.V. A cluster of virus-encoded microRNAs accelerates acute systemic Epstein-Barr virus infection but does not significantly enhance virus-induced oncogenesis in vivo. *J. Virol.* **2013**, *87*, 5437–5446. [CrossRef] [PubMed]

171. Majoros, W.H.; Lekprasert, P.; Mukherjee, N.; Skalsky, R.L.; Corcoran, D.L.; Cullen, B.R.; Ohler, U. MicroRNA target site identification by integrating sequence and binding information. *Nat. Methods* **2013**, *10*, 630–633. [CrossRef] [PubMed]

172. Skalsky, R.L.; Cullen, B.R. EBV Noncoding RNAs. *Curr. Top. Microbiol. Immunol.* **2015**, *391*, 181–217. [PubMed]

173. Song, M.S.; Salmena, L.; Pandolfi, P.P. The functions and regulation of the PTEN tumour suppressor. *Nat. Rev. Mol. Cell Biol.* **2012**, *13*, 283–296. [CrossRef] [PubMed]

174. Warnatz, H.J.; Schmidt, D.; Manke, T.; Piccini, I.; Sultan, M.; Borodina, T.; Balzereit, D.; Wruck, W.; Soldatov, A.; Vingron, M.; et al. The BTB and CNC homology 1 (BACH1) target genes are involved in the oxidative stress response and in control of the cell cycle. *J. Biol. Chem.* **2011**, *286*, 23521–23532. [CrossRef] [PubMed]

175. Castellini, L.; Moon, E.J.; Razorenova, O.V.; Krieg, A.J.; von Eyben, R.; Giaccia, A.J. KDM4B/JMJD2B is a p53 target gene that modulates the amplitude of p53 response after DNA damage. *Nucleic Acids Res.* **2017**, *45*, 3674–3692. [CrossRef] [PubMed]

176. Gilligan, K.J.; Rajadurai, P.; Lin, J.C.; Busson, P.; Abdel-Hamid, M.; Prasad, U.; Tursz, T.; Raab-Traub, N. Expression of the Epstein-Barr virus BamHI A fragment in nasopharyngeal carcinoma: Evidence for a viral protein expressed in vivo. *J. Virol.* **1991**, *65*, 6252–6259. [PubMed]

177. Hitt, M.M.; Allday, M.J.; Hara, T.; Karran, L.; Jones, M.D.; Busson, P.; Tursz, T.; Ernberg, I.; Griffin, B.E. EBV gene expression in an NPC-related tumour. *EMBO J.* **1989**, *8*, 2639–2651. [PubMed]

178. Sadler, R.H.; Raab-Traub, N. Structural analyses of the Epstein-Barr virus BamHI A transcripts. *J. Virol.* **1995**, *69*, 1132–1141. [PubMed]

179. Smith, P.R.; Gao, Y.; Karran, L.; Jones, M.D.; Snudden, D.; Griffin, B.E. Complex nature of the major viral polyadenylated transcripts in Epstein-Barr virus-associated tumors. *J. Virol.* **1993**, *67*, 3217–3225. [PubMed]

180. Zhang, J.; Chen, H.; Weinmaster, G.; Hayward, S.D. Epstein-Barr virus BamHi-a rightward transcript-encoded RPMS protein interacts with the CBF1-associated corepressor CIR to negatively regulate the activity of EBNA2 and NotchIC. *J. Virol.* **2001**, *75*, 2946–2956. [CrossRef] [PubMed]

181. Smith, P.R.; de Jesus, O.; Turner, D.; Hollyoake, M.; Karstegl, C.E.; Griffin, B.E.; Karran, L.; Wang, Y.; Hayward, S.D.; Farrell, P.J. Structure and coding content of CST (BART) family RNAs of Epstein-Barr virus. *J. Virol.* **2000**, *74*, 3082–3092. [CrossRef] [PubMed]

182. Kusano, S.; Raab-Traub, N. An Epstein-Barr virus protein interacts with Notch. *J. Virol.* **2001**, *75*, 384–395. [CrossRef] [PubMed]

183. Al-Mozaini, M.; Bodelon, G.; Karstegl, C.E.; Jin, B.; Al-Ahdal, M.; Farrell, P.J. Epstein-Barr virus BART gene expression. *J. Gen. Virol.* **2009**, *90*, 307–316. [CrossRef] [PubMed]

184. Van Beek, J.; Brink, A.A.; Vervoort, M.B.; van Zijp, M.J.; Meijer, C.J.; van den Brule, A.J.; Middeldorp, J.M. In vivo transcription of the Epstein-Barr virus (EBV) BamHI-A region without associated in vivo BARF0 protein expression in multiple EBV-associated disorders. *J. Gen. Virol.* **2003**, *84*, 2647–2659. [CrossRef] [PubMed]

185. Bornkamm, G.W.; Delius, H.; Zimber, U.; Hudewentz, J.; Epstein, M.A. Comparison of Epstein-Barr virus strains of different origin by analysis of the viral DNAs. *J. Virol.* **1980**, *35*, 603–618. [PubMed]

186. Raab-Traub, N.; Dambaugh, T.; Kieff, E. DNA of Epstein-Barr virus VIII: B95–8, the previous prototype, is an unusual deletion derivative. *Cell* **1980**, *22*, 257–267. [CrossRef]

187. Vereide, D.T.; Seto, E.; Chiu, Y.F.; Hayes, M.; Tagawa, T.; Grundhoff, A.; Hammerschmidt, W.; Sugden, B. Epstein-Barr virus maintains lymphomas via its miRNAs. *Oncogene* **2014**, *33*, 1258–1264. [CrossRef] [PubMed]

188. Dolken, L.; Malterer, G.; Erhard, F.; Kothe, S.; Friedel, C.C.; Suffert, G.; Marcinowski, L.; Motsch, N.; Barth, S.; Beitzinger, M.; et al. Systematic analysis of viral and cellular microRNA targets in cells latently infected with human gamma-herpesviruses by RISC immunoprecipitation assay. *Cell Host Microbe* **2010**, *7*, 324–334. [CrossRef] [PubMed]

189. Kang, H.S.; Ock, J.; Lee, H.J.; Lee, Y.J.; Kwon, B.M.; Hong, S.H. Early growth response protein 1 upregulation and nuclear translocation by 2′-benzoyloxycinnamaldehyde induces prostate cancer cell death. *Cancer Lett.* **2013**, *329*, 217–227. [CrossRef] [PubMed]

190. Bellot, G.; Cartron, P.F.; Er, E.; Oliver, L.; Juin, P.; Armstrong, L.C.; Bornstein, P.; Mihara, K.; Manon, S.; Vallette, F.M. TOM22, a core component of the mitochondria outer membrane protein translocation pore, is a mitochondrial receptor for the proapoptotic protein Bax. *Cell Death Differ.* **2007**, *14*, 785–794. [CrossRef] [PubMed]

191. Chi, S.W.; Zang, J.B.; Mele, A.; Darnell, R.B. Argonaute HITS-CLIP decodes microRNA-mRNA interaction maps. *Nature* **2009**, *460*, 479–486. [CrossRef] [PubMed]

192. Riley, K.J.; Rabinowitz, G.S.; Yario, T.A.; Luna, J.M.; Darnell, R.B.; Steitz, J.A. EBV and human microRNAs co-target oncogenic and apoptotic viral and human genes during latency. *EMBO J.* **2012**, *31*, 2207–2221. [CrossRef] [PubMed]

193. Lin, Z.; Xu, G.; Deng, N.; Taylor, C.; Zhu, D.; Flemington, E.K. Quantitative and qualitative RNA-Seq-based evaluation of Epstein-Barr virus transcription in type I latency Burkitt's lymphoma cells. *J. Virol.* **2010**, *84*, 13053–13058. [CrossRef] [PubMed]

194. Moss, W.N.; Steitz, J.A. Genome-wide analyses of Epstein-Barr virus reveal conserved RNA structures and a novel stable intronic sequence RNA. *BMC Genom.* **2013**, *14*, 543. [CrossRef] [PubMed]

195. Thorley-Lawson, D.A. EBV Persistence—Introducing the Virus. *Curr. Top. Microbiol. Immunol.* **2015**, *390*, 151–209. [PubMed]

196. Vrzalikova, K.; Vockerodt, M.; Leonard, S.; Bell, A.; Wei, W.; Schrader, A.; Wright, K.L.; Kube, D.; Rowe, M.; Woodman, C.B.; et al. Down-regulation of BLIMP1alpha by the EBV oncogene, LMP-1, disrupts the plasma cell differentiation program and prevents viral replication in B cells: Implications for the pathogenesis of EBV-associated B-cell lymphomas. *Blood* **2011**, *117*, 5907–5917. [CrossRef] [PubMed]

197. Sinclair, A.J. Epigenetic control of Epstein-Barr virus transcription—Relevance to viral life cycle? *Front. Genet.* **2013**, *4*, 161. [CrossRef] [PubMed]

198. Niller, H.H.; Wolf, H.; Minarovits, J. Epigenetic dysregulation of the host cell genome in Epstein-Barr virus-associated neoplasia. *Semin. Cancer Biol.* **2009**, *19*, 158–164. [CrossRef] [PubMed]

199. Kalla, M.; Schmeinck, A.; Bergbauer, M.; Pich, D.; Hammerschmidt, W. AP-1 homolog BZLF1 of Epstein-Barr virus has two essential functions dependent on the epigenetic state of the viral genome. *Proc. Natl. Acad. Sci. USA* **2010**, *107*, 850–855. [CrossRef] [PubMed]

200. Woellmer, A.; Arteaga-Salas, J.M.; Hammerschmidt, W. BZLF1 governs CpG-methylated chromatin of Epstein-Barr Virus reversing epigenetic repression. *PLoS Pathog.* **2012**, *8*, e1002902. [CrossRef] [PubMed]

201. Gruffat, H.; Sergeant, A. Characterization of the DNA-binding site repertoire for the Epstein-Barr virus transcription factor R. *Nucleic Acids Res.* **1994**, *22*, 1172–1178. [CrossRef] [PubMed]

202. Gutsch, D.E.; Marcu, K.B.; Kenney, S.C. The Epstein-Barr virus BRLF1 gene product transactivates the murine and human c-myc promoters. *Cell. Mol. Biol.* **1994**, *40*, 747–760. [PubMed]

203. Ragoczy, T.; Miller, G. Autostimulation of the Epstein-Barr virus BRLF1 promoter is mediated through consensus Sp1 and Sp3 binding sites. *J. Virol.* **2001**, *75*, 5240–5251. [CrossRef] [PubMed]

204. Hammerschmidt, W.; Sugden, B. Identification and characterization of oriLyt, a lytic origin of DNA replication of Epstein-Barr virus. *Cell* **1988**, *55*, 427–433. [CrossRef]

205. Kawanishi, M. Epstein-Barr virus induces fragmentation of chromosomal DNA during lytic infection. *J. Virol.* **1993**, *67*, 7654–7658. [PubMed]

206. Morrison, T.E.; Kenney, S.C. BZLF1, an Epstein-Barr virus immediate-early protein, induces p65 nuclear translocation while inhibiting p65 transcriptional function. *Virology* **2004**, *328*, 219–232. [CrossRef] [PubMed]

207. Zuo, J.; Thomas, W.A.; Haigh, T.A.; Fitzsimmons, L.; Long, H.M.; Hislop, A.D.; Taylor, G.S.; Rowe, M. Epstein-Barr virus evades CD4+ T cell responses in lytic cycle through BZLF1-mediated downregulation of CD74 and the cooperation of vBcl-2. *PLoS Pathog.* **2011**, *7*, e1002455. [CrossRef] [PubMed]

208. Williams, L.R.; Quinn, L.L.; Rowe, M.; Zuo, J. Induction of the Lytic Cycle Sensitizes Epstein-Barr Virus-Infected B Cells to NK Cell Killing That Is Counteracted by Virus-Mediated NK Cell Evasion Mechanisms in the Late Lytic Cycle. *J. Virol.* **2015**, *90*, 947–958. [CrossRef] [PubMed]

209. Inman, G.J.; Binne, U.K.; Parker, G.A.; Farrell, P.J.; Allday, M.J. Activators of the Epstein-Barr Virus Lytic Program Concomitantly Induce Apoptosis, but Lytic Gene Expression Protects from Cell Death. *J. Virol.* **2001**, *75*, 2400–2410. [CrossRef] [PubMed]

210. Oussaief, L.; Hippocrate, A.; Clybouw, C.; Rampanou, A.; Ramirez, V.; Desgranges, C.; Vazquez, A.; Khelifa, R.; Joab, I. Activation of the lytic program of the Epstein-Barr virus in Burkitt's lymphoma cells leads to a two steps downregulation of expression of the proapoptotic protein BimEL, one of which is EBV-late-gene expression dependent. *Virology* **2009**, *387*, 41–49. [CrossRef] [PubMed]

211. Wen, W.; Iwakiri, D.; Yamamoto, K.; Maruo, S.; Kanda, T.; Takada, K. Epstein-Barr virus BZLF1 gene, a switch from latency to lytic infection, is expressed as an immediate-early gene after primary infection of B lymphocytes. *J. Virol.* **2007**, *81*, 1037–1042. [CrossRef] [PubMed]

212. Zeidler, R.; Eissner, G.; Meissner, P.; Uebel, S.; Tampe, R.; Lazis, S.; Hammerschmidt, W. Downregulation of TAP1 in B lymphocytes by cellular and Epstein-Barr virus-encoded interleukin-10. *Blood* **1997**, *90*, 2390–2397. [PubMed]

213. Kim, H.; Choi, H.; Lee, S.K. Epstein-Barr Virus MicroRNA miR-BART20–5p Suppresses Lytic Induction by Inhibiting BAD-Mediated caspase-3-Dependent Apoptosis. *J. Virol.* **2015**, *90*, 1359–1368. [CrossRef] [PubMed]

214. Yuan, J.; Cahir-McFarland, E.; Zhao, B.; Kieff, E. Virus and cell RNAs expressed during Epstein-Barr virus replication. *J. Virol.* **2006**, *80*, 2548–2565. [CrossRef] [PubMed]

215. Nikitin, P.A.; Price, A.M.; McFadden, K.; Yan, C.M.; Luftig, M.A. Mitogen-induced B-cell proliferation activates Chk2-dependent G1/S cell cycle arrest. *PLoS ONE* **2014**, *9*, e87299. [CrossRef] [PubMed]

216. Nikitin, P.A.; Yan, C.M.; Forte, E.; Bocedi, A.; Tourigny, J.P.; White, R.E.; Allday, M.J.; Patel, A.; Dave, S.S.; Kim, W.; et al. An ATM/Chk2-mediated DNA damage-responsive signaling pathway suppresses Epstein-Barr virus transformation of primary human B cells. *Cell Host Microbe* **2010**, *8*, 510–522. [CrossRef] [PubMed]

217. McFadden, K.; Hafez, A.Y.; Kishton, R.; Messinger, J.E.; Nikitin, P.A.; Rathmell, J.C.; Luftig, M.A. Metabolic stress is a barrier to Epstein-Barr virus-mediated B-cell immortalization. *Proc. Natl. Acad. Sci. USA* **2016**, *113*, E782–E790. [CrossRef] [PubMed]

218. Allday, M.J.; Sinclair, A.; Parker, G.; Crawford, D.H.; Farrell, P.J. Epstein-Barr virus efficiently immortalizes human B cells without neutralizing the function of p53. *EMBO J.* **1995**, *14*, 1382–1391. [PubMed]

219. Bernasconi, M.; Ueda, S.; Krukowski, P.; Bornhauser, B.C.; Ladell, K.; Dorner, M.; Sigrist, J.A.; Campidelli, C.; Aslandogmus, R.; Alessi, D.; et al. Early gene expression changes by Epstein-Barr virus infection of B-cells indicate CDKs and survivin as therapeutic targets for post-transplant lymphoproliferative diseases. *Int. J. Cancer* **2013**, *133*, 2341–2350. [CrossRef] [PubMed]

220. Szekely, L.; Pokrovskaja, K.; Jiang, W.Q.; Selivanova, G.; Lowbeer, M.; Ringertz, N.; Wiman, K.G.; Klein, G. Resting B-cells, EBV-infected B-blasts and established lymphoblastoid cell lines differ in their Rb, p53 and EBNA-5 expression patterns. *Oncogene* **1995**, *10*, 1869–1874. [PubMed]

221. Levine, A.J. The common mechanisms of transformation by the small DNA tumor viruses: The inactivation of tumor suppressor gene products: P53. *Virology* **2009**, *384*, 285–293. [CrossRef] [PubMed]

222. Forte, E.; Luftig, M.A. MDM2-dependent inhibition of p53 is required for Epstein-Barr virus B-cell growth transformation and infected-cell survival. *J. Virol.* **2009**, *83*, 2491–2499. [CrossRef] [PubMed]

223. Shumilov, A.; Tsai, M.H.; Schlosser, Y.T.; Kratz, A.S.; Bernhardt, K.; Fink, S.; Mizani, T.; Lin, X.; Jauch, A.; Mautner, J.; et al. Epstein-Barr virus particles induce centrosome amplification and chromosomal instability. *Nat. Commun.* **2017**, *8*, 14257. [CrossRef] [PubMed]

224. Jha, H.C.; Yang, K.; El-Naccache, D.W.; Sun, Z.; Robertson, E.S. EBNA3C regulates p53 through induction of Aurora kinase B. *Oncotarget* **2015**, *6*, 5788–5803. [CrossRef] [PubMed]

225. Saha, A.; Bamidele, A.; Murakami, M.; Robertson, E.S. EBNA3C attenuates the function of p53 through interaction with inhibitor of growth family proteins 4 and 5. *J. Virol.* **2011**, *85*, 2079–2088. [CrossRef] [PubMed]

226. Kashuba, E.; Yurchenko, M.; Yenamandra, S.P.; Snopok, B.; Szekely, L.; Bercovich, B.; Ciechanover, A.; Klein, G. Epstein-Barr virus-encoded EBNA-5 forms trimolecular protein complexes with MDM2 and p53 and inhibits the transactivating function of p53. *Int. J. Cancer* **2011**, *128*, 817–825. [CrossRef] [PubMed]

227. Strasser, A.; Cory, S.; Adams, J.M. Deciphering the rules of programmed cell death to improve therapy of cancer and other diseases. *EMBO J.* **2011**, *30*, 3667–3683. [CrossRef] [PubMed]

228. Kelly, G.L.; Strasser, A. The essential role of evasion from cell death in cancer. *Adv. Cancer Res.* **2011**, *111*, 39–96. [PubMed]

229. Gregory, C.D.; Dive, C.; Henderson, S.; Smith, C.A.; Williams, G.T.; Gordon, J.; Rickinson, A.B. Activation of Epstein-Barr virus latent genes protects human B cells from death by apoptosis. *Nature* **1991**, *349*, 612–614. [CrossRef] [PubMed]

230. Henderson, E.; Miller, G.; Robinson, J.; Heston, L. Efficiency of transformation of lymphocytes by Epstein-Barr virus. *Virology* **1977**, *76*, 152–163. [CrossRef]

231. Sugden, B.; Mark, W. Clonal transformation of adult human leukocytes by Epstein-Barr virus. *J. Virol.* **1977**, *23*, 503–508. [PubMed]

232. Kanda, T.; Furuse, Y.; Oshitani, H.; Kiyono, T. Highly Efficient CRISPR/Cas9-Mediated Cloning and Functional Characterization of Gastric Cancer-Derived Epstein-Barr Virus Strains. *J. Virol.* **2016**, *90*, 4383–4393. [CrossRef] [PubMed]

233. Harris-Arnold, A.; Arnold, C.P.; Schaffert, S.; Hatton, O.; Krams, S.M.; Esquivel, C.O.; Martinez, O.M. Epstein-Barr virus modulates host cell microRNA-194 to promote IL-10 production and B lymphoma cell survival. *Am. J. Transplant.* **2015**, *15*, 2814–2824. [CrossRef] [PubMed]

234. Hatton, O.; Lambert, S.L.; Phillips, L.K.; Vaysberg, M.; Natkunam, Y.; Esquivel, C.O.; Krams, S.M.; Martinez, O.M. Syk-induced phosphatidylinositol-3-kinase activation in Epstein-Barr virus posttransplant lymphoproliferative disorder. *Am. J. Transplant.* **2013**, *13*, 883–890. [CrossRef] [PubMed]

235. Ghigna, M.R.; Reineke, T.; Rince, P.; Schuffler, P.; El Mchichi, B.; Fabre, M.; Jacquemin, E.; Durrbach, A.; Samuel, D.; Joab, I.; et al. Epstein-Barr virus infection and altered control of apoptotic pathways in posttransplant lymphoproliferative disorders. *Pathobiology* **2013**, *80*, 53–59. [CrossRef] [PubMed]

236. Magrath, I. Epidemiology: Clues to the pathogenesis of Burkitt lymphoma. *Br. J. Haematol.* **2012**, *156*, 744–756. [CrossRef] [PubMed]

237. Levine, P.H.; Kamaraju, L.S.; Connelly, R.R.; Berard, C.W.; Dorfman, R.F.; Magrath, I.; Easton, J.M. The American Burkitt's Lymphoma Registry: Eight years' experience. *Cancer* **1982**, *49*, 1016–1022. [CrossRef]

238. Araujo, I.; Foss, H.D.; Bittencourt, A.; Hummel, M.; Demel, G.; Mendonca, N.; Herbst, H.; Stein, H. Expression of Epstein-Barr virus-gene products in Burkitt's lymphoma in Northeast Brazil. *Blood* **1996**, *87*, 5279–5286. [PubMed]

239. Queiroga, E.M.; Gualco, G.; Weiss, L.M.; Dittmer, D.P.; Araujo, I.; Klumb, C.E.; Harrington, W.J., Jr.; Bacchi, C.E. Burkitt lymphoma in Brazil is characterized by geographically distinct clinicopathologic features. *Am. J. Clin. Pathol.* **2008**, *130*, 946–956. [CrossRef] [PubMed]

240. Manolov, G.; Manolova, Y. Marker band in one chromosome 14 from Burkitt lymphomas. *Nature* **1972**, *237*, 33–34. [CrossRef] [PubMed]

241. Zech, L.; Haglund, U.; Nilsson, K.; Klein, G. Characteristic chromosomal abnormalities in biopsies and lymphoid-cell lines from patients with Burkitt and non-Burkitt lymphomas. *Int. J. Cancer* **1976**, *17*, 47–56. [CrossRef] [PubMed]

242. Dalla-Favera, R.; Bregni, M.; Erikson, J.; Patterson, D.; Gallo, R.C.; Croce, C.M. Human c-myc onc gene is located on the region of chromosome 8 that is translocated in Burkitt lymphoma cells. *Proc. Natl. Acad. Sci. USA* **1982**, *79*, 7824–7827. [CrossRef] [PubMed]

243. Adams, J.M.; Gerondakis, S.; Webb, E.; Corcoran, L.M.; Cory, S. Cellular myc oncogene is altered by chromosome translocation to an immunoglobulin locus in murine plasmacytomas and is rearranged similarly in human Burkitt lymphomas. *Proc. Natl. Acad. Sci. USA* **1983**, *80*, 1982–1986. [CrossRef] [PubMed]

244. Adams, J.M.; Harris, A.W.; Pinkert, C.A.; Corcoran, L.M.; Alexander, W.S.; Cory, S.; Palmiter, R.D.; Brinster, R.L. The c-myc oncogene driven by immunoglobulin enhancers induces lymphoid malignancy in transgenic mice. *Nature* **1985**, *318*, 533–538. [CrossRef] [PubMed]

245. Schmidt, E.V. The role of c-myc in cellular growth control. *Oncogene* **1999**, *18*, 2988–2996. [CrossRef] [PubMed]

246. Pelengaris, S.; Khan, M.; Evan, G. c-MYC: More than just a matter of life and death. *Nat. Rev. Cancer* **2002**, *2*, 764–776. [CrossRef] [PubMed]

247. Gaidano, G.; Ballerini, P.; Gong, J.Z.; Inghirami, G.; Neri, A.; Newcomb, E.W.; Magrath, I.T.; Knowles, D.M.; Dalla-Favera, R. p53 mutations in human lymphoid malignancies: Association with Burkitt lymphoma and chronic lymphocytic leukemia. *Proc. Natl. Acad. Sci. USA* **1991**, *88*, 5413–5417. [CrossRef] [PubMed]

248. Farrell, P.J.; Allan, G.J.; Shanahan, F.; Vousden, K.H.; Crook, T. p53 is frequently mutated in Burkitt's lymphoma cell lines. *EMBO J.* **1991**, *10*, 2879–2887. [PubMed]

249. Vousden, K.H.; Crook, T.; Farrell, P.J. Biological activities of p53 mutants in Burkitt's lymphoma cells. *J. Gen. Virol.* **1993**, *74*, 803–810. [CrossRef] [PubMed]

250. Cherney, B.W.; Bhatia, K.G.; Sgadari, C.; Gutierrez, M.I.; Mostowski, H.; Pike, S.E.; Gupta, G.; Magrath, I.T.; Tosato, G. Role of the p53 tumor suppressor gene in the tumorigenicity of Burkitt's lymphoma cells. *Cancer Res.* **1997**, *57*, 2508–2515. [PubMed]

251. Eischen, C.M.; Weber, J.D.; Roussel, M.F.; Sherr, C.J.; Cleveland, J.L. Disruption of the ARF-Mdm2-p53 tumor suppressor pathway in Myc-induced lymphomagenesis. *Genes Dev.* **1999**, *13*, 2658–2669. [CrossRef] [PubMed]

252. Lindstrom, M.S.; Klangby, U.; Wiman, K.G. p14ARF homozygous deletion or MDM2 overexpression in Burkitt lymphoma lines carrying wild type p53. *Oncogene* **2001**, *20*, 2171–2177. [CrossRef] [PubMed]

253. Egle, A.; Harris, A.W.; Bouillet, P.; Cory, S. Bim is a suppressor of Myc-induced mouse B cell leukemia. *Proc. Natl. Acad. Sci. USA* **2004**, *101*, 6164–6169. [CrossRef] [PubMed]

254. Eischen, C.M.; Woo, D.; Roussel, M.F.; Cleveland, J.L. Apoptosis triggered by Myc-induced suppression of Bcl-X(L) or Bcl-2 is bypassed during lymphomagenesis. *Mol. Cell. Biol.* **2001**, *21*, 5063–5070. [CrossRef] [PubMed]

255. Maclean, K.H.; Keller, U.B.; Rodriguez-Galindo, C.; Nilsson, J.A.; Cleveland, J.L. c-Myc augments gamma irradiation-induced apoptosis by suppressing Bcl-XL. *Mol. Cell. Biol.* **2003**, *23*, 7256–7270. [CrossRef] [PubMed]

256. Juin, P.; Hunt, A.; Littlewood, T.; Griffiths, B.; Swigart, L.B.; Korsmeyer, S.; Evan, G. c-Myc functionally cooperates with Bax to induce apoptosis. *Mol. Cell. Biol.* **2002**, *22*, 6158–6169. [CrossRef] [PubMed]

257. Mitchell, K.O.; Ricci, M.S.; Miyashita, T.; Dicker, D.T.; Jin, Z.; Reed, J.C.; El-Deiry, W.S. Bax is a transcriptional target and mediator of c-myc-induced apoptosis. *Cancer Res.* **2000**, *60*, 6318–6325. [PubMed]

258. Michalak, E.M.; Jansen, E.S.; Happo, L.; Cragg, M.S.; Tai, L.; Smyth, G.K.; Strasser, A.; Adams, J.M.; Scott, C.L. Puma and to a lesser extent Noxa are suppressors of Myc-induced lymphomagenesis. *Cell Death Differ.* **2009**, *16*, 684–696. [CrossRef] [PubMed]

259. Happo, L.; Cragg, M.S.; Phipson, B.; Haga, J.M.; Jansen, E.S.; Herold, M.J.; Dewson, G.; Michalak, E.M.; Vandenberg, C.J.; Smyth, G.K.; et al. Maximal killing of lymphoma cells by DNA damage-inducing therapy requires not only the p53 targets Puma and Noxa, but also Bim. *Blood* **2010**, *116*, 5256–5267. [CrossRef] [PubMed]

260. Garrison, S.P.; Jeffers, J.R.; Yang, C.; Nilsson, J.A.; Hall, M.A.; Rehg, J.E.; Yue, W.; Yu, J.; Zhang, L.; Onciu, M.; et al. Selection against PUMA gene expression in Myc-driven B-cell lymphomagenesis. *Mol. Cell. Biol.* **2008**, *28*, 5391–5402. [CrossRef] [PubMed]

261. Piazza, R.; Magistroni, V.; Mogavero, A.; Andreoni, F.; Ambrogio, C.; Chiarle, R.; Mologni, L.; Bachmann, P.S.; Lock, R.B.; Collini, P.; et al. Epigenetic silencing of the proapoptotic gene BIM in anaplastic large cell lymphoma through an MeCP2/SIN3a deacetylating complex. *Neoplasia* **2013**, *15*, 511–522. [CrossRef] [PubMed]

262. Abate, F.; Ambrosio, M.R.; Mundo, L.; Laginestra, M.A.; Fuligni, F.; Rossi, M.; Zairis, S.; Gazaneo, S.; De Falco, G.; Lazzi, S.; et al. Distinct Viral and Mutational Spectrum of Endemic Burkitt Lymphoma. *PLoS Pathog.* **2015**, *11*, e1005158. [CrossRef] [PubMed]

263. Adhikary, S.; Eilers, M. Transcriptional regulation and transformation by Myc proteins. *Nat. Rev. Mol. Cell Biol.* **2005**, *6*, 635–645. [CrossRef] [PubMed]

264. Dang, C.V.; O'Donnell, K.A.; Juopperi, T. The great MYC escape in tumorigenesis. *Cancer Cell* **2005**, *8*, 177–178. [CrossRef] [PubMed]

265. Love, C.; Sun, Z.; Jima, D.; Li, G.; Zhang, J.; Miles, R.; Richards, K.L.; Dunphy, C.H.; Choi, W.W.; Srivastava, G.; et al. The genetic landscape of mutations in Burkitt lymphoma. *Nat. Genet.* **2012**, *44*, 1321–1325. [CrossRef] [PubMed]

266. Hemann, M.T.; Bric, A.; Teruya-Feldstein, J.; Herbst, A.; Nilsson, J.A.; Cordon-Cardo, C.; Cleveland, J.L.; Tansey, W.P.; Lowe, S.W. Evasion of the p53 tumour surveillance network by tumour-derived MYC mutants. *Nature* **2005**, *436*, 807–811. [CrossRef] [PubMed]

267. Kaymaz, Y.; Oduor, C.I.; Yu, H.; Otieno, J.A.; Ong'echa, J.M.; Moormann, A.M.; Bailey, J.A. Comprehensive Transcriptome and Mutational Profiling of Endemic Burkitt Lymphoma Reveals EBV Type-Specific Differences. *Mol. Cancer Res.* **2017**, *15*, 563–576. [CrossRef] [PubMed]

268. Schmitz, R.; Young, R.M.; Ceribelli, M.; Jhavar, S.; Xiao, W.; Zhang, M.; Wright, G.; Shaffer, A.L.; Hodson, D.J.; Buras, E.; et al. Burkitt lymphoma pathogenesis and therapeutic targets from structural and functional genomics. *Nature* **2012**, *490*, 116–120. [CrossRef] [PubMed]

269. Richter, J.; Schlesner, M.; Hoffmann, S.; Kreuz, M.; Leich, E.; Burkhardt, B.; Rosolowski, M.; Ammerpohl, O.; Wagener, R.; Bernhart, S.H.; et al. Recurrent mutation of the ID3 gene in Burkitt lymphoma identified by integrated genome, exome and transcriptome sequencing. *Nat. Genet.* **2012**, *44*, 1316–1320. [CrossRef] [PubMed]

270. Rowe, M.; Fitzsimmons, L.; Bell, A.I. Epstein-Barr virus and Burkitt lymphoma. *Chin. J. Cancer* **2014**, *33*, 609–619. [CrossRef] [PubMed]

271. Kennedy, G.; Komano, J.; Sugden, B. Epstein-Barr virus provides a survival factor to Burkitt's lymphomas. *Proc. Natl. Acad. Sci. USA* **2003**, *100*, 14269–14274. [CrossRef] [PubMed]

272. Nasimuzzaman, M.; Kuroda, M.; Dohno, S.; Yamamoto, T.; Iwatsuki, K.; Matsuzaki, S.; Mohammad, R.; Kumita, W.; Mizuguchi, H.; Hayakawa, T.; et al. Eradication of Epstein-Barr virus episome and associated inhibition of infected tumor cell growth by adenovirus vector-mediated transduction of dominant-negative EBNA1. *Mol. Ther.* **2005**, *11*, 578–590. [CrossRef] [PubMed]

273. Amato, T.; Abate, F.; Piccaluga, P.; Iacono, M.; Fallerini, C.; Renieri, A.; De Falco, G.; Ambrosio, M.R.; Mourmouras, V.; Ogwang, M.; et al. Clonality Analysis of Immunoglobulin Gene Rearrangement by Next-Generation Sequencing in Endemic Burkitt Lymphoma Suggests Antigen Drive Activation of BCR as Opposed to Sporadic Burkitt Lymphoma. *Am. J. Clin. Pathol.* **2016**, *145*, 116–127. [CrossRef] [PubMed]

274. Bellan, C.; Lazzi, S.; Hummel, M.; Palummo, N.; de Santi, M.; Amato, T.; Nyagol, J.; Sabattini, E.; Lazure, T.; Pileri, S.A.; et al. Immunoglobulin gene analysis reveals 2 distinct cells of origin for EBV-positive and EBV-negative Burkitt lymphomas. *Blood* **2005**, *106*, 1031–1036. [CrossRef] [PubMed]

275. Capello, D.; Scandurra, M.; Poretti, G.; Rancoita, P.M.; Mian, M.; Gloghini, A.; Deambrogi, C.; Martini, M.; Rossi, D.; Greiner, T.C.; et al. Genome wide DNA-profiling of HIV-related B-cell lymphomas. *Br. J. Haematol.* **2010**, *148*, 245–255. [CrossRef] [PubMed]

276. Shiramizu, B.; McGrath, M.S. Molecular pathogenesis of AIDS-associated non-Hodgkin's lymphoma. *Hematol. Oncol. Clin. N. Am.* **1991**, *5*, 323–330.

277. Pelicci, P.G.; Knowles, D.M., 2nd; Magrath, I.; Dalla-Favera, R. Chromosomal breakpoints and structural alterations of the c-myc locus differ in endemic and sporadic forms of Burkitt lymphoma. *Proc. Natl. Acad. Sci. USA* **1986**, *83*, 2984–2988. [CrossRef] [PubMed]

278. Vaux, D.L.; Cory, S.; Adams, J.M. Bcl-2 gene promotes haemopoietic cell survival and cooperates with c-myc to immortalize pre-B cells. *Nature* **1988**, *335*, 440–442. [CrossRef] [PubMed]

279. Shimizu, N.; Tanabe-Tochikura, A.; Kuroiwa, Y.; Takada, K. Isolation of Epstein-Barr virus (EBV)-negative cell clones from the EBV-positive Burkitt's lymphoma (BL) line Akata: Malignant phenotypes of BL cells are dependent on EBV. *J. Virol.* **1994**, *68*, 6069–6073. [PubMed]

280. Komano, J.; Sugiura, M.; Takada, K. Epstein-Barr virus contributes to the malignant phenotype and to apoptosis resistance in Burkitt's lymphoma cell line Akata. *J. Virol.* **1998**, *72*, 9150–9156. [PubMed]

281. Chodosh, J.; Holder, V.P.; Gan, Y.J.; Belgaumi, A.; Sample, J.; Sixbey, J.W. Eradication of latent Epstein-Barr virus by hydroxyurea alters the growth-transformed cell phenotype. *J. Infect. Dis.* **1998**, *177*, 1194–1201. [CrossRef] [PubMed]

282. Kirchmaier, A.L.; Sugden, B. Dominant-negative inhibitors of EBNA-1 of Epstein-Barr virus. *J. Virol.* **1997**, *71*, 1766–1775. [PubMed]

283. Vereide, D.; Sugden, B. Proof for EBV's sustaining role in Burkitt's lymphomas. *Semin. Cancer Biol.* **2009**, *19*, 389–393. [CrossRef] [PubMed]

284. Vereide, D.T.; Sugden, B. Lymphomas differ in their dependence on Epstein-Barr virus. *Blood* **2011**, *117*, 1977–1985. [CrossRef] [PubMed]

285. Fitzsimmons, L.; Boyce, A.J.; Wei, W.; Chang, C.; Croom-Carter, D.; Tierney, R.J.; Herold, M.J.; Bell, A.I.; Strasser, A.; Kelly, G.L.; et al. Coordinated repression of BIM and PUMA by Epstein-Barr virus latent genes maintains the survival of Burkitt lymphoma cells. *Cell Death Differ.* **2017**. [CrossRef] [PubMed]

286. Ruf, I.K.; Rhyne, P.W.; Yang, H.; Borza, C.M.; Hutt-Fletcher, L.M.; Cleveland, J.L.; Sample, J.T. Epstein-barr virus regulates c-MYC, apoptosis, and tumorigenicity in Burkitt lymphoma. *Mol. Cell. Biol.* **1999**, *19*, 1651–1660. [CrossRef] [PubMed]

287. Komano, J.; Maruo, S.; Kurozumi, K.; Oda, T.; Takada, K. Oncogenic role of Epstein-Barr virus-encoded RNAs in Burkitt's lymphoma cell line Akata. *J. Virol.* **1999**, *73*, 9827–9831. [PubMed]

288. Ruf, I.K.; Rhyne, P.W.; Yang, C.; Cleveland, J.L.; Sample, J.T. Epstein-Barr virus small RNAs potentiate tumorigenicity of Burkitt lymphoma cells independently of an effect on apoptosis. *J. Virol.* **2000**, *74*, 10223–10228. [CrossRef] [PubMed]

289. Fukuda, M.; Longnecker, R. Latent membrane protein 2A inhibits transforming growth factor-beta 1-induced apoptosis through the phosphatidylinositol 3-kinase/Akt pathway. *J. Virol.* **2004**, *78*, 1697–1705. [CrossRef] [PubMed]

290. Bieging, K.T.; Amick, A.C.; Longnecker, R. Epstein-Barr virus LMP2A bypasses p53 inactivation in a MYC model of lymphomagenesis. *Proc. Natl. Acad. Sci. USA* **2009**, *106*, 17945–17950. [CrossRef] [PubMed]

291. Bieging, K.T.; Swanson-Mungerson, M.; Amick, A.C.; Longnecker, R. Epstein-Barr virus in Burkitt's lymphoma: A role for latent membrane protein 2A. *Cell Cycle* **2010**, *9*, 901–908. [CrossRef] [PubMed]

292. Bell, A.I.; Groves, K.; Kelly, G.L.; Croom-Carter, D.; Hui, E.; Chan, A.T.; Rickinson, A.B. Analysis of Epstein-Barr virus latent gene expression in endemic Burkitt's lymphoma and nasopharyngeal carcinoma tumour cells by using quantitative real-time PCR assays. *J. Gen. Virol.* **2006**, *87*, 2885–2890. [CrossRef] [PubMed]

293. Tao, Q.; Robertson, K.D.; Manns, A.; Hildesheim, A.; Ambinder, R.F. Epstein-Barr virus (EBV) in endemic Burkitt's lymphoma: Molecular analysis of primary tumor tissue. *Blood* **1998**, *91*, 1373–1381. [PubMed]

294. Xue, S.A.; Labrecque, L.G.; Lu, Q.L.; Ong, S.K.; Lampert, I.A.; Kazembe, P.; Molyneux, E.; Broadhead, R.L.; Borgstein, E.; Griffin, B.E. Promiscuous expression of Epstein-Barr virus genes in Burkitt's lymphoma from the central African country Malawi. *Int. J. Cancer* **2002**, *99*, 635–643. [CrossRef] [PubMed]

295. Choy, E.Y.; Siu, K.L.; Kok, K.H.; Lung, R.W.; Tsang, C.M.; To, K.F.; Kwong, D.L.; Tsao, S.W.; Jin, D.Y. An Epstein-Barr virus-encoded microRNA targets PUMA to promote host cell survival. *J. Exp. Med.* **2008**, *205*, 2551–2560. [CrossRef] [PubMed]

296. Marquitz, A.R.; Mathur, A.; Nam, C.S.; Raab-Traub, N. The Epstein-Barr Virus BART microRNAs target the pro-apoptotic protein Bim. *Virology* **2011**, *412*, 392–400. [CrossRef] [PubMed]

297. Pimienta, G.; Fok, V.; Haslip, M.; Nagy, M.; Takyar, S.; Steitz, J.A. Proteomics and Transcriptomics of BJAB Cells Expressing the Epstein-Barr Virus Noncoding RNAs EBER1 and EBER2. *PLoS ONE* **2015**, *10*, e0124638. [CrossRef] [PubMed]

298. Coloff, J.L.; Mason, E.F.; Altman, B.J.; Gerriets, V.A.; Liu, T.; Nichols, A.N.; Zhao, Y.; Wofford, J.A.; Jacobs, S.R.; Ilkayeva, O.; et al. Akt requires glucose metabolism to suppress puma expression and prevent apoptosis of leukemic T cells. *J. Biol. Chem.* **2011**, *286*, 5921–5933. [CrossRef] [PubMed]

299. Wu, B.; Guo, B.; Kang, J.; Deng, X.; Fan, Y.; Zhang, X.; Ai, K. Downregulation of Smurf2 ubiquitin ligase in pancreatic cancer cells reversed TGF-beta-induced tumor formation. *Tumour Biol.* **2016**. [CrossRef] [PubMed]

300. Spender, L.C.; Carter, M.J.; O'Brien, D.I.; Clark, L.J.; Yu, J.; Michalak, E.M.; Happo, L.; Cragg, M.S.; Inman, G.J. Transforming Growth Factor-β directly induces PUMA during the rapid induction of apoptosis in Myc-driven B-cell lymphomas. *J. Biol. Chem.* **2013**, *288*, 5198–5219. [CrossRef] [PubMed]

301. Flavell, J.R.; Baumforth, K.R.; Wood, V.H.; Davies, G.L.; Wei, W.; Reynolds, G.M.; Morgan, S.; Boyce, A.; Kelly, G.L.; Young, L.S.; et al. Down-regulation of the TGF-beta target gene, PTPRK, by the Epstein-Barr virus encoded EBNA1 contributes to the growth and survival of Hodgkin lymphoma cells. *Blood* **2008**, *111*, 292–301. [CrossRef] [PubMed]

302. Kelly, G.L.; Milner, A.E.; Baldwin, G.S.; Bell, A.I.; Rickinson, A.B. Three restricted forms of Epstein-Barr virus latency counteracting apoptosis in c-myc-expressing Burkitt lymphoma cells. *Proc. Natl. Acad. Sci. USA* **2006**, *103*, 14935–14940. [CrossRef] [PubMed]

303. Kelly, G.; Bell, A.; Rickinson, A. Epstein-Barr virus-associated Burkitt lymphomagenesis selects for downregulation of the nuclear antigen EBNA2. *Nat. Med.* **2002**, *8*, 1098–1104. [CrossRef] [PubMed]

304. Obexer, P.; Hagenbuchner, J.; Rupp, M.; Salvador, C.; Holzner, M.; Deutsch, M.; Porto, V.; Kofler, R.; Unterkircher, T.; Ausserlechner, M.J. p16INK4A sensitizes human leukemia cells to FAS- and glucocorticoid-induced apoptosis via induction of BBC3/Puma and repression of MCL1 and BCL2. *J. Biol. Chem.* **2009**, *284*, 30933–30940. [CrossRef] [PubMed]

305. Garibal, J.; Hollville, E.; Bell, A.I.; Kelly, G.L.; Renouf, B.; Kawaguchi, Y.; Rickinson, A.B.; Wiels, J. Truncated form of the Epstein-Barr virus protein EBNA-LP protects against caspase-dependent apoptosis by inhibiting protein phosphatase 2A. *J. Virol.* **2007**, *81*, 7598–7607. [CrossRef] [PubMed]

306. Cancer Genome Atlas Research Network. Comprehensive molecular characterization of gastric adenocarcinoma. *Nature* **2014**, *513*, 202–209.

307. Ribeiro, J.; Oliveira, C.; Malta, M.; Sousa, H. Epstein-Barr virus gene expression and latency pattern in gastric carcinomas: A systematic review. *Future Oncol.* **2017**, *13*, 567–579. [CrossRef] [PubMed]

308. Hu, L.; Lin, Z.; Wu, Y.; Dong, J.; Zhao, B.; Cheng, Y.; Huang, P.; Xu, L.; Xia, T.; Xiong, D.; et al. Comprehensive profiling of EBV gene expression in nasopharyngeal carcinoma through paired-end transcriptome sequencing. *Front. Med.* **2016**, *10*, 61–75. [CrossRef] [PubMed]

309. Zhu, S.; Sun, P.; Zhang, Y.; Yan, L.; Luo, B. Expression of c-myc and PCNA in Epstein-Barr virus-associated gastric carcinoma. *Exp. Ther. Med.* **2013**, *5*, 1030–1034. [CrossRef] [PubMed]

310. Zur Hausen, A.; Brink, A.A.; Craanen, M.E.; Middeldorp, J.M.; Meijer, C.J.; van den Brule, A.J. Unique transcription pattern of Epstein-Barr virus (EBV) in EBV-carrying gastric adenocarcinomas: Expression of the transforming BARF1 gene. *Cancer Res.* **2000**, *60*, 2745–2748. [PubMed]

311. Cabras, G.; Decaussin, G.; Zeng, Y.; Djennaoui, D.; Melouli, H.; Broully, P.; Bouguermouh, A.M.; Ooka, T. Epstein-Barr virus encoded BALF1 gene is transcribed in Burkitt's lymphoma cell lines and in nasopharyngeal carcinoma's biopsies. *J. Clin. Virol.* **2005**, *34*, 26–34. [CrossRef] [PubMed]

312. Strockbine, L.D.; Cohen, J.I.; Farrah, T.; Lyman, S.D.; Wagener, F.; DuBose, R.F.; Armitage, R.J.; Spriggs, M.K. The Epstein-Barr virus BARF1 gene encodes a novel, soluble colony-stimulating factor-1 receptor. *J. Virol.* **1998**, *72*, 4015–4021. [PubMed]

313. Hoebe, E.K.; Le Large, T.Y.; Tarbouriech, N.; Oosterhoff, D.; De Gruijl, T.D.; Middeldorp, J.M.; Greijer, A.E. Epstein-Barr virus-encoded BARF1 protein is a decoy receptor for macrophage colony stimulating factor and interferes with macrophage differentiation and activation. *Viral. Immunol.* **2012**, *25*, 461–470. [CrossRef] [PubMed]

314. Dolcetti, R. Cross-talk between Epstein-Barr virus and microenvironment in the pathogenesis of lymphomas. *Semin. Cancer Biol.* **2015**, *34*, 58–69. [CrossRef] [PubMed]

315. Concha, M.; Wang, X.; Cao, S.; Baddoo, M.; Fewell, C.; Lin, Z.; Hulme, W.; Hedges, D.; McBride, J.; Flemington, E.K. Identification of new viral genes and transcript isoforms during Epstein-Barr virus reactivation using RNA-Seq. *J. Virol.* **2012**, *86*, 1458–1467. [CrossRef] [PubMed]

316. Fox, C.P.; Haigh, T.A.; Taylor, G.S.; Long, H.M.; Lee, S.P.; Shannon-Lowe, C.; O'Connor, S.; Bollard, C.M.; Iqbal, J.; Chan, W.C.; et al. A novel latent membrane 2 transcript expressed in Epstein-Barr virus-positive NK- and T-cell lymphoproliferative disease encodes a target for cellular immunotherapy. *Blood* **2010**, *116*, 3695–3704. [CrossRef] [PubMed]

317. Yuen, K.S.; Chan, C.P.; Kok, K.H.; Jin, D.Y. Mutagenesis and Genome Engineering of Epstein-Barr Virus in Cultured Human Cells by CRISPR/Cas9. *Methods Mol. Biol.* **2017**, *1498*, 23–31. [PubMed]

318. Pujals, A.; Favre, L.; Pioche-Durieu, C.; Robert, A.; Meurice, G.; Le Gentil, M.; Chelouah, S.; Martin-Garcia, N.; Le Cam, E.; Guettier, C.; et al. Constitutive autophagy contributes to resistance to TP53-mediated apoptosis in Epstein-Barr virus-positive latency III B-cell lymphoproliferations. *Autophagy* **2015**, *11*, 2275–2287. [CrossRef] [PubMed]

319. Lee, D.Y.; Sugden, B. The latent membrane protein 1 oncogene modifies B-cell physiology by regulating autophagy. *Oncogene* **2008**, *27*, 2833–2842. [CrossRef] [PubMed]

320. Brune, W.; Andoniou, C.E. Die Another Day: Inhibition of Cell Death Pathways by Cytomegalovirus. *Viruses* **2017**, *9*, 249. [CrossRef] [PubMed]

321. Veyer, D.L.; Carrara, G.; Maluquer de Motes, C.; Smith, G.L. Vaccinia virus evasion of regulated cell death. *Immunol. Lett.* **2017**, *186*, 68–80. [CrossRef] [PubMed]

322. Kofahi, H.M.; Taylor, N.G.; Hirasawa, K.; Grant, M.D.; Russell, R.S. Hepatitis C Virus Infection of Cultured Human Hepatoma Cells Causes Apoptosis and Pyroptosis in Both Infected and Bystander Cells. *Sci. Rep.* **2016**, *6*, 37433. [CrossRef] [PubMed]

![viruses logo] *viruses*

MDPI

Review

Die Another Day: Inhibition of Cell Death Pathways by Cytomegalovirus

Wolfram Brune [1],* 🔵 and Christopher E. Andoniou [2,3],*

1   Heinrich Pette Institute, Leibniz Institute for Experimental Virology, Hamburg 20251, Germany
2   Immunology and Virology Program, Centre for Ophthalmology and Visual Science,
    the University of Western Australia, Crawley 6009, WA, Australia
3   Centre for Experimental Immunology, Lions Eye Institute, Nedlands 6009, WA, Australia
*   Correspondence: wolfram.brune@leibniz-hpi.de (W.B.); candoniou@lei.org.au (C.E.A.);
    Tel.: +49-40-48051351 (W.B.); +61-8-9381-0701 (C.E.A.)

Academic Editor: Marc Kvansakul
Received: 26 July 2017; Accepted: 28 August 2017; Published: 2 September 2017

Abstract: Multicellular organisms have evolved multiple genetically programmed cell death pathways that are essential for homeostasis. The finding that many viruses encode cell death inhibitors suggested that cellular suicide also functions as a first line of defence against invading pathogens. This theory was confirmed by studying viral mutants that lack certain cell death inhibitors. Cytomegaloviruses, a family of species-specific viruses, have proved particularly useful in this respect. Cytomegaloviruses are known to encode multiple death inhibitors that are required for efficient viral replication. Here, we outline the mechanisms used by the host cell to detect cytomegalovirus infection and discuss the methods employed by the cytomegalovirus family to prevent death of the host cell. In addition to enhancing our understanding of cytomegalovirus pathogenesis we detail how this research has provided significant insights into the cross-talk that exists between the various cell death pathways.

Keywords: apoptosis; necrosis; necroptosis; HCMV; MCMV; vMIA; vICA; vIRA; vIBO

1. Introduction

The capacity to eliminate damaged or unwanted cells via regulated cell death programmes is essential for the wellbeing of multicellular organisms. Indeed, defective regulation of cell death pathways can result in the development of pathological conditions such as cancer, autoimmunity and inflammatory disease [1,2]. In addition to a role in embryonic development and tissue homeostasis [3], programmed cell death has the capacity to act as defence mechanism against invading pathogens [4,5]. The finding that many pathogens encode cell death inhibitors suggests that the ability to circumvent this process is essential for optimal replication and/or transmission of many pathogens.

Human cytomegalovirus (HCMV) is a large DNA virus that is common in the human population. After the initial acute infection, that is typically subclinical, HCMV establishes a latent infection that lasts for the lifetime of the host [6,7]. HCMV is known to employ an array of strategies designed to interfere with the host immune response and thereby prevent viral clearance. Since HCMV infection is strictly species-specific, animal models are often employed to study how virally-encoded genes contribute to viral pathogenesis in vivo. Murine CMV (MCMV), a natural mouse pathogen, shares a high degree of sequence homology and biology with HCMV making it an excellent model for HCMV infection. Targeted disruption of selected viral open reading frames has established that efficient viral replication depends on the capacity to inhibit the host cell death pathways. In this review we describe how CMV infection triggers a death response in the host cell, outline the countermeasures employed by CMV to prevent death, and detail how these contribute to in vivo viral pathogenesis.

2. Multiple Pathways to Death

Several genetically controlled cell death pathways are now recognised. Apoptosis is characterised by cell shrinkage, nuclear condensation, DNA fragmentation, and the formation of apoptotic bodies. The process is generally non-inflammatory since phagocytic cells rapidly engulf and digest the apoptotic bodies. Additional pathways to death such as necroptosis and pyroptosis have also been described. Unlike apoptosis, these necrotic forms of death are accompanied by cellular swelling, and ultimately rupture of the cell membrane, resulting in the release of cellular components into the extracellular space. Significant cross-talk between the different cell death pathways exist, allowing the distinct pathways to operate in a coordinated manner in order to promote immune responses to invading pathogens.

2.1. Apoptosis

The process of apoptosis may be initiated by either extrinsic or intrinsic signals. The intrinsic pathway is triggered in response to stimuli such as DNA damage, growth factor deprivation, or endoplasmic reticulum (ER) stress. Bcl-2 family proteins are the principle regulators of the intrinsic pathway that function by regulating the integrity of the mitochondrial outer membrane (MOM) [8]. MOM permeabilization (MOMP) results in the release of proteins such as cytochrome *c*, Smac/DIABLO, and Htr2/Omi that promotes the activation of caspases, a family of cysteine proteases responsible for mediating cellular destruction [9–12]. The Bcl-2 family is composed of three functional subgroups, BH3-only proteins that act as stress sensors and initiate apoptosis, the effector proteins Bax and Bak that mediate MOMP, and pro-survival proteins that maintain mitochondrial membrane integrity. Pro-survival Bcl-2 proteins have the capacity to bind to Bax and Bak and thus prevent their activation [8]. All pro-survival proteins appear capable of inhibiting Bax, while only Mcl-1, Bcl-x$_L$ and A1 seem capable of holding Bak in check [13–15]. BH3-only proteins initiate apoptosis by binding to pro-survival Bcl-2 proteins and thereby releasing Bax and Bak [13,14]. Alternatively, some BH3-only proteins have the capacity to interact with Bax and Bak and directly catalyse their activation [16–20]. In healthy cells Bax and Bak exist as inert monomers, but as apoptosis proceeds the proteins undergo conformational changes resulting in the formation of large homo-oligomers that permeabilize the MOM. The Bcl-2 pathway therefore determines cell fate by regulating the activity of Bax and Bak. The importance of Bax and Bak to the apoptotic cascade was confirmed by the finding that cells isolated from Bax/Bak$^{-/-}$ mice are highly resistant to many forms of apoptosis [21,22].

Death receptors (DR) may promote cell survival or death depending on the composition of the signalling complexes formed after receptor activation. DR are a subset of the tumour necrosis factor (TNF) receptor family characterised by a cytoplasmic domain of approximately 80 amino acids termed the death domain (DD). Fas-associated DD protein (FADD) is an adaptor protein that is critical for apoptotic signalling downstream of DR. Following activation of the DR Fas, DR4 or DR5, FADD is recruited to the receptor, via homotypic DD interactions, and FADD in turn, recruits initiator caspase-8 or caspase-10 [23,24]. Recruitment of the initiator caspases promotes the formation of dimers resulting in autocatalytic caspase activation [25]. Once activated, the initiator caspases promote apoptosis by cleaving effector caspases, such as caspase-3 and caspase-7, that then degrade critical cellular components. This process can be inhibited by the cellular FLICE inhibitor protein (cFLIP), a non-catalytic paralogue of caspase-8 (FLICE), that forms a heterodimer with caspase-8, thus preventing autocatalytic activation [26]. Initiator caspase activation, and hence apoptosis, may also result after TNF receptor 1 (TNFR1) activation. Ligand binding by TNFR1 results in the recruitment of the adaptors TNFR1-associated DD protein (TRADD) and TNFR-associated factor 2 (TRAF2), the receptor-interacting protein 1 (RIP1) kinase, and the cellular inhibitor of apoptosis protein (cIAP) into a signalling structure termed complex I. Formation of complex I promotes cell survival by activating the NF-κB pathway and inducing the production of pro-survival proteins [27]. Following internalization of complex I, complex II, a cytoplasmic signalling unit that includes the FADD adaptor and caspase-8 is formed. Several forms of complex II have been described with the stability and constituents of the complex regulated by an intricate series of ubiquitination event and phosphorylation events. Formation of complex II results in apoptosis only if autocatalytic processing of caspase-8 is able to take place, for example when cFLIP levels are low [28].

2.2. Necroptosis

Receptor interacting protein kinase-1 and -3 (RIPK1 and RIPK3) are key mediators of necroptosis with the study of signalling events downstream of TNFR1 being critical in understanding how these kinases promote necrotic death [29]. As outlined above, complex II is assembled in response to TNFR1 activation. Complex II, also called the "necrosome", is composed of a heterodimer of caspase-8 and cFLIP along with FADD, RIPK1 and RIPK3. The presence of cFLIP in the complex prevents the autocatalytic activation of caspase-8, thus sparing the cell from apoptotic death. While caspase-8 is unable to undergo full catalytic activation when bound to cFLIP, a basal protease activity is present, and inhibition of caspase-8 activity has long been recognised as a method to promote necroptosis [30,31]. The finding that RIPK1 and RIPK3 are substrates of caspase-8 suggests that caspase-8 acts as a negative regulator of necroptosis by degrading and hence silencing the pro-necroptotic kinases [32,33]. Under circumstances where caspase-8 activity is silenced, RIPK1 recruits and activates RIPK3 through a RIP homotypic interaction motif (RHIM) [34–36]. Once activated, RIPK3 phosphorylates the mixed lineage kinase domain-like (MLKL) protein causing a conformational change within the MLKL that promotes its oligomerization. The oligomeric form of MLKL translocates to the plasma membrane where it interacts with the cell membrane and causes membrane rupture [37–40]. RIPK1-independent mechanisms for the activation of RIPK3 have also been described: the intracellular nucleotide sensor DNA-dependent activator of IFN-regulatory factors (DAI), also known as Z-DNA-binding protein 1(ZBP1), and the TIR-domain-containing adaptor inducing interferon-β (TRIF), both of which carry an RHIM, are capable of engaging RIPK3 and activating the necroptotic pathway [41,42].

2.3. Pyroptosis

The inflammatory caspases, that includes caspase-1, are critical for mediating innate defence. In response to cellular stress, inflammasomes are assembled that act as platforms for the activation of caspase-1. Activated caspase-1 promotes inflammation by cleaving pro-interleukin (IL)-1β and pro-IL-18 into their active forms, and initiates pyroptotic death by cleaving the gasdermin D protein. Once cleaved, the N-terminus of gasdermin D forms pores within the cell membrane causing the cell to swell and eventually results in membrane rupture [43–45]. The inflammasome is formed when pattern recognition receptors (PRRs), such as members of the Nod-like receptor family (NLR) or the absent in melanoma 2 (AIM2) protein, oligomerize in response to the detection of danger- or pathogen-associated molecules [46]. Clustering of the PRR results in the recruitment of the apoptosis-associated speck-like adaptor protein containing a CARD (ASC) that binds pro-caspase-1. The recruitment of pro-caspase-1 into the multi-protein assembly triggers proteolytic cleavage of pro-caspase-1 into its active form allowing the processing of downstream substrates. It should be noted that while caspase-1 is essential for the processing of pro-IL-1β and pro-IL-18, activation of caspase-11 can induce pyoptosis in some settings [47]. The inflammatory caspases therefore link the processes of inflammation and cell death induced following infection.

2.4. Cross-Talk between Death Pathways

The preceding description of the cell death pathways may give the impression that cell death results from the activation of linear signalling cascades. In fact, significant cross-talk between the various pathways exist that determines if a cell will die, and if so, by what mechanism. The interconnection of the pathways is exemplified by caspase-8 whose activation initiates apoptosis, while basal activity of uncleaved casapse-8 is essential to prevent necroptosis. Similarly, instances where activation of RIPK1 causes apoptosis, or promotes cell survival, rather than activating necroptosis have been reported. Antagonism of the ubiquitin ligases cIAP1 and cIAP2 is sufficient to result in the formation of the "Ripoptosome" a complex composed of FADD, caspase-8 and RIPK1 that assembles independently of DR signalling [48,49]. The kinase activity of RIPK1 is required for formation of the Ripoptosome with the complex mediating caspase-8 dependent apoptosis. The finding that RIPK1-deficent mice die shortly after birth suggested that in addition to its role in promoting death RIPK1 has a pro-survival function [50]. Inactivation of both caspase-8 and RIPK3 (or FADD and RIPK1) was required to prevent the lethality associated with RIPK1, indicating that FADD and caspase-8 promote survival by suppressing RIPK1 and RIPK3-mediated necroptosis during development [51–53]. Moreover, RIPK1 also has a kinase-independent pro-survival function under certain conditions [54–56].

The processes of apoptosis and pyroptosis are also linked. Procasapse-8 can be recruited to the inflammasome by interacting with the ASC adaptor protein leading to apoptotic death [57]. These observations demonstrate some of the ways in which the cell death pathways are linked. A feature of the interlocking regulation of these pathways is that inhibiting one cell death mechanism often results in the activation of an alternative pathway. An extension of this observation then is that ability of pathogens to inhibit the innate death response will require multiple inhibitory proteins.

3. Activation of Cell Death Pathways by CMV

In order for cell death to function as a defence mechanism, the host cell must have an effective means for detecting the presence of an invading pathogen. Cell death can be triggered when the host cell detects pathogen-associated molecular patterns (PAMPs), or can occur in response to cellular stress that is induced by the replication of intracellular pathogens. The following section describes some of the known mechanisms used by the host to detect CMV and how these can activate a cell death response.

3.1. Direct Recognition of CMV Infection

PRRs are a class of receptors that recognise molecules expressed, or produced by pathogens. Toll-like receptors (TLRs) were the first members of the PRR family to be identified with several implicated in the detection of CMV. A heterodimer of TLR1 and TRL2 was found to bind glycoproteins B and H produced by HCMV [58]. Whether MCMV can be detected by this mechanism has not been determined, but TLR2 knockout mice were found to be more susceptible to MCMV infection suggesting this pathway could be relevant during in vivo infection [59]. Increased titres of MCMV were also detected in mice lacking TLR3, which recognises double-stranded RNA (dsRNA), or TLR9, a receptor for unmethylated DNA [60,61]. TLR signalling typically results in the production of pro-inflammatory cytokines and chemokines. More recently, though, activation of many TLRs was found to result in the formation of signalling complexes capable of inducing RIPK-dependent necroptosis or caspase-8-dependent apoptosis [42,49,62].

In addition to TLRs, the DAI/ZBP1 and AIM2 sensors detect MCMV infection. Both of these PRR recognise cytoplasmic dsDNA, and both are activated during MCMV infection. Upon activation, DAI/ZBP1 can interact with RIPK3, promoting its activation that ultimately results in necroptosis [41]. More recent studies have shown that DAI/ZBP1 activation during MCMV infection requires viral gene transcription, but not viral DNA replication, and that DAI/ZBP1 is activated by the dsRNA in the Z-conformation (Z-RNA) rather than by the Z-DNA [63,64]. By contrast, the DNA sensor AIM2 suppresses MCMV replication via activation of the inflammasome [65]. While AIM2 was found to be required for the production of anti-viral cytokines in this context, the potential contribution of pyroptosis to the control of MCMV was not addressed. Activation of dsRNA-dependent protein kinase (PKR) is an additional mechanism by which CMV infection is detected [66,67]. PKR forms a homodimer and undergoes autophosphorylation after binding an RNA target resulting in catalytic activation of PKR. Once activated, PKR inhibits viral replication via several mechanisms, the best characterised of which is the inhibition of cellular protein translation by phosphorylation of the eukaryotic translation initiation factor 2α (eIF2α) [68]. How PKR activation mediates cell death has not been completely elucidated, but activation of NF-κB and/or FADD-dependent caspase-8 activation have been implicated [69,70]. PKR induces cell death in response to infection by influenza or poxviruses, but whether this process occurs following CMV infection has not been determined.

3.2. Cellular Stress Responses

Viral replication places a significant burden on the host cell, which responds by activating stress response pathways. CMV replication requires large amounts of protein to be synthesised. As a consequence, unfolded proteins accumulate within the ER resulting in the activation of a stress response after CMV infection [71,72]. The unfolded protein response is designed to restore ER homeostasis, however, if ER stress is not resolved cell death can result [73]. In response to ER stress the PKR-like ER kinase (PERK) phosphorylates eIF2α, resulting in the attenuation of mRNA translation, and in doing so, preventing the influx of additional protein into the stressed ER. Activation of PERK also allows for the preferential translation of proteins that contribute to the resolution of ER stress, or the induction of apoptosis, including the transcription factor C/EBP homologous protein (CHOP). A second mediator of ER stress responses is inositol-requiring enzyme 1 (IRE1), which undergoes oligomerisation in response to the accumulation of unfolded proteins within the ER. Once activated, IRE1 stimulates expression of ER-associated degradation factors and chaperones in order to alleviate ER stress. However, IRE1 can also promote apoptosis by activating a TRAF2-ASK1-JNK signalling cascade [73], by promoting caspase-12 or caspase-2 activation [74,75], or by interacting with Bax and Bak [76].

A second form of stress response triggered by CMV infection is the DNA damage response. Following detection of DNA damage, pathways that inhibit cell cycle progression are activated allowing for repair of the damaged DNA, or if the damage is too severe, apoptosis ensues [77]. Replication of the HCMV genome is achieved by a rolling-circle mechanism, and the resulting concatameric genomes are then cleaved to unit-length genomes [7]. This mechanism produces multiple exposed ends that can be recognised as dsDNA breaks. The ataxia-telangiectasia mutated (ATM) kinase pathway that responds to dsDNA breaks is activated following HCMV infection [78]. ATM has the capacity to activate p53 that can then induce pro-apoptotic proteins including Bax, Fas and the p53-induced protein with a DD (PIDD). HCMV replication also triggers a second type of DNA damage response activated in response to stalled replication forks [79]. The ATM-Rad3-related kinase (ATR) is the main transducer of this process, with p53 again acting as one of the downstream effectors for this pathway. Hence, replication of the viral genome, a process that is essential for viral propagation, activates several host cell defence mechanisms.

In summary, the host cell has the capacity to activate multiple cell death pathways either by directly detecting CMV constituents or indirectly by stress pathways activated by viral replication.

4. CMV-Encoded Death Inhibitors

Given the array of mechanisms that can lead to cell death in response to viral infection, a slow growing virus such as CMV would be expected to encode multiple death inhibitors. The ability to rapidly generate viral mutants lacking specific viral open reading frames has resulted in the identification of numerous CMV-encoded death inhibitors and aided in understanding how these proteins contribute to viral replication.

4.1. Inhibition of MOMP

A key step in the execution of intrinsic apoptosis is the permeabilization of the MOM by Bax and/or Bak. HCMV prevents MOMP by targeting Bax, and possibly Bak. The protein product of UL37 exon 1 (UL37x1) of HCMV, termed viral mitochondria-localized inhibitor of apoptosis (vMIA), binds and sequesters Bax at the mitochondrial membrane [80–82] (Figure 1). These early reports defined vMIA as a Bax-specific inhibitor, and suggested that activation of Bak requires the action of Bax in some cell types [81]. Thus, the inhibition of Bax by vMIA may be sufficient to prevent MOMP during HCMV infection. Several later studies have demonstrated an association between vMIA and Bak, implying that vMIA is capable of inhibiting both pro-apoptotic effector proteins [83,84]. If vMIA inhibits Bak, in addition to Bax, during viral infection remains to be confirmed. Several additional functions have been ascribed to vMIA. Expression of vMIA is sufficient to induce the release of calcium stores from the ER, a process that may modulate the apoptotic response [85]. In addition to its anti-apoptotic role vMIA inhibits anti-viral signalling downstream of the mitochondrial antiviral-signaling protein (MAVS) at mitochondria and peroxisomes [86,87]. vMIA therefore has the potential to promote viral replication via several mechanisms.

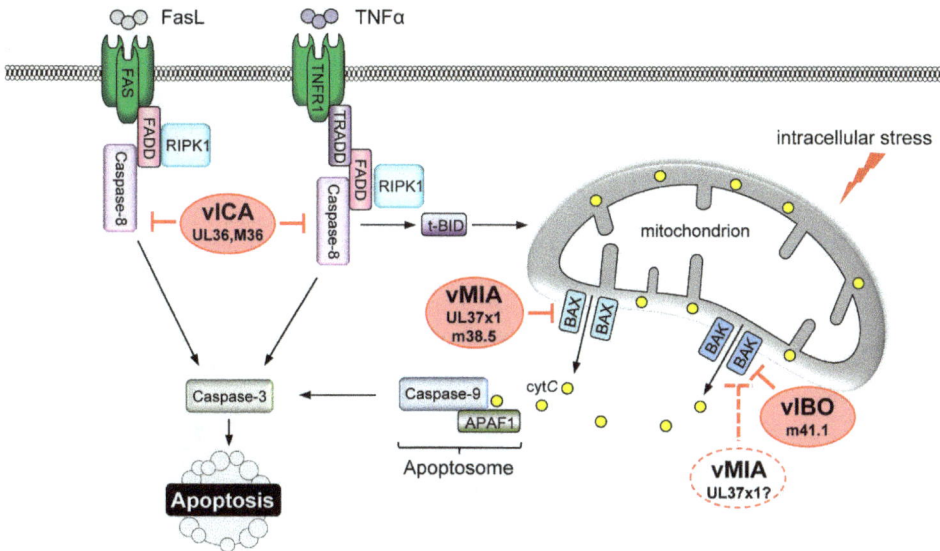

Figure 1. Inhibition of apoptosis by cytomegalovirus (CMV). The viral mitochondria-localized inhibitor of apoptosis (vMIA) and the viral inhibitor of BAK oligomerization (vIBO) inhibit mitochondrial outer membrane permablization and release of proapoptotic factors (such as cytochrome *C* (cytC)) by interacting with BAX and BAK, respectively. While murine CMV (MCMV) encodes two specific inhibitors, m38.5 and m41.1, human CMV (HCMV) has only one inhibitor, UL37x1. Whether the UL37x1 protein is BAX-specific or inhibits both BAX and BAK remains controversial. The extrinsic apoptosis pathway initiated by death receptor stimulation is blocked by the viral inhibitor of caspase-8 activation (vICA), which is encoded by the HCMV *UL36* and the MCMV *M36* gene, respectively. FasL: Fas ligand; FADD: Fas-associated death domain protein; RIPK1: receptor interacting protein kinase-1; TNFα: tumour necrosis factor α; TNFR1: TNF receptor 1; TRADD: TNFR1-associated death domain protein; t-BID: truncated BH3-interacting domain death agonist; APAF1: apoptotic protease activating factor 1.

Unlike HCMV, MCMV encodes distinct inhibitors of Bax and Bak. The m38.5 protein of MCMV localises to mitochondria where it binds Bax and prevents its activation [84,88–90]. Although the MCMV m38.5 and HCMV UL37x1 proteins share little sequence similarity, they are very similar in their functions and their genes are located at analogous positions within the viral genomes. Therefore, m38.5 is also referred to as the vMIA of MCMV (Figure 1). A second MCMV-derived inhibitor, m41.1, associates with Bak at the mitochondrial membrane and acts as a viral inhibitor of Bak oligomerisation (vIBO) [91] (Figure 1). Cells infected in vitro with MCMV mutants lacking either m38.5 or m41.1 are sensitive to apoptosis induced by a range of stimuli [88,89,91,92]. Since activation of either Bax or Bak is sufficient to induce apoptosis, it is surprising that the in vivo growth characteristics of an Δm38.5 mutant differed from that observed when m41.1 was absent. Replication of a Δm41.1 mutant was attenuated in the liver and lungs, while deletion of m38.5 had no impact on viral replication at these sites [89,92–94]. By contrast, MCMV replication in leukocytes was reduced to a similar extent when either m38.5 or m41.1 was absent [89,92,93]. Optimal replication of MCMV therefore depends upon m38.5 and m41.1, whose combined activities maintain mitochondrial integrity. Overall the data suggest that inhibition of the intrinsic apoptotic pathway is an important requirement for CMV replication.

The perturbation of mitochondrial metabolism that occurs during viral infection can induce apoptosis. HCMV prevents cell death induced by oxidative stress by producing large amounts of a 2.7-kilobase non-coding RNA. During infection the β2.7 RNA interacts with complex I of the respiratory transport chain, resulting in maintenance of mitochondrial membrane potential [95]. Genes associated with retinoid/interferon-induced mortality (GRIM)-19 is an essential component of complex I that relocalises to a perinuclear region in response to oxidative stress [96]. β2.7 interacts with GRIM-19 and prevents its relocalisation from the mitochondria, allowing oxidative phosphorylation to continue and preventing oxidative stress-induced death [95].

4.2. Suppression of the ER Stress Response

The survival of CMV-infected cells depends on the ability to modulate the ER stress response. HCMV counteracts this process, in part, via the production of UL38. Cells infected with a HCMV mutant lacking UL38 die prematurely with cells displaying morphological changes consistent with the induction of apoptosis [97]. UL38 is a multifunctional protein with expression of the N-terminal 239 amino acids sufficient to suppress apoptosis [98,99]. Expression of UL38 is associated with accumulation of the activating transcription factor 4 (ATF4) and suppression of JNK activity [98]. The ATF4 transcription factor helps to resolve ER stress by inducing the production of proteins that facilitate protein folding within the ER. The inhibition of JNK activation prevents phosphorylation of Bcl-2 and Bim and so maintains the integrity of the mitochondrial membrane. Importantly, overexpression of ATF4 or inhibition of JNK activity reduced the death of cells infected with a pUL38-deficient virus demonstrating the functional relevance of these changes to the suppression of apoptosis [98]. The *M38* gene of MCMV shares significant homology with *UL38*, suggesting that MCMV might have conserved this mechanism for inhibiting ER stress-induced death, however, this has not yet been formally tested.

A second mechanism for manipulating the ER stress response by CMV is downregulation of IRE1 protein levels. The M50 protein of MCMV interacts with IRE1, causing its degradation at late times post infection [100]. UL50, the HCMV homologue of M50, was found to have a similar impact on IRE1 expression [100]. By inducing the degradation of IRE1 the M50/UL50 proteins should restrict all IRE1 signalling events, including the activation of apoptosis-inducing pathways. However, the impact of M50/UL50 on ER stress-induced apoptosis has not yet been investigated.

4.3. Inhibition of DR-Mediated Apoptosis

Multiple proteins involved in immune recognition including DR are downregulated following infection with CMV. Cell surface expression of the Fas and TNF receptors are reduced following infection with HCMV or MCMV, respectively [101,102]. Surprisingly, the UL138 protein encoded by HCMV increases TNFR1 levels at the cell surface [103,104]. While these changes have been noted following in vitro infection how they impact on viral replication in vivo is unclear. The impact of CMV infection on the expression of the TNF-related apoptosis-inducing ligand (TRAIL) receptors has been more extensively characterised. The MCMV m166 open reading frame inhibits the cell surface expression of the TRAIL receptor [105]. Importantly, the in vivo replication of an m166 deletion virus was compromised, an effect that could be overcome by depleting NK cells. Direct targeting of the TRAIL receptor therefore allows MCMV-infected cells to avoid elimination by innate immune effector cells. A similar pathway is likely to exist in humans since the UL141 glycoprotein of HCMV is capable of binding to TRAIL receptors and promoting receptor retention within the cell [106].

CMV-encoded proteins that interfere with DR signalling also contribute to viral pathogenesis. HCMV encodes the viral inhibitor of caspase-8 activation (vICA), the protein product of the *UL36* gene. HCMV vICA inhibits Fas-induced cell death by binding to pro-caspase-8 and preventing its activation [107] (Figure 1). Homologues of vICA have been identified in genomes of CMVs from different species implying that the capacity to inhibit DR signalling is an important requirement for CMV pathogenesis [108]. M36 is the vICA gene of MCMV. It is dispensable for viral replication in fibroblasts, but required for optimal replication in macrophages in vitro [109]. The replication defect of an MCMV M36 deletion mutant in macrophages was fully rescued by expression of HCMV UL36 or overexpression of a dominant-negative FADD [110,111] and partially rescued by expression of MC159, a viral FLIP of the molluscum contagiosum virus [112]. Remarkably, the in vivo growth defect of MCMV mutants lacking M36 was rescued by depletion of macrophages [113]. This finding is consistent with in vitro experiments where TNFα produced by macrophages induced apoptosis in cells infected with a ΔM36 virus. Thus, M36 allows MCMV-infected cells to resist the action of DR ligands produced by the host in response to infection. Similar to what has been observed with MCMV M36, HCMV UL36 was also required for cell death inhibition and viral replication in monocyte-derived macrophages [114]. However, cell death induced by a UL36-deficient virus could be inhibited by the broad-spectrum caspase inhibitor zVAD-fmk only at early but not at late times of monocyte-to-macrophage differentiation. This finding was interpreted as an indication that UL36 inhibits caspase-dependent as well as caspase-independent cell death programs in monocyte-derived macrophages [114].

4.4. Replication of CMV Requires Suppression of Necroptosis

As mentioned previously, the necroptotic death pathway can be activated under circumstances where caspase-8 activity is suppressed. The action of vICA, while offering protection from DR-mediated apoptosis, could conceivably sensitise cells to necroptosis. Detection of MCMV infection by host PRRs constitutes an additional mechanism by which the necroptotic pathway could be activated. Analysis of MCMV mutants has established that the capacity to inhibit necroptosis is essential for viral replication in vivo. Screening of a random transposon library identified the M45 protein as being important for preventing the death of infected endothelial cells and macrophages in vitro, and that the activity of M45 is essential for in vivo replication [115,116]. M45 interacts with both RIPK1 and RIPK3 and prevents necroptosis initiated by Fas or TNFR activation in vitro [117,118]. Therefore, M45 has been termed viral inhibitor of RIP activation (vIRA) (Figure 2). Construction of an MCMV mutant bearing an inactivating mutation within M45 clarified the physiological relevance of these findings [119]. Knockdown of RIPK1 expression or pharmacological inhibition of RIPK1 activity was unable to prevent necroptosis induced by the M45 mutant virus in vitro establishing that MCMV infection activates necroptosis via a RIPK1-independent mechanism [119]. By contrast, growth of the M45 mutant virus was equivalent to that of wild type (WT) virus in RIPK3 knockout mice [119]. The growth defect of the M45 mutant could also be rescued by infecting mice lacking DAI/ZBP1 [41] or expressing a Z-RNA binding-deficient DAI/ZBP1 protein [63]. This elegant series of observations established that the DAI/ZBP1 PRR senses Z-RNA produced during MCMV infection resulting in activation of RIPK3-dependent necroptosis (Figure 2), and that successful replication of MCMV depends upon suppression of this process by M45. Similarly, M45 can block TLR3 and TLR4-induced necroptosis by inhibiting the RHIM-dependent activation of RIP3 by the adaptor protein TRIF [42] (Figure 2).

Besides inhibiting necroptosis, M45 also modulates TNFR-dependent activation of transcription factor NF-κB and p38 mitogen-activated protein kinase by interacting with RIPK1 [117,120]. Moreover, M45 also interacts with the NF-κB essential modulator (NEMO) and redirects it to autophagosomes for degradation [121]. By blocking all canonical NF-κB activating pathways, M45 inhibits NF-κB-dependent expression of proinflammatory cytokines and survival factors. Thus, M45 has multiple functions that affect cell death, survival, and inflammation.

How the processes of apoptosis and necroptosis interact during viral infection was recently investigated using an MCMV mutant lacking functional M36 and M45. Macrophages infected with the double mutant activated caspase-8 to drive apoptosis early, with cells then progressing to secondary RIPK3-dependant necroptosis [122]. This form of cell death was highly inflammatory, resulting in an improved T cell response after in vivo infection. This data suggests that the simultaneous suppression of apoptosis and necroptosis not only prevents the premature death of infected cells, but also serves as a means of restricting the inflammatory response.

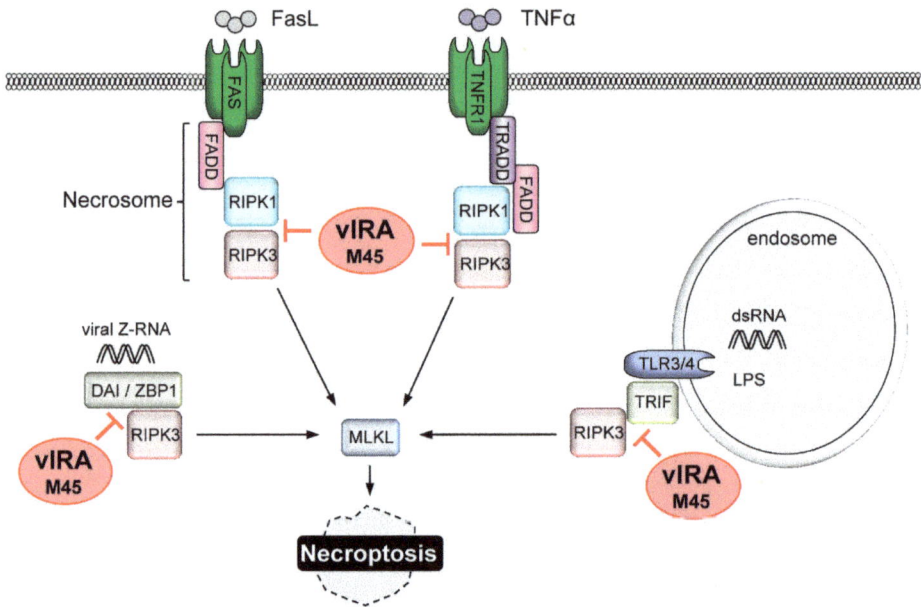

Figure 2. Inhibition of necroptosis by MCMV. Induction of programmed necrosis (necroptosis) involves activation of RIPK3 and mixed lineage kinase domain-like (MLKL). The viral inhibitor of RIP activation (vIRA), encoded by the MCMV *M45* gene, contains a RIP homotypic interaction motif (RHIM) and inhibits RHIM-dependent activation of RIPK3 by RIPK1, DAI/ZBP1, or TRIF. Death receptor-induced necroptosis requires caspase-8 inhibition, e.g., by vICA. LPS: lipopolysaccharide.

HCMV infection is also known to block the induction of necroptosis [123]. In contrast to MCMV, inhibition of necroptosis by HCMV occurs after the activation of RIPK3 and phosphorylation of MLKL. The viral protein responsible for conferring protection is regulated by IE1, but its identity has not been determined [123]. Interestingly, HCMV UL45 differs from its MCMV homologue, M45, in that it does not contain a RHIM. In contrast, the homologous proteins in herpes simplex virus (HSV) type 1 and 2 carry a RHIM [124]. The HSV1 ICP6 protein inhibits necroptosis in human cells [125], but surprisingly activates necroptosis in murine cells [126,127]. These results have established that necroptosis is an evolutionarily conserved response to herpesvirus infection. RHIM-mediated inhibition of necroptosis is a conserved strategy of both MCMV and HSV (at least in cells of their natural host), but MCMV and HCMV have developed distinct methods for antagonizing this process.

4.5. Interference with Pyroptosis?

CMV infection is known to activate the inflammasome [65]. If CMV can suppress this pathway is unclear, but some evidence from in vitro experiments suggests this may be the case. Expression of HCMV UL83 was sufficient to cause a reduction in the expression of AIM2 and inhibit the processing of IL-1β [128]. A reduction in the expression level of pro-IL-1β following MCMV infection has also been noted [129]. While tantalising, further work is required to determine if CMV has any significant capacity to interfere with pyroptosis.

5. Concluding Remarks

Mammalian cells have evolved an array of mechanisms to detect the presence of intracellular pathogens, with many of these pathways culminating in the activation of the cell death response. The study of CMV mutants has clearly established that successful viral replication requires the capacity to curb multiple host cell death pathways. While many of the viral inhibitors are conserved between HCMV and MCMV, some differences in how the respective viruses prevent cell death have been noted. For example, both viruses prevent necroptosis but they do so by targeting different points in the pathway. Indeed, the divergent anti-death strategies used by the various CMVs are proposed as one of the reasons why these viruses are so highly species specific [130].

Inhibition of cell death not only affords CMV an opportunity to complete the replication cycle, but also restricts the inflammatory response required for the generation of the adaptive immune response. Preventing cell death therefore serves several functions that contribute to viral pathogenesis. CMV-derived inhibitors of all the major cell death pathways have been described, with the exception of pyroptosis. If CMV inhibits the process of pyroptosis to any significant extent, and how this contributes to viral infection, is one of the significant questions that remains to be addressed. A second unresolved issue is if CMV latency or reactivation requires the inhibition of cell death. In support of this possibility, CD34$^+$ progenitors harbouring a latent HCMV infection exhibit an increased resistance to apoptosis [131]. The elucidation of the mechanisms used by CMV to prevent cell death has not only enhanced our understanding of viral pathogenesis but provided insights into how host cell death responses are regulated. An exciting possibility is that this knowledge could be used to develop improved treatments for HCMV infections, a virus that causes significant morbidity and mortality in immunosuppressed patients.

Acknowledgments: Work in the authors' laboratories was supported by the Deutsche Forschungsgemeinschaft (BR1730/3-2), the Australian Research Council (DP150102437), and the National Health and Medical Research Council of Australia (APP1125357 and APP1109288).

Conflicts of Interest: The authors declare no conflict of interest. The funding sponsors had no role in the writing of the manuscript or in the decision to publish it.

References

1. Dillon, C.P.; Green, D.R. Molecular Cell Biology of Apoptosis and Necroptosis in Cancer. *Adv. Exp. Med. Biol.* **2016**, *930*, 1–23. [PubMed]
2. Nagata, S.; Tanaka, M. Programmed cell death and the immune system. *Nat. Rev. Immunol.* **2017**, *17*, 333–340. [CrossRef] [PubMed]
3. Tuzlak, S.; Kaufmann, T.; Villunger, A. Interrogating the relevance of mitochondrial apoptosis for vertebrate development and postnatal tissue homeostasis. *Genes Dev.* **2016**, *30*, 2133–2151. [CrossRef] [PubMed]
4. Jorgensen, I.; Rayamajhi, M.; Miao, E.A. Programmed cell death as a defence against infection. *Nat. Rev. Immunol.* **2017**, *17*, 151–164. [CrossRef] [PubMed]
5. Sridharan, H.; Upton, J.W. Programmed necrosis in microbial pathogenesis. *Trends Microbiol.* **2014**, *22*, 199–207. [CrossRef] [PubMed]
6. Ho, M. The history of cytomegalovirus and its diseases. *Med. Microbiol. Immunol.* **2008**, *197*, 65–73. [CrossRef] [PubMed]
7. Mocarski, E.S.; Shenk, T.; Griffiths, P.D.; Pass, R.F. Cytomegaloviruses. In *Fields Virology*; Knipe, D.M., Howley, P.M., Cohen, J.I., Griffin, D.E., Lamb, R.A., Martin, M.A., Racaniello, V.R., Roizman, B., Eds.; Lippincott, Williams and Wilkins: Philadelphia, PA, USA, 2013; pp. 1960–2014.
8. Tait, S.W.; Green, D.R. Mitochondria and cell death: Outer membrane permeabilization and beyond. *Nat. Rev. Mol. Cell Biol.* **2010**, *11*, 621–632. [CrossRef] [PubMed]
9. Kluck, R.M.; Bossy-Wetzel, E.; Green, D.R.; Newmeyer, D.D. The release of cytochrome c from mitochondria: A primary site for Bcl-2 regulation of apoptosis. *Science* **1997**, *275*, 1132–1136. [CrossRef] [PubMed]

10. Li, P.; Nijhawan, D.; Budihardjo, I.; Srinivasula, S.M.; Ahmad, M.; Alnemri, E.S.; Wang, X. Cytochrome c and dATP-dependent formation of Apaf-1/caspase-9 complex initiates an apoptotic protease cascade. *Cell* **1997**, *91*, 479–489. [CrossRef]

11. Verhagen, A.M.; Ekert, P.G.; Pakusch, M.; Silke, J.; Connolly, L.M.; Reid, G.E.; Moritz, R.L.; Simpson, R.J.; Vaux, D.L. Identification of DIABLO, a mammalian protein that promotes apoptosis by binding to and antagonizing IAP proteins. *Cell* **2000**, *102*, 43–53. [CrossRef]

12. Du, C.; Fang, M.; Li, Y.; Li, L.; Wang, X. Smac, a mitochondrial protein that promotes cytochrome c-dependent caspase activation by eliminating IAP inhibition. *Cell* **2000**, *102*, 33–42. [CrossRef]

13. Willis, S.N.; Chen, L.; Dewson, G.; Wei, A.; Naik, E.; Fletcher, J.I.; Adams, J.M.; Huang, D.C. Proapoptotic Bak is sequestered by Mcl-1 and Bcl-x$_L$, but not Bcl-2, until displaced by BH$_3$-only proteins. *Genes Dev.* **2005**, *19*, 1294–1305. [CrossRef] [PubMed]

14. Willis, S.N.; Fletcher, J.I.; Kaufmann, T.; van Delft, M.F.; Chen, L.; Czabotar, P.E.; Ierino, H.; Lee, E.F.; Fairlie, W.D.; Bouillet, P.; et al. Apoptosis initiated when BH3 ligands engage multiple Bcl-2 homologs, not Bax or Bak. *Science* **2007**, *315*, 856–859. [CrossRef] [PubMed]

15. Fletcher, J.I.; Meusburger, S.; Hawkins, C.J.; Riglar, D.T.; Lee, E.F.; Fairlie, W.D.; Huang, D.C.; Adams, J.M. Apoptosis is triggered when prosurvival Bcl-2 proteins cannot restrain Bax. *Proc. Natl. Acad. Sci. USA* **2008**, *105*, 18081–18087. [CrossRef] [PubMed]

16. Wei, M.C.; Lindsten, T.; Mootha, V.K.; Weiler, S.; Gross, A.; Ashiya, M.; Thompson, C.B.; Korsmeyer, S.J. tBID, a membrane-targeted death ligand, oligomerizes BAK to release cytochrome c. *Genes Dev.* **2000**, *14*, 2060–2071. [PubMed]

17. Letai, A.; Bassik, M.C.; Walensky, L.D.; Sorcinelli, M.D.; Weiler, S.; Korsmeyer, S.J. Distinct BH3 domains either sensitize or activate mitochondrial apoptosis, serving as prototype cancer therapeutics. *Cancer Cell* **2002**, *2*, 183–192. [CrossRef]

18. Kim, H.; Tu, H.C.; Ren, D.; Takeuchi, O.; Jeffers, J.R.; Zambetti, G.P.; Hsieh, J.J.; Cheng, E.H. Stepwise activation of BAX and BAK by tBID, BIM, and PUMA initiates mitochondrial apoptosis. *Mol. Cell* **2009**, *36*, 487–499. [CrossRef] [PubMed]

19. Czabotar, P.E.; Westphal, D.; Dewson, G.; Ma, S.; Hockings, C.; Fairlie, W.D.; Lee, E.F.; Yao, S.; Robin, A.Y.; Smith, B.J.; et al. Bax crystal structures reveal how BH3 domains activate Bax and nucleate its oligomerization to induce apoptosis. *Cell* **2013**, *152*, 519–531. [CrossRef] [PubMed]

20. Moldoveanu, T.; Grace, C.R.; Llambi, F.; Nourse, A.; Fitzgerald, P.; Gehring, K.; Kriwacki, R.W.; Green, D.R. BID-induced structural changes in BAK promote apoptosis. *Nat. Struct. Mol. Biol.* **2013**, *20*, 589–597. [CrossRef] [PubMed]

21. Lindsten, T.; Ross, A.J.; King, A.; Zong, W.X.; Rathmell, J.C.; Shiels, H.A.; Ulrich, E.; Waymire, K.G.; Mahar, P.; Frauwirth, K.; et al. The combined functions of proapoptotic Bcl-2 family members Bak and Bax are essential for normal development of multiple tissues. *Mol. Cell* **2000**, *6*, 1389–1399. [CrossRef]

22. Rathmell, J.C.; Lindsten, T.; Zong, W.X.; Cinalli, R.M.; Thompson, C.B. Deficiency in Bak and Bax perturbs thymic selection and lymphoid homeostasis. *Nat. Immunol.* **2002**, *3*, 932–939. [CrossRef] [PubMed]

23. Boldin, M.P.; Varfolomeev, E.E.; Pancer, Z.; Mett, I.L.; Camonis, J.H.; Wallach, D. A novel protein that interacts with the death domain of Fas/APO1 contains a sequence motif related to the death domain. *J. Biol. Chem.* **1995**, *270*, 7795–7798. [CrossRef] [PubMed]

24. Chinnaiyan, A.M.; O'Rourke, K.; Tewari, M.; Dixit, V.M. FADD, a novel death domain-containing protein, interacts with the death domain of Fas and initiates apoptosis. *Cell* **1995**, *81*, 505–512. [CrossRef]

25. Boatright, K.M.; Renatus, M.; Scott, F.L.; Sperandio, S.; Shin, H.; Pedersen, I.M.; Ricci, J.E.; Edris, W.A.; Sutherlin, D.P.; Green, D.R.; et al. A unified model for apical caspase activation. *Mol. Cell* **2003**, *11*, 529–541. [CrossRef]

26. Irmler, M.; Thome, M.; Hahne, M.; Schneider, P.; Hofmann, K.; Steiner, V.; Bodmer, J.L.; Schroter, M.; Burns, K.; Mattmann, C.; et al. Inhibition of death receptor signals by cellular FLIP. *Nature* **1997**, *388*, 190–195. [CrossRef] [PubMed]

27. Van Antwerp, D.J.; Martin, S.J.; Kafri, T.; Green, D.R.; Verma, I.M. Suppression of TNF-α-induced apoptosis by NF-κB. *Science* **1996**, *274*, 787–789. [CrossRef] [PubMed]

28. Micheau, O.; Tschopp, J. Induction of TNF receptor I-mediated apoptosis via two sequential signaling complexes. *Cell* **2003**, *114*, 181–190. [CrossRef]

29. Grootjans, S.; Vanden Berghe, T.; Vandenabeele, P. Initiation and execution mechanisms of necroptosis: An overview. *Cell Death Differ.* **2017**, *24*, 1184–1195. [CrossRef] [PubMed]

30. Pop, C.; Oberst, A.; Drag, M.; van Raam, B.J.; Riedl, S.J.; Green, D.R.; Salvesen, G.S. FLIP_L induces caspase 8 activity in the absence of interdomain caspase 8 cleavage and alters substrate specificity. *Biochem. J.* **2011**, *433*, 447–457. [CrossRef] [PubMed]

31. Vercammen, D.; Beyaert, R.; Denecker, G.; Goossens, V.; van Loo, G.; Declercq, W.; Grooten, J.; Fiers, W.; Vandenabeele, P. Inhibition of caspases increases the sensitivity of L929 cells to necrosis mediated by tumor necrosis factor. *J. Exp. Med.* **1998**, *187*, 1477–1485. [CrossRef] [PubMed]

32. Lin, Y.; Devin, A.; Rodriguez, Y.; Liu, Z.G. Cleavage of the death domain kinase RIP by caspase-8 prompts TNF-induced apoptosis. *Genes Dev.* **1999**, *13*, 2514–2526. [CrossRef] [PubMed]

33. Feng, S.; Yang, Y.; Mei, Y.; Ma, L.; Zhu, D.E.; Hoti, N.; Castanares, M.; Wu, M. Cleavage of RIP3 inactivates its caspase-independent apoptosis pathway by removal of kinase domain. *Cell. Signal.* **2007**, *19*, 2056–2067. [CrossRef] [PubMed]

34. Cho, Y.S.; Challa, S.; Moquin, D.; Genga, R.; Ray, T.D.; Guildford, M.; Chan, F.K. Phosphorylation-driven assembly of the RIP1-RIP3 complex regulates programmed necrosis and virus-induced inflammation. *Cell* **2009**, *137*, 1112–1123. [CrossRef] [PubMed]

35. He, S.; Wang, L.; Miao, L.; Wang, T.; Du, F.; Zhao, L.; Wang, X. Receptor interacting protein kinase-3 determines cellular necrotic response to TNF-α. *Cell* **2009**, *137*, 1100–1111. [CrossRef] [PubMed]

36. Zhang, D.W.; Shao, J.; Lin, J.; Zhang, N.; Lu, B.J.; Lin, S.C.; Dong, M.Q.; Han, J. RIP3, an energy metabolism regulator that switches TNF-induced cell death from apoptosis to necrosis. *Science* **2009**, *325*, 332–336. [CrossRef] [PubMed]

37. Wang, H.; Sun, L.; Su, L.; Rizo, J.; Liu, L.; Wang, L.F.; Wang, F.S.; Wang, X. Mixed lineage kinase domain-like protein MLKL causes necrotic membrane disruption upon phosphorylation by RIP3. *Mol. Cell* **2014**, *54*, 133–146. [CrossRef] [PubMed]

38. Cai, Z.; Jitkaew, S.; Zhao, J.; Chiang, H.C.; Choksi, S.; Liu, J.; Ward, Y.; Wu, L.G.; Liu, Z.G. Plasma membrane translocation of trimerized MLKL protein is required for TNF-induced necroptosis. *Nat. Cell Biol.* **2014**, *16*, 55–65. [CrossRef] [PubMed]

39. Chen, X.; Li, W.; Ren, J.; Huang, D.; He, W.T.; Song, Y.; Yang, C.; Li, W.; Zheng, X.; Chen, P.; et al. Translocation of mixed lineage kinase domain-like protein to plasma membrane leads to necrotic cell death. *Cell Res.* **2014**, *24*, 105–121. [CrossRef] [PubMed]

40. Dondelinger, Y.; Declercq, W.; Montessuit, S.; Roelandt, R.; Goncalves, A.; Bruggeman, I.; Hulpiau, P.; Weber, K.; Sehon, C.A.; Marquis, R.W.; et al. MLKL compromises plasma membrane integrity by binding to phosphatidylinositol phosphates. *Cell Rep.* **2014**, *7*, 971–981. [CrossRef] [PubMed]

41. Upton, J.W.; Kaiser, W.J.; Mocarski, E.S. DAI/ZBP1/DLM-1 complexes with RIP3 to mediate virus-induced programmed necrosis that is targeted by murine cytomegalovirus vIRA. *Cell Host Microbe* **2012**, *11*, 290–297. [CrossRef] [PubMed]

42. Kaiser, W.J.; Sridharan, H.; Huang, C.; Mandal, P.; Upton, J.W.; Gough, P.J.; Sehon, C.A.; Marquis, R.W.; Bertin, J.; Mocarski, E.S. Toll-like receptor 3-mediated necrosis via TRIF, RIP3, and MLKL. *J. Biol. Chem.* **2013**, *288*, 31268–31279. [CrossRef] [PubMed]

43. He, W.T.; Wan, H.; Hu, L.; Chen, P.; Wang, X.; Huang, Z.; Yang, Z.H.; Zhong, C.Q.; Han, J. Gasdermin D is an executor of pyroptosis and required for interleukin-1β secretion. *Cell Res.* **2015**, *25*, 1285–1298. [CrossRef] [PubMed]

44. Kayagaki, N.; Stowe, I.B.; Lee, B.L.; O'Rourke, K.; Anderson, K.; Warming, S.; Cuellar, T.; Haley, B.; Roose-Girma, M.; Phung, Q.T.; et al. Caspase-11 cleaves gasdermin D for non-canonical inflammasome signalling. *Nature* **2015**, *526*, 666–671. [CrossRef] [PubMed]

45. Shi, J.; Zhao, Y.; Wang, K.; Shi, X.; Wang, Y.; Huang, H.; Zhuang, Y.; Cai, T.; Wang, F.; Shao, F. Cleavage of GSDMD by inflammatory caspases determines pyroptotic cell death. *Nature* **2015**, *526*, 660–665. [CrossRef] [PubMed]

46. Guo, H.; Callaway, J.B.; Ting, J.P. Inflammasomes: Mechanism of action, role in disease, and therapeutics. *Nat. Med.* **2015**, *21*, 677–687. [CrossRef] [PubMed]

47. Kayagaki, N.; Warming, S.; Lamkanfi, M.; Vande Walle, L.; Louie, S.; Dong, J.; Newton, K.; Qu, Y.; Liu, J.; Heldens, S.; et al. Non-canonical inflammasome activation targets caspase-11. *Nature* **2011**, *479*, 117–121. [CrossRef] [PubMed]

48. Tenev, T.; Bianchi, K.; Darding, M.; Broemer, M.; Langlais, C.; Wallberg, F.; Zachariou, A.; Lopez, J.; MacFarlane, M.; Cain, K.; et al. The Ripoptosome, a signaling platform that assembles in response to genotoxic stress and loss of IAPs. *Mol. Cell* **2011**, *43*, 432–448. [CrossRef] [PubMed]

49. Feoktistova, M.; Geserick, P.; Kellert, B.; Dimitrova, D.P.; Langlais, C.; Hupe, M.; Cain, K.; MacFarlane, M.; Hacker, G.; Leverkus, M. cIAPs block Ripoptosome formation, a RIP1/caspase-8 containing intracellular cell death complex differentially regulated by cFLIP isoforms. *Mol. Cell* **2011**, *43*, 449–463. [CrossRef] [PubMed]

50. Kelliher, M.A.; Grimm, S.; Ishida, Y.; Kuo, F.; Stanger, B.Z.; Leder, P. The death domain kinase RIP mediates the TNF-induced NF-κB signal. *Immunity* **1998**, *8*, 297–303. [CrossRef]

51. Oberst, A.; Dillon, C.P.; Weinlich, R.; McCormick, L.L.; Fitzgerald, P.; Pop, C.; Hakem, R.; Salvesen, G.S.; Green, D.R. Catalytic activity of the caspase-8-FLIP$_L$ complex inhibits RIPK3-dependent necrosis. *Nature* **2011**, *471*, 363–367. [CrossRef] [PubMed]

52. Kaiser, W.J.; Upton, J.W.; Long, A.B.; Livingston-Rosanoff, D.; Daley-Bauer, L.P.; Hakem, R.; Caspary, T.; Mocarski, E.S. RIP3 mediates the embryonic lethality of caspase-8-deficient mice. *Nature* **2011**, *471*, 368–372. [CrossRef] [PubMed]

53. Zhang, H.; Zhou, X.; McQuade, T.; Li, J.; Chan, F.K.; Zhang, J. Functional complementation between FADD and RIP1 in embryos and lymphocytes. *Nature* **2011**, *471*, 373–376. [CrossRef] [PubMed]

54. Dillon, C.P.; Weinlich, R.; Rodriguez, D.A.; Cripps, J.G.; Quarato, G.; Gurung, P.; Verbist, K.C.; Brewer, T.L.; Llambi, F.; Gong, Y.N.; et al. RIPK1 blocks early postnatal lethality mediated by caspase-8 and RIPK3. *Cell* **2014**, *157*, 1189–1202. [CrossRef] [PubMed]

55. Rickard, J.A.; O'Donnell, J.A.; Evans, J.M.; Lalaoui, N.; Poh, A.R.; Rogers, T.; Vince, J.E.; Lawlor, K.E.; Ninnis, R.L.; Anderton, H.; et al. RIPK1 regulates RIPK3-MLKL-driven systemic inflammation and emergency hematopoiesis. *Cell* **2014**, *157*, 1175–1188. [CrossRef] [PubMed]

56. Kaiser, W.J.; Daley-Bauer, L.P.; Thapa, R.J.; Mandal, P.; Berger, S.B.; Huang, C.; Sundararajan, A.; Guo, H.; Roback, L.; Speck, S.H.; et al. RIP1 suppresses innate immune necrotic as well as apoptotic cell death during mammalian parturition. *Proc. Natl. Acad. Sci. USA* **2014**, *111*, 7753–7758. [CrossRef] [PubMed]

57. Sagulenko, V.; Thygesen, S.J.; Sester, D.P.; Idris, A.; Cridland, J.A.; Vajjhala, P.R.; Roberts, T.L.; Schroder, K.; Vince, J.E.; Hill, J.M.; et al. AIM2 and NLRP3 inflammasomes activate both apoptotic and pyroptotic death pathways via ASC. *Cell Death Differ.* **2013**, *20*, 1149–1160. [CrossRef] [PubMed]

58. Boehme, K.W.; Guerrero, M.; Compton, T. Human cytomegalovirus envelope glycoproteins B and H are necessary for TLR2 activation in permissive cells. *J. Immunol.* **2006**, *177*, 7094–7102. [CrossRef] [PubMed]

59. Szomolanyi-Tsuda, E.; Liang, X.; Welsh, R.M.; Kurt-Jones, E.A.; Finberg, R.W. Role for TLR2 in NK cell-mediated control of murine cytomegalovirus in vivo. *J. Virol.* **2006**, *80*, 4286–4291. [CrossRef] [PubMed]

60. Tabeta, K.; Georgel, P.; Janssen, E.; Du, X.; Hoebe, K.; Crozat, K.; Mudd, S.; Shamel, L.; Sovath, S.; Goode, J.; et al. Toll-like receptors 9 and 3 as essential components of innate immune defense against mouse cytomegalovirus infection. *Proc. Natl. Acad. Sci. USA* **2004**, *101*, 3516–3521. [CrossRef] [PubMed]

61. Krug, A.; French, A.R.; Barchet, W.; Fischer, J.A.; Dzionek, A.; Pingel, J.T.; Orihuela, M.M.; Akira, S.; Yokoyama, W.M.; Colonna, M. TLR9-dependent recognition of MCMV by IPC and DC generates coordinated cytokine responses that activate antiviral NK cell function. *Immunity* **2004**, *21*, 107–119. [CrossRef] [PubMed]

62. Estornes, Y.; Toscano, F.; Virard, F.; Jacquemin, G.; Pierrot, A.; Vanbervliet, B.; Bonnin, M.; Lalaoui, N.; Mercier-Gouy, P.; Pacheco, Y.; et al. dsRNA induces apoptosis through an atypical death complex associating TLR3 to caspase-8. *Cell Death Differ.* **2012**, *19*, 1482–1494. [CrossRef] [PubMed]

63. Maelfait, J.; Liverpool, L.; Bridgeman, A.; Ragan, K.B.; Upton, J.W.; Rehwinkel, J. Sensing of viral and endogenous RNA by ZBP1/DAI induces necroptosis. *EMBO J.* **2017**. [CrossRef] [PubMed]

64. Sridharan, H.; Ragan, K.B.; Guo, H.; Gilley, R.P.; Landsteiner, V.J.; Kaiser, W.J.; Upton, J.W. Murine cytomegalovirus IE3-dependent transcription is required for DAI/ZBP1-mediated necroptosis. *EMBO Rep.* **2017**, *18*, 1429–1441. [CrossRef] [PubMed]

65. Rathinam, V.A.; Jiang, Z.; Waggoner, S.N.; Sharma, S.; Cole, L.E.; Waggoner, L.; Vanaja, S.K.; Monks, B.G.; Ganesan, S.; Latz, E.; et al. The AIM2 inflammasome is essential for host defense against cytosolic bacteria and DNA viruses. *Nat. Immunol.* **2010**, *11*, 395–402. [CrossRef] [PubMed]

66. Valchanova, R.S.; Picard-Maureau, M.; Budt, M.; Brune, W. Murine cytomegalovirus m142 and m143 are both required to block protein kinase R-mediated shutdown of protein synthesis. *J. Virol.* **2006**, *80*, 10181–10190. [CrossRef] [PubMed]

67. Marshall, E.E.; Bierle, C.J.; Brune, W.; Geballe, A.P. Essential role for either TRS1 or IRS1 in human cytomegalovirus replication. *J. Virol.* **2009**, *83*, 4112–4120. [CrossRef] [PubMed]

68. Dauber, B.; Wolff, T. Activation of the Antiviral Kinase PKR and Viral Countermeasures. *Viruses* **2009**, *1*, 523–544. [CrossRef] [PubMed]

69. Gil, J.; Alcami, J.; Esteban, M. Induction of apoptosis by double-stranded-RNA-dependent protein kinase (PKR) involves the α subunit of eukaryotic translation initiation factor 2 and NF-κB. *Mol. Cell Biol.* **1999**, *19*, 4653–4663. [CrossRef] [PubMed]

70. Balachandran, S.; Kim, C.N.; Yeh, W.C.; Mak, T.W.; Bhalla, K.; Barber, G.N. Activation of the dsRNA-dependent protein kinase, PKR, induces apoptosis through FADD-mediated death signaling. *EMBO J.* **1998**, *17*, 6888–6902. [CrossRef] [PubMed]

71. Isler, J.A.; Skalet, A.H.; Alwine, J.C. Human cytomegalovirus infection activates and regulates the unfolded protein response. *J. Virol.* **2005**, *79*, 6890–6899. [CrossRef] [PubMed]

72. Qian, Z.; Xuan, B.; Chapa, T.J.; Gualberto, N.; Yu, D. Murine cytomegalovirus targets transcription factor ATF4 to exploit the unfolded-protein response. *J. Virol.* **2012**, *86*, 6712–6723. [CrossRef] [PubMed]

73. Tabas, I.; Ron, D. Integrating the mechanisms of apoptosis induced by endoplasmic reticulum stress. *Nat. Cell Biol.* **2011**, *13*, 184–190. [CrossRef] [PubMed]

74. Yoneda, T.; Imaizumi, K.; Oono, K.; Yui, D.; Gomi, F.; Katayama, T.; Tohyama, M. Activation of caspase-12, an endoplastic reticulum (ER) resident caspase, through tumor necrosis factor receptor-associated factor 2-dependent mechanism in response to the ER stress. *J. Biol. Chem.* **2001**, *276*, 13935–13940. [CrossRef] [PubMed]

75. Upton, J.P.; Wang, L.; Han, D.; Wang, E.S.; Huskey, N.E.; Lim, L.; Truitt, M.; McManus, M.T.; Ruggero, D.; Goga, A.; et al. IRE1α cleaves select microRNAs during ER stress to derepress translation of proapoptotic Caspase-2. *Science* **2012**, *338*, 818–822. [CrossRef] [PubMed]

76. Hetz, C.; Bernasconi, P.; Fisher, J.; Lee, A.H.; Bassik, M.C.; Antonsson, B.; Brandt, G.S.; Iwakoshi, N.N.; Schinzel, A.; Glimcher, L.H.; et al. Proapoptotic BAX and BAK modulate the unfolded protein response by a direct interaction with IRE1α. *Science* **2006**, *312*, 572–576. [CrossRef] [PubMed]

77. Xiaofei, E.; Kowalik, T.F. The DNA damage response induced by infection with human cytomegalovirus and other viruses. *Viruses* **2014**, *6*, 2155–2185. [PubMed]

78. Gaspar, M.; Shenk, T. Human cytomegalovirus inhibits a DNA damage response by mislocalizing checkpoint proteins. *Proc. Natl. Acad. Sci. USA* **2006**, *103*, 2821–2826. [CrossRef] [PubMed]

79. Luo, M.H.; Rosenke, K.; Czornak, K.; Fortunato, E.A. Human cytomegalovirus disrupts both ataxia telangiectasia mutated protein (ATM)- and ATM-Rad3-related kinase-mediated DNA damage responses during lytic infection. *J. Virol.* **2007**, *81*, 1934–1950. [CrossRef] [PubMed]

80. Goldmacher, V.S.; Bartle, L.M.; Skaletskaya, A.; Dionne, C.A.; Kedersha, N.L.; Vater, C.A.; Han, J.; Lutz, R.J.; Watanabe, S.; McFarland, E.D.; et al. A cytomegalovirus-encoded mitochondria-localized inhibitor of apoptosis structurally unrelated to Bcl-2. *Proc. Natl. Acad. Sci. USA* **1999**, *96*, 12536–12541. [CrossRef] [PubMed]

81. Arnoult, D.; Bartle, L.M.; Skaletskaya, A.; Poncet, D.; Zamzami, N.; Park, P.U.; Sharpe, J.; Youle, R.J.; Goldmacher, V.S. Cytomegalovirus cell death suppressor vMIA blocks Bax- but not Bak-mediated apoptosis by binding and sequestering Bax at mitochondria. *Proc. Natl. Acad. Sci. USA* **2004**, *101*, 7988–7993. [CrossRef] [PubMed]

82. Poncet, D.; Larochette, N.; Pauleau, A.L.; Boya, P.; Jalil, A.A.; Cartron, P.F.; Vallette, F.; Schnebelen, C.; Bartle, L.M.; Skaletskaya, A.; et al. An anti-apoptotic viral protein that recruits Bax to mitochondria. *J. Biol. Chem.* **2004**, *279*, 22605–22614. [CrossRef] [PubMed]

83. Karbowski, M.; Norris, K.L.; Cleland, M.M.; Jeong, S.Y.; Youle, R.J. Role of Bax and Bak in mitochondrial morphogenesis. *Nature* **2006**, *443*, 658–662. [CrossRef] [PubMed]

84. Norris, K.L.; Youle, R.J. Cytomegalovirus proteins vMIA and m38.5 link mitochondrial morphogenesis to Bcl-2 family proteins. *J. Virol.* **2008**, *82*, 6232–6243. [CrossRef] [PubMed]

85. Sharon-Friling, R.; Goodhouse, J.; Colberg-Poley, A.M.; Shenk, T. Human cytomegalovirus pUL37x1 induces the release of endoplasmic reticulum calcium stores. *Proc. Natl. Acad. Sci. USA* **2006**, *103*, 19117–19122. [CrossRef] [PubMed]

86. Castanier, C.; Garcin, D.; Vazquez, A.; Arnoult, D. Mitochondrial dynamics regulate the RIG-I-like receptor antiviral pathway. *EMBO Rep.* **2010**, *11*, 133–138. [CrossRef] [PubMed]

87. Magalhaes, A.C.; Ferreira, A.R.; Gomes, S.; Vieira, M.; Gouveia, A.; Valenca, I.; Islinger, M.; Nascimento, R.; Schrader, M.; Kagan, J.C.; et al. Peroxisomes are platforms for cytomegalovirus' evasion from the cellular immune response. *Sci. Rep.* **2016**, *6*, 26028. [CrossRef] [PubMed]

88. Jurak, I.; Schumacher, U.; Simic, H.; Voigt, S.; Brune, W. Murine cytomegalovirus m38.5 protein inhibits Bax-mediated cell death. *J. Virol.* **2008**, *82*, 4812–4822. [CrossRef] [PubMed]

89. Manzur, M.; Fleming, P.; Huang, D.C.; Degli-Esposti, M.A.; Andoniou, C.E. Virally mediated inhibition of Bax in leukocytes promotes dissemination of murine cytomegalovirus. *Cell Death Differ.* **2009**, *16*, 312–320. [CrossRef] [PubMed]

90. Arnoult, D.; Skaletskaya, A.; Estaquier, J.; Dufour, C.; Goldmacher, V.S. The murine cytomegalovirus cell death suppressor m38.5 binds Bax and blocks Bax-mediated mitochondrial outer membrane permeabilization. *Apoptosis* **2008**, *13*, 1100–1110. [CrossRef] [PubMed]

91. Çam, M.; Handke, W.; Picard-Maureau, M.; Brune, W. Cytomegaloviruses inhibit Bak- and Bax-mediated apoptosis with two separate viral proteins. *Cell Death Differ.* **2010**, *17*, 655–665. [CrossRef] [PubMed]

92. Fleming, P.; Kvansakul, M.; Voigt, V.; Kile, B.T.; Kluck, R.M.; Huang, D.C.; Degli-Esposti, M.A.; Andoniou, C.E. MCMV-mediated inhibition of the pro-apoptotic Bak protein is required for optimal in vivo replication. *PLoS Pathog.* **2013**, *9*, e1003192. [CrossRef] [PubMed]

93. Crosby, L.N.; McCormick, A.L.; Mocarski, E.S. Gene products of the embedded m41/m41.1 locus of murine cytomegalovirus differentially influence replication and pathogenesis. *Virology* **2013**, *436*, 274–283. [CrossRef] [PubMed]

94. Handke, W.; Luig, C.; Popovic, B.; Krmpotic, A.; Jonjic, S.; Brune, W. Viral inhibition of BAK promotes murine cytomegalovirus dissemination to salivary glands. *J. Virol.* **2013**, *87*, 3592–3596. [CrossRef] [PubMed]

95. Reeves, M.B.; Davies, A.A.; McSharry, B.P.; Wilkinson, G.W.; Sinclair, J.H. Complex I binding by a virally encoded RNA regulates mitochondria-induced cell death. *Science* **2007**, *316*, 1345–1348. [CrossRef] [PubMed]

96. Lu, H.; Cao, X. GRIM-19 is essential for maintenance of mitochondrial membrane potential. *Mol. Biol. Cell* **2008**, *19*, 1893–1902. [CrossRef] [PubMed]

97. Terhune, S.; Torigoi, E.; Moorman, N.; Silva, M.; Qian, Z.; Shenk, T.; Yu, D. Human cytomegalovirus UL38 protein blocks apoptosis. *J. Virol.* **2007**, *81*, 3109–3123. [CrossRef] [PubMed]

98. Xuan, B.; Qian, Z.; Torigoi, E.; Yu, D. Human cytomegalovirus protein pUL38 induces ATF4 expression, inhibits persistent JNK phosphorylation, and suppresses endoplasmic reticulum stress-induced cell death. *J. Virol.* **2009**, *83*, 3463–3474. [CrossRef] [PubMed]

99. Qian, Z.; Xuan, B.; Gualberto, N.; Yu, D. The human cytomegalovirus protein pUL38 suppresses endoplasmic reticulum stress-mediated cell death independently of its ability to induce mTORC1 activation. *J. Virol.* **2011**, *85*, 9103–9113. [CrossRef] [PubMed]

100. Stahl, S.; Burkhart, J.M.; Hinte, F.; Tirosh, B.; Mohr, H.; Zahedi, R.P.; Sickmann, A.; Ruzsics, Z.; Budt, M.; Brune, W. Cytomegalovirus downregulates IRE1 to repress the unfolded protein response. *PLoS Pathog.* **2013**, *9*, e1003544. [CrossRef] [PubMed]

101. Seirafian, S.; Prod'homme, V.; Sugrue, D.; Davies, J.; Fielding, C.; Tomasec, P.; Wilkinson, G.W. Human cytomegalovirus suppresses Fas expression and function. *J. Gen. Virol.* **2014**, *95*, 933–939. [CrossRef] [PubMed]

102. Popkin, D.L.; Virgin, H.W., 4th. Murine cytomegalovirus infection inhibits tumor necrosis factor α responses in primary macrophages. *J. Virol.* **2003**, *77*, 10125–10130. [CrossRef] [PubMed]

103. Le, V.T.; Trilling, M.; Hengel, H. The cytomegaloviral protein pUL138 acts as potentiator of tumor necrosis factor (TNF) receptor 1 surface density to enhance ULb'-encoded modulation of TNF-α signaling. *J. Virol.* **2011**, *85*, 13260–13270. [CrossRef] [PubMed]

104. Montag, C.; Wagner, J.A.; Gruska, I.; Vetter, B.; Wiebusch, L.; Hagemeier, C. The latency-associated UL138 gene product of human cytomegalovirus sensitizes cells to tumor necrosis factor α (TNF-α) signaling by upregulating TNF-α receptor 1 cell surface expression. *J. Virol.* **2011**, *85*, 11409–11421. [CrossRef] [PubMed]

105. Verma, S.; Loewendorf, A.; Wang, Q.; McDonald, B.; Redwood, A.; Benedict, C.A. Inhibition of the TRAIL death receptor by CMV reveals its importance in NK cell-mediated antiviral defense. *PLoS Pathog.* **2014**, *10*, e1004268. [CrossRef] [PubMed]

106. Smith, W.; Tomasec, P.; Aicheler, R.; Loewendorf, A.; Nemcovicova, I.; Wang, E.C.; Stanton, R.J.; Macauley, M.; Norris, P.; Willen, L.; et al. Human cytomegalovirus glycoprotein UL141 targets the TRAIL death receptors to thwart host innate antiviral defenses. *Cell Host Microbe* **2013**, *13*, 324–335. [CrossRef] [PubMed]

107. Skaletskaya, A.; Bartle, L.M.; Chittenden, T.; McCormick, A.L.; Mocarski, E.S.; Goldmacher, V.S. A cytomegalovirus-encoded inhibitor of apoptosis that suppresses caspase-8 activation. *Proc. Natl. Acad. Sci. USA* **2001**, *98*, 7829–7834. [CrossRef] [PubMed]

108. McCormick, A.L.; Skaletskaya, A.; Barry, P.A.; Mocarski, E.S.; Goldmacher, V.S. Differential function and expression of the viral inhibitor of caspase 8-induced apoptosis (vICA) and the viral mitochondria-localized inhibitor of apoptosis (vMIA) cell death suppressors conserved in primate and rodent cytomegaloviruses. *Virology* **2003**, *316*, 221–233. [CrossRef] [PubMed]

109. Menard, C.; Wagner, M.; Ruzsics, Z.; Holak, K.; Brune, W.; Campbell, A.E.; Koszinowski, U.H. Role of murine cytomegalovirus US22 gene family members in replication in macrophages. *J. Virol.* **2003**, *77*, 5557–5570. [CrossRef] [PubMed]

110. Chaudhry, M.; Kasmapour, B.; Plaza, C.; Bajagic, M.; Casalegno Garduno, R.; Borkner, L.; Lenac Roviš, T.; Scrima, A.; Jonjic, S.; Schmitz, I.; et al. UL36 Rescues Apoptosis Inhibition and In Vivo Replication of A Chimeric MCMV Lacking the M36 Gene. *Front. Microbiol.* **2017**, *7*, 312. [CrossRef] [PubMed]

111. Cicin-Sain, L.; Ruzsics, Z.; Podlech, J.; Bubic, I.; Menard, C.; Jonjic, S.; Reddehase, M.J.; Koszinowski, U.H. Dominant-negative FADD rescues the in vivo fitness of a cytomegalovirus lacking an antiapoptotic viral gene. *J. Virol.* **2008**, *82*, 2056–2064. [CrossRef] [PubMed]

112. Hüttmann, J.; Krause, E.; Schommartz, T.; Brune, W. Functional Comparison of Molluscum Contagiosum Virus vFLIP MC159 with Murine Cytomegalovirus M36/vICA and M45/vIRA Proteins. *J. Virol.* **2015**, *90*, 2895–2905. [CrossRef] [PubMed]

113. Ebermann, L.; Ruzsics, Z.; Guzman, C.A.; van Rooijen, N.; Casalegno-Garduno, R.; Koszinowski, U.; Cicin-Sain, L. Block of death-receptor apoptosis protects mouse cytomegalovirus from macrophages and is a determinant of virulence in immunodeficient hosts. *PLoS Pathog.* **2012**, *8*, e1003062. [CrossRef] [PubMed]

114. McCormick, A.L.; Roback, L.; Livingston-Rosanoff, D.; St Clair, C. The human cytomegalovirus *UL36* gene controls caspase-dependent and -independent cell death programs activated by infection of monocytes differentiating to macrophages. *J. Virol.* **2010**, *84*, 5108–5123. [CrossRef] [PubMed]

115. Brune, W.; Ménard, C.; Heesemann, J.; Koszinowski, U.H. A ribonucleotide reductase homolog of cytomegalovirus and endothelial cell tropism. *Science* **2001**, *291*, 303–305. [CrossRef] [PubMed]

116. Lembo, D.; Donalisio, M.; Hofer, A.; Cornaglia, M.; Brune, W.; Koszinowski, U.; Thelander, L.; Landolfo, S. The ribonucleotide reductase R1 homolog of murine cytomegalovirus is not a functional enzyme subunit but is required for pathogenesis. *J. Virol.* **2004**, *78*, 4278–4288. [CrossRef] [PubMed]

117. Mack, C.; Sickmann, A.; Lembo, D.; Brune, W. Inhibition of proinflammatory and innate immune signaling pathways by a cytomegalovirus RIP1-interacting protein. *Proc. Natl. Acad. Sci. USA* **2008**, *105*, 3094–3099. [CrossRef] [PubMed]

118. Upton, J.W.; Kaiser, W.J.; Mocarski, E.S. Cytomegalovirus M45 cell death suppression requires receptor-interacting protein (RIP) homotypic interaction motif (RHIM)-dependent interaction with RIP1. *J. Biol. Chem.* **2008**, *283*, 16966–16970. [CrossRef] [PubMed]

119. Upton, J.W.; Kaiser, W.J.; Mocarski, E.S. Virus inhibition of RIP3-dependent necrosis. *Cell Host Microbe* **2010**, *7*, 302–313. [CrossRef] [PubMed]

120. Krause, E.; de Graaf, M.; Fliss, P.M.; Dölken, L.; Brune, W. Murine cytomegalovirus virion-associated protein M45 mediates rapid NF-κB activation after infection. *J. Virol.* **2014**, *88*, 9963–9975. [CrossRef] [PubMed]

121. Fliss, P.M.; Jowers, T.P.; Brinkmann, M.M.; Holstermann, B.; Mack, C.; Dickinson, P.; Hohenberg, H.; Ghazal, P.; Brune, W. Viral mediated redirection of NEMO/IKKγ to autophagosomes curtails the inflammatory cascade. *PLoS Pathog.* **2012**, *8*, e1002517. [CrossRef] [PubMed]

122. Daley-Bauer, L.P.; Roback, L.; Crosby, L.N.; McCormick, A.L.; Feng, Y.; Kaiser, W.J.; Mocarski, E.S. Mouse cytomegalovirus M36 and M45 death suppressors cooperate to prevent inflammation resulting from antiviral programmed cell death pathways. *Proc. Natl. Acad. Sci. USA* **2017**, *114*, E2786–E2795. [CrossRef] [PubMed]

123. Omoto, S.; Guo, H.; Talekar, G.R.; Roback, L.; Kaiser, W.J.; Mocarski, E.S. Suppression of RIP3-dependent necroptosis by human cytomegalovirus. *J. Biol. Chem.* **2015**, *290*, 11635–11648. [CrossRef] [PubMed]

124. Lembo, D.; Brune, W. Tinkering with a viral ribonucleotide reductase. *Trends Biochem. Sci.* **2009**, *34*, 25–32. [CrossRef] [PubMed]

125. Guo, H.; Omoto, S.; Harris, P.A.; Finger, J.N.; Bertin, J.; Gough, P.J.; Kaiser, W.J.; Mocarski, E.S. Herpes simplex virus suppresses necroptosis in human cells. *Cell Host Microbe* **2015**, *17*, 243–251. [CrossRef] [PubMed]

126. Wang, X.; Li, Y.; Liu, S.; Yu, X.; Li, L.; Shi, C.; He, W.; Li, J.; Xu, L.; Hu, Z.; et al. Direct activation of RIP3/MLKL-dependent necrosis by herpes simplex virus 1 (HSV-1) protein ICP6 triggers host antiviral defense. *Proc. Natl. Acad. Sci. USA* **2014**, *111*, 15438–15443. [CrossRef] [PubMed]

127. Huang, Z.; Wu, S.Q.; Liang, Y.; Zhou, X.; Chen, W.; Li, L.; Wu, J.; Zhuang, Q.; Chen, C.; Li, J.; et al. RIP1/RIP3 binding to HSV-1 ICP6 initiates necroptosis to restrict virus propagation in mice. *Cell Host Microbe* **2015**, *17*, 229–242. [CrossRef] [PubMed]

128. Huang, Y.; Ma, D.; Huang, H.; Lu, Y.; Liao, Y.; Liu, L.; Liu, X.; Fang, F. Interaction between HCMV pUL83 and human AIM2 disrupts the activation of the AIM2 inflammasome. *Virol. J.* **2017**, *14*, 34. [CrossRef] [PubMed]

129. Sester, D.P.; Sagulenko, V.; Thygesen, S.J.; Cridland, J.A.; Loi, Y.S.; Cridland, S.O.; Masters, S.L.; Genske, U.; Hornung, V.; Andoniou, C.E.; et al. Deficient NLRP3 and AIM2 Inflammasome Function in Autoimmune NZB Mice. *J. Immunol.* **2015**, *195*, 1233–1241. [CrossRef] [PubMed]

130. Jurak, I.; Brune, W. Induction of apoptosis limits cytomegalovirus cross-species infection. *EMBO J.* **2006**, *25*, 2634–2642. [CrossRef] [PubMed]

131. Poole, E.; McGregor Dallas, S.R.; Colston, J.; Joseph, R.S.; Sinclair, J. Virally induced changes in cellular microRNAs maintain latency of human cytomegalovirus in CD34$^+$ progenitors. *J. Gen. Virol.* **2011**, *92*, 1539–1549. [CrossRef] [PubMed]

viruses

MDPI

Review

Investigations of Pro- and Anti-Apoptotic Factors Affecting African Swine Fever Virus Replication and Pathogenesis

Linda K. Dixon [1,*], Pedro J. Sánchez-Cordón [1,†], Inmaculada Galindo [2] and Covadonga Alonso [2]

[1]   The Pirbright Institute, Ash Road, Pirbright, Woking, Surrey GU24 0NF, UK; pedro.sanchez-crdon@apha.gsi.gov.uk

[2]   Department of Biotechnology, Instituto Nacional de Investigación y Tecnología Agraria y Alimentaria, INIA, Ctra. de la Coruña Km 7.5, 28040 Madrid, Spain; galindo@inia.es (I.G.); calonso@inia.es (C.A.)

*   Correspondence: linda.dixon@pirbright.ac.uk; Tel.: +44-1483-231009

†   Current Address: Pathology Department, Animal and Plant Health Agency (APHA-Weybridge), Woodham Lane, New Haw, Addlestone KT15 3NB, UK.

Academic Editor: Marc Kvansakul
Received: 3 August 2017; Accepted: 21 August 2017; Published: 25 August 2017

Abstract: African swine fever virus (ASFV) is a large DNA virus that replicates predominantly in the cell cytoplasm and is the only member of the *Asfarviridae* family. The virus causes an acute haemorrhagic fever, African swine fever (ASF), in domestic pigs and wild boar resulting in the death of most infected animals. Apoptosis is induced at an early stage during virus entry or uncoating. However, ASFV encodes anti-apoptotic proteins which facilitate production of progeny virions. These anti-apoptotic proteins include A179L, a Bcl-2 family member; A224L, an inhibitor of apoptosis proteins (IAP) family member; EP153R a C-type lectin; and DP71L. The latter acts by inhibiting activation of the stress activated pro-apoptotic pathways pro-apoptotic pathways. The mechanisms by which these proteins act is summarised. ASF disease is characterised by massive apoptosis of uninfected lymphocytes which reduces the effectiveness of the immune response, contributing to virus pathogenesis. Mechanisms by which this apoptosis is induced are discussed.

Keywords: African swine fever virus; apoptosis; A179L; A224L pathogenesis

1. Introduction to African Swine Fever Virus

African swine fever virus (ASFV) is a large double-stranded DNA virus that replicates predominantly in the cell cytoplasm. The virus causes an acute haemorrhagic fever, African swine fever (ASF), in domestic pigs and wild boar with lethality approaching 100%. In its long-term reservoir hosts in East Africa, warthogs, bushpigs and soft ticks of the *Ornithodoros* species, ASFV causes long-term persistent infections without significant clinical signs.

The disease has a high socio-economic impact upon affected countries in sub-Saharan Africa, Sardinia, Russia and Eastern Europe. Since it spread to Georgia in 2007, ASF has extended through the Trans-Caucasus, Russian Federation and Eastern Europe including EU countries in the Baltic States and Poland [1–3]. Most recently, ASF spread in June 2017 to the Czech Republic and in August to Romania [4].

ASFV is the only member of the *Asfarviridae* family. Several large DNA viruses that infect amoeba, including Faustovirus, Kaumoebavirus and Pacmanvirus, are distantly related to *ASFV* and share about 30 conserved genes. These have genomes of approximately 400 kbp, considerably larger in comparison to the ASFV genome of 170 to 193 kbp [5–7].

The ASFV genome encodes many non-essential proteins that have important roles in evading host defences. These include proteins that inhibit type I interferon responses, the main early innate antiviral response, and proteins that inhibit apoptosis. The target cells for ASFV replication are mononuclear phagocyte system cells with key roles in activation of innate and adaptive responses. Manipulation of the function of these cells can profoundly affect the host's response to infection.

In this review, we describe different impacts of ASFV infection on apoptosis. These include the inhibition of apoptosis in infected cells to facilitate virus replication. We also review the massive induction of apoptosis in uninfected cells, particularly lymphocytes, which is a characteristic of acute ASF disease [8].

2. Induction of Apoptosis in Infected Cells

The induction of apoptosis in infected cells is an important mechanism by which host cells restrict virus replication. Activation of this process can prevent viruses from completing their replication cycle and thus reduce production of infectious progeny viruses. In common with other viruses, ASFV infection of cells was shown to induce apoptosis, as it induces caspase 3 activation (Figure 1). Other caspases that are activated previous to execution caspase 3 are caspase 9, which is characteristic of the mitochondrial pathway of apoptosis and caspase 12, which is associated with endoplasmic reticulum (ER) stress [9]. In fact, caspase 3 activation after infection occurs in the absence of virus protein synthesis or DNA replication [10]. Inhibition of endosomal acidification blocked the induction of apoptosis as did UV-inactivation of virions. These results suggested that a step including fusion of the viral membrane with the endosomal membrane or virus uncoating, could be involved in initial induction of apoptosis following ASFV infection [10]. Perturbation of membranes as a consequence of fusion or disruption can initiate signalling pathways that lead to cell death [11,12]. Another mechanism for induction of apoptosis involves the interaction of the ASFV structural protein E183L/p54 with the light chain of dynein (DLC8), the microtubule motor protein [13]. The binding site of E183L/p54 to DLC8 dynein is similar to that by which the pro-apoptotic Bcl-2 family member Bim-3 binds. It was suggested that E183L/p54 induces apoptosis by displacement of Bim-3 from microtubules [13,14]. The latter would account for apoptosis induction by the mitochondrial pathway, however, recent evidence has shown that ER stress plays an important role in apoptosis induction after ASFV infection [9]. ER stress might be elicited by the large amounts of viral proteins that are synthesized and accumulate in infected cells potentially overloading the ER protein folding capacity. ER chaperones calnexin and calreticulin are markedly increased 16 hours post-infection (hpi) and protein disulfide isomerase (PDI) at later infection time points (48 hpi). Also, there is a marked increase in caspase 12 activation which is characteristic of ER stress and induces apoptosis. This apoptosis induction might be beneficial for viral spread. In fact, there is a marked activation of ATF6 which was translocated to the nucleus to activate transcription of chaperone-encoding genes and ATF4 only at 48 hpi. It was reported that inhibition of ATF6 action results both in inhibition of all caspases activation and viral production [9].

Inhibition of apoptosis will favour virus replication during the process of progeny virion production. Conversely, at the later stages of infection, it may be advantageous for viruses to induce apoptosis. This would facilitate virus spread by increasing virus release from the cell but avoiding the induction of inflammatory signals that could activate an immune response to clear infection. Apoptosis and the presence of ASFV particles in apoptotic bodies have been observed at late stages of infection [15]. Uptake of these apoptotic bodies into macrophages mediated by phosphatidyl serine receptors could be another route for infection. Induction of apoptosis has been observed in cell culture in ASFV-infected macrophages and in vascular endothelial cells [16].

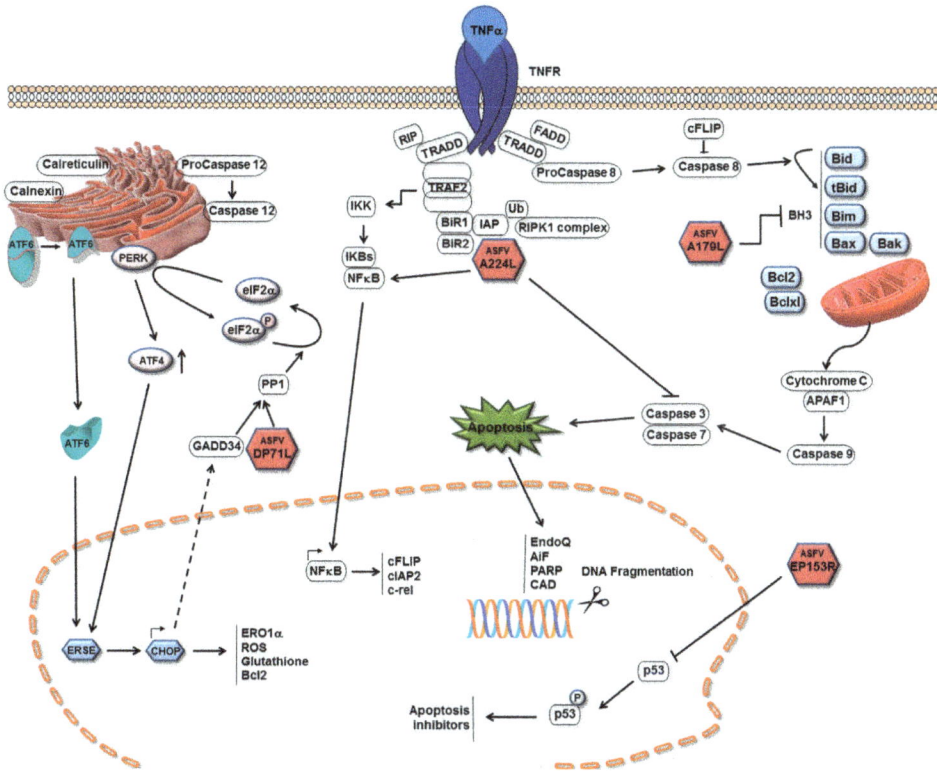

Figure 1. Mechanisms of apoptosis inhibition by African swine fever virus (ASFV). Pathways by which ASFV inhibits induction of apoptosis in infected cells and ASFV proteins are shown as red hexagons with the name of the protein inside. The ASFV A179L Bcl-2 family protein binds to and inhibits several BH3 only domain pro-apoptotic proteins. The A224L IAP-family protein binds to and inhibits caspase 3 and activates nuclear factor kappa-light-chain-enhancer of activated B cells (NF-κB) signalling, thus increasing expression of anti-apoptotic genes including *cFLIP*, *cIAP2* and *c-rel*. The DP71L protein recruits protein phosphatase 1 to dephosphorylate eIF2a, restoring global protein synthesis and inhibiting transcriptional activation of pro-apoptotic CCAAT-enhancer-binding protein homologous protein (CHOP). The EP153R protein inhibits activation of the p53 protein.

3. Inhibiting Apoptosis in the Infected Cell

In the early 1990s *ASFV* genes with similarities to known apoptosis inhibitory families were identified. These included one protein with similarity to Bcl-2 family members and one with similarity to IAP family members [17–19]. These were proposed to inhibit the induction of apoptosis in infected cells, thus promoting cell survival and favouring virus replication. Subsequently, additional ASFV proteins have been shown to regulate cell death pathways (see Figure 1).

3.1. Inhibitors of Apoptosis in ASFV-Infected Cells

The ASFV Bcl-2 Family Member A179L

Early studies focused on defining the functions of the predicted ASFV-encoded apoptosis inhibitors, including a Bcl-2 family member. The Bcl-2 family proteins contain up to four Bcl-2 homology regions (BH1-4) that are key to their functions as either anti- or pro-apoptotic. The apoptosis inducers include BH3-only proteins which sense cellular damage and initiate the death process and the Bax and Bak proteins that act downstream of BH3-only proteins to permeabilise the mitochondrial outer membrane [20,21]. Bax is primarily cytosolic, translocating to the mitochondrial outer membrane (MOM) after an apoptotic stimulus. Bax and Bak activation is induced by the expression of BH3-only proteins [22]. The BH3-only proteins include Bim, Bid, Puma, Noxa, Bmf, Bik, Bad and Hrk, and function either by directly activating Bak and Bax, or sequestering and neutralizing the pro-survival Bcl-2 members. BH3-only proteins engage the canonical ligand-binding groove on the pro-survival proteins as an α-helix [23].

The A179L protein sequence contains domains similar to all BH domains including a well conserved BH3 domain in the centre [24]. The A179L protein is very well conserved in different ASFV isolates which share between 94% and 99% amino acid identity across 179 amino acids encoded by the entire protein. The similarity with cellular Bcl-2 proteins varies between 33% in mouse and humans, 34% in bovines and 30% in zebrafish. The sequence is most conserved with cellular Bcl2-family proteins between residues 70 to 138 which contain the BH1 and BH2 predicted domains, however A179L lacks the transmembrane domain of cellular Bcl-2 (see Figure 2). The A179L protein was shown to be expressed at both early and late times post-infection in macrophages as an 18 kDa protein [17]. In later studies, A179L was shown to localise at the mitochondria or endoplasmic reticulum [25]. This protein suppressed apoptotic cell death in different cellular systems. For example, it suppressed the strong apoptosis induced by the double-stranded RNA-activated protein kinase (p68) in HeLa and BSC-40 cells [26] and the apoptosis induced by inhibitors of macromolecular synthesis in the human myeloid leukemia cell line K562 [27], demonstrating it was a functional member of the Bcl-2 family. Expression of A179L in insect cells from recombinant baculovirus extended the survival time of infected insect cells when grown as a monolayer but not in suspension suggesting the anti-apoptotic activity may be cell-anchorage dependent [28]. This observation has not been followed up using mammalian cells or porcine macrophages but may be relevant to the understanding of ASFV replication in vivo.

Using a yeast two-hybrid assay, A179L was shown to bind to several BH-3 only proteins, including the activated truncated forms of the Bid protein. Co-precipitation of A179L with active truncated Bid-(p13 and p15) and binding of A179L to other BH-3 domain proteins and pro-apoptotic Bak and Bax (Figure 1) was observed in pull-down assays but not to full length Bid and Noxa [29]. In addition to its role in apoptosis, A179L modulates autophagy via interaction with Beclin-1, and inhibits autophagosome formation under starvation conditions [25]. This was further investigated by determining the kinetics of binding of BH-3 motif peptides to A179L [30]. The results confirmed high affinity binding with several pro-apoptotic BH-3 motifs including Bid, Bim, Puma and also with Bak and Bax. Lower affinity binding was detected to Bmf, Bik and Bad. Hrk and Noxa bound with much lower affinities [30]. The crystal structure of A179L bound to Bid and Bax BH-3 motifs was determined to understand the basis for the unusual promiscuity of A179L. The configuration of the A179L ligand binding groove and some specific interactions suggested a mechanism for the broad specificity. The region corresponding to the α3 helix in Bcl-2 proteins that forms one side of the A179L ligand-binding groove is not helical, and adapts an extended configuration. Elevation of B-factors in this region suggested considerably flexibility in binding, as would be required to engage such a broad range of pro-death Bcl-2 ligands [30]. However, further studies are required to investigate the possible influence of other homology regions in binding specificity. As yet, there are few studies on the role of A179L protein in virus infection and the importance of its interaction with different binding partners. It is possible that the promiscuity in A179L binding is required for its function in both arthropod and mammalian cells.

```
                                                                          BH4
BHRF1    1 --------MAYSTRE ILLALC IRD--------------------------------SRVH
A179L    1 -MEGEELIYHNIINE ILVGYI KYYI-----------------------------------
HS_Bcl-x 1 ------MS--QSNRE LVVDFL SYKLSQKGYSWSQFSDVEENRTEAPEGTESEMETPSAIN
HS_Bcl-2 1 MAHAGRTG--YDNRE IVMKYI HYKLSQRGYEWDAGDVGAA-----PPGAAPA---PGIFS
                            .*::::

                                                                          BH3
BHRF1    GNGTLHP--------VLELAARETP-------LRLSPEDTVVLR------YHVLLEEII
A179L    ------------------------NDISEHELSPYQQ------QIKKILTYYDECLN
HS_Bcl-x GNPSWH--------------LADSPAVNGATGHSSSLDAREVIPMAAVKQ ALREAGDEFE
HS_Bcl-2 SQPGHTPHPAASRDPVARTSPLQTPAAPGAAA---GPALSPV--PPVVHLT RQAGDDFS
                                                            .          : :

                                              BH1
BHRF1    ERNSETFTETWNRFITHTEHVDLDFNSVFL EIFHRGDPSLGRALAWMAWC MHACRTLCCN
A179L    KQVTITFSL------TSVQEIKTQFTGVVT ELFKDL-INWGRICGFIVFS AKMAKYC-KD
HS_Bcl-x LRYRRAFSDLTSQLHITPGTAYQSFEQVVN ELFRDG-VNWGRIVAFFSFG GALCVES-VD
HS_Bcl-2 RRYRRDFAEMSSQLHLTPFTARGRFATVVE ELFRDG-VNWGRIVAFFEFG GVMCVES-VN
             :    *:           *  *. *:*:    . **  .::: .         :

                                        BH2
BHRF1    QSTPYYVVDLSVRGMLEASEGLD GWIHQQGGWSTLIEDNI PGSRR---F------SWTLF
A179L    AN-NHLESTVITTAYNFMKHNLL PWMISHGGQEEFLAFSL HSDMYSVIFNIKYFLSKFCN
HS_Bcl-x KEMQVLVSRIAAWMATYLNDHLE PWIQENGGWDTFVELYG NNAAAESRKGQERFNRWFLT
HS_Bcl-2 REMSPLVDNIALWMTEYLNRHLH TWIQDNGGWDAFVELYG PSMRPLFDF------SWL-S
             .          :    .   * *:. :** . :  .

BHRF1    LAGLT—LSLLVICSYLFISRGRH    191
A179L    HMFFRSCVQLLRNCNLI-------    179
HS_Bcl-x GMTV-A------GVVLLGSLFSRK    233
HS_Bcl-2 LKTL-LSLALVGACITLGAYLGHK    240
```

Figure 2. Sequence comparison of ASFV A179L protein with other Bcl-2 family proteins. The ASFV A179L (Interpro annotation P42485) protein sequence was compared with those from EBV BHRF1 (P03182), human Bcl-xL (Q07817) and Bcl-2 (P10415). BH domains are shown as coloured backgrounds. The BH4 domain is shown in yellow, BH3 in grey, BH1 in green and BH2 in turquoise. Amino acid identities between the sequences are shown as asterisks * and similarities as double (:) or single (.) dots. Amended from [30] http://jvi.asm.org/content/91/6/e02228-16.full?sid=d5398579-2b8b-47c3-8fbc-c68ac6dc3ddf.

3.2. The ASFV IAP-Family Member

3.2.1. Roles of Cellular IAPs

The IAP inhibitor of the apoptosis protein family were first identified in baculovirus and shown to inhibit cell death in insect cells [31]. Recently, the baculovirus inhibitor of apoptosis Op-IAP3 was shown to bind and stabilise an insect cellular IAP, preventing the virus-induced degradation of this protein. Formation of this complex was shown to be critical for the anti-apoptotic activity of Op-IAP3 [32]. Mammalian cellular IAPs (cIAPs) were first identified as proteins which bound indirectly to TNFR2 through TRAF1 and TRAF2 (see [33] for review). These proteins shared similarity in multiple BIR (Baculoviral IAP repeat) motifs and RING fingers, although the functions of these domains were unknown at that time. It was later established that IAPs bind to TRAFs (Figure 1) via their BIR1 domains [34,35]. Structural analysis showed that a single IAP molecule binds to a TRAF2 trimer [36,37]. The cellular XIAP protein was shown to bind to and inhibit caspase 3 and caspase-mediated apoptosis [38,39]. The BIR2 domain was required to inhibit both caspase 3 and 7 [40]. The BIR2 domain is required to bind processed caspase 3 and the inhibitory domain shown to lie between BIR1 and BIR2. An important development was the demonstration that some cellular IAPs have ubiquitin ligase activity mediated through the RING finger domains. This ubiquitin ligase activity is required for the ubiquitinylation of the RIPK1 complex (see [33] for review). Additional roles for cIAPs were discovered in inhibiting the pro-inflammatory cell death pathways, necroptosis and inflammasome activation leading to the pyroptosis cell death pathway. Consequences of inflammasome activation include the induction of caspase 1- and 11-driven lytic inflammatory cell death by pyropotosis or activation of inflammation mediated by cleavage of IL-1β precursor. Well beyond the inhibition of cell death pathways, XIAP proteins also have roles in regulating signalling from innate immune receptors that depend on their ubiquitin ligase function. The domain structure of BIR motif-containing proteins is very varied as described in the Interpro databases (available online: http://www.ebi.ac.uk/interpro/entry/IPR001370).

3.2.2. The ASFV IAP Protein A224L

Studies on the function of the ASFV IAP-like protein, A224Lwere reported prior to 2002 lacking knowledge about the more recent evidence describing the role of IAPs in different cell death pathways and signalling. Consequently, information is lacking on the potential broader roles of A224L. The A224L protein is 224 amino acids long and well conserved in different ASFV isolates sharing 90% to 99% amino acid identity. Comparison with other members of the IAP family identified between 25% and 34% amino acid identity compared to Baculovirus IAP family proteins between region 17 to 97 amino acids (see Figure 3). The highest amino acid identity with cellular proteins is 32% with Drosophila between residues 19 to 97. This region encodes the A224L protein BIR repeat, which is between residues 29 to 92. A canonical RING motif is not present in A224L protein. The RING motif is present in many other IAP proteins and is required for the ubiquitin ligase activity, suggesting that A224L lacks this function. A224L contains a predicted zinc finger of the 4 cysteine type near the C-terminus (residues 189 to 207) in contrast to ring fingers of the C3HC4 type found in some other IAP (see Figure 3). Stable expression of the A224L IAP-like protein in cells substantially inhibited caspase 3 activity (Figure 1) and cell death induced by treatment with tumour necrosis factor α [41]. When transiently overexpressed, A224L inhibited cell death induced by cycloheximide or staurosporine. Proteolytic cleavage of caspase 3 was increased at later times in Vero cells infected with an A224L deletion mutant compared to wild-type virus. Thus, the A224L protein was indicated to promote cell survival, although the yield of infectious progeny virus was not affected by deletion of A224L. The data suggested that A224L may directly interact with the processed fragment of caspase 3 [41]. Expression of the A224L protein was detected late during infection and the protein was incorporated into virus particles, supporting a role for the protein late in infection or early after entry of the virus particle [42]. A possible alternative mechanism by which A224L may inhibit cell death was suggested from studies which showed that transient expression of A224L activated an NF-kB-dependent reporter. In cell lines stably expressing A224L PMA and ionomycin, stimulation induced greater levels of *c-rel*, an NF-kB dependent gene, than in control cells. This NF-kB inducing activity was abrogated by an IKK-2-dominant negative mutant and enhanced by expression of TNF receptor-associated factor 2 [43]. The activation of NF-kB mediated by TNF-R2 can inhibit apoptotic cell death by activating transcription of a number of anti-apoptotic genes including *IAP* and *Bcl-2* family members. Activation of NF-kB also drives expression of *cFLIP*, an inactive caspase 8 homolog that inhibits its activity (see [33] for review). Deletion of the A224L (*4CL*) gene from the virulent Malawi isolate did not affect levels of virus replication in porcine macrophage cell cultures or the infected macrophage survival time and the induction and magnitude of apoptosis. Moreover, the deletion of this gene from this virulent isolate did not reduce virulence in infected pigs [18]. Thus, although this A224L IAP-like protein was shown to be functional in mammalian cells in inhibiting cell death, no obvious phenotype associated with deletion of the gene was identified either during virus replication in macrophages or infection of pigs [18]. It is possible that the loss of the gene is compensated for by other cell-death inhibitors encoded by the virus. Given the more recently discovered roles of IAP proteins in other cell-death pathways and in cell signalling and inflammation, further investigation should be undertaken of the role of A224L protein in regulating these processes.

```
A224L      --------------------------------------------------------------
opIAP    1 MSSRAIGAPQEGADMKNKAARLGTYTNWPVQFLEPSRMAASGFYYLGRGDEVRCAFCKVE


A224L    1 --------------------MYPKINTID----TYISLRLFEVKPKYAGYSSVDARNKS
opIAP      ITNWVRGDDPETDHKRWAPQCPFVRNNAHDTPHDRAPPARSAAAHPQYAT---EAARLRT
                           : : *: *          *     .:*:**    ** ::

                                BIR2
A224L      FA---IHDIKNYEKFSNAGLFYTSP-TEITCYCCGMKFCNWLYEKHPLQVHGFWSRNCGF
opIAP      FAEWPRGLKQRPEELAEAGFFYTGQGDKTRCFCCDGGLKDWEPDDAPWQQHARWYDRCEY
           **        :. *:::**:***.    :    *:**.   : :*   :. * * *. *   .* :

A224L      MRATLGIIGLKKMIDSYNDYFTHEVSVKHKNRVYTHKRLEDMGFSKCFMRFILANAFMPP
opIAP      VLLVKGRDFVQRVMTEACVVR-DADNEPHIERPAVEAEVADDRL----------------
           :  . *    ::::: .    .   . * :*   ..  .: *   :.

                                                            Zn
A224L      YRKYIHKIILNERYFTFKFVAYLLSFHKVKLDNQTTYC-MTCGIE----QINKDENFCSA
opIAP      -----------------------CKICLGAEKTVCFVPCGHVVACGKCAAGVTTCPV
                                  *: *. :.* * : **       :    . . * .

A224L      CKTLNYKYYKMLNFSIKL 224
opIAP      CRGQLDKAVRMYQV---- 268
           *:    *   :* :.
```

Figure 3. Comparison of ASFV A224L and opIAP protein sequences. The sequences of the ASFV BA71V isolate A224L protein (AOAOCAZXO) and Baculovirus opIAP (P41437) were aligned using Clustal Omega. The positions of domains in the proteins are indicated as coloured backgrounds. The BIR1 repeat (yellow) is present only in the opIAP protein. The BIR2 repeat is in both A224L and OpIAP sequences. At the C-terminus, A224L has a predicted C4 Zn binding domain (grey). The OpIAP protein contains a RING finger domain (turquoise). Identical amino acids are shown as asterisks (*) and similarities as double (:) or single (.) dots.

3.3. ASFV Inhibition of the Stress-Activated Apoptosis Pathway

The pro-apoptotic CCAAT-enhancer-binding protein homologous protein (CHOP) pathway is a cell-death pathway that is activated in cells in response to stress signals including virus infection and the unfolded protein response. This pathway can be induced following phosphorylation of the translation initiation factor eIF2-α by a family of stress-activated protein kinases. These include the double-stranded RNA-activated protein kinase, PKR, which is activated following infection with many viruses including poxvirus but not ASFV. The endoplasmic reticulum resident protein kinase, PERK, is activated during the unfolded protein response (Figure 1). Phosphorylation of eIF2-α on serine 51 leads to reduction of global protein synthesis due to the increased affinity of eIF2-α for the guanine nucleotide exchange factor, eIF2B, which limits formation of the pre-initiation complex required for protein translation. A small subset of proteins can still be translated when eIF2-α is phosphorylated. These include the transcription factor ATF4 and downstream targets including the pro-apoptotic transcription factor CHOP [44,45]. CHOP decreases transcription of Bcl2, depletes cellular glutathione and increases production of reactive oxygen species, sensitising the cell to ER stress and apoptosis [46]. CHOP participates in oxidative stress-mediated apoptosis though the induction of ER oxidase 1α (ERO1α), hyperoxidising the lumen which may result in leakage of hydrogen peroxide into the cytoplasm [47].

The ASFV DP71L protein shares similarity in a C-terminal domain with the HSV ICP34.5 protein and the host GADD34 and CreP proteins. This protein is encoded by most analysed ASFV isolates as a short form of 71 or 72 amino acids (DP71Ls). In some isolates (for example, genotype VIII from Malawi), a longer form of the protein with an amino-terminal extension of about 112 amino acids

(DP71Ll) is encoded. The DP71L protein is expressed late during the replication cycle. The DP71Ls proteins share 94% to 100% amino acid identity with each other. The DP71Ll proteins share 61% identity with DP71Ls over the domains 7 to 71 in DP71Ls and 118 to 185 in DP71Ll (see Figure 4). Amino acid identities of DP71Ls over this domain are approximately 30% to 40% with ICP34.5, GADD34 and CreP proteins; All of these proteins recruit protein phosphatase 1 to dephosphorylate eIF-2 α and restore global protein synthesis [45,48–52] (Figure 1). ASFV activates PP1 and promotes the expression of GADD34 [9]. As a consequence, exogenously expressed DP71L protein can inhibit the stress-induced induction and activation of the pro-apoptotic CHOP protein. Induction of CHOP is inhibited in ASFV infected cells, even those infected with ASFV lacking the *DP71L* gene. This suggests that ASFV may encode other inhibitors of this pathway [50]. The deletion of the DP71L protein from one isolate (E70) reduced virus virulence in pigs, whereas from another isolate (Malawi LIL20/1) no reduction was observed [53,54]. It is possible that the different virus gene complements may explain these differences in results.

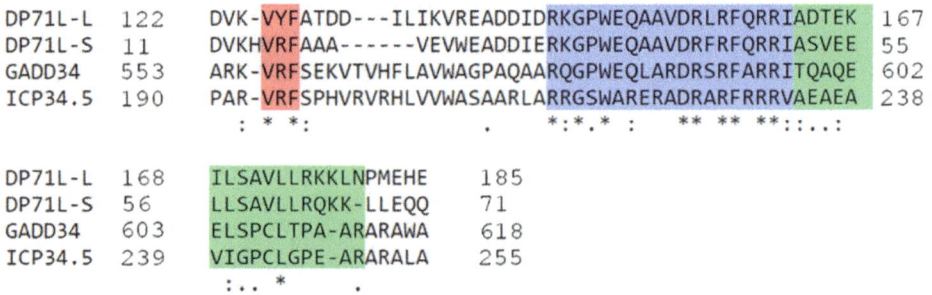

```
DP71L-L  122  DVK-VYFATDD---ILIKVREADDIDRKGPWEQAAVDRLRFQRRIADTEK  167
DP71L-S   11  DVKHVRFAAA------VEVWEADDIERKGPWEQAAVDRFRFQRRIASVEE   55
GADD34   553  ARK-VRFSEKVTVHFLAVWAGPAQAARQGPWEQLARDRSRFARRITQAQE  602
ICP34.5  190  PAR-VRFSPHVRVRHLVVWASAARLARRGSWARERADRARFRRRVAEAEA  238
              : * *:            .    *:*.* :    ** ** **::..:


DP71L-L  168  ILSAVLLRKKLNPMEHE        185
DP71L-S   56  LLSAVLLRQKK-LLEQQ         71
GADD34   603  ELSPCLTPA-ARARAWA        618
ICP34.5  239  VIGPCLGPE-ARARALA        255
              :.. *         .
```

Figure 4. Alignment of ASFV DP71L with GADD34 and ICP34.5 of HSV-1. The long and short forms of DP71L share significant homology with the C-terminal domain of ICP34.5 of HSV-1 and GADD34. Within the C-terminal region of ICP34.5, residues 233–248 (shaded green) have been identified as the eIF2α binding domain [55]. The LSAVL motif within this was identified as critical for function [48]. The eIF2α binding motif described [56] in GADD34 is highlighted in blue. Identical amino acid residues shard between the sequences are shown with an asterisk (*) and similarities are shown as double (:) or single (.) dots.

3.4. The ASFV C-Type Lectin Domain Containing Protein

The ASFV protein, EP153R, was indicated to inhibit the induction of apoptosis. Increased caspase 3 activity and cell death were observed in cells infected with an *EP153R* gene-deletion mutant as compared with infection with the parental BA71V strain. Both transient and stable expression of the *EP153R* gene in cells resulted in a partial protection of the cells from apoptosis induced in response to virus infection or external stimuli. EP153R reduced the transactivating activity of the cellular protein p53 following induction of apoptosis (Figure 1). Since p53 activates transcription of a number of apoptosis inhibitors, this could explain the mechanism of EP153R activation [57,58].

4. Role of Apoptosis in Pathology and Immune Responses

Apoptosis in Tissue Samples from ASFV Infected Pigs

In lethal forms of ASF, early leukopenia is frequently described due to a decrease in the number of circulating monocytes, B- and T- lymphocytes [59,60]. In addition, pronounced depletion of lymphoid tissues is a hallmark of ASF that has been associated with lymphopenia. Other changes observed include severe vascular lesions affecting different organs and body cavities such as hyperemic splenomegaly, haemorrhages or edemas, thrombocytopenia and the induction of disseminated intravascular coagulation. Here, we summarize studies on apoptosis in lymphoid and non-lymphoid organs of pigs infected with ASFV isolates.

ASFV replicates in cells of the mononuclear phagocyte system, predominantly monocytes and fixed-tissue macrophages [8,61,62]. However several other cell types were shown to be infected, especially in the later stages of the disease [60,63]. Infection induces apoptosis of infected macrophages in vivo and the massive apoptosis of bystander lymphocytes (Figure 5) is one of the hallmarks of the acute disease [8,61,62]. Widespread cell death was observed in infected cells of the mononuclear phagocyte system due to programmed cell death. However, at later stages of the disease, cytophatic effects with morphological changes of necrosis are identified in ultrastructural studies [61–65].

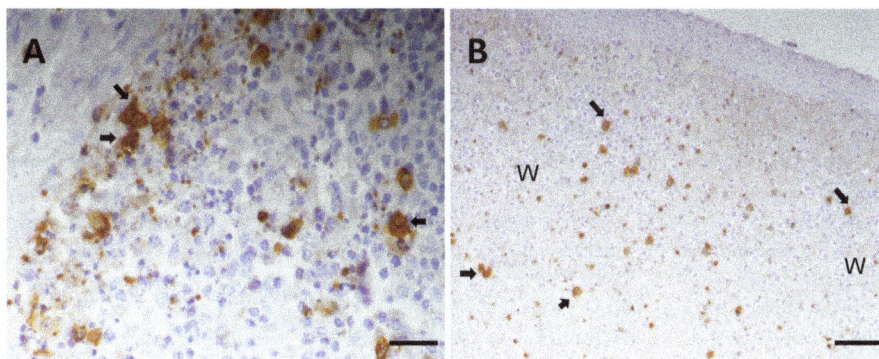

Figure 5. Immunohistochemical detection of ASFV protein P30 on wax-embedded tissue sections from pigs inoculated with the highly virulent ASFV isolate OURT88/1 and euthanized at day 5 post-infection. (**A**) Tonsil, Bar 40 µm. Lymphoid follicle with severe lymphoid depletion. Observe the presence of infected macrophages (arrows) close to areas where lymphocytes show characteristic features of apoptosis such as reduced size and hyperchromatic nuclei. Cell debris and apoptotic bodies, many of them immunolabeled, are also observed; (**B**) spleen, Bar 80 µm. Note the presence of infected cells, mainly macrophages (arrows), along with pyknotic cells, cell debris and apoptotic bodies immunolabeled against P30 in white pulp areas (WP) with severe lymphoid depletion.

Both lymphoid depletion in primary and secondary lymphoid organs and the death of infiltrate associated with the lymphocytes in non-lymphoid organs such as liver and kidney have been attributed to massive apoptosis of lymphocyte subsets [8,61,62]. The mechanisms by which this apoptosis is induced are poorly understood. Lymphocyte destruction was observed mainly in T areas of retropharyngeal lymph nodes, gastrohepatic lymph nodes [64] and tonsils [65] from pigs infected with a highly virulent isolate. Many apoptotic lymphocytes and apoptotic bodies appeared to be phagocytosed by macrophages. There is no reported evidence for ASFV virus replication in cells of lymphoid origin.

The massive lymphocyte apoptosis that affected both B and T areas of lymphoid tissues was not infection of these cells by ASFV. However, the presence of infected macrophages close to areas with intense apoptotic phenomena suggested that the infected cells may have an indirect effect on uninfected lymphocytes such as secretion of chemical mediators by these macrophages [61,64].

A comparison of infections with a virulent isolate in susceptible domestic pigs or bushpigs, which do not develop clinical signs of acute ASF suggested lower virus replication in bushpigs was correlated with reduced secretion of cytokines or vasoactive substances, and less lymphocyte apoptosis [66]. This could in part explain why bushpigs survive infections with virulent isolates of ASF. Further research is required to support this tentative hypothesis.

Gómez del Moral et al. [67] demonstrated that TNFα containing supernatants from macrophage cultures infected with the ASFV virulent isolate Spain-75 induced apoptosis in uninfected lymphocytes. This effect was partially abrogated by pre-incubation with the anti-TNFα specific antibody. TNFα transcripts were detectable at 2–3 dpi in the liver, spleen and lymph nodes and correlated with viral protein expression. Elevated TNFα concentrations in serum were correlated to the onset of clinical signs in pigs. This confirmed a role for TNFα and probably additional pro-apoptotic factors in induction of lymphocyte apoptosis. TNFα-producing cells and infected cells were both identified as macrophages by immunohistochemistry in frozen samples of spleen and lymph nodes from infected pigs but not in non-infected controls [67].

Systematic analysis of pigs infected with the highly virulent isolate Spain-70 showed an increase in serum levels of TNFα and IL-1β from day 2 post-infection (dpi) [68]. From 3 dpi, pigs also displayed a severe leukopenia due to a decrease of circulating lymphocytes and monocytes [69]. Lymphoid depletion that affected both B and T areas was evident from 3 dpi (spleen, lymph nodes, tonsils), 4 dpi (thymus) and correlated with the presence of massive apoptotic phenomena. [68–71]. Infected cells, mainly macrophages, were detected close to apoptotic areas from 1 dpi in the spleen and lymph nodes and from 3 dpi in the thymus and tonsils. Immunohistochemistry showed significant increase of macrophages secreting cytokines. TNFα was secreted at a higher and more constant rate, while secretion of IL-1α was only detected at an early stage. A similar sequence was observed in the liver, where apoptosis in cell infiltrates was described from 5 dpi [72].

In summary, apoptosis affected both macrophage target cells and uninfected lymphocytes from initial stages of disease. An increase of macrophage counts in different areas of lymphoid organs was observed that coincided with the appearance of infected cells, mainly macrophages, and preceded massive lymphocyte apoptosis and lymphoid depletion typically associated with lethal forms of ASF. Macrophage activation was associated with the release of cytokines, mainly TNFα and IL-1α capable of inducing lymphoid tissue destruction by apoptosis. So, the presence of the virus might induce an increase of cytokine secretion in non-infected adjacent cells as a result of an autocrine effect. Lymphocyte apoptosis was correlated with the penetration of ASFV into organs and structures. Controlled apoptosis in lymphocytes would result in a diminished immune response enabling ASFV-infected cells to evade the immune system and replicate.

5. Conclusions and Future Work

As with other viruses, host cells respond to ASFV infection by initiation of apoptosis to limit virus replication. ASFV infection is sensed at the stage of virus entry before the onset of viral protein synthesis to activate caspase 3 and initiate apoptosis. Also, caspase 9 and 12 play an important role in apoptosis induction both by the mitochondrial pathway and the extrinsic pathway of ER stress. However, several viruses-encoded proteins block induction of apoptosis to enable replication of progeny virions. The induction of apoptosis observed at late stages of ASFV infection would favour "silent" virus spread, avoiding the activation of inflammatory responses that are induced by other cell-death pathways including necroptosis and pyroptosis. The activation of inflammatory responses could result in virus clearance by cells of the innate response, limiting virus replication.

ASFV encodes two anti-apoptotic proteins with similarity to cellular protein families. The A179L Bcl-2 like protein has an unusually broad specificity of binding to pro-apoptotic BH3-domain containing proteins. A179L is presumed to function by neutralising the pro-apoptotic function of these cellular proteins. The preferred binding partners for A179L have been indicated by direct binding and interactions in uninfected cells. The binding partners of A179L in infected cells and the impact of expression on infection in cells and in animals remain to be confirmed. Evaluating the role of A179L in virus persistence in wild suids in Africa and soft tick vectors would be of great interest.

The ASFV A224L IAP-like protein acts to inhibit apoptosis by two mechanisms, direct binding and inhibition of caspase 3 and activation of the NF-kB transcription factor and of the anti-apoptotic genes it controls. A224L lacks the RING finger domain that is present in some cellular IAP proteins, suggesting that it lacks ubiquitin ligase function, but this has yet to be confirmed. Cellular IAP proteins have a role in inhibition of necroptosis and pyroptosis and signalling. Potential functions of A224L in these pathways should also be investigated. Additional modulators of cell-death pathways in infected cells undoubtedly remain to be identified. In this context, it will also be interesting to investigate the role of non-coding RNAs in regulating apoptosis during ASFV infection. For example, micro RNAs have been shown to negatively regulate mRNAs for proteins in a number of cellular pathways including apoptosis.

Mechanisms leading to the massive apoptosis of non-infected lymphocytes in lymphoid and non-lymphoid tissues observed in acute ASFV infections are poorly understood. The evidence indicates that factors secreted from ASFV-infected macrophages are involved and TNF-α is suggested to be at least one mediator of this process. Further investigation is needed to establish the key mediators of lymphocyte apoptosis in tissues. The lymphopenia observed in blood during acute ASFV infection is likewise poorly understood. This may involve, in addition to direct effects of ASFV-infected cells, indirect effects which influence homeostasis of lymphocyte populations.

A better understanding of the factors influencing cell death during ASFV infection will contribute to the understanding of disease pathogenesis and the development of effective vaccine strategies.

Acknowledgments: We thank our colleagues Drs Dave Chapman, Ana Reis, Maria Montoya, Chris Netherton, for helpful discussions and BBSRC, UK (BB/L004267/1) and Ministerio de Economia, Industria y Competitividad of Spain (AGL2015-69598-R) for funding.

Conflicts of Interest: The authors have no conflicts of interest.

References

1. Smietanka, K.; Wozniakowski, G.; Kozak, E.; Niemczuk, K.; Fraczyk, M.; Bocian, L.; Kowalczyk, A.; Pejsak, Z. African Swine Fever Epidemic, Poland, 2014–2015. *Emerg. Infect. Dis.* **2016**, *22*, 1201–1207. [CrossRef] [PubMed]

2. Wozniakowski, G.; Kozak, E.; Kowalczyk, A.; Lyjak, M.; Pomorska-Mol, M.; Niemczuk, K.; Pejsak, Z. Current status of African swine fever virus in a population of wild boar in eastern Poland (2014–2015). *Arch. Virol.* **2016**, *161*, 189–195. [CrossRef] [PubMed]

3. Iglesias, I.; Munoz, M.J.; Montes, F.; Perez, A.; Gogin, A.; Kolbasov, D.; de la Torre, A. Reproductive Ratio for the Local Spread of African Swine Fever in Wild Boars in the Russian Federation. *Transbound. Emerg. Dis.* **2016**, *63*, E237–E245. [CrossRef] [PubMed]

4. OIE. World Animal Health Information Database (WAHID) 2017, World 731 Organisation for Animal Health (OIE). Available online: http://www.oie.int/wahis_2/public/wahid.php/Diseaseinformation/diseasehome (accessed on 2 August 2017).

5. Andreani, J.; Bou Khalil, J.Y.; Sevvana, M.; Benamar, S.; Di Pinto, F.; Bitam, I.; Colson, P.; Klose, T.; Rossmann, M.G.; Raoult, D.; et al. Pacmanvirus, a new giant icosahedral virus at the crossroads between Asfarviridae and Faustoviruses. *J. Virol.* **2017**. [CrossRef] [PubMed]

6. Bajrai, L.H.; Benamar, S.; Azhar, E.I.; Robert, C.; Levasseur, A.; Raoult, D.; la Scola, B. Kaumoebavirus, a New Virus That Clusters with Faustoviruses and Asfarviridae. *Viruses* **2016**, *8*, 278. [CrossRef] [PubMed]

7. Reteno, D.G.; Benamar, S.; Khalil, J.B.; Andreani, J.; Armstrong, N.; Klose, T.; Rossmann, M.; Colson, P.; Raoult, D.; La Scola, B. Faustovirus, an Asfarvirus-Related New Lineage of Giant Viruses Infecting Amoebae. *J. Virol.* **2015**, *89*, 6585–6594. [CrossRef] [PubMed]

8. RamiroIbanez, F.; Ortega, A.; RuizGonzalvo, F.; Escribano, J.M.; Alonso, C. Modulation of immune cell populations and activation markers in the pathogenesis of African swine fever virus infection. *Virus Res.* **1997**, *47*, 31–40. [CrossRef]

9. Galindo, I.; Hernaez, B.; Munoz-Moreno, R.; Cuesta-Geijo, M.A.; Dalmau-Mena, I.; Alonso, C. The ATF6 branch of unfolded protein response and apoptosis are activated to promote African swine fever virus infection. *Cell Death Dis.* **2012**, *3*, e341. [CrossRef] [PubMed]

10. Carrascosa, A.L.; Bustos, M.J.; Nogal, M.L.; de Buitrago, G.G.; Revilla, Y. Apoptosis induced in an early step of African swine fever virus entry into Vero cells does not require virus replication. *Virology* **2002**, *294*, 372–382. [CrossRef] [PubMed]

11. Danthi, P. Enter the kill zone: Initiation of death signaling during virus entry. *Virology* **2011**, *411*, 316–324. [CrossRef] [PubMed]

12. Danthi, P. Viruses and the Diversity of Cell Death. In *Annual Review of Virology*; Enquist, L.W., Ed.; Annual Reviews: Palo Alto, CA, USA, 2016; Volume 3, pp. 533–553.

13. Alonso, C.; Miskin, J.; Hernaez, B.; Fernandez-Zapatero, P.; Soto, L.; Canto, C.; Rodriguez-Crespo, I.; Dixon, L.; Escribano, J.M. African swine fever virus protein p54 interacts with the microtubular motor complex through direct binding to light-chain dynein. *J. Virol.* **2001**, *75*, 9819–9827. [CrossRef] [PubMed]

14. Hernaez, B.; Diaz-Gil, G.; Garcia-Gallo, M.; Quetglas, J.I.; Rodriguez-Crespo, I.; Dixon, L.; Escribano, J.M.; Alonso, C. The African swine fever virus dynein-binding protein p54 induces infected cell apoptosis. *FEBS Lett.* **2004**, *569*, 224–228. [CrossRef] [PubMed]

15. Hernaez, B.; Escribano, J.M.; Alonso, C. Visualization of the African swine fever virus infection in living cells by incorporation into the virus particle of green fluorescent protein-p54 membrane protein chimera. *Virology* **2006**, *350*, 1–14. [CrossRef] [PubMed]

16. Vallee, I.; Tait, S.W.G.; Powell, P.P. African swine fever virus infection of porcine aortic endothelial cells leads to inhibition of inflammatory responses, activation of the thrombotic state, and apoptosis. *J. Virol.* **2001**, *75*, 10372–10382. [CrossRef] [PubMed]

17. Neilan, J.G.; Lu, Z.; Afonso, C.L.; Kutish, G.F.; Sussman, M.D.; Rock, D.L. An African Swine Fever Virus Gene with Similarity to the Protooncogene Bcl-2 and the Epstein-Barr-Virus Gene BHRF1. *J. Virol.* **1993**, *67*, 4391–4394. [PubMed]

18. Neilan, J.G.; Lu, Z.; Kutish, G.F.; Zsak, L.; Burrage, T.G.; Borca, M.V.; Carrillo, C.; Rock, D.L. A BIR motif containing gene of African swine fever virus, 4CL, is nonessential for growth in vitro and viral virulence. *Virology* **1997**, *230*, 252–264. [CrossRef] [PubMed]

19. Yanez, R.J.; Rodriguez, J.M.; Nogal, M.L.; Yuste, L.; Enriquez, C.; Rodriguez, J.F.; Vinuela, E. Analysis of the Complete Nucleotide-Sequence of African Swine Fever Virus. *Virology* **1995**, *208*, 249–278. [CrossRef] [PubMed]

20. Bouillet, P.; Strasser, A. Bax and Bak: Back-bone of T cell death. *Nat. Immunol.* **2002**, *3*, 893–894. [CrossRef] [PubMed]

21. Strasser, A. The role of BH3-only proteins in the immune system. *Nat. Rev. Immunol.* **2005**, *5*, 189–200. [CrossRef] [PubMed]

22. Youle, R.J.; Strasser, A. The BCL-2 protein family: Opposing activities that mediate cell death. *Nat. Rev. Mol. Cell Biol.* **2008**, *9*, 47–59. [CrossRef] [PubMed]

23. Kvansakul, M.; Hinds, M.G. The structural biology of BH3-only proteins. *Methods Enzymol.* **2014**, *544*, 49–74. [PubMed]

24. Afonso, C.L.; Neilan, J.G.; Kutish, G.F.; Rock, D.L. An African swine fever virus Bcl-2 homolog, 5-HL, suppresses apoptotic cell death. *J. Virol.* **1996**, *70*, 4858–4863. [PubMed]

25. Hernaez, B.; Cabezas, M.; Munoz-Moreno, R.; Galindo, I.; Cuesta-Geijo, M.A.; Alonso, C. A179L, a New Viral Bcl2 Homolog Targeting Beclin 1 Autophagy Related Protein. *Curr. Mol. Med.* **2013**, *13*, 305–316. [CrossRef] [PubMed]

26. Brun, A.; Rivas, C.; Esteban, M.; Escribano, J.M.; Alonso, C. African swine fever virus gene *A179L*, a viral homologue of Bcl-2, protects cells from programmed cell death. *Virology* **1996**, *225*, 227–230. [CrossRef] [PubMed]

27. Revilla, Y.; Cebrian, A.; Baixeras, E.; Martinez, C.; Vinuela, E.; Salas, M.L. Inhibition of apoptosis by the African swine fever virus Bcl-2 homologue: Role of the BH1 domain. *Virology* **1997**, *228*, 400–404. [CrossRef] [PubMed]

28. Brun, A.; Rodriguez, F.; Escribano, J.M.; Alonso, C. Functionality and cell anchorage dependence of the African swine fever virus gene *A179L*, a viral Bcl-2 homolog, in insect cells. *J. Virol.* **1998**, *72*, 10227–10233. [PubMed]

29. Galindo, I.; Hernaez, B.; Diaz-Gil, G.; Escribano, J.M.; Alonso, C. A 179L, a viral Bcl-2 homologue, targets the core Bcl-2 apoptotic machinery and its upstream BH3 activators with selective binding restrictions for Bid and Noxa. *Virology* **2008**, *375*, 561–572. [CrossRef] [PubMed]

30. Banjara, S.; Caria, S.; Dixon, L.K.; Hinds, M.G.; Kvansakul, M. Structural Insight into African Swine Fever Virus A179L-Mediated Inhibition of Apoptosis. *J. Virol.* **2017**, *91*, e02228-16. [CrossRef] [PubMed]

31. Clem, R.J. Viral IAPs, then and now. *Semin. Cell Dev. Biol.* **2015**, *39*, 72–79. [CrossRef] [PubMed]

32. Byers, N.M.; Vandergaast, R.L.; Friesen, P.D. Baculovirus Inhibitor-of-Apoptosis Op-IAP3 Blocks Apoptosis by Interaction with and Stabilization of a Host Insect Cellular IAP. *J. Virol.* **2016**, *90*, 533–544. [CrossRef] [PubMed]

33. Silke, J.; Vince, J. IAPs and Cell Death. In *Apoptotic and Non-Apoptotic Cell Death*; Nagata, S., Nakano, H., Eds.; Springer International Publishing AG: Cham, Switzerland, 2017; Volume 403, pp. 95–117.

34. Varfolomeev, E.; Blankenship, J.W.; Wayson, S.M.; Fedorova, A.V.; Kayagaki, N.; Garg, P.; Zobel, K.; Dynek, J.N.; Elliott, L.O.; Wallweber, H.J.A.; et al. IAP antagonists induce autoubiquitination of c-IAPs, NF-κB activation, and TNFα-dependent apoptosis. *Cell* **2007**, *131*, 669–681. [CrossRef] [PubMed]

35. Varfolomeev, E.; Wayson, S.M.; Dixit, V.M.; Fairbrother, W.J.; Vucic, D. The inhibitor of apoptosis protein fusion c-IAP2 center dot MALT1 stimulates NF-κB activation independently of TRAF1 and TRAF2. *J. Biol. Chem.* **2006**, *281*, 29022–29029. [CrossRef] [PubMed]

36. Mace, P.D.; Smits, C.; Vaux, D.L.; Silke, J.; Day, C.L. Asymmetric Recruitment of cIAPs by TRAF2. *J. Mol. Biol.* **2010**, *400*, 8–15. [CrossRef] [PubMed]

37. Zheng, C.; Kabaleeswaran, V.; Wang, Y.; Cheng, G.; Wu, H. Crystal Structures of the TRAF2: cIAP2 and the TRAF1: TRAF2: cIAP2 Complexes: Affinity, Specificity, and Regulation. *Mol. Cell* **2010**, *38*, 101–113. [CrossRef] [PubMed]

38. Deveraux, Q.L.; Leo, E.; Stennicke, H.R.; Welsh, K.; Salvesen, G.S.; Reed, J.C. Cleavage of human inhibitor of apoptosis protein XIAP results in fragments with distinct specificities for caspases. *EMBO J.* **1999**, *18*, 5242–5251. [CrossRef] [PubMed]

39. Deveraux, Q.L.; Stennicke, H.R.; Salvesen, G.S.; Reed, J.C. Endogenous inhibitors of caspases. *J. Clin. Immunol.* **1999**, *19*, 388–398. [CrossRef] [PubMed]

40. Takahashi, R.; Deveraux, Q.; Tamm, I.; Welsh, K.; Assa-Munt, N.; Salvesen, G.S.; Reed, J.C. A single BIR domain of XIAP sufficient for inhibiting caspases. *J. Biol. Chem.* **1998**, *273*, 7787–7790. [CrossRef] [PubMed]

41. Nogal, M.L.; de Buitrago, G.G.; Rodriguez, C.; Cubelos, B.; Carrascosa, A.L.; Salas, M.L.; Revilla, Y. African swine fever virus IAP homologue inhibits caspase activation and promotes cell survival in mammalian cells. *J. Virol.* **2001**, *75*, 2535–2543. [CrossRef] [PubMed]

42. Chacon, M.R.; Almazan, F.; Nogal, M.L.; Vinuela, E.; Rodriguez, J.F. The African swine fever virus IAP homolog is a late structural polypeptide. *Virology* **1995**, *214*, 670–674. [CrossRef] [PubMed]

43. Rodriguez, C.I.; Nogal, M.L.; Carrascosa, A.L.; Salas, M.L.; Fresno, M.; Revilla, Y. African swine fever virus IAP-like protein induces the activation of nuclear factor κB. *J. Virol.* **2002**, *76*, 3936–3942. [CrossRef] [PubMed]

44. Harding, H.P.; Novoa, I.; Zhang, Y.H.; Zeng, H.Q.; Wek, R.; Schapira, M.; Ron, D. Regulated translation initiation controls stress-induced gene expression in mammalian cells. *Mol. Cell* **2000**, *6*, 1099–1108. [CrossRef]

45. Novoa, I.; Zeng, H.Q.; Harding, H.P.; Ron, D. Feedback inhibition of the unfolded protein response by GADD34-mediated dephosphorylation of eIF2 α. *J. Cell Biol.* **2001**, *153*, 1011–1021. [CrossRef] [PubMed]

46. McCullough, K.D.; Martindale, J.L.; Klotz, L.O.; Aw, T.Y.; Holbrook, N.J. Gadd153 sensitizes cells to endoplasmic reticulum stress by down-regulating Bcl2 and perturbing the cellular redox state. *Mol. Cell Biol.* **2001**, *21*, 1249–1259. [CrossRef] [PubMed]

47. Marciniak, S.J.; Yun, C.Y.; Oyadomari, S.; Novoa, I.; Zhang, Y.; Jungreis, R.; Nagata, K.; Harding, H.P.; Ron, D. CHOP induces death by promoting protein synthesis and oxidation in the stressed endoplasmic reticulum. *Genes Dev.* **2004**, *18*, 3066–3077. [CrossRef] [PubMed]

48. Barber, C.; Netherton, C.; Goatley, L.; Moon, A.; Goodbourn, S.; Dixon, L. Identification of residues within the African swine fever virus DP71L protein required for dephosphorylation of translation initiation factor eIF2 α and inhibiting activation of pro-apoptotic CHOP. *Virology* **2017**, *504*, 107–113. [CrossRef] [PubMed]

49. Rivera, J.; Abrams, C.; Hernaez, B.; Alcazar, A.; Escribano, J.M.; Dixon, L.; Alonso, C. The MyD116 African swine fever virus homologue interacts with the catalytic subunit of protein phosphatase 1 and activates its phosphatase activity. *J. Virol.* **2007**, *81*, 2923–2929. [CrossRef] [PubMed]

50. Zhang, F.; Moon, A.; Childs, K.; Goodbourn, S.; Dixon, L.K. The African Swine Fever Virus DP71L Protein Recruits the Protein Phosphatase 1 Catalytic Subunit To Dephosphorylate eIF2 α and Inhibits CHOP Induction but Is Dispensable for These Activities during Virus Infection. *J. Virol.* **2010**, *84*, 10681–10689. [CrossRef] [PubMed]

51. Wilcox, D.R.; Longnecker, R. The Herpes Simplex Virus Neurovirulence Factor gamma 34.5: Revealing Virus-Host Interactions. *PLoS Pathog.* **2016**, *12*, e1005449. [CrossRef] [PubMed]

52. Jousse, C.; Oyadomari, S.; Novoa, I.; Lu, P.; Zhang, Y.H.; Harding, H.P.; Ron, D. Inhibition of a constitutive translation initiation factor 2 α phosphatase CReP, promotes survival of stressed cells. *J. Cell Biol.* **2003**, *163*, 767–775. [CrossRef] [PubMed]

53. Zsak, L.; Lu, Z.; Kutish, G.F.; Neilan, J.G.; Rock, D.L. An African swine fever virus virulence-associated gene NL-S with similarity to the herpes simplex virus *ICP34.5* gene. *J. Virol.* **1996**, *70*, 8865–8871. [PubMed]

54. Afonso, C.L.; Zsak, L.; Carrillo, C.; Borca, M.V.; Rock, D.L. African swine fever virus *NL* gene is not required for virus virulence. *J. Gen. Virol.* **1998**, *79*, 2543–2547. [CrossRef] [PubMed]

55. Rojas, M.; Vasconcelos, G.; Dever, T.E. An eIF2 α-binding motif in protein phosphatase 1 subunit GADD34 and its viral orthologs is required to promote dephosphorylation of eIF2 α. *Proc. Natl. Acad. Sci. USA* **2015**, *112*, E3466–E3475. [CrossRef] [PubMed]

56. Li, Y.; Zhang, C.; Chen, X.; Yu, J.; Wang, Y.; Yang, Y.; Du, M.; Jin, H.; Ma, Y.; He, B.; et al. ICP34.5 Protein of Herpes Simplex Virus Facilitates the Initiation of Protein Translation by Bridging Eukaryotic Initiation Factor 2 α (eIF2 α) and Protein Phosphatase 1. *J. Biol. Chem.* **2011**, *286*, 24785–24792. [CrossRef] [PubMed]

57. Granja, A.G.; Nogal, M.L.; Hurtado, C.; Salas, J.; Salas, M.L.; Carrascosa, A.L.; Revilla, Y. Modulation of p53 cellular function and cell death by African swine fever virus. *J. Virol.* **2004**, *78*, 7165–7174. [CrossRef] [PubMed]

58. Hurtado, C.; Granja, A.G.; Bustos, M.J.; Nogal, M.L.; de Buitrago, G.G.; de Yebenes, V.G.; Salas, M.L.; Revilla, Y.; Carrascosa, A.L. The C-type lectin homologue gene (*EP153R*) of African swine fever virus inhibits apoptosis both in virus infection and in heterologous expression. *Virology* **2004**, *326*, 160–170. [CrossRef] [PubMed]

59. Blome, S.; Gabriel, C.; Beer, M. Pathogenesis of African swine fever in domestic pigs and European wild boar. *Virus Res.* **2013**, *173*, 122–130. [CrossRef] [PubMed]

60. Gomez-Villamandos, J.C.; Bautista, M.J.; Sanchez-Cordon, P.J.; Carrasco, L. Pathology of African swine fever: The role of monocyte-macrophage. *Virus Res.* **2013**, *173*, 140–149. [CrossRef] [PubMed]

61. Gomezvillamandos, J.C.; Hervas, J.; Mendez, A.; Carrasco, L.; Delasmulas, J.M.; Villeda, C.J.; Wilkinson, P.J.; Sierra, M.A. Experimental African Swine Fever—Apoptosis of Lymphocytes and Virus-Replication in Other Cells. *J. Gen. Virol.* **1995**, *76*, 2399–2405. [CrossRef] [PubMed]

62. RamiroIbanez, F.; Ortega, A.; Brun, A.; Escribano, J.M.; Alonso, C. Apoptosis: A mechanism of cell killing and lymphoid organ impairment during acute African swine fever virus infection. *J. Gen. Virol.* **1996**, *77*, 2209–2219. [CrossRef] [PubMed]

63. Zsak, L.; Neilan, J.G. Regulation of apoptosis in African swine fever virus-infected macrophages. *Sci. World J.* **2002**, *2*, 1186–1195. [CrossRef] [PubMed]

64. Carrasco, L.; deLara, F.C.M.; delasMulas, J.M.; GomezVillamandos, J.C.; Perez, J.; Wilkinson, P.J.; Sierra, M.A. Apoptosis in lymph nodes in acute African swine fever. *J. Comp. Pathol.* **1996**, *115*, 415–428. [CrossRef]

65. GomezVillamandos, J.C.; Hervas, J.; Moreno, C.; Carrasco, L.; Bautista, M.J.; Caballero, J.M.; Wilkinson, P.J.; Sierra, M.A. Subcellular changes in the tonsils of pigs infected with acute African swine fever virus. *Vet. Res.* **1997**, *28*, 179–189.

66. Oura, C.A.L.; Powell, P.P.; Anderson, E.; Parkhouse, R.M.E. The pathogenesis of African swine fever in the resistant bushpig. *J. Gen. Virol.* **1998**, *79*, 1439–1443. [CrossRef] [PubMed]

67. Gomez Del Moral, M.; Ortuno, E.; Fernandez-Zapatero, P.; Alonso, F.; Alonso, C.; Ezquerra, A.; Dominguez, J. African swine fever virus infection induces tumor necrosis factor α production: Implications in pathogenesis. *J. Virol.* **1999**, *73*, 2173–2180. [PubMed]

68. Salguero, F.J.; Ruiz-Villamor, E.; Bautista, M.J.; Sanchez-Cordon, P.J.; Carrasco, L.; Gomez-Villamandos, J.C. Changes in macrophages in spleen and lymph nodes during acute African swine fever: Expression of cytokines. *Vet. Immunol. Immunopathol.* **2002**, *90*, 11–22. [CrossRef]

69. Salguero, F.J.; Sanchez-Cordon, P.J.; Sierra, M.A.; Jover, A.; Nunez, A.; Gomez-Villamandos, J.C. Apoptosis of thymocytes in experimental African swine fever virus infection. *Histol. Histopathol.* **2004**, *19*, 77–84. [PubMed]

70. Fernandez de Marco, M.; Salguero, F.J.; Bautista, M.J.; Nunez, A.; Sanchez-Cordon, P.J.; Gomez-Villamandos, J.C. An immunohistochemical study of the tonsils in pigs with acute African swine fever virus infection. *Res. Vet. Sci.* **2007**, *83*, 198–203. [CrossRef] [PubMed]

71. Salguero, F.J.; Sanchez-Cordon, P.J.; Nunez, A.; de Marco, M.F.; Gomez-Villamandos, J.C. Proinflammatory cytokines induce lymphocyte apoptosis in acute African swine fever infection. *J. Comp. Pathol.* **2005**, *132*, 289–302. [CrossRef] [PubMed]
72. Sanchez-Cordon, P.; Lorenzo Romero-Trevejo, J.; Pedrera, M.; Manuel Sanchez-Vizcaino, J.; Jose Bautista, M.; Carlos Gomez-Villamandos, J. Role of hepatic macrophages during the viral haemorrhagic fever induced by African Swine Fever Virus. *Histol. Histopathol.* **2008**, *23*, 683–691. [PubMed]

Review

Influenza Virus Infection, Interferon Response, Viral Counter-Response, and Apoptosis

Jung Min Shim [1], Jinhee Kim [1], Tanel Tenson [2], Ji-Young Min [1] and Denis E. Kainov [1,2,3,*]

[1] Institut Pasteur Korea, Gyeonggi-do 13488, Korea; jungmin.shim@ip-korea.org (J.M.S.); jinhee.kim@ip-korea.org (J.K.); jiyoung.min@ip-korea.org (J.-Y.M.)
[2] Institute of Technology, University of Tartu, Tartu 50090, Estonia; tanel.tenson@ut.ee
[3] Department of Clinical and Molecular Medicine, Norwegian University of Science and Technology, Trondheim 7028, Norway
[*] Correspondence: denikaino@gmail.com; Tel.: +358-50-415-5460

Academic Editor: Marc Kvansakul
Received: 30 June 2017; Accepted: 8 August 2017; Published: 12 August 2017

Abstract: Human influenza A viruses (IAVs) cause global pandemics and epidemics, which remain serious threats to public health because of the shortage of effective means of control. To combat the surge of viral outbreaks, new treatments are urgently needed. Developing new virus control modalities requires better understanding of virus-host interactions. Here, we describe how IAV infection triggers cellular apoptosis and how this process can be exploited towards the development of new therapeutics, which might be more effective than the currently available anti-influenza drugs.

Keywords: influenza virus; apoptosis; antiviral agent; innate immunity; host response

1. Introduction

Influenza A and B viruses are common causes of seasonal epidemics. Infected individuals display mild symptoms like cough, sore throat, nasal discharge, fever, headache, and muscle pain [1]. However, the symptoms can be more severe and lead to serious complications like bronchitis and pneumonia. Globally, influenza viruses are the culprits in 3–5 million annual cases of hospitalization and 250,000–500,000 deaths [2,3].

Influenza A virus (IAV) in particular is a potential threat to global health. In contrast to influenza B virus which is only found in humans, IAV can cause pandemic outbreaks when a novel subtype emerges, typically from an animal origin [4]. In the 20th century alone, four influenza pandemics were recorded. The most severe pandemic "Spanish Flu" swept the continents in 1918–1919, affected 500 million people, and caused over 30 million deaths [5]. The most recent pandemic in 2009 emerged when the swine-origin virus, so called "Swine flu", began to infect humans [6]. In addition, "Avian Flu" represents an ongoing threat that may result in devastating consequences if not controlled.

Anti-influenza drugs that target influenza neuraminidase (NA) have been used to prevent and treat influenza virus infections for many years. In particular, oseltamivir, zanamivir, and peramivir exert antiviral effects [7], but certain amino acid changes in NA give rise to drug-resistant IAV strains [8,9]. Due to increasing cases of drug-resistance, and thus reduced efficacy of current treatment, a critical question remains: what will be the next generation of anti-influenza drugs that is less likely to lead to a selection of drug-resistant virus variants?

Developing new virus control modalities requires better understanding of virus-host interactions. Here, we attempt to summarize our knowledge in virus-host cell interactions with a particular focus on programmed death of infected cells (apoptosis). We propose a concept of using apoptosis-inducing drugs as a new class of potential anti-influenza agents. These small molecules can facilitate apoptosis of infected cells, without affecting non-infected cells and, therefore, limit IAV replication and spread. The concept can be expanded to other viral diseases.

2. Influenza A Virus Structure and Replication Cycle

IAV belongs to the *Orthomyxoviridae* family [10]. Its genome is comprised of eight single-stranded viral RNA segments (vRNA) of negative polarity. Two gene segments encode pre-mRNAs that are alternatively spliced to produce nonstructural protein 1 (NS1)/nuclear export protein (NEP) and matrix M1/proton channel M2 proteins, whereas six others encode mRNAs which are translated into nucleoprotein (NP), polymerase subunit PA, PB1 or PB2, hemagglutinin (HA), and NA. Two of the six mRNAs, however, can be translated using different start/stop codons to produce PA-X/N40 and PB1-F2 [10].

In the virions, NP and three viral polymerase subunits bind to vRNA to make eight viral ribonucleoproteins (vRNPs). Eight vRNPs are surrounded by M1 and a lipid membrane, derived from the host cell. The membrane is embedded with HA, NA, and M2. NS1, NEP, PB1-F2, PA-X, and N40 are only expressed in the infected cells and not present in the virion.

IAVs are divided into subtypes based on the structure of virus surface glycoproteins HA and NA. Currently, there are 18 known subtypes of HA (H1-18) and 11 of NA (N1-11) [11]. Only a limited number of IAV subtypes including H1N1 and H3N2 are capable of infecting humans.

The replication cycle of IAV begins when the HA bind to sialic acids on the surface of epithelial cells of the respiratory tract, dendritic cells, type II pneumocytes, alveolar macrophages, or retinal epithelial cells (Figure 1A) [12–14]. Viruses are internalized by endocytosis and then transported to late endosomes [15]. The acidic environment in the late endosomes facilitates HA-mediated fusion of the viral and endosomal membrane, followed by degradation of M1 and release of vRNPs in the cytoplasm [16,17]. The vRNPs enter the nucleus [18]. In the nucleus, negative-sense vRNA is transcribed into positive-sense mRNA using viral polymerase [19,20]. The polymerase snatches 5′ caps from cellular RNA and 3′ RNA is polyadenylated in order to make viral pre-mRNA. The viral proteins are translated from mRNA in the cytoplasm by ribosomes in a cap-dependent manner. Some viral proteins are imported into the nucleus to replicate vRNA. Replication of vRNA occurs in two steps: (i) synthesis of positive-sense complementary RNA (cRNA); (ii) copying of cRNA into new negative-sense vRNAs. Newly assembled vRNPs and viral proteins are transported to the apical side of the cell plasma membrane, where virions are assembled and released by NA [21].

Approximately 0.18–0.21% of amino acids in IAV proteins mutate every year due to the error-prone nature of viral polymerase [22]. Some of these mutations cause antigenic drift, which allows emerging viruses to evade host immunity developed from previous IAV infections or vaccinations. The viruses can also undergo reassortment of genetic segments to generate even greater variations and sometimes antigenic shift. The genetic shifts and drifts are potential causes of epidemic and pandemic outbreaks [10].

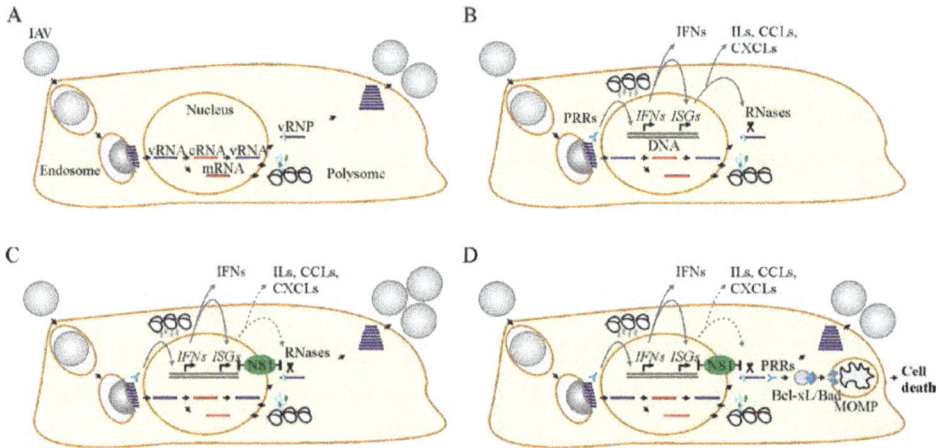

Figure 1. Influenza A virus (IAV) replication cycle, interferon (IFN) response, viral counter-response, and apoptosis. (**A**) IAV replication cycle consists of entry through endocytosis into the host cell and uncoating of viral ribonucleoproteins (vRNPs), import of vRNPs into the nucleus, transcription and replication of the viral genome, translation of viral proteins in the cytoplasm, assembly of vRNPs in the nucleus, export of the vRNPs from the nucleus, and assembly and budding of virions at the host cell plasma membrane. (**B**) When IAV enters the cell, pathogen recognition receptors (PRRs) sense viral RNA (vRNA) and initiate the transcription of interferon (*IFN*) genes. Once transcribed, *IFNs* mediate the expression of IFN-stimulated genes (*ISGs*) in self or, when secreted, in neighboring non-infected cells. *ISGs* encode different antiviral proteins including RNases, which degrade vRNA in infected cells. *ISGs* also encode interleukins (ILs), C-X-C and C-C motif chemokines (CXCLs and CCLs) and other cytokines to recruit immune cells to the site of infection. (**C**) IAV nonstructural protein 1 (NS1) hinders the cellular *IFN-ISG* response by binding with cellular DNA, vRNA, or other cellular factors. The viral replication cycle continues. (**D**) Apoptosis is initiated in response to a large amount of vRNA or its replication intermediates. PRRs recognize vRNA and transduce signals to anti-apoptotic B-cell lymphoma 2 (Bcl-2) proteins. Bcl-2 proteins release pro-apoptotic proteins to initiate mitochondrial outer membrane permiabilization (MoMP), ATP degradation and caspase 3 activation. This results in cell death.

3. Cellular Factors Essential for Influenza A Virus Replication

Partly due to the simplicity of the genome, IAVs complete successful replication by relying on multiple cellular proteins [23–30]. Cellular clathrin, epsin-1 Ras-related GTPases, and COPI are important for virus dynamin-dependent endocytic uptake. Cellular vATPase acidifies the interior of late endosomes. This activates cellular serine proteases, which cleave HA and mediate fusion of viral and endosomal membranes and the release of vRNPs surrounded by M1. The aggresome formation and disassembly machinery degrades the M1 shell and uncoats vRNPs. Subsequently, cytoplasmic importins mediate nuclear import of vRNPs through the nuclear pore complex (NPC). Cellular hCLE, cyclin T1, CDK9, ANP32A, and pol II are required for vRNA transcription. PTBP1, NHP2L1, SNRP70, SF3B1, SF3A1, CLK1, UAP56, p14, and PRPF8 are necessary to splice NS1/NEP and M1/M2 pre-mRNAs. NPC, with the help of cellular NXF1, E1B-AP5, Rae1, and p15, transport viral mRNAs into the cytoplasm. In the cytoplasm, a translation apparatus translates viral mRNAs into proteins and GRSF1 stimulates this process. Subsequently, quality control of newly synthesized viral proteins is carried out by cellular chaperones and chaperonins. In addition, ISGylation, SUMOylation, and phosphorylation processes mediated by cellular machineries could modify novel viral proteins. Importins and HSP90 assist in the translocation of viral polymerase, NP, and NEP via NPC back to the nucleus where they form NEP-vRNP complexes. Crm1, HRB, hNup98, and Raf–MEK–ERK are required for transport of NEP-vRNPs into the cytoplasm through NPC. In the cytoplasm, microtubules and Rab11 bring the complexes to the plasma membrane. Newly synthesized M1, M2, HA, and NA are also transported to the plasma membrane through the trans-Golgi network with the help of COPI and Rab8. β-actin, CK2 and Rab11 are cellular proteins required for the budding and release of new virions.

IAV also actively exploits cell metabolism for the production of viral RNA, proteins, and lipids [24,26,31–36]. Free NTPs are used by viral polymerase which produces vRNA and its replication intermediates. In addition, IAV utilizes amino acids to synthesize viral proteins by hijacking the PI3K–mTor–Akt-mediated autophagy. Virus assembly and budding depends on lipid metabolism (including fatty acid biosynthesis, phospholipid metabolism, de novo synthesis of cholesterol). Finally, virus replication is sensitive to the cellular redox state, which is essential for maturation of HA and for the quality of released viral particles. These are only a few examples of cellular factors essential for virus replication.

4. Cellular Factors that Limit Virus Replication and Spread

Apart from cellular factors that support viral replication, there are dozens of those which restrict this process. When IAV enters the cells, stimulus-specific signals are transduced along the interferon signaling pathway to activate antiviral responses (Figure 1B) [37]. Pattern recognition receptors (PRRs), such as TLR3, TLR7, IRF7, MDA5, and RIGI sense incoming viruses and activate transcription of interferon (IFN) genes, such as *IFNB1*, *IL28A*, *IL29*, *IL28B*, *IFNW1*, *IFNA7*, *IFNA14*, *IFNA10*, *IFNA13*, *IFNA16*, *IFNA8*, *IFNA1*, *IFNG*, *IFNA2*, and *IFNA21* [38]. IFNs launch the expression of IFN-stimulated genes (*ISGs*) in infected cells as well as in nearby non-infected cells, protecting them from potential viral invasion (Figure S1) [39–41].

The ISGs encode a variety of antiviral proteins with diverse modes of action. These include IFITM1 and SAMD9, which prevent fusion between viral and endosome membranes; HERC5, HERC6, USP18, ISG15, TRIM22, and ISG20, which mark viral proteins for degradation and, thereby, mediate vRNA uncoating; IFIT1, IFIT2, OASL, IRF7, DDX60, DDX58/RIG-I, IFIH1/MDA5, and EIF2AK2/PKR, which recognize vRNA, and OAS1, OAS2, and OAS3 which degrade vRNA; ZBP1, PARP1, PARP9, PARP14, and PRIC285, which inhibit transcription and translation of vRNA and activating expression of cellular antiviral genes; lipid raft-disturbing factor RSAD2 which prevents coating of vRNPs with host membrane; and cholesterol-depleting factor IFITM3 which inactivates budding viruses (Figure 2B) [42–47]. ISGs also encode IFI27 and XAF1 for regulation of apoptosis; IDO, COX2, and CH25H for production of neuro- and immuno-modulators; cytokines and chemokines for activation and recruitment of immune cells to the site of infection; MX1, MX2, GBP1, GBP2, GBP3, GBP5, IFI44, GMPR, and NT5C3 for GTP catabolism and cytokine processing; STAT1 for amplification of autocrine *ISG* expression, as well as many other antiviral factors. As a result, *ISG* products can inhibit viral replication in infected cells, alert non-infected cells for potential infections, attract immune cells, and trigger an alarm in the central nervous system about the ongoing infection.

In counter-response to cellular IFNs, IAV utilizes non-structural protein NS1 (Figure 1C) [48]. NS1 is produced within a few hours of infection [49]. NS1 can block the transcription of innate antiviral genes by directly binding with cellular DNA [50]. In addition, NS1 interacts with vRNA and its replication intermediates to prevent its recognition by cellular PRRs and RNAses [51–54]. It can also bind TRIM25, ISG15, GBP1, and other *ISG* products to inhibit their functions at transcriptional, post-transcriptional, translational, and post-translational stages [55]. Thus, the levels of innate immune mediators are regulated by viruses to ensure IAV replication and to avoid excessive IFN responses, which are often associated with severe disease [56,57].

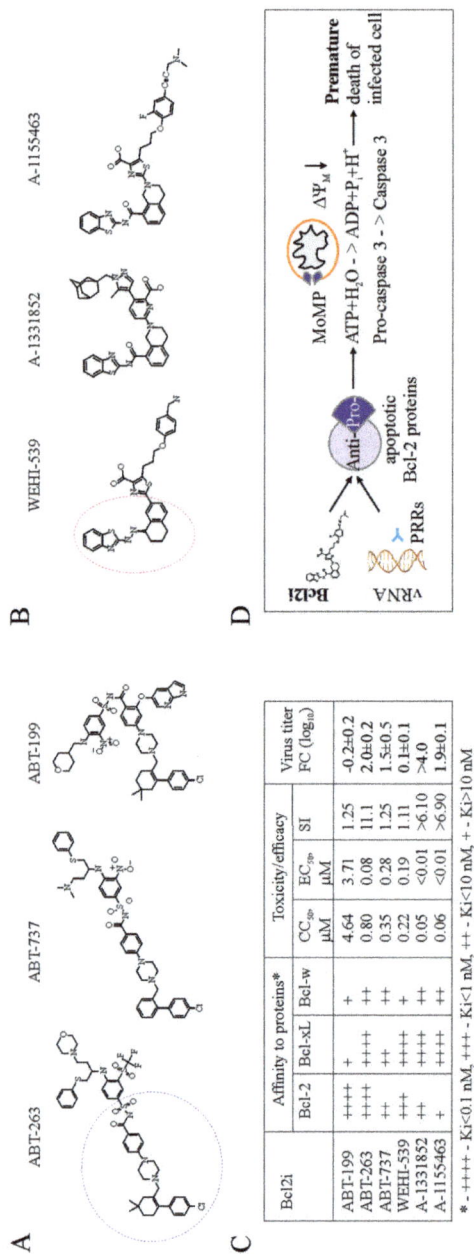

Figure 2. Bcl-2 inhibitors (Bcl2i) facilitate Bcl-2-dependent apoptosis in cells containing viral RNA. (**A,B**) Structures of ABT-263, ABT-737, ABT-199, WEHI-539, A-1331852, and A-1155463 revealed that these molecules fall into two distinct classes. Core structures are highlighted. (**C**) Table showing Bcl2i antiviral activities and affinities for three Bcl-2 proteins. "+" indicates inhibitory effect. Increased inhibition is marked by a higher "+" designation. (**D**) Schematic diagram showing how chemical inhibitors of Bcl-2 proteins induce premature death of cells containing viral nucleic acids. Bcl: B-cell lymphoma; CC_{50}: half-maximum cytotoxic concentration; EC_{50}: half-maximum efficacy concentration; SI: selectivity index; FC: fold-change; PRRs: pattern recognition receptors.

5. Apoptosis Is a Cellular Process That Restricts Virus Replication and Spread

When the IFN responses fail to control IAV replication, cells may activate a secondary antiviral response via programmed death called apoptosis (Figure 1D). This is particularly important when IAV escapes the IFN responses through the action of NS1. During this process, PRRs, including RIG-I, MDA5, PKR (encoded by ISGs: *IFIH1*, *DDX58*, and *EIF2AK2*), recognize accumulating vRNA and activate apoptotic machinery that directs the fate of IAV-infected cells [58]. The anti-apoptotic (Bcl-2, Bcl-xL, and Bcl-w) and pro-apoptotic (Bax, Bak, Bad, Bim, Bid, Puma, and Noxa) Bcl-2 proteins associate or dissociate to start a cascade of reactions resulting first in mitochondria membrane permeabilization (MoMP). This is followed by cytochrome c release, apoptosome activation, ATP degradation, and eventually cell death [59–62]. As the initial trigger of this process, the concentration of vRNA is, therefore, a critical rate-limiting factor. Alternatively, if the viral load is high enough, apoptosis could be initiated during virus entry.

All Bcl-2 proteins contain Bcl2-homology 3 (BH3) domains, which are essential for their protein-protein interactions and functions [63]. Cellular proteins including UACA, PAWR, FLII, Trim21, IMMT, 14-3-3, EFHD2, DHX9, DDX3, NLRP3, and LRRFIP2 as well as viral factors M2, PB1-F2, NS1, HA, and NP may stabilize or disrupt the interactions of BH3-domain proteins in infected cells [62,64–66]. However, further studies are required to verify their specific functions in apoptosis.

6. Apoptosis-Inducing Small Molecules

Bcl-2 dependent apoptosis represents a potential target for antiviral drug development. In particular, anticancer Bcl-2 inhibitors (Bcl2i) may be repurposed to treat viral diseases. The first anticancer Bcl-2 inhibitor, ABT-737, was engineered based on the structure of Bad bound to Bcl-xL in order to mimic Bad BH3-peptide [67,68]. Several derivatives have been developed to have improved pharmacokinetic properties, and the resulting product, ABT-263, is currently in clinical trials, and ABT-199 is approved to treat multiple lymphoid malignancies (Figure 2A) [63,69–71]. Another group of Bcl2i with anticancer properties was discovered using high-throughput screening [72]. This includes WEHI-539 and its derivatives, A-1331852, and A-1155463 (Figure 2B). Also, other Bcl-2 inhibitors (such as TW-37, gossypol, UMI-77, A-1210477 and BDA-366) that are structurally distinct from ABT-737 and WEHI-539 have been developed. All these compounds have different affinities for Bcl-2 proteins [58].

Importantly, ABT-737, ABT-263, ABT-199, WEHI-539, A-1331852, and A-1155463, but not TW-37, gossypol, UMI-77, A-1210477, and BDA-366, can universally induce premature death of IAV-infected cells at concentrations not toxic for non-infected cells (Figure 2C) [62]. However, only ABT-263, A-1331852, and A-1155463 could effectively limit viral replication and spread (unpublished data [73]).

We propose a model for this effect in Figure 2D. PRRs recognize vRNA or its replication intermediates and send signals to anti-apoptotic Bcl-xL. Bcl-xL releases its pro-apoptotic partners to initiate MoMP, ATP degradation, and caspase-3 activation. This results in cell death. ABT-263, A-1331852, or A-1155463 act synergistically with viral RNA and thereby facilitate the cell death.

ABT-263, unlike A-1155463, causes irreversible thrombocytopenia [74,75], which makes A-1155463 a better candidate for antiviral testing in animals. Moreover, half-maximum efficacy concentration (EC_{50}) for A-1155463 is lower than that for ABT-263. In addition, half-maximum cytotoxic concentration (CC_{50}) value of A-1155463 is higher than that of A-1331852, whereas EC_{50} of both are lower than that for ABT-263 (unpublished data [73]). Thus, A-1155463 could represent an antiviral lead candidate, which would reinforce the necessary therapeutic arsenal for the treatment of influenza and perhaps other viral diseases.

7. Accelerating Apoptosis of Infected Cells: A Novel Antiviral Strategy

The typical approach in antiviral drug discovery has been to identify virus inhibitors that target various stages of virus replication and to preserve infected cells from death

(Figure 3) [23,24,30,44,76–83]. Examples of such antiviral drugs are DAS181, JNJ872, ribavirin, verdinexor, CH65, C05, SaliPhe, nucleozin, geldanamycin, 17-AAG, LJ001, SA-19, fattiviracin, TBHQ, 4C, gemcitabine, ASN2, bortezamib, carfilzomib, C75, 25HC, SNS-032, and MK2206 (Figure 3) [23,24,44,79–84]. As an alternative to the traditional method, there is the use of Bcl2i. The Bcl2i selectively causes apoptosis in only virus infected cells, leaving virus-free cells intact. Therefore, Bcl2i represents a novel class of antiviral compounds with potential that is worth exploring.

However, Bcl2i must be used as a prophylactic rather than a therapeutic drug because of the following issues. Although the induction of apoptosis has been shown to be selective for infected cells in vitro, inhibition of Bcl2 proteins may have off-target effects in vivo [74,75]. Our preliminary results also indicate that treatment with Bcl2i of IAV-infected mice may affect cytokine expression and, therefore, may prevent development of innate and adaptive immune responses [62]. In addition, Bcl2i may have adverse effects in acute virus infection. The viral dose is likely to be high, infecting a large number of cells. Inducing apoptosis may result in extensive tissue damage in this case.

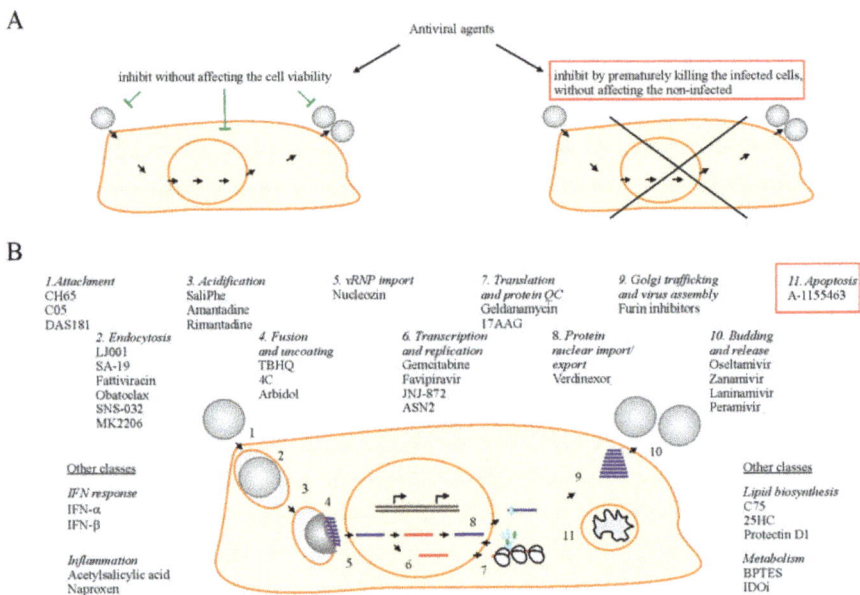

Figure 3. Two strategies of antiviral drug development. (**A**) One strategy is focused on discovery of antivirals to inhibit viral infection without affecting the viability of infected cells, whereas another exploits small molecules to inhibit viral replication by specifically killing only the virus-infected cells. (**B**) Examples of existing and emerging anti-IAV drugs. Existing and emerging drugs that target certain stages of virus replication cycle are shown. Bcl2 inhibitors (Bcl2i) are shown in a red box.

8. Conclusions

Cellular antiviral responses including IFN response and apoptosis are employed in order to inhibit virus replication and spread. IAV has evolved to gain mechanisms to disconcert these responses to ensure its replication. Based on our knowledge on host-virus interaction, we can explore ways to develop pharmacological interventions to control IAV infections. In particular, our advance in understanding apoptosis has shown potential in developing apoptosis-inducing molecules as antiviral drugs against flu. A-1155463 could serve as a lead compound in this process. Prophylactic treatment with A-1155463 may prevent development of severe disease. Successful prevention of flu using Bcl2i could provide an alternative therapeutic option for IAV, against which current treatment is limited.

Having wider treatment options could reduce the use of drugs targeting virus proteins, and thus slow down the rise of drug-resistant virus strain through evolutionary selection pressure. Timely use of Bcl2i may also reduce the use of antibiotics, which are utilized for treatment of secondary bacterial infections. This will limit the development of emerging antibiotic-resistant bacteria. Exploring a new class of antiviral drugs is crucial, and further investigations on the antiviral properties of Bcl2i could lead to development of new drugs to prevent other viral diseases, associated with HIV, ZIKV, HBV, and VZV (29–31).

Supplementary Materials: The following are available online at http://www.mdpi.com/1999-4915/9/8/223/s1. Figure S1. IAV infection and interferon response in human macrophages and RPE cells.

Acknowledgments: We thank Alexandr Ianevsky and Dmitry Guschin for critical reading of the manuscript. This work was funded by Mobilitas pluss top researcher grant (contract No MOBTT39, to D.K.), IP-Korea and Bio & Medical Technology Development Program of the National Research Foundation and the Ministry of science and information technology, Republic of Korea (grant No. 2014K1A4A7A01074646, to J.Y.M.).

Author Contributions: All authors analyzed the literature and wrote the manuscript.

Conflicts of Interest: The authors declare no conflict of interest.

Abbreviations

vATPase: vacuolar ATPase; hCLE: human homolog of chicken CLE; CDK9: cyclin-dependent kinase 9; ANP32A: acidic leucine-rich nuclear phosphoprotein 32 family member A; pol II: DNA-directed RNA polymerase 2; PTBP1: polypyrimidine tract-binding protein 1; NHP2L1: NHP2-like protein 1; SNRP70: small nuclear ribonucleoprotein U1 subunit 70; SF3B1: splicing factor 3B subunit 1; SF3A1: splicing factor 3A subunit 1; CLK1: dual specificity protein kinase 1; UAP56: spliceosome RNA helicase DDX39B ; p14: pre-mRNA branch site protein p14; PRPF8: pre-mRNA-processing-splicing factor 8; NXF1: nuclear RNA export factor 1; E1B-AP5: heterogeneous nuclear ribonucleoprotein U-like protein 1; Rae1: mRNA export factor 1; p15: mRNA export factor; GRSF1: G-rich sequence factor 1; Crm1: chromosome region maintenance 1; HRB: human immunodeficiency virus rev-binding; hNup98: human nucleoporin 98; Raf: Raf proto-oncogene serine/threonine-protein kinase; MEK: dual specificity mitogen-activated protein kinase kinase 1; Erk: extracellular signal–regulated kinases; Rab11: Ras-related protein 11; COPI: coatomer 1 vesicular transport complex; Rab8: Ras-related protein 8; CK2: casein kinase 2; PI3K: phosphatidylinositol 3-kinase; mTor: mechanistic target of rapamycin; Akt: protein kinase B; TLR3: Toll-like receptor 3; TLR7: Toll-like receptor 7; IRF7: Interferon regulatory factor 7; MDA5: interferon-induced helicase C domain-containing protein 1; RIGI: probable ATP-dependent RNA helicase DDX58; HERC5: E3 ISG15–protein ligase 5; HERC6: E3 ISG15–protein ligase 6; USP18: Ubl carboxyl-terminal hydrolase 18; ISG15: ubiquitin-like protein 15; TRIM22: E3 ubiquitin-protein ligase 22; ISG20: ubiquitin-like protein 22; IFIT1: interferon-induced protein with tetratricopeptide repeats 1; IFIT2: interferon-induced protein with tetratricopeptide repeats 2; OASL: 2′-5′-oligoadenylate synthase-like protein; DDX60: probable ATP-dependent RNA helicase 60; EIF2AK2: Interferon-induced, double-stranded RNA-activated protein kinase; OAS1: 2′-5′-oligoadenylate synthase 1; OAS2: 2′-5′-oligoadenylate synthase 2; OAS3: 2′-5′-oligoadenylate synthase 3; ZBP1: Z-DNA-binding protein 1; PARP1: poly [ADP-ribose] polymerase 1; PARP9: poly [ADP-ribose] polymerase 9; PARP14: poly [ADP-ribose] polymerase 14; PRIC285: helicase with zinc finger domain 2; RSAD2: radical S-adenosyl methionine domain-containing protein 2; IDO: indoleamine 2,3-dioxygenase 1; COX2: Prostaglandin G/H synthase 2; CH25H: Cholesterol 25-hydroxylase; MX1: interferon-induced GTP-binding protein 1; MX2: interferon-induced GTP-binding protein 2; GBP1: guanylate-binding protein 1; GBP2: guanylate-binding protein 2; GBP3: guanylate-binding protein 3; GBP5: guanylate-binding protein 5; IFI44: interferon-induced protein 44; GMPR: GMP reductase 2; NT5C3: cytosolic 5′-nucleotidase 3A; GTP: guanosine triphosphate; Bax: apoptosis regulator BAX; Bak: Bcl-2 homologous antagonist/killer; Bad: Bcl2-associated agonist of cell death; Bim: Bcl-2-like protein 11; Bid: BH3-interacting domain death agonist; Puma: Bcl-2-binding component 3; Noxa: phorbol-12-myristate-13-acetate-induced protein 1; 25HC: 25-Hydroxycholesterol; HIV: human immunodeficiency virus; ZIKV: Zika virus; HBV: hepatitis B virus; VZV: Varicella zoster virus.

References

1. WHO Influenza (Seasonal). Available online: http://wwwwhoint/mediacentre/factsheets/fs211/en/ (accessed on 8 July 2017).

2. Global Burden of Disease Study 2013 Collaborators. Global, regional, and national incidence, prevalence, and years lived with disability for 301 acute and chronic diseases and injuries in 188 countries, 1990–2013: A systematic analysis for the Global Burden of Disease Study 2013. *Lancet* **2015**, *386*, 743–800.

3. Lafond, K.E.; Nair, H.; Rasooly, M.H.; Valente, F.; Booy, R.; Rahman, M.; Kitsutani, P.; Yu, H.; Guzman, G.; Coulibaly, D.; et al. Global Role and Burden of Influenza in Pediatric Respiratory Hospitalizations, 1982–2012: A Systematic Analysis. *PLoS Med.* **2016**, *13*, e1001977. [CrossRef] [PubMed]

4. Horimoto, T.; Kawaoka, Y. Influenza: Lessons from past pandemics, warnings from current incidents. *Nat. Rev. Microbiol.* **2005**, *3*, 591–600. [CrossRef] [PubMed]
5. Taubenberger, J.K.; Morens, D.M. 1918 Influenza: The mother of all pandemics. *Emerg. Infect. Dis.* **2006**, *12*, 15–22. [CrossRef] [PubMed]
6. Fineberg, H.V. Pandemic preparedness and response—Lessons from the H1N1 influenza of 2009. *N. Engl. J. Med.* **2014**, *370*, 1335–1342. [CrossRef] [PubMed]
7. CDC Influenza Antiviral Medications: Summary for Clinicians. Available online: https://wwwcdcgov/flu/professionals/antivirals/summary-clinicianshtm (accessed on 8 July 2017).
8. Spanakis, N.; Pitiriga, V.; Gennimata, V.; Tsakris, A. A review of neuraminidase inhibitor susceptibility in influenza strains. *Expert Rev. Anti-Infect. Ther.* **2014**, *12*, 1325–1336. [CrossRef] [PubMed]
9. Hussain, M.; Galvin, H.D.; Haw, T.Y.; Nutsford, A.N.; Husain, M. Drug resistance in influenza A virus: The epidemiology and management. *Infect. Drug Resist.* **2017**, *10*, 121–134. [CrossRef] [PubMed]
10. Bouvier, N.M.; Palese, P. The biology of influenza viruses. *Vaccine* **2008**, *26* (Suppl. 4), D49–D53. [CrossRef] [PubMed]
11. Pinsent, A.; Fraser, C.; Ferguson, N.M.; Riley, S. A systematic review of reported reassortant viral lineages of influenza A. *BMC Infect. Dis.* **2016**, *16*, 3. [CrossRef] [PubMed]
12. Werner, J.L.; Steele, C. Innate receptors and cellular defense against pulmonary infections. *J. Immunol.* **2014**, *193*, 3842–3850. [CrossRef] [PubMed]
13. Mansour, D.E.; El-Shazly, A.A.; Elawamry, A.I.; Ismail, A.T. Comparison of ocular findings in patients with H1N1 influenza infection versus patients receiving influenza vaccine during a pandemic. *Ophthalmic Res.* **2012**, *48*, 134–138. [CrossRef] [PubMed]
14. Michaelis, M.; Geiler, J.; Klassert, D.; Doerr, H.W.; Cinatl, J., Jr. Infection of human retinal pigment epithelial cells with influenza A viruses. *Investig. Ophthalmol. Vis. Sci.* **2009**, *50*, 5419–5425. [CrossRef] [PubMed]
15. Edinger, T.O.; Pohl, M.O.; Stertz, S. Entry of influenza A virus: Host factors and antiviral targets. *J. Gen. Virol.* **2014**, *95*, 263–277. [CrossRef] [PubMed]
16. White, J.M.; Whittaker, G.R. Fusion of Enveloped Viruses in Endosomes. *Traffic* **2016**, *17*, 593–614. [CrossRef] [PubMed]
17. Banerjee, I.; Miyake, Y.; Nobs, S.P.; Schneider, C.; Horvath, P.; Kopf, M.; Matthias, P.; Helenius, A.; Yamauchi, Y. Influenza A virus uses the aggresome processing machinery for host cell entry. *Science* **2014**, *346*, 473–477. [CrossRef] [PubMed]
18. Pumroy, R.A.; Ke, S.; Hart, D.J.; Zachariae, U.; Cingolani, G. Molecular determinants for nuclear import of influenza A PB2 by importin alpha isoforms 3 and 7. *Structure* **2015**, *23*, 374–384. [CrossRef] [PubMed]
19. Te Velthuis, A.J.; Fodor, E. Influenza virus RNA polymerase: Insights into the mechanisms of viral RNA synthesis. *Nat. Rev. Microbiol.* **2016**, *14*, 479–493. [CrossRef] [PubMed]
20. Reguera, J.; Gerlach, P.; Cusack, S. Towards a structural understanding of RNA synthesis by negative strand RNA viral polymerases. *Curr. Opin. Struct. Biol.* **2016**, *36*, 75–84. [CrossRef] [PubMed]
21. Lakdawala, S.S.; Fodor, E.; Subbarao, K. Moving On Out: Transport and Packaging of Influenza Viral RNA into Virions. *Annu. Rev. Virol.* **2016**, *3*, 411–427. [CrossRef] [PubMed]
22. Belanov, S.S.; Bychkov, D.; Benner, C.; Ripatti, S.; Ojala, T.; Kankainen, M.; Kai Lee, H.; Wei-Tze Tang, J.; Kainov, D.E. Genome-Wide Analysis of Evolutionary Markers of Human Influenza A(H1N1)pdm09 and A(H3N2) Viruses May Guide Selection of Vaccine Strain Candidates. *Genome Biol. Evol.* **2015**, *7*, 3472–3483. [CrossRef] [PubMed]
23. Muller, K.H.; Kakkola, L.; Nagaraj, A.S.; Cheltsov, A.V.; Anastasina, M.; Kainov, D.E. Emerging cellular targets for influenza antiviral agents. *Trends Pharmacol. Sci.* **2012**, *33*, 89–99. [CrossRef] [PubMed]
24. Soderholm, S.; Fu, Y.; Gaelings, L.; Belanov, S.; Yetukuri, L.; Berlinkov, M.; Cheltsov, A.V.; Anders, S.; Aittokallio, T.; Nyman, T.A.; et al. Multi-Omics Studies towards Novel Modulators of Influenza A Virus-Host Interaction. *Viruses* **2016**, *8*, E269. [CrossRef] [PubMed]
25. Shaw, M.L.; Stertz, S. Role of Host Genes in Influenza Virus Replication. *Curr. Top. Microbiol. Immunol.* **2017**, 1–99. [CrossRef]
26. Powell, J.D.; Waters, K.M. Influenza-Omics and the Host Response: Recent Advances and Future Prospects. *Pathogens* **2017**, *6*, E25. [CrossRef] [PubMed]
27. Yen, H.L. Current and novel antiviral strategies for influenza infection. *Curr. Opin. Virol.* **2016**, *18*, 126–134. [CrossRef] [PubMed]

28. Watanabe, T.; Kawaoka, Y. Influenza virus-host interactomes as a basis for antiviral drug development. *Curr. Opin. Virol.* **2015**, *14*, 71–78. [CrossRef] [PubMed]

29. Tripathi, S.; Batra, J.; Lal, S.K. Interplay between influenza A virus and host factors: Targets for antiviral intervention. *Arch. Virol.* **2015**, *160*, 1877–1891. [CrossRef] [PubMed]

30. Keener, A.B. Host with the most: Targeting host cells instead of pathogens to fight infectious disease. *Nat. Med.* **2017**, *23*, 528–531. [CrossRef] [PubMed]

31. Cui, L.; Zheng, D.; Lee, Y.H.; Chan, T.K.; Kumar, Y.; Ho, W.E.; Chen, J.Z.; Tannenbaum, S.R.; Ong, C.N. Metabolomics Investigation Reveals Metabolite Mediators Associated with Acute Lung Injury and Repair in a Murine Model of Influenza Pneumonia. *Sci. Rep.* **2016**, *6*, 26076. [CrossRef] [PubMed]

32. Paul, P.; Munz, C. Autophagy and Mammalian Viruses: Roles in Immune Response, Viral Replication, and Beyond. *Adv. Virus Res.* **2016**, *95*, 149–195. [PubMed]

33. Chlanda, P.; Zimmerberg, J. Protein-lipid interactions critical to replication of the influenza A virus. *FEBS Lett.* **2016**, *590*, 1940–1954. [CrossRef] [PubMed]

34. Tisoncik-Go, J.; Gasper, D.J.; Kyle, J.E.; Eisfeld, A.J.; Selinger, C.; Hatta, M.; Morrison, J.; Korth, M.J.; Zink, E.M.; Kim, Y.M.; et al. Integrated Omics Analysis of Pathogenic Host Responses during Pandemic H1N1 Influenza Virus Infection: The Crucial Role of Lipid Metabolism. *Cell Host Microbe* **2016**, *19*, 254–266. [CrossRef] [PubMed]

35. Proia, R.L.; Hla, T. Emerging biology of sphingosine-1-phosphate: Its role in pathogenesis and therapy. *J. Clin. Investig.* **2015**, *125*, 1379–1387. [CrossRef] [PubMed]

36. Fu, Y.; Gaelings, L.; Soderholm, S.; Belanov, S.; Nandania, J.; Nyman, T.A.; Matikainen, S.; Anders, S.; Velagapudi, V.; Kainov, D.E. JNJ872 inhibits influenza A virus replication without altering cellular antiviral responses. *Antivir. Res.* **2016**, *133*, 23–31. [CrossRef] [PubMed]

37. Killip, M.J.; Fodor, E.; Randall, R.E. Influenza virus activation of the interferon system. *Virus Res.* **2015**, *209*, 11–22. [CrossRef] [PubMed]

38. Bowie, A.G.; Unterholzner, L. Viral evasion and subversion of pattern-recognition receptor signalling. *Nat. Rev. Immunol.* **2008**, *8*, 911–922. [CrossRef] [PubMed]

39. Liu, S.Y.; Sanchez, D.J.; Aliyari, R.; Lu, S.; Cheng, G. Systematic identification of type I and type II interferon-induced antiviral factors. *Proc. Natl. Acad. Sci. USA* **2012**, *109*, 4239–4244. [CrossRef] [PubMed]

40. Schoggins, J.W.; Wilson, S.J.; Panis, M.; Murphy, M.Y.; Jones, C.T.; Bieniasz, P.; Rice, C.M. A diverse range of gene products are effectors of the type I interferon antiviral response. *Nature* **2011**, *472*, 481–485. [CrossRef] [PubMed]

41. Soderholm, S.; Anastasina, M.; Islam, M.M.; Tynell, J.; Poranen, M.M.; Bamford, D.H.; Stenman, J.; Julkunen, I.; Sauliene, I.; De Brabander, J.K.; et al. Immuno-modulating properties of saliphenylhalamide, SNS-032, obatoclax, and gemcitabine. *Antivir. Res.* **2016**, *126*, 69–80. [CrossRef] [PubMed]

42. Melchjorsen, J.; Kristiansen, H.; Christiansen, R.; Rintahaka, J.; Matikainen, S.; Paludan, S.R.; Hartmann, R. Differential regulation of the *OASL* and *OAS1* genes in response to viral infections. *J. Interferon Cytokine Res.* **2009**, *29*, 199–207. [CrossRef] [PubMed]

43. Ludwig, S.; Wolff, T. Influenza A virus TRIMs the type I interferon response. *Cell Host Microbe* **2009**, *5*, 420–421. [CrossRef] [PubMed]

44. Gaelings, L.; Soderholm, S.; Bugai, A.; Fu, Y.; Nandania, J.; Schepens, B.; Lorey, M.B.; Tynell, J.; Vande Ginste, L.; Le Goffic, R.; et al. Regulation of kynurenine biosynthesis during influenza virus infection. *FEBS J.* **2017**, *284*, 222–236. [CrossRef] [PubMed]

45. Dudek, S.E.; Nitzsche, K.; Ludwig, S.; Ehrhardt, C. Influenza A viruses suppress cyclooxygenase-2 expression by affecting its mRNA stability. *Sci. Rep.* **2016**, *6*, 27275. [CrossRef] [PubMed]

46. Meunier, E.; Broz, P. Interferon-inducible GTPases in cell autonomous and innate immunity. *Cell. Microbiol.* **2016**, *18*, 168–180. [CrossRef] [PubMed]

47. Gold, E.S.; Diercks, A.H.; Podolsky, I.; Podyminogin, R.L.; Askovich, P.S.; Treuting, P.M.; Aderem, A. 25-Hydroxycholesterol acts as an amplifier of inflammatory signaling. *Proc. Natl. Acad. Sci. USA* **2014**, *111*, 10666–10671. [CrossRef] [PubMed]

48. Ayllon, J.; Garcia-Sastre, A. The NS1 protein: a multitasking virulence factor. *Curr. Top. Microbiol. Immunol.* **2015**, *386*, 73–107. [PubMed]

49. Anastasina, M.; Schepens, B.; Soderholm, S.; Nyman, T.A.; Matikainen, S.; Saksela, K.; Saelens, X.; Kainov, D.E. The C terminus of NS1 protein of influenza A/WSN/1933(H1N1) virus modulates antiviral responses in infected human macrophages and mice. *J. Gen. Virol.* **2015**, *96*, 2086–2091. [CrossRef] [PubMed]

50. Anastasina, M.; Le May, N.; Bugai, A.; Fu, Y.; Soderholm, S.; Gaelings, L.; Ohman, T.; Tynell, J.; Kyttanen, S.; Barboric, M.; et al. Influenza virus NS1 protein binds cellular DNA to block transcription of antiviral genes. *Biochim. Biophys. Acta* **2016**, *1859*, 1440–1448. [CrossRef] [PubMed]

51. Bornholdt, Z.A.; Prasad, B.V. X-ray structure of NS1 from a highly pathogenic H5N1 influenza virus. *Nature* **2008**, *456*, 985–988. [CrossRef] [PubMed]

52. Min, J.Y.; Li, S.; Sen, G.C.; Krug, R.M. A site on the influenza A virus NS1 protein mediates both inhibition of PKR activation and temporal regulation of viral RNA synthesis. *Virology* **2007**, *363*, 236–243. [CrossRef] [PubMed]

53. Min, J.Y.; Krug, R.M. The primary function of RNA binding by the influenza A virus NS1 protein in infected cells: Inhibiting the 2′-5′ oligo (A) synthetase/RNase L pathway. *Proc. Natl. Acad. Sci. USA* **2006**, *103*, 7100–7105. [CrossRef] [PubMed]

54. Li, S.; Min, J.Y.; Krug, R.M.; Sen, G.C. Binding of the influenza A virus NS1 protein to PKR mediates the inhibition of its activation by either PACT or double-stranded RNA. *Virology* **2006**, *349*, 13–21. [CrossRef] [PubMed]

55. Hale, B.G.; Randall, R.E.; Ortin, J.; Jackson, D. The multifunctional NS1 protein of influenza A viruses. *J. Gen. Virol.* **2008**, *89*, 2359–2376. [CrossRef] [PubMed]

56. Baskin, C.R.; Bielefeldt-Ohmann, H.; Tumpey, T.M.; Sabourin, P.J.; Long, J.P.; Garcia-Sastre, A.; Tolnay, A.E.; Albrecht, R.; Pyles, J.A.; Olson, P.H.; et al. Early and sustained innate immune response defines pathology and death in nonhuman primates infected by highly pathogenic influenza virus. *Proc. Natl. Acad. Sci. USA* **2009**, *106*, 3455–3460. [CrossRef] [PubMed]

57. Kash, J.C.; Tumpey, T.M.; Proll, S.C.; Carter, V.; Perwitasari, O.; Thomas, M.J.; Basler, C.F.; Palese, P.; Taubenberger, J.K.; Garcia-Sastre, A.; et al. Genomic analysis of increased host immune and cell death responses induced by 1918 influenza virus. *Nature* **2006**, *443*, 578–581. [CrossRef] [PubMed]

58. Ashkenazi, A.; Fairbrother, W.J.; Leverson, J.D.; Souers, A.J. From basic apoptosis discoveries to advanced selective Bcl-2 family inhibitors. *Nat. Rev. Drug Discov.* **2007**, *16*, 273–284. [CrossRef] [PubMed]

59. Tran, A.T.; Cortens, J.P.; Du, Q.; Wilkins, J.A.; Coombs, K.M. Influenza virus induces apoptosis via BAD-mediated mitochondrial dysregulation. *J. Virol.* **2013**, *87*, 1049–1060. [CrossRef] [PubMed]

60. McLean, J.E.; Datan, E.; Matassov, D.; Zakeri, Z.F. Lack of Bax prevents influenza A virus-induced apoptosis and causes diminished viral replication. *J. Virol.* **2009**, *83*, 8233–8246. [CrossRef] [PubMed]

61. Hinshaw, V.S.; Olsen, C.W.; Dybdahl-Sissoko, N.; Evans, D. Apoptosis: A mechanism of cell killing by influenza A and B viruses. *J. Virol.* **1994**, *68*, 3667–3673. [PubMed]

62. Kakkola, L.; Denisova, O.V.; Tynell, J.; Viiliainen, J.; Ysenbaert, T.; Matos, R.C.; Nagaraj, A.; Ohman, T.; Kuivanen, S.; Paavilainen, H.; et al. Anticancer compound ABT-263 accelerates apoptosis in virus-infected cells and imbalances cytokine production and lowers survival rates of infected mice. *Cell Death Dis.* **2013**, *4*, e742. [CrossRef] [PubMed]

63. Delbridge, A.R.; Grabow, S.; Strasser, A.; Vaux, D.L. Thirty years of BCL-2: Translating cell death discoveries into novel cancer therapies. *Nat. Rev. Cancer* **2016**, *16*, 99–109. [CrossRef] [PubMed]

64. Ong, J.D.; Mansell, A.; Tate, M.D. Hero turned villain: NLRP3 inflammasome-induced inflammation during influenza A virus infection. *J. Leukoc. Biol.* **2017**, *101*, 863–874. [CrossRef] [PubMed]

65. Herold, S.; Ludwig, S.; Pleschka, S.; Wolff, T. Apoptosis signaling in influenza virus propagation, innate host defense, and lung injury. *J. Leukoc. Biol.* **2012**, *92*, 75–82. [CrossRef] [PubMed]

66. Subramanian, T.; Vijayalingam, S.; Kuppuswamy, M.; Chinnadurai, G. Interaction of cellular proteins with Bcl-xL targeted to cytoplasmic inclusion bodies in adenovirus infected cells. *Virology* **2015**, *483*, 21–31. [CrossRef] [PubMed]

67. Lee, E.F.; Czabotar, P.E.; Smith, B.J.; Deshayes, K.; Zobel, K.; Colman, P.M.; Fairlie, W.D. Crystal structure of ABT-737 complexed with Bcl-xL: Implications for selectivity of antagonists of the Bcl-2 family. *Cell Death Differ.* **2007**, *14*, 1711–1713. [CrossRef] [PubMed]

68. Kvansakul, M.; Hinds, M.G. The Bcl-2 family: Structures, interactions and targets for drug discovery. *Apoptosis* **2015**, *20*, 136–150. [CrossRef] [PubMed]

69. Souers, A.J.; Leverson, J.D.; Boghaert, E.R.; Ackler, S.L.; Catron, N.D.; Chen, J.; Dayton, B.D.; Ding, H.; Enschede, S.H.; Fairbrother, W.J.; et al. ABT-199, a potent and selective Bcl-2 inhibitor, achieves antitumor activity while sparing platelets. *Nat. Med.* **2013**, *19*, 202–208. [CrossRef] [PubMed]

70. Vandenberg, C.J.; Cory, S. ABT-199, a new Bcl-2-specific BH3 mimetic, has in vivo efficacy against aggressive Myc-driven mouse lymphomas without provoking thrombocytopenia. *Blood* **2013**, *121*, 2285–2288. [CrossRef] [PubMed]

71. Roberts, A.W.; Huang, D. Targeting BCL2 with BH3 Mimetics: Basic Science and Clinical Application of Venetoclax in Chronic Lymphocytic Leukemia and Related B Cell Malignancies. *Clin. Pharmacol. Ther.* **2017**, *101*, 89–98. [CrossRef] [PubMed]

72. Lessene, G.; Czabotar, P.E.; Sleebs, B.E.; Zobel, K.; Lowes, K.N.; Adams, J.M.; Baell, J.B.; Colman, P.M.; Deshayes, K.; Fairbrother, W.J.; et al. Structure-guided design of a selective Bcl-xL inhibitor. *Nat. Chem. Biol.* **2013**, *9*, 390–397. [CrossRef] [PubMed]

73. Bulanova, D.; Ianevski1, A.; Bugai, A.; Akimov, E.; Kuivanen, S.; Paavilainen, H.; Kakkola, L.; Nandania, J.; Turunen, L.; Ohman, T.; et al. Antiviral properties of anticancer Bcl-2 inhibitors. *Mol. Microbiol.* **2017**, submitted.

74. Leverson, J.D.; Phillips, D.C.; Mitten, M.J.; Boghaert, E.R.; Diaz, D.; Tahir, S.K.; Belmont, L.D.; Nimmer, P.; Xiao, Y.; Ma, X.M.; et al. Exploiting selective Bcl-2 family inhibitors to dissect cell survival dependencies and define improved strategies for cancer therapy. *Sci. Transl. Med.* **2015**, *7*, 279ra40. [CrossRef] [PubMed]

75. Tao, Z.F.; Hasvold, L.; Wang, L.; Wang, X.; Petros, A.M.; Park, C.H.; Boghaert, E.R.; Catron, N.D.; Chen, J.; Colman, P.M.; et al. Discovery of a Potent and Selective Bcl-xL Inhibitor with in Vivo Activity. *ACS Med. Chem. Lett.* **2014**, *5*, 1088–1093. [CrossRef] [PubMed]

76. Martinez, J.P.; Sasse, F.; Bronstrup, M.; Diez, J.; Meyerhans, A. Antiviral drug discovery: broad-spectrum drugs from nature. *Nat. Prod. Rep.* **2015**, *32*, 29–48. [CrossRef] [PubMed]

77. Vigant, F.; Santos, N.C.; Lee, B. Broad-spectrum antivirals against viral fusion. *Nat. Rev. Microbiol.* **2015**, *13*, 426–437. [CrossRef] [PubMed]

78. Lou, Z.; Sun, Y.; Rao, Z. Current progress in antiviral strategies. *Trends Pharmacol. Sci.* **2014**, *35*, 86–102. [CrossRef] [PubMed]

79. Loregian, A.; Mercorelli, B.; Nannetti, G.; Compagnin, C.; Palu, G. Antiviral strategies against influenza virus: Towards new therapeutic approaches. *Cell. Mol. Life Sci.* **2014**, *71*, 3659–3683. [CrossRef] [PubMed]

80. Vanderlinden, E.; Naesens, L. Emerging antiviral strategies to interfere with influenza virus entry. *Med. Res. Rev.* **2014**, *34*, 301–339. [CrossRef] [PubMed]

81. Zumla, A.; Rao, M.; Wallis, R.S.; Kaufmann, S.H.; Rustomjee, R.; Mwaba, P.; Vilaplana, C.; Yeboah-Manu, D.; Chakaya, J.; Ippolito, G.; et al. Host-directed therapies for infectious diseases: current status, recent progress, and future prospects. *Lancet Infect. Dis.* **2016**, *16*, e47–e63. [CrossRef]

82. McKimm-Breschkin, J.L.; Fry, A.M. Meeting report: 4th ISIRV antiviral group conference: Novel antiviral therapies for influenza and other respiratory viruses. *Antivir. Res.* **2016**, *129*, 21–38. [CrossRef] [PubMed]

83. Soderholm, S.; Kainov, D.E.; Ohman, T.; Denisova, O.V.; Schepens, B.; Kulesskiy, E.; Imanishi, S.Y.; Corthals, G.; Hintsanen, P.; Aittokallio, T.; et al. Phosphoproteomics to Characterize Host Response During Influenza A Virus Infection of Human Macrophages. *Mol. Cell. Proteom.* **2016**, *15*, 3203–3219. [CrossRef] [PubMed]

84. Holthausen, D.J.; Lee, S.H.; Kumar, V.T.; Bouvier, N.M.; Krammer, F.; Ellebedy, A.H.; Wrammert, J.; Lowen, A.C.; George, S.; Pillai, M.R.; et al. An Amphibian Host Defense Peptide Is Virucidal for Human H1 Hemagglutinin-Bearing Influenza Viruses. *Immunity* **2017**, *46*, 587–595. [CrossRef] [PubMed]

![viruses logo] *viruses*

MDPI

Article

Upregulation of miRNA-4776 in Influenza Virus Infected Bronchial Epithelial Cells Is Associated with Downregulation of NFKBIB and Increased Viral Survival

Sreekumar Othumpangat [1,*], Nicole B. Bryan [2], Donald H. Beezhold [1] and John D. Noti [1]

1   Allergy and Clinical Immunology Branch, Health Effects Laboratory Division, National Institute for Occupational Safety and Health, Centers for Disease Control and Prevention, Morgantown, WV 26505, USA; zec1@cdc.gov (D.H.B.); ivr2@cdc.gov (J.D.N.)
2   School of Medicine, West Virginia University, Morgantown, WV 26506, USA; nbryan2@hsc.wvu.edu
*   Correspondence: seo8@cdc.gov; Tel.: +1-304-285-5839

Academic Editor: Marc Kvansakul
Received: 2 March 2017; Accepted: 20 April 2017; Published: 27 April 2017

Abstract: Influenza A virus (IAV) infection remains a significant cause of morbidity and mortality worldwide. One key transcription factor that is activated upon IAV infection is nuclear factor Kappa B (NF-κB). NF-κB regulation involves the inhibitor proteins NF-κB inhibitor beta (NFKBIB), (also known as IκBβ), which form complexes with NF-κB to sequester it in the cytoplasm. In this study, microarray data showed differential expression of several microRNAs (miRNAs) on exposure to IAV. Target scan analysis revealed that miR-4776, miR-4514 and miR-4742 potentially target NFKBIB messenger RNA (mRNA). Time-course analysis of primary bronchial epithelial cells (HBEpCs) showed that miR-4776 expression is increased within 1 h of infection, followed by its downregulation 4 h post-exposure to IAV. NFKBIB upregulation of miR-4776 correlated with a decrease in NFKBIB expression within 1 h of infection and a subsequent increase in NFKBIB expression 4 h post-infection. In addition, miRNA ago-immunoprecipitation studies and the three prime untranslated region (3′ UTR) luciferase assay confirmed that miR-4776 targets NFKBIB mRNA. Furthermore, uninfected HBEpCs transfected with miR-4776 mimic showed decreased expression of NFKBIB mRNA. Overexpression of NFKBIB protein in IAV infected cells led to lower levels of IAV. Taken together, our data suggest that miRNA-4776 modulates IAV production in infected cells through NFKBIB expression, possibly through the modulation of NF-κB.

Keywords: NFKBIB; influenza virus; bronchial epithelial cells; NF-κB; virus survival

1. Introduction

Influenza infection is a significant cause of morbidity and mortality causing an estimated 3–5 million infections per year [1]. Upon infection, the virus hijacks the host's cellular machinery for its survival and replication. One mechanism by which influenza may influence host gene expression is through differential expression of host cell microRNAs (miRNA), which are endogenously produced ~22 nucleotide single-strand RNAs that interact with the three prime untranslated region (3′ UTR) of messenger RNAs (mRNAs) and destabilize the transcripts or degrade them to repress translation [2]. These molecules have been shown to play a key role in the regulation of a diverse array of cellular responses including inflammation and cell death. Since their initial discovery, miRNAs have been developed as both diagnostic tools and therapeutic targets. A number of miRNAs identified are predicted to target key elements of the immune response and cell survival pathways [3].

Nuclear factor Kappa B (NF-κB) plays an important role in regulating several genes that are necessary for cell proliferation, apoptosis and cell survival [4]. NF-κB also activates pro-inflammatory genes in cells at the sites of inflammation in several diseases including infectious diseases [5,6]. In addition, NF-κB itself is activated upon influenza infection [7–9]. It was previously believed that activation of NF-κB served as a cellular defense mechanism upon influenza viral infection with production of inflammatory cytokines, including interferon β (IFNβ) [7,10]. Studies conducted in A549 and U1752 cells showed that low NF-κB activity were resistant to influenza A virus (IAV) infection, but cells became susceptible to IAV upon activation of NF-κB. One suggested role of NF-κB signaling is in the endocytosis pathway that is necessary for efficient influenza virus infection (9). Some of the viral proteins are involved in the activation of NF-κB [11]. The viral non-structural protein NS1 acts as a suppressor of NF-κB activation [12]. However, the antagonistic activity of NS1 is not exhaustive for complete blocking of NF-κB [13]. Moreover, other viral components such as hemagglutinin [14], nucleoprotein and matrix proteins can induce NF-κB activation [11]. In vitro studies showed that the NF-κB specific inhibitor, SC75741, expressed in bronchial epithelial cells decreases the propagation of IAV [15]. This effect was subsequently confirmed in an in vivo mouse model in which SC75741 was able to effectively protect mice from the deadly effects of three highly pathogenic influenza strains. Based on these studies, it is hypothesized that IAV has evolved a strategy to activate NF-κB to prevent apoptosis of the host cell and facilitate viral replication.

Regulation of NF-κB is a complex process involving both internal and external cellular stimuli [16]. NF-κB regulation involves a family of inhibitor proteins, NF-κB inhibitor beta protein (NFKBIB, also known as IκBβ) that forms a complex with NF-κB and sequesters it into the cytoplasm [17]. This effectively prevents nuclear translocation of NF-κB and subsequent transcription of its target genes [18]. Phosphorylation of the serine residues on NFKBIB proteins directs them for degradation and releases NF-κB from the complex resulting in its translocation to the nucleus [19]. Given that NF-κB plays a role in influenza infection, our aim was to determine the role of specific miRNAs in regulating the NF-κB pathway. In gastric cancer cell lines, miR-20a was shown to target NFKBIB [20], and miR-182-5p was shown to target NFKBIB in breast cancer cells [21]. However, possibly due to tissue specific expression, we saw no significant change in these miRNAs in IAV infected alveolar lung epithelial (A549) cells. In our miRNA microarray analysis, we show that miR-4514, miR-4742, and miR-4776 are significantly upregulated in IAV infected A549 cells and target scan [22] analysis showed that these miRNAs may target NFKBIB mRNA. We show that these three miRNAs are differentially expressed in IAV infected human primary bronchial epithelial cells (HBEpCs) and that miR-4776 specifically downregulates expression of NFKBIB that leads to activation of NF-κB and increased survival of IAV in HBEpCs cells.

2. Materials and Methods

2.1. Cell Culture

HBEpCs were purchased from PromoCell GmbH (Heidelberg, Germany) and sub-cultured in media and growth factors recommended by the supplier. A549 cells (CCL-34, American Type Culture Collection (ATCC), Manassas, VA, USA) were cultured in standard F12K medium with heat inactivated 10% fetal bovine serum (FBS), 100 IU/mL penicillin and 100 μg/mL streptomycin sulfate. Madin–Darby canine kidney (MDCK) cells were used for the propagation of influenza virus. MDCK cells were cultured in Minimum Essential Medium (MEM) (ATCC) supplemented with 10% FBS, 100 IU/mL penicillin and 100 μg/mL streptomycin sulfate [22].

2.2. Viruses and Their Infections

Influenza virus A/WSN/33 (H1N1) was a kind gift from Prof. Robert A. Lamb (Northwestern University, Chicago, IL, USA) and cultivation and maintenance of the virus was carried out as described earlier [23].

All infections of airway epithelial cells were performed in six-well plates at a dose of 1.0 multiplicity of infection (MOI) unless otherwise specified. Controls referred are the cells that were mock infected. Six-well plates were seeded with 5×10^5 cells per well and grown to 80% confluence. Cells were rinsed with phosphate buffered saline (PBS), and then the virus diluted in modified Hank's Balanced Salt Solution (HBSS) was added to each well. After 45 min incubation at 37 °C, excess virus was washed off with PBS. Fresh F12 media was added containing 1 μg/mL of tosyl phenylalanyl chloromethyl ketone (TPCK)-trypsin (Sigma-Aldrich, St Louis, MO, USA) and incubated at 37 °C and 5% CO_2. TPCK was not added to A/WSN/33 virus. Cells were harvested at different time intervals and used for protein and RNA studies.

2.3. Real-Time Reverse Transcription Polymerase Chain Reaction

After the experimental treatment of the HBEpCs and A549 cells total RNA was extracted from these cells with the RNeasy plus Mini Kit (Qiagen, Germantown, MD, USA). RNA was quantified with a NanoDrop spectrophotometer (Thermofisher Scientific, Foster City, CA, USA). Total RNA was then reverse transcribed with the High-Capacity complementary DNA (cDNA) Reverse Transcription Kit (Applied Biosystems, Foster City, CA, USA). All reverse transcription-polymerase chain reactions (RT-PCRs) were carried out using standard TaqMan primers for NFKBIB (assay id # Hs00182115_m1), NF-κB (assay ID# Hs00765730_m1), and glyceraldehyde phosphate dehydrogenase (GAPDH) (assay ID # Hs03929097_g1), which were purchased from Applied Biosystems (Thermofisher Scientific). Fold change in expression was determined using the ΔΔct method after normalizing to GAPDH [22].

miRNA was isolated using the miReasy Kit (Qiagen) and analyzed by RT-PCR with the TaqMan MicroRNA Reverse Transcription Kit (Lifetechnologies, Foster City, CA, USA) and specific primers for miR-4776 (assay # 462695), 4514 (assay # 462737), 4742 (assay # 463053) and the U6 (assay # 001973) control. Influenza matrix gene expression was quantified and reported as influenza copy number. RT-PCR was performed using TaqMan assay with matrix-specific primers, as reported earlier [24]. The M segment of the RT-PCR was specific for viral RNA. A standard curve was generated from the cloned influenza IAV matrix gene by RT-PCR for IAV quantification.

2.4. miRNA Microarray

A549 cells were infected with IAV for 3 h and the RNA was extracted using the Exiqon miRCURY locked nucleic acid (LNA) miRNA extraction kit (Exiqon, Vedbaek, Denmark). The quality of the total RNA was verified in an Agilent 2100 Bioanalyzer profile (Agilent Technologies, Inc., Santa Clara, CA, USA). In addition, 750 ng total RNA from both sample and reference was labeled with Hy3 ™ and Hy5 ™ fluorescent label, respectively, using the miRCURY LNA ™ microRNA Hi-Power Labeling Kit, Hy3 ™/Hy5 ™ (Exiqon) according to the procedure described by the manufacturer. The Hy3 ™-labeled samples and a Hy5 ™-labeled reference RNA sample were mixed pairwise and hybridized to the miRCURY LNA ™ microRNA Array 7th Gen (Exiqon), which contains capture probes targeting all miRNAs for human, mouse or rat registered in the miRBASE 18.0. The hybridization was performed according to the miRCURY LNA ™ microRNA Array Instruction manual using a Tecan HS4800 ™ hybridization station (Tecan, Grodig, Austria). After hybridization, the microarray slides were scanned and stored in an ozone free environment to prevent potential bleaching of the fluorescent dyes. The miRCURY LNA ™ microRNA array slides were scanned using the Agilent G2565BA Microarray Scanner System (Agilent Technologies) and the image analysis was carried out using the ImaGene ®9 (miRCURY LNA ™ microRNA Array Analysis Software, Exiqon). The quantified signals were background corrected and normalized using the global Lowess (LOcally WEighted

Scatterplot Smoothing) regression algorithm. The data obtained were subjected to statistical analysis and differentially regulated miRNAs in infected and uninfected cells were identified and reported by Exiqon.

2.5. Transfection Studies

HBEpCs were transfected with a miRNA-4776 or 4514 inhibitor oligonucleotide (complementary strand to miRNAs) or a miRNA-4776 or 4514 mimic oligonucleotide (corresponding to the miRNA-sequence) (Life Technologies, Carlsbad, CA, USA) using the lipid-based Lipofectamine 2000 reagent diluted in Opti-MEM-I reduced serum medium (Life Technologies) according to the protocol provided by the supplier. Briefly, HBEpCs were grown to 80% confluence in six-well plates. Transfection complexes were directly applied to the cells (final concentration of 50 nM) and the plates were incubated in a humidified chamber with 5% CO_2 at 37 °C. After 6 h of transfection, the medium was replaced with fresh complete medium. As negative control, cells were transfected with a scrambled oligonucleotide (Life Technologies). To evaluate the effects of miRNA-4514 and 4776 in the context of viral infection, cells after 48 h of transfection were infected with IAV at a MOI of 1. Virus were allowed to attach to the cells for 45 min at 37 °C, then the excess virus was washed off and fresh medium containing 1 μg/mL of TPCK trypsin (Sigma-Aldrich) was added. Cells were then incubated for another 4 h. Following the incubation, cells were harvested at different time intervals and used for protein and RNA studies.

2.6. Imaging with Confocal Microscopy

HBEpCs cells were grown on chamber slides overnight (Chamber slide ™, Lab-TekII, Thermo Fisher Scientific, Rochester, NY, USA) to 80–90% confluence. Cells were then exposed to IAV for 4 h. Subsequently, cells were washed with PBS and fixed with 4% methanol-free formaldehyde (Polysciences Inc., Warrington, PA, USA). Immunofluorescent staining was done as described earlier [23] and stained with an antibody that recognizes the phosphorylated form of NF-κBp65, rabbit anti-phospho-NF-κB p65 antibody (Millipore, Billerica, MA, USA) for 1 h, followed by Alexa-488 conjugated anti-rabbit secondary antibody (Life Technologies). HBEpCs overexpressing NFKBIB were infected with IAV and then stained with NFKBIB antibody (Cell signaling, Danvers, MA, USA) and NS1 antibody (Invitrogen, Foster City, CA, USA), followed by appropriate secondary antibodies (Alexa 488 and Alexa-555). The glass slides were mounted with 4′,6-diamidino-2-phenylindole (DAPI)-Prolong Gold anti-fade reagent (Life Technologies) and protected with cover slips. Photomicrographs were made using a Zeiss Laser Scanning Microscopy (LSM)-510 (Carl Zeiss AG, Obertochen, Germany) confocal microscope.

2.7. Overexpression of NFKBIB

HBEpCs cells were transiently transfected with the open reading frame (ORF) of NFKBIB cloned in pCMV6-Entry vector from Origene (Origene, Rockville, MD, USA) and lipofectamine 2000 (Invitrogen). Cells transfected with an empty pCMV6 vector were used as control. Following transfection for 48 h, cells were infected with IAV at 1 MOI for another 10 h. Cells were harvested at different time points as described in Results and Discussion. RNA was isolated and RT-PCR for the viral matrix copy number as well as the NFKBIB transcripts were done as described earlier.

2.7.1. Validation of miRNA by Ago Immunoprecipitation

Immunoprecipitation (IP) of the miRNA was done according to the instructions from the kit available from Active Motif (Active Motif, Carlsbad, CA, USA). Briefly, HBEpCs were grown to 80% confluency in six-well plates and were transfected with 25 nM mimics of miR-4776 or miR-4514 or a negative control (scrambled oligonucleotide of mimic or inhibitor) for 24 h. An equal number of cells were taken for the IP to minimize variability. IP uses G-coupled magnetic beads and pan-Ago antibody that recognizes Ago 1, Ago 2 and Ago 3 to precipitate the miRNA/mRNA complex. An isotype antibody control was also run in parallel. IP was done as described in the manufacturer's protocol. The precipitated complex was collected and the RNA purified from the complex using trizol reagent. The RNA was converted to cDNA with the High Capacity cDNA Reverse Transfection Kit (Applied Biosystems) and specific primers for NFKBIB were used for RT-PCR. The data was analyzed by comparing the cells transfected with mimic miRNA or negative control oligonucleotide and the fold enrichment of NFKBIB was calculated from the Ago and isotype antibody preparations as described by the manufacturer.

2.7.2. Luciferase Assay

The 3′ UTR sequence of NFKBIB was cloned downstream of the firefly luciferase gene. The transcript level is regulated by its interaction with miRNAs resulting in reduced luciferase activity that is measured by the Promega Dual Luciferase Assay (Madison, WI, USA). NFKBIB 3′ UTR reporter plasmid was synthesized by Origene (Origene). HBEpCs cells were transfected on a 96-well plate with a complex containing transfection agent (Lipofectamine 2000, Life Technologies), NFKBIB UTR reporter plasmid, and 25 nM of either the miR-4776 mimic or negative control (a scrambled oligonucleotide that is not a target of any gene). After 18 h of transfection, the medium was replaced with complete medium and the cells were incubated for an additional 16–18 h. Cells were then collected and lysed in lysis buffer supplied by the manufacturer. Luciferase activity was assessed in microtiter plates using the Promega Dual Luciferase Reporter Assay system according to the manufacturer's instructions (Dual-Glo ®Luciferase Assay, Promega). The plates were read on a Glomax ®96-well microplate Luminometer (Promega). The relative Renilla luciferase (Promega) activity was calculated by normalizing transfection efficiency to the firefly luciferase activity.

2.8. Western Immunoblotting

HBEpCs were transfected with miR-4776 mimic and then mock infected or infected at a MOI of 1 IAV. Cells were also transfected with a negative control of mimic. Four hours post infection, the cells were lysed using a cytoplasmic protein extraction buffer containing protease inhibitors (Thermofisher Scientific, Foster City, CA, USA). The isolated proteins were electrophoresed on a 10% SDS-polyacrylamide gel (sodium dodecyl sulfate-polyacrylamide gel electrophoresis (SDS-PAGE), Pre-cast gel, BioRad (Hercules, CA, USA). Separated proteins were transferred to a nitrocellulose membrane (Millipore) and the membrane was blocked using Odyssey Blocking Buffer (LI-COR Biosciences, Lincoln, NE, USA). Western blot analysis was carried out using mouse monoclonal anti-phospho-NF-κB p65 (Millipore) and mouse monoclonal anti-GAPDH (Abcam, Cambridge, MA, USA) and NS1 of IAV (Invitrogen). Appropriate mouse and rabbit IR Dye 680 or 800 secondary antibodies (LI-COR Biosciences, Lincoln, NE, USA) were used. Near-infrared fluorescence detection was performed on the Odyssey Imaging System (LI-COR Biosciences).

2.9. Viral Plaque Assay

MDCK cells were cultured in six-well tissue culture plate and incubated at 35 °C in a humidified CO_2 incubator. Confluent monolayers of the cells were washed with PBS and inoculated with 800 μL of serially diluted cell culture supernatant collected from the HBEpC cells exposed to IAV and incubated for 45 min at 35 °C. Immediately after incubation, the cells were washed twice with 2 mL of PBS, overlaid with supplemented Dulbecco's Modified Eagle's medium (DMEM)/F12 containing 0.6% agarose (Oxoid Ltd., Basingstoke , Hampshire, UK) and incubated at 35 °C for another 60 h. Cells were then fixed with 10% formalin and the agarose removed with tap water and stained with 1.0% crystal violet and plaque forming units (PFU) were counted.

2.10. Statistical Analysis

One-way analysis of variance (ANOVA) was used to analyze RT-PCR data and post-hoc pairwise multiple comparisons between means were performed using the Holm–Sidak method with a *p*-value of <0.05 considered statistically significant using Sigma stat version 11.0 for Windows (Systat Software, Chicago, IL, USA).

3. Results and Discussion

3.1. Microarray Analysis

The virus host paradigm provides a mechanism by which RNA viruses can regulate several host genes through the modulation of host-specific miRNAs [25–27]. To identify influenza-induced changes in host miRNA expression, a microarray analysis of host miRNAs differentially regulated following IAV infection was performed. A549 cells exposed to IAV showed differential expression of miRNAs as shown in Figure 1. We focused on the regulatory mechanisms for NF-κB expression, as it may have a role in the replication and survival of IAV [9,10]. Little is known regarding miRNAs that regulate the expression of NF-κB negative regulator, NFKBIB, and recent studies indicate that miR-20a targets NFKBIB in gastric cancer cells [20]. However, miR-20a did not show any significant differential expression in A549 cells infected with IAV. In contrast, miR-4514 exhibited a slight upregulation and miR-4776 was downregulated in A549 cells infected with IAV (Figure 1) and target scan analysis revealed that miR-4514 and miR-4776 are putative targets of NFKBIB (Figure 2). We added miR-4742-3p to our validation study since it showed significant upregulation at 18 h exposure to IAV (data not shown), and target scan predicted it as a target of NFKBIB. miRNA-4776 was of particular interest in that two putative binding sites for this miRNA are present in the NFKBIB 3′ UTR region, one at position 162–168 and another at position 648–654. Even though several microarray studies have shown a wide range of miRNAs being modulated by IAV infection, very few miRNAs were studied in detail to understand the target and role in IAV replications. Moreover, the induction of miRNA also depends on the cell type, incubation time and the MOI of virus used. Previous studies also have shown that many target genes of miRNAs contain several miRNA binding sites and the level of translational repression exponentially increases with the number of miRNA binding sites in the 3′ UTR region [28]. Therefore, this potential increase in binding of miRNA-4776 to the 3′ UTR of NFKBIB may lead to reduced expression of this repressor and trigger increased activity of NF-κB. We further examined the role and specificity of these miRNAs in NFKBIB expression (Table S1 shows the complete microarray data).

Annotation	AvgHy3	Control	Treatment	logFC
hsa-miR-1260a	8.610425	-1.14164	-0.34654	0.795098
hsa-miR-1260b	14.2583	-0.91229	-0.13554	0.776744
hsa-miR-1280	14.47228	-0.96412	-0.23068	0.733437
hsa-miR-4695-3p	8.733011	-0.96841	-0.267	0.701414
hsa-miR-4443	11.70945	-1.13694	-0.50818	0.628756
hsa-miR-720	11.04486	-0.08444	0.435426	0.519871
hsa-miR-1264	7.16946	0.341482	0.835945	0.494464
hsa-miR-29b-3p	7.758527	0.408334	0.894104	0.48577
hsa-miR-4417	5.854004	-1.0368	-0.568	0.4688
hsa-miR-4286	7.405508	-0.35991	0.062421	0.422326
hsa-miR-3182	10.54429	0.48656	0.900717	0.414157
hsa-miR-4797-5p	5.256111	-0.76644	-0.36416	0.402285
hsa-miR-629-3p	5.139955	-0.79141	-0.40945	0.381966
hsa-miR-3646	8.747996	-0.79603	-0.41561	0.38042
hsa-miR-5704	9.710873	-0.31407	0.059809	0.373883
hsa-miR-4456	12.78199	-0.07369	0.297498	0.371187
hsa-miR-32-5p	5.272177	0.299411	0.670507	0.371096
hsa-miR-4454	14.68543	0.39734	0.765259	0.367919
hsa-miR-4514	5.362792	-1.54513	-1.31048	0.234644
hsa-miR-4742-3p	6.76656	-1.73109	-1.72774	0.00335
hsa-miR-548an	6.140453	-0.94959	-1.21936	-0.26977
hsa-miR-4508	6.040203	-0.88739	-1.15923	-0.27184
hsa-miR-4776-3p	5.144972	-0.92846	-1.20811	-0.27966
hsa-miR-4516	6.949994	-1.73942	-2.03221	-0.29279
hsa-let-7a-5p	7.746293	0.462963	0.1642	-0.29876
hsa-let-7d-5p	6.053223	0.354016	0.050855	-0.30316
hsa-let-7g-5p	7.333395	0.617302	0.247358	-0.36994
hsa-miR-4674	5.367789	-1.04093	-1.41291	-0.37198
hsa-miR-210	5.604469	0.373951	-0.06	-0.43395
hsa-miR-4521	6.033983	-0.22751	-0.97363	-0.74612

Figure 1. Influenza A virus (IAV) infection induced differential expression of microRNAs (miRNAs). Microarray analysis for miRNA expression was performed with RNA extracted from IAV infected A549 cells for 3 h. The log fold changes in expression of miRNAs that are up or downregulated on exposure to IAV are shown from the microarray. AvgHy3-Average signal intensity, logFC—log fold change, Control-Mock; Treatment-Infected with H1N1.

Position 385-392 of NFKBIB 3' UTR

hsa-miR-4742   5' ...GUCAGAGAACCCAGCAAUACAGA...

                  ||| | ||| |

      3'   GACGUCCGUUUCCUCUUAUGUCU

Position 1000-1006 of NFKBIB 3' UTR

hsa-miR-4514   5' ...GCCGCCUGAGCCCCU------CUGCCUGG...

                 |||| |||||||

      3'   AAGGGGUUAGGACGGACA

Position 648-654 of NFKBIB 3' UTR

hsa-miR-4776   5' ...ACCAGCCUGGCCAACAUGGCAAG...

                  |||||||

      3'   UACGUCACCUGGUCCUACCGUUC

Position 162-168 of NFKBIB 3' UTR

hsa-miR-4776   5' ...GAAAAACUAAGGCAGUGGCAAAG...

                  ||||||

      3'   UACGUCACCUGGUCCUACCGUUC

Figure 2. Target scan analysis of nuclear factor Kappa B inhibitor beta (NFKBIB) three prime untranslated region (3' UTR). A search of the sequences for miRNAs miR-4514, 4742, and 4776 showed putative target binding sites in NFKBIB messenger RNA (mRNA).

3.2. miRNA and NFKBIB Expression

Our earlier studies [23] showed that the host cellular response (for cell survival) in response to the viral invasion occurs at the early stages of infection (up to 3 h). Thus, it is appropriate to look for changes that occur at early hours of infection, before the virus take control of the cell metabolism and signaling. To validate the initial microarray analysis, A549 cells were infected with IAV for 1 to 4 h and miRNA expression was analyzed by RT-PCR. MiR-4514 expression was significantly decreased at 1 h exposure to IAV compared to mock infected control (Figure 3A), increased at 2 h to the level observed at 3 h in the microarray analysis, but then decreased again at 3 h and 4 h. The data from RT-PCR is more sensitive and reliable than microarray analysis and may account for the apparent discrepancy in expression between these assays at 2–3 h. Expression of miR-4776 in infected cells increased significantly throughout the (1–3 h) infection, and expression peaked (12 fold increase) by 3 h and then sharply decreased by 4 h (Figure 3B), a similar trend was seen by array analysis with the difference in incubation time of 1 h, the difference in time again could be due to the increased viral copies in array. The array data showed a decrease at 3 h, and the difference in time with validation and array could be attributed to the difference in MOI. Expression of miR-4742 decreased in the initial 3 h of infection and then increased by 4 h to that level in the mock infected control (Figure 3C). This return to basal level of miR-4742 after 4 h of infection was also seen after 3 h of infection in the initial microarray analysis, which may be due to the higher MOI of 3 used in the microarray analysis. The high MOI used in the array analysis was to ensure that every cell in culture was infected with one or more virus, whereas in the validation studies, we reduced the MOI to 1, so that the overburden of the virus on the culture cells could be avoided and the host cells response to viral infection could be measured. A549 cells are transformed lung epithelial cells that could also produce variation in expression compared to primary normal lung cells, hence we used human primary bronchial epithelial cells to confirm the repeatability of the results.

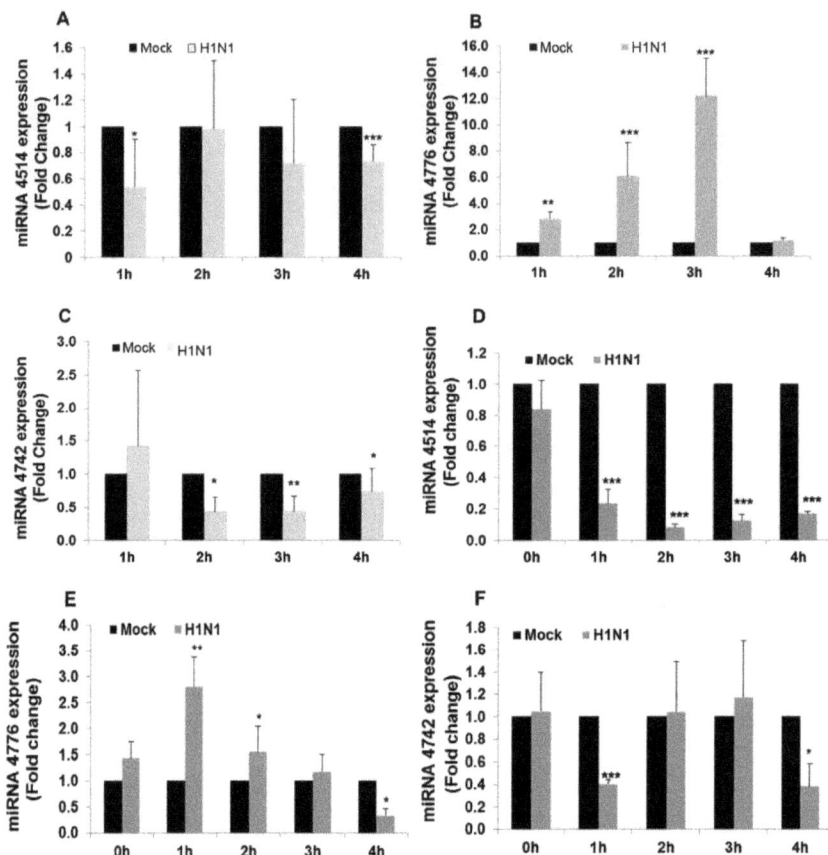

Figure 3. miRNA expression profile on exposure to IAV compared with mock and infected cells. (**A**) expression pattern of miR-4514. A549 cells were infected with IAV at multiplicity of infection (MOI) of 1 for 4 h and samples were taken every hour, miRNAs extracted and analyzed by reverse transcription-polymerase chain reaction (RT-PCR). *** $p < 0.001$; (**B**) expression pattern of miR-4776. A549 cells were infected with IAV at MOI of 1 for 4 h and samples were taken every hour, miRNAs extracted and analyzed by RT-PCR. ** $p < 0.01$, *** $p < 0.001$; (**C**) expression pattern of miR-4742. A549 cells were infected with IAV at MOI of 1 for 4 h and samples were taken every hour, miRNAs extracted and analyzed by RT-PCR. Data is from three independent experiments. * $p < 0.05$, ** $p < 0.01$; (**D**) expression pattern of miR-4514. Human bronchial epithelial cells (HBEpCs) infected with IAV at MOI 1 for 4 h and samples were taken every hour, miRNAs extracted and analyzed by RT-PCR. Data is from three independent experiments. *** $p < 0.001$; (**E**) expression pattern of miR-4776. HBEpCs were infected with IAV at MOI of 1 for 4 h and samples were taken every hour, miRNAs extracted and analyzed by RT-PCR. Data is from three independent experiments. * $p < 0.05$, ** $p < 0.01$; and (**F**) expression pattern of miR-4742. HBEpCs were infected with IAV at MOI of 1 for 4 h and samples were taken every hour, miRNAs extracted and analyzed by RT-PCR. Data is from three independent experiments. * $p < 0.05$, *** $p < 0.001$. Expression of each miRNA was normalized to expression of U6 miRNA.

HBEpCs cells were infected with IAV at MOI of 1 for 1–4 h and miRNA expression was analyzed by RT-PCR. MiR-4514 expression was significantly decreased throughout the 4 h exposure compared to mock infected control (Figure 3D). Expression of miR-4776 showed a significant increase at 1 h of IAV infection (2.8 fold) but then gradually decreased to that of the mock control after 3 h and decreased further by 4 h (Figure 3E). Expression of miR-4742 initially decreased after 1 h of infection and then returned to that of the mock control after 2 h and 3 h, followed again by a sharp reduction in expression after 4 h (Figure 3F). The patterns of expression of the three miRNAs in HBEpCs was similar (but not exactly) as was found in A549 infected cells with expression of miR-4776 showing the most consistency, although expression of this miRNA peaked earlier in the infection in HBEpCs.

We compared the expression levels of NFKBIB mRNA with those of the three miRNAs in HBEpC cells. Expression of NFKBIB mRNA significantly decreased 1.6-fold after 1 h of infection, returned to the level of the mock infected control after 2 h and 3 h, and then significantly increased (2-fold) after 4 h exposure to IAV (Figure 4A). When the expression of the three miRNAs (miR-4415, 4742, 4776) was compared with that of NFKBIB mRNA, only miRNA-4776 showed an inverse correlation, as would be expected if NFKBIB mRNA expression is regulated by the specific miRNA. The influenza copy number gradually increased after 4 h of exposure to IAV (Figure 4B). A classical approach to identify the actual target mRNA of these miRNA is by introducing mimics and inhibitors of the specific miRNA and evaluating the target gene expression [29,30]. We further examined the role and specificity of miR-4776 and miR-4514 miRNAs to NFKBIB but did not study miR-4742-3p, as it did not appear to have a pattern of expression, suggesting it is involved in NFKBIB expression.

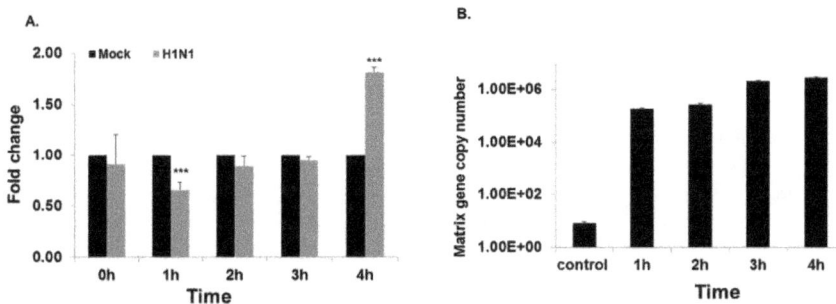

Figure 4. Influenza infection modulates NFKBIB expression: (**A**) HBEpCs grown to 80% confluency were infected with IAV at an MOI of 1 or mock for 4 h and samples were taken every hour. RNA was extracted and complementary DNA (cDNA) was synthesized, and used for RT-PCR analysis. NFKBIB expression was analyzed and normalized to glyceraldehyde phosphate dehydrogenase (GAPDH), and data are expressed as ± standard error of the mean (SEM), *** $p < 0.001$, $n = 3$ independent experiments; (**B**) influenza virus matrix gene copy number was analyzed from the RNA extracted from HBEpCs infected with IAV for 4 h; $n = 3$ independent experiments analyzed in duplicates. Control is the mock infected.

3.3. Functional Analysis of miR-4776 and miR-4514

To examine the functionality of these miRNAs, HBEpCs were transfected with the mimic or inhibitors of miR-4514 and miR-4776 and the changes in NFKBIB mRNA expression were determined. The miR-4776 inhibitor increased the NFKBIB mRNA expression 1.75-fold and the exposure to miR-4776 mimic decreased the NFKBIB mRNA expression approximately 50% in uninfected cells transfected with the negative control (Figure 5A). In contrast, the change in NFKBIB expression in the presence of miR-4514 mimic was not consistent with a role in regulating NFKBIB mRNA (Figure 5B). HBEpCs transfected with miR-4776 inhibitor and subsequently infected with IAV showed an increase in NFKBIB expression, and a decrease with mimic (Figure 5C), whereas miR-4514 mimic increased NFKBIB and inhibitor did not significantly alter NFKBIB expression in IAV infected cells (Figure 5D). This apparent lack of expected inverse correlation between miR-4514 expression and NFKBIB expression may be due to an indirect effect on the NFKBIB pathway, as miRNA target multiple mRNAs.

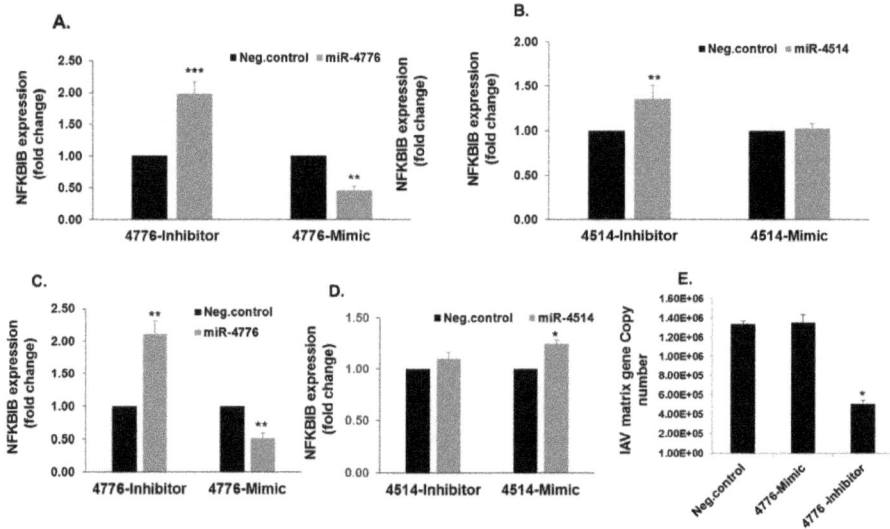

Figure 5. Downregulation of miR-4776 induce NFKBIB expression: (**A**) HBEpCs were transfected with miR-4776 inhibitor or mimic or negative control. After 48 h of incubation, RNA was extracted and converted to cDNA and used for RT-PCR. GAPDH was used as the housekeeping gene. Data are expressed as ± SEM, *** $p < 0.001$, ** $p < 0.01$ ($n = 3$); (**B**) HBEpCs were transfected with miR-4514 inhibitor or mimic or the negative control. After 48 h of incubation, RNA was extracted and converted to cDNA and used for RT-PCR. The relative abundance of NFKBIB was measured in cells and the data normalized to the abundance of GAPDH. Data are expressed as ± SEM, * $p < 0.05$, ** $p < 0.01$ ($n = 3$); (**C**) HBEpCs were transfected with miR-4776 mimic or inhibitor or a negative control. After 48 h of incubation, the transfected cells were either mock infected or infected with 1 MOI IAV for an additional 4 h. Relative expression of NFKBIB was analyzed by RT-PCR. Data are expressed as ± SEM, ** $p < 0.01$ compared the cells transfected with the negative control, ($n = 3$); (**D**) HBEpCs transfected with miR-4514 mimic or inhibitor or a negative control. After 48 h of incubation, the cells were infected with 1 MOI of IAV for another 4 h. After 4 h of incubation, RNA was extracted and converted to cDNA and used for RT-PCR. The relative NFKBIB abundance was measured in infected cells, and GAPDH was used as the internal control. Data are expressed as ± SEM, * $p < 0.05$ compared the cells transfected with the negative control, ($n = 3$); (**E**) abundance of NFKBIB modulates viral replication. HBEpCs were transfected with miR-4776 mimic or inhibitor or a negative control. Following infection, the transfected cells were infected with 1 MOI of IAV for an additional 4 h. After 4 h of incubation, RNA was extracted and converted to cDNA and used for RT-PCR. Matrix copy number in cells transfected with either miR-4776 mimic or inhibitor or negative control is shown. Data are expressed as ± SEM, ** $p < 0.001$ ($n = 4$).

The increased expression of NFKBIB resulting from transfection of miR-4776 inhibitor in HBEpCs also is associated with a significant decrease in IAV matrix gene copy number as evidenced in the 65% decrease in copy number (Figure 5E). This result is consistent with the concept that IAV is dependent on the NF-κB pathway [10] and the increase in NFKBIB reduced the level of active NF-κB available for inducing the transcription of viral genes. In contrast, further decreasing the level of NFKBIB mRNA with miR-4776 mimic did not alter the matrix copy number in infected cells, suggesting that there was a sufficient number of NF-κB molecules to support IAV replication (Figure 5E). miRNAs are known to induce changes in the cellular physiology by different mechanisms, and each miRNA has multiple gene targets [29,31,32]. The exogenous administration of miRNA inhibitor to antagonize influenza replication in airway epithelial cells could be a potential therapeutic strategy to explore for controlling influenza infections.

Unstimulated cells contain a pool of NF-κB dimers in the cytosol that are stoichiometrically bound with the inhibitors [16]. NFKBIB is complexed with NF-κB but upon infection or external stimuli activated NF-κB is released [17], and subsequently translocated to the nucleus where it induces transcription of genes. HBEpCs transfected with mimic miR-4776 followed with IAV infection for 4 h were monitored for the expression of NF-κB by confocal microscopy (Figure 6A) and compared to the negative control. The cells were stained with an antibody to phosphorylated NF-κBp65 that specifically binds to the p65 component of active NF-κB (green), and the nucleus was stained with DAPI (blue). Active phosphorylated NF-κB was significantly higher in miR-4776 transfected cells compared to the negative control mimic transfected cells infected with IAV (Figure 6B). This data suggests that mimic miR-4776 lowers the levels of NFKBIB mRNA and the amount of NFKBIB protein available to bind with NF-κB results in an increased level of active NF-κB. This is consistent with previous studies that showed NFKBIB, once released from the NF-κB/NFKBIB complex, undergoes degradation and leads to the phosphorylation and activation of NF-κB [18,33]. We further confirmed the results obtained by confocal microscopy with Western blot analysis. Data presented in Figure 6B shows that phosho-NF-κB p65 is overexpressed in cells transfected with the miR-4776 mimic compared to the negative control mimic transfected cells. The 4 h time point of infection was selected based on the results (Figures 3E and 4A) wherein 4 h exposure to IAV showed a reduction in miR-4776 expression and upregulation of NFKBIB transcripts. Moreover, the miR-4776 mimic transfected cells showed a reduction in NFKBIB protein compared to the control cells transfected with negative control (Figure 6C). Taken together, the results from this study clearly indicates that miR-4776 has a significant role in regulating NFKBIB expression to alter the level of active NF-κB.

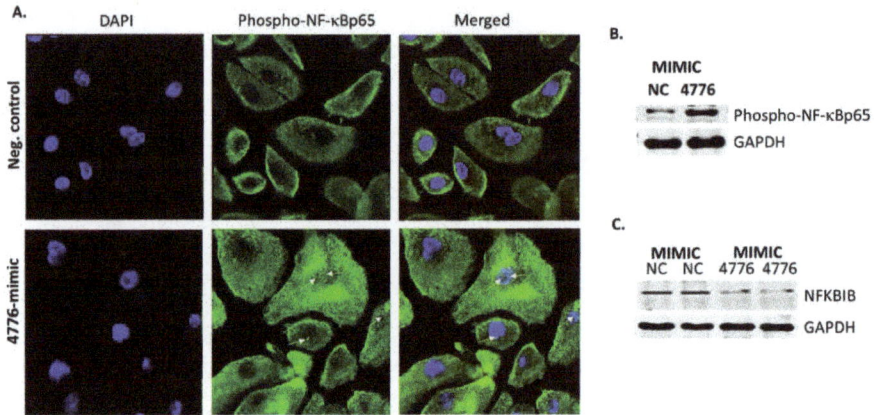

Figure 6. Immunofluorescence of IAV infected cells showing the expression of NF-κB. (**A**) HBEpCs were transfected with miR-4776 mimic or a negative control. After 48 h of incubation, the cells were infected with 1 MOI of IAV for another 4 h. (NC represents the negative control and is a scrambled oligonucleotide) (top panel), miRNA-4776 mimic (bottom panel). The expression of activated phospho-NF-κB p65 protein (green) was determined by immunofluorescence. 4′,6-diamidino-2-phenylindole (DAPI) (blue) stained the nucleus of the cells. Arrows indicate the NF-kB translocated into the nucleus; (**B**) HBEpCs transfected with miR-4776 mimic or a negative control. After 48 h of incubation, the cells were infected with 1 MOI of IAV for another 4 h. Western blot analysis of cells transfected with mimic miR-4776 showing increased phospho-NF-κB p65 protein; (**C**) immunoblot showing the decreased expression of NFKBIB protein in cells transfected with miR-4776 and infected with IAV. HBEpCs transfected with miR-4776 mimic or a negative control. After 48 h of incubation, the cells were infected with 1 MOI of IAV for another 4 h. GAPDH is used as the housekeeping protein.

To validate the role of NFKBIB, we used two classic methods: silencing of NFKBIB by small interfering RNA (siRNA) and overexpression of NFKBIB by plasmid DNA. Four NFKBIB siRNA oligonucleotides were obtained from Dharmacon (GE, Pittsburgh, PA, USA) but were not effective in reducing the expression of NFKBIB mRNA in HBEpCs. All four siRNA oligonucleotides individually or in combination failed to produce a reduction of NFKBIB transcripts (data not shown). Plasmids carrying the ORF of NFKBIB were transiently transfected in HBEpCs. After 24–36 h of transfection, cells were infected with IAV (1 MOI) for up to 10 h. The results (Figure 7A) indicate that there was an increase (11–21 fold) in NFKBIB expression in cells transfected with the ORF of NFKBIB. When HBEpCs were transfected with NFKBIB ORF plasmids, a significant reduction in IAV copy numbers was observed (Figure 7B). Confocal analysis of the cells overexpressing NFKBIB (Red) infected with IAV (Figure 7C) shows reduced expression of influenza NS1 protein (green) demonstrating the low levels of IAV protein in cells. We selected influenza NS1, as it has been shown to have a significant impact on host cell gene expression and a central role in inhibiting interferon and activating NF-κB and PI3-K pathways [34,35]. The results from these studies show that overexpression of NFKBIB blocks NF-κB activity needed for viral replication. The essential role of NF-κB in viral replication has been well documented [6]. The results corroborate the effect observed with the miR-4776 inhibitor, which increased NFKBIB expression and led to a reduction in matrix gene copy numbers. Figure 7D showed a reduction in viable IAV in cells overexpressing NFKBIB protein compared to the control cells infected with IAV.

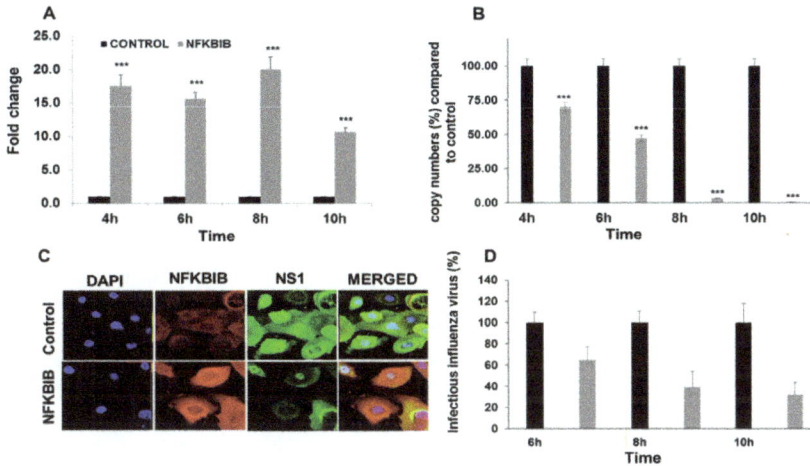

Figure 7. NFKBIB overexpression leads to reduction in IAV M gene copy number. (**A**) HBEpCs grown to 80% confluency were transfected with NFKBIB plasmid or control plasmid for 36–48 h. NFKBIB overexpressing HBEpCs were infected with 1 MOI of IAV for another 10 h, and the expression of NFKBIB was analyzed using RT-PCR. *** $p < 0.001$ ($n = 3$); (**B**) NFKBIB overexpressing HBEpCs were infected with 1 MOI of IAV for another for 10 h, and the IAV matrix gene copy numbers were analyzed using the matrix gene primers, and a standard curve for the matrix gene. *** $p < 0.001$ (Black—Control plasmid, Grey—NFKBIB gene overexpressing); (**C**) HBEpCs overexpressing NFKBIB gene were cultured on slides and infected with IAV of 1 MOI. Cells were then fixed and stained with NFKBIB (red) and NS1 (influenza) antibodies (green), followed by respective fluorescent secondary antibodies. Nucleus (blue) was stained with DAPI. Slides were observed under the Laser Scanning Microscopy (LSM)-510 confocal microscope using an oil 63X objective; (**D**) HBEpCs grown to 80% confluency were transfected with NFKBIB plasmid or control plasmid for 36–48 h. NFKBIB overexpressing HBEpCs were infected with 1 MOI of IAV for another 10 h. Culture supernatant were collected at different time points. Percent infectious influenza virus was determined by viral plaque assay from culture supernatant. Percent calculations were normalized to control cells infected with IAV.

3.4. miRNA Target Confirmation Studies

miRNA is incorporated with Argonaute (Ago) proteins into the RNA-induced silencing complex [36,37], where miRNA guides the complex to partial complementary binding sites located in the 3′ UTR regions [37] of target mRNAs to induce translational repression or degradation of mRNAs. To confirm the specificity of miR-4776 to NFKBIB mRNA, we approached this in two ways, using an Ago antibody pulldown method, and using a 3′ UTR luciferase assay. The Ago protein binds the miRNA to enable the miRNA to base pair with the target mRNA [36], which can be immunoprecipitated (IP) and separated for target identification. Ago was precipitated using a pan-ago antibody (specific for ago-1, ago-2 and ago-3) and protein G magnetic beads (Active Motif kit). HBEpCs were transfected with either miR-4776 or miR-4514 mimic or the negative control mimic. The results presented in Figure 8A show a 10 fold increase in enrichment of NFKBIB mRNA obtained from the IP of miR-4776 mimic transfected cells compared to the negative control mimic transfected cells. Moreover, transfection of miRNA-4514 did not lead to an enrichment of NFKBIB mRNA, suggesting that NFKBIB mRNA binds to miR-4776 and not with miR-4514.

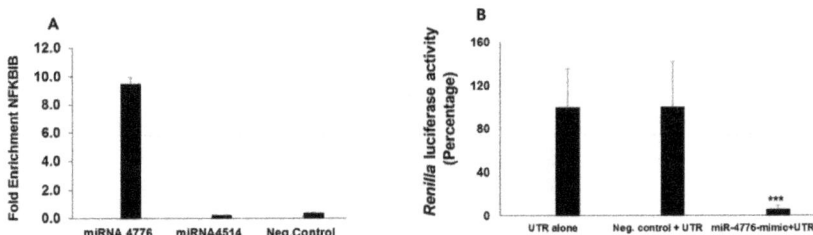

Figure 8. Confirmation of miR-4776 target by Argonaut Immunoprecipitation and luciferase 3′ UTR assay. (**A**) HBEpCs were transfected with miR-4776 mimic, miR-4514 mimic or negative control mimic, and, after 24–36 h. the cells were collected and immunoprecipitated using the Ago or isotype antibody. RNA was isolated and the fold enrichment of NFKBIB transcript was calculated. Independent experiments (*n* = 2) were analyzed in duplicate; (**B**) HBEpCs grown to 80% confluency were co-transfected with the luciferase reporter construct and with either mimic miR-4776 or negative control and after 36 h, cells were infected with IAV at MOI of 1 for 4 h. Luciferase activity was measured using the Dual luciferase assay kit as described by the manufacturer. *** *p* < 0.001, Error bars represent the standard error from three independent experiments.

We then addressed the specificity of miR-4776 to target NFKBIB mRNA using the 3′ UTR Luciferase Reporter Assay. Luciferase UTR experiments are considered the confirmatory experiments in miRNA target identification in various studies utilizing miRNA [38,39]. The 3′ UTR of NFKBIB predicted to interact with miR-4776 was cloned into a reporter luciferase vector and co-transfected with miR-4776 mimic into HBEpC cells. Transfection of miR-4776 mimic resulted in 80% reduction in luciferase activity compared to that from cells transfected with the negative control (Figure 8B). These results demonstrate that miR-4776 regulates NFKBIB mRNA expression through binding of the 3′ UTR.

Influenza viruses are a major concern due to the emergence in recent years of new and divergent strains that are potentially dangerous including H7N9 and the pandemic 2009 H1N1. There are reports that the virus also becomes drug resistant and the number of anti-viral drugs are limited. Therefore, understanding the host cell's regulatory machinery used during influenza infection has the potential for identifying host cell targets in anti-influenza therapies. Regulation of NF-κB is a complex process, and its regulation involves a family of inhibitor proteins (NFKBIB), which form complexes with NF-κB, and prevents its translocation to the nucleus where activated NF-κB can affect the transcription of target genes. We also noted that the decreased expression of miR-4776 in bronchial epithelial cells was followed by an increase in NFKBIB mRNA and that subsequently resulted in an inactive NF-κB protein. IAV manipulates the NF-κB pathway for efficient replication of the virus and differentially regulates viral RNA synthesis [33]. Our data suggests that NF-κB is an important signaling molecule that plays a critical role in the replication of IAV and is modulated by the NFKBIB protein, which, in turn, is regulated by miR-4776.

4. Conclusions

In conclusion, the results presented here provide the first evidence that miR-4776 can modulate the expression of NFKBIB, the regulatory protein for a key transcription factor (NF-κB) that is necessary for cell proliferation, apoptosis and cell survival. Downregulation of NFKBIB mRNA by miR-4776 led to the production of active NF-κB that possibly modulates the increased viral survival. Identifying novel miRNAs for their target genes provides an alternative approach to reduce IAV infections and block influenza transmission. Validation of specific miRNAs as potential anti-viral agents will require further testing in animal models.

Supplementary Materials: The following are available online at www.mdpi.com/1999-4915/9/5/94/s1, Table S1: Microarray data on miRNA expression on A549 cells exposed to IAV for 3 h. The log fold changes in expression of miRNAs that are up or down regulated on exposure to IAV are shown. AvgHy3-Average signal intensity, logFC-log fold change, Control-Mock; Treatment- Infected with H1N1.

Author Contributions: S.O., D.H.B. and J.D.N. conceived and designed the experiments; S.O. and N.B.B. performed the experiments; S.O. analyzed the data; J.D.N. contributed reagents/materials/analysis tools; and S.O., J.D.N. and D.H.B. wrote the paper.

Conflicts of Interest: The findings and the conclusions in this report are those of the authors and do not necessarily represent the views of the National Institute for Occupational Safety and Health. The authors declare no conflict of interest.

References

1. Thompson, W.W.; Shay, D.K.; Weintraub, E.; Brammer, L.; Bridges, C.B.; Cox, N.J.; Fukuda, K. Influenza-associated hospitalizations in the United States. *JAMA* **2004**, *292*, 1333–1340. [CrossRef] [PubMed]

2. Friedman, R.C.; Farh, K.K.; Burge, C.B.; Bartel, D.P. Most mammalian mRNAs are conserved targets of microRNAs. *Genome Res.* **2009**, *19*, 92–105. [CrossRef] [PubMed]

3. Song, L.; Liu, H.; Gao, S.; Jiang, W.; Huang, W. Cellular microRNAs inhibit replication of the H1N1 influenza A virus in infected cells. *J. Virol.* **2010**, *84*, 8849–8860. [CrossRef] [PubMed]

4. Perkins, N.D.; Gilmore, T.D. Good cop, bad cop: The different faces of NF-kappaB. *Cell Death Differ.* **2006**, *13*, 759–772. [CrossRef] [PubMed]

5. Ben-Neriah, Y.; Karin, M. Inflammation meets cancer, with NF-kappaB as the matchmaker. *Nat. Immunol.* **2011**, *12*, 715–723. [CrossRef] [PubMed]

6. Tak, P.P.; Firestein, G.S. NF-kappaB: A key role in inflammatory diseases. *J. Clin. Investig.* **2001**, *107*, 7–11. [CrossRef] [PubMed]

7. Wurzer, W.J.; Planz, O.; Ehrhardt, C.; Giner, M.; Silberzahn, T.; Pleschka, S.; Ludwig, S. Caspase 3 activation is essential for efficient influenza virus propagation. *EMBO J.* **2003**, *22*, 2717–2728. [CrossRef] [PubMed]

8. Wei, L.; Sandbulte, M.R.; Thomas, P.G.; Webby, R.J.; Homayouni, R.; Pfeffer, L.M. NFkappaB negatively regulates interferon-induced gene expression and anti-influenza activity. *J. Biol. Chem.* **2006**, *281*, 11678–11684. [CrossRef] [PubMed]

9. Nimmerjahn, F.; Dudziak, D.; Dirmeier, U.; Hobom, G.; Riedel, A.; Schlee, M.; Staudt, L.M.; Rosenwald, A.; Behrends, U.; Bornkamm, G.W.; et al. Active NF-kappaB signalling is a prerequisite for influenza virus infection. *J. Gen. Virol.* **2004**, *85*, 2347–2356. [CrossRef] [PubMed]

10. Wurzer, W.J.; Ehrhardt, C.; Pleschka, S.; Berberich-Siebelt, F.; Wolff, T.; Walczak, H.; Planz, O.; Ludwig, S. NF-κB-dependent induction of tumor necrosis factor-related apoptosis-inducing ligand (TRAIL) and Fas/FasL is crucial for efficient influenza virus propagation. *J. Biol. Chem.* **2004**, *279*, 30931–30937. [CrossRef] [PubMed]

11. Flory, E.; Kunz, M.; Scheller, C.; Jassoy, C.; Stauber, R.; Rapp, U.R.; Ludwig, S. Influenza virus-induced NF-κB-dependent gene expression is mediated by overexpression of viral proteins and involves oxidative radicals and activation of IkappaB kinase. *J. Biol. Chem.* **2000**, *275*, 8307–8314. [CrossRef] [PubMed]

12. Wang, X.; Li, M.; Zheng, H.; Muster, T.; Palese, P.; Beg, A.A.; Garcia-Sastre, A. Influenza A virus NS1 protein prevents activation of NF-kappaB and induction of alpha/beta interferon. *J. Virol.* **2000**, *74*, 11566–11573. [CrossRef] [PubMed]

13. Ludwig, S.; Pleschka, S.; Planz, O.; Wolff, T. Ringing the alarm bells: Signalling and apoptosis in influenza virus infected cells. *Cell Microbiol.* **2006**, *8*, 375–386. [CrossRef] [PubMed]

14. Pahl, H.L.; Baeuerle, P.A. Expression of influenza virus hemagglutinin activates transcription factor NF-κB. *J. Virol.* **1995**, *69*, 1480–1484. [PubMed]

15. Ehrhardt, C.; Ruckle, A.; Hrincius, E.R.; Haasbach, E.; Anhlan, D.; Ahmann, K.; Banning, C.; Reiling, S.J.; Kuhn, J.; Strobl, S.; et al. The NF-κB inhibitor SC75741 efficiently blocks influenza virus propagation and confers a high barrier for development of viral resistance. *Cell Microbiol.* **2013**, *15*, 1198–1211. [CrossRef] [PubMed]

16. Hoffmann, A.; Natoli, G.; Ghosh, G. Transcriptional regulation via the NF-κB signaling module. *Oncogene* **2006**, *25*, 6706–6716. [CrossRef] [PubMed]

17. Tsui, R.; Kearns, J.D.; Lynch, C.; Vu, D.; Ngo, K.A.; Basak, S.; Ghosh, G.; Hoffmann, A. IkappaBbeta enhances the generation of the low-affinity NF-κB/RelA homodimer. *Nat. Commun.* **2015**, *6*, 7068. [CrossRef] [PubMed]

18. Ghosh, S.; May, M.J.; Kopp, E.B. NF-κB and Rel proteins: Evolutionarily conserved mediators of immune responses. *Annu. Rev. Immunol.* **1998**, *16*, 225–260. [CrossRef] [PubMed]

19. Palombella, V.J.; Rando, O.J.; Goldberg, A.L.; Maniatis, T. The ubiquitin-proteasome pathway is required for processing the NF-κB1 precursor protein and the activation of NF-kappa B. *Cell* **1994**, *78*, 773–785. [CrossRef]

20. Du, Y.; Zhu, M.; Zhou, X.; Huang, Z.; Zhu, J.; Xu, J.; Cheng, G.; Shu, Y.; Liu, P.; Zhu, W.; et al. miR-20a enhances cisplatin resistance of human gastric cancer cell line by targeting NFKBIB. *Tumour Biol.* **2016**, *37*, 1261–1269. [CrossRef] [PubMed]

21. Krishnan, K.; Steptoe, A.L.; Martin, H.C.; Wani, S.; Nones, K.; Waddell, N.; Mariasegaram, M.; Simpson, P.T.; Lakhani, S.R.; Gabrielli, B.; et al. MicroRNA-182-5p targets a network of genes involved in DNA repair. *RNA* **2013**, *19*, 230–242. [CrossRef] [PubMed]

22. Othumpangat, S.; Noti, J.D.; Beezhold, D.H. Lung epithelial cells resist influenza A infection by inducing the expression of cytochrome C oxidase VIc which is modulated by miRNA 4276. *Virology* **2014**, *468–470*, 256–264. [CrossRef] [PubMed]

23. Othumpangat, S.; Noti, J.D.; Blachere, F.M.; Beezhold, D.H. Expression of non-structural-1A binding protein in lung epithelial cells is modulated by miRNA-548an on exposure to influenza A virus. *Virology* **2013**, *447*, 84–94. [CrossRef] [PubMed]

24. Spackman, E.; Senne, D.A.; Myers, T.J.; Bulaga, L.L.; Garber, L.P.; Perdue, M.L.; Lohman, K.; Daum, L.T.; Suarez, D.L. Development of a Real-Time Reverse Transcriptase PCR Assay for Type A Influenza Virus and the Avian H5 and H7 Hemagglutinin Subtypes. *J. Clin. Microbiol.* **2002**, *40*, 3256–3260. [CrossRef] [PubMed]

25. Fabian, M.R.; Sonenberg, N.; Filipowicz, W. Regulation of mRNA translation and stability by microRNAs. *Annu. Rev. Biochem.* **2010**, *79*, 351–379. [CrossRef] [PubMed]

26. Gottwein, E.; Cullen, B.R. Viral and cellular microRNAs as determinants of viral pathogenesis and immunity. *Cell Host Microbe* **2008**, *3*, 375–387. [CrossRef] [PubMed]

27. Grundhoff, A.; Sullivan, C.S. Virus-encoded microRNAs. *Virology* **2011**, *411*, 325–343. [CrossRef] [PubMed]

28. Doench, J.G.; Petersen, C.P.; Sharp, P.A. siRNAs can function as miRNAs. *Genes Dev.* **2003**, *17*, 438–442. [CrossRef] [PubMed]

29. Bartel, D.P. MicroRNAs: Genomics, biogenesis, mechanism, and function. *Cell* **2004**, *116*, 281–297. [CrossRef]

30. Othumpangat, S.; Walton, C.; Piedimonte, G. MicroRNA-221 modulates RSV replication in human bronchial epithelium by targeting NGF expression. *PLoS ONE* **2012**, *7*, e30030. [CrossRef] [PubMed]

31. Ambros, V. MicroRNA pathways in flies and worms: Growth, death, fat, stress, and timing. *Cell* **2003**, *113*, 673–676. [CrossRef]

32. He, L.; Hannon, G.J. MicroRNAs: Small RNAs with a big role in gene regulation. *Nat. Rev. Genet.* **2004**, *5*, 522–531. [CrossRef] [PubMed]

33. Kumar, N.; Xin, Z.T.; Liang, Y.; Ly, H. NF-κB signaling differentially regulates influenza virus RNA synthesis. *J. Virol.* **2008**, *82*, 9880–9889. [CrossRef] [PubMed]

34. Geiss, G.K.; Salvatore, M.; Tumpey, T.M.; Carter, V.S.; Wang, X.; Basler, C.F.; Taubenberger, J.K.; Bumgarner, R.E.; Palese, P.; Katze, M.G.; et al. Cellular transcriptional profiling in influenza A virus-infected lung epithelial cells: The role of the nonstructural NS1 protein in the evasion of the host innate defense and its potential contribution to pandemic influenza. *Proc. Natl. Acad. Sci. USA* **2002**, *99*, 10736–10741. [CrossRef] [PubMed]

35. Shin, Y.K.; Liu, Q.; Tikoo, S.K.; Babiuk, L.A.; Zhou, Y. Effect of the phosphatidylinositol 3-kinase/Akt pathway on influenza A virus propagation. *J. Gen. Virol.* **2007**, *88*, 942–950. [CrossRef] [PubMed]

36. Matranga, C.; Tomari, Y.; Shin, C.; Bartel, D.P.; Zamore, P.D. Passenger-strand cleavage facilitates assembly of siRNA into Ago2-containing RNAi enzyme complexes. *Cell* **2005**, *123*, 607–620. [CrossRef] [PubMed]

37. Gregory, R.I.; Chendrimada, T.P.; Cooch, N.; Shiekhattar, R. Human RISC couples microRNA biogenesis and posttranscriptional gene silencing. *Cell* **2005**, *123*, 631–640. [CrossRef] [PubMed]

Viruses **2017**, *9*, 94

38. Jin, Y.; Chen, Z.; Liu, X.; Zhou, X. Evaluating the microRNA targeting sites by luciferase reporter gene assay. *Methods Mol. Biol.* **2013**, *936*, 117–127. [PubMed]

39. Krek, A.; Grun, D.; Poy, M.N.; Wolf, R.; Rosenberg, L.; Epstein, E.J.; MacMenamin, P.; da Piedade, I.; Gunsalus, K.C.; Stoffel, M.; et al. Combinatorial microRNA target predictions. *Nat. Genet.* **2005**, *37*, 495–500. [CrossRef] [PubMed]

Review

Regulation of Apoptosis during Flavivirus Infection

Toru Okamoto *, Tatsuya Suzuki, Shinji Kusakabe, Makoto Tokunaga, Junki Hirano, Yuka Miyata and Yoshiharu Matsuura

Department of Molecular Virology, Research Institute for Microbial Diseases, Osaka University, Osaka 565-0871, Japan; tsuzuki@biken.osaka-u.ac.jp (T.S.); shinji@biken.osaka-u.ac.jp (S.K.); mtoku@biken.osaka-u.ac.jp (M.T.); kamati1101@biken.osaka-u.ac.jp (J.H.); yuppi_ramune0309@yahoo.co.jp (Y.Mi.); matsuura@biken.osaka-u.ac.jp (Y.Ma.)
* Correspondence: toru@biken.osaka-u.ac.jp; Tel.: +81-6-6879-8343; Fax: +81-6-6879-8269

Academic Editor: Marc Kvansakul
Received: 20 July 2017; Accepted: 25 August 2017; Published: 28 August 2017

Abstract: Apoptosis is a type of programmed cell death that regulates cellular homeostasis by removing damaged or unnecessary cells. Its importance in host defenses is highlighted by the observation that many viruses evade, obstruct, or subvert apoptosis, thereby blunting the host immune response. Infection with Flaviviruses such as Japanese encephalitis virus (JEV), Dengue virus (DENV) and West Nile virus (WNV) has been shown to activate several signaling pathways such as endoplasmic reticulum (ER)-stress and AKT/PI3K pathway, resulting in activation or suppression of apoptosis in virus-infected cells. On the other hands, expression of some viral proteins induces or protects apoptosis. There is a discrepancy between induction and suppression of apoptosis during flavivirus infection because the experimental situation may be different, and strong links between apoptosis and other types of cell death such as necrosis may make it more difficult. In this paper, we review the effects of apoptosis on viral propagation and pathogenesis during infection with flaviviruses.

Keywords: apoptosis; B-cell lymphoma 2 (BCL2); flavivirus; dengue virus; Japanese encephalitis virus; West Nile virus

1. Introduction

1.1. Overview of Apoptosis Signaling

Apoptosis, or programmed cell death, is an evolutionarily conserved process essential for the removal of damaged, infected or excess amounts of cells [1,2]. It is required for normal development, tissue homeostasis, and countering infection. Whether a cell lives or dies in response to diverse developmental cues or cellular stresses is largely determined by the interactions of three members of this protein family, Bcl-2 homology 3 (BH3)-only proteins, BCL2 proteins and Bcl-2-associated X protein (BAX)/Bcl-2 homologous antagonist/killer (BAK) effectors [1]. To promote apoptosis, the BH3-only proteins (e.g., Bcl-2 interacting mediator (BIM), Bcl-2 associated death promoter (BAD), NADPH oxidase activator (NOXA) sense cellular damage, but the critical downstream mediators of apoptosis are BAX and BAK, and their combined absence abolishes most apoptotic responses [1,2]. When activated, BAX and BAK permeabilize the outer mitochondrial membrane, releasing proapoptogenic factors, such as cytochrome c, which then promote the activation of the cysteine proteases (caspases) that mediate cellular destruction [3]. The activation of BAX and BAK is opposed by the prosurvival proteins, including BCL2, B-cell lymphoma-extra -large (BCLXL), BCL2-like 2 protein (BCLW), myeloid cell lymphoma 1 (MCL1), and Bcl-2 related protein A1 (A1). These proteins are inactivated by the insertion of BH3 domains of the BH3-only proteins into a groove on the prosurvival proteins to allow the activation of BAX and BAK [1,4]. Activated BAX and BAK then permeabilize the outer mitochondrial membrane to release proapoptotic molecules, such as cytochrome c and second mitochondria-derived activator of caspases (SMAC). Cytochrome c interacts with apoptotic protease-activating factor 1 (APAF1) to activate the effector caspases, including caspase 9 [5]. At same time, SMAC block the caspase inhibitor, X-linked inhibitor of apoptosis proteins (XIAP). Death receptors, such as FAS/CD95 and tumor necrosis factor receptors (TNFR), are activated by their ligands (FASL and TNF, respectively) and then recruit the FAS-associated death domain protein (FADD) to cytoplasmic domains of receptors, resulting in the activation of caspase 8 and the effector caspases [6]. To link mitochondria-mediated apoptosis and death-receptor-mediated apoptosis, caspase 8 is activated by the death receptors and cleaves BH3 interacting domain death agonist (BID) to generate truncated BID, inducing mitochondria-mediated apoptosis [7,8] (Figure 1).

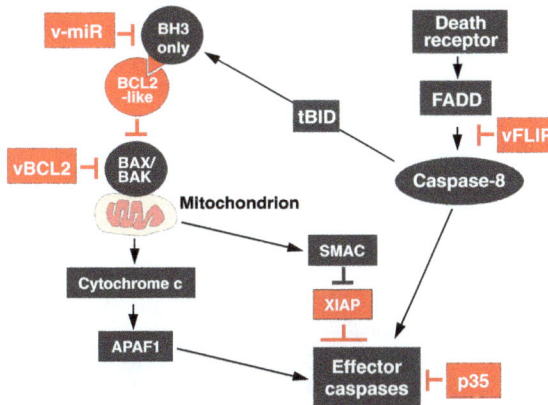

Figure 1. Pathways of apoptosis. Prosurvival proteins (BCL2, BCLW, BCLX$_L$, MCL1, and A1) sequester BH3-only proteins to prevent the activation of BAX and BAK under normal conditions. In the presence of apoptotic stimuli, such as the deprivation of cytokines or cellular damage, BH3-only proteins are activated and displace the prosurvival proteins from their interaction with BAX and BAK, disrupting the mitochondrial outer membrane and thus releasing cytochrome c, second mitochondria-derived activator of caspases (SMAC), and so on. The cytochrome c released interacts with apoptotic protease-activating factor 1 (APAF1) to activate effector caspases, including caspase 9. At same time, SMAC blocks the caspase inhibitor X-linked inhibitor of apoptosis proteins (XIAP). Death receptors, such as FAS/CD95 or tumor necrosis factor receptors (TNFR), are activated by their ligands (FASL or TNF, respectively) and then recruit FAS-associated death domain protein (FADD) to cytoplasmic domains of receptors, resulting in the activation of caspase 8 and effector caspases. Activated caspase 8 cleaves BID to generate truncated BID (tBID), inducing mitochondria-mediated apoptosis. Some DNA viruses encode several genes to control apoptosis (shown in red). The black arrows indicated signaling pathways and T bars shows suppression of activity of indicated proteins.

1.2. Application of Mitochondria Mediated Apoptosis to Cancer Therapy

Small-molecule inhibitors ABT-737 [9] (or ABT-263 (navitoclax), an orally available clinical derivative of ABT-737 [10]), ABT-199 (venetoclax) [11], and A-1331852 [12] have been developed as cancer therapeutics. Whereas ABT-737/ABT-263 targets BCL2, BCLW, and BCLX$_L$, ABT-199 and A-1331852 target BCL2 and BCLX$_L$, respectively. The U.S. Food and Drug Administration has approved ABT-199 for the treatment of 17-p-deleted chronic lymphocytic leukemia (CLL) [13]. It should be noted that the induction of cell death by targeting BCL2 proteins is a promising therapeutic strategy based on the removal of unnecessary cells.

1.3. Apoptosis Regulation by RNA Virus: Lessons from DNA Virus

The consequences of effects of apoptosis during infection with RNA virus remain unclear. However, apart from RNA virus, DNA viruses have evolved a capacity to control cell death [14]. For example, adenovirus, Karposi's sarcoma-associated herpesvirus (KSHV), Epstein-Barr virus (EBV), and murine gamma-herpesvirus 68 (γHV68) encode homologues of mammalian BCL2: ADE1B19K, KSHV vBCL2, BHRF-1, and HV68 M11, respectively. These homologues act as prosurvival proteins by inhibiting the activation of BAX and BAK. Some viruses also encode microRNAs that target proapoptotic host proteins [15–17]. For example, EBV encodes a microRNA, miR-BART5, that targets BH3-only proteins, including p53 upregulated modulator of apoptosis (PUMA) [18]. Furthermore, the majority of poxviruses encode a viral BCL2, which has comparable functions to the BCL2 protein family, despite of the lack of sequence similarity [19,20]. Importantly, these viral proteins are also associated with viral virulence. The p35 protein encoded by the baculoviruses is a viral inhibitor of apoptosis that inhibits a broad range of caspases [21]. Furthermore, virus-encoded Fas-associated death domain-like interleukin-1β (IL-1β)-converting enzyme inhibitory proteins (vFLIP) inhibit the activation of caspase 8 by targeting a complex of FADD and caspase 8 [22–24]. The CrmB protein encoded by orthopoxviruses is a soluble protein that tightly binds TNF to inhibit TNF/TNFR-mediated apoptosis [25]. These data suggest that DNA viruses have acquired strategies to inhibit cell death in order to prolong infection and enhance the production of progeny viruses.

1.4. Life Cycle of Flavivirus Infection

The family *Flaviviridae* contains viruses that are global human health concerns, including *Dengue virus* (DENV), *West Nile virus* (WNV), *Japanese encephalitis virus* (JEV), and *Zika virus* (ZIKV) [26,27]. The viruses of the family *Flaviviridae* have a positive-sense single-stranded RNA genome. The viral RNA is translated to a single polyprotein (of about 3000 amino acids) that is cleaved by host proteases and specific viral proteases, producing at least 10 viral proteins (Figure 2).

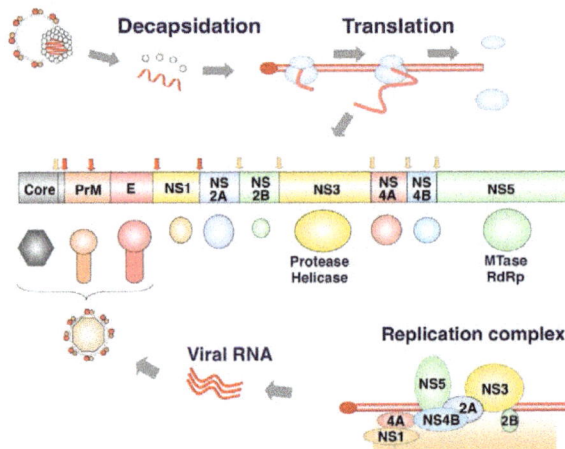

Figure 2. Structure of the flavivirus polyprotein. After decapsidation, the *Flavivirus* genome acts as an mRNA and is directly translated to a polyprotein in a Cap-dependent manner. The polyprotein is cleaved by host signal peptidase (SP) and viral protease (NS3). Core, prM, and E are components of the viral particles. Nonstructural (NS) proteins form a complex on the ER membrane and produce viral RNA. NS5 has methyltransferase (MTase) and RNA-dependent RNA polymerase (RdRp) activities. Red arrow shows host proteases cleave viral proteins while yellow arrow shows cleavages of viral proteins by viral proteases.

The core protein is the first protein to be translated from the viral genome and cleaved by the viral NS2B/3 protease and then the host signal peptidase. The core protein forms a dimer and the viral capsid. The precursor membrane (PrM) and envelope (E) are translated into the ER. The PreM assists the proper folding of E, and plays a role in shielding the fusion peptide of E protein. The E protein is a protein representing on the surface of virions, and is important for receptor binding and membrane fusion. Seven nonstructural proteins are components of the viral replication complex (VRC) [28,29]. The NS1 glycoprotein is translated into the ER, and the intracellular dimer form of NS1 plays roles in viral RNA replication, where the NS1 hexamer is secreted from mammalian cells to play a role in evasion of humoral immune responses. The NS2B and NS3 (NS2B-3) is a serine protease. The NS5 possesses activity of methyltransferase (MTase) and RNA-dependent RNA polymerase (RdRp). The life cycle of the flaviviruses is summarized in Figure 3.

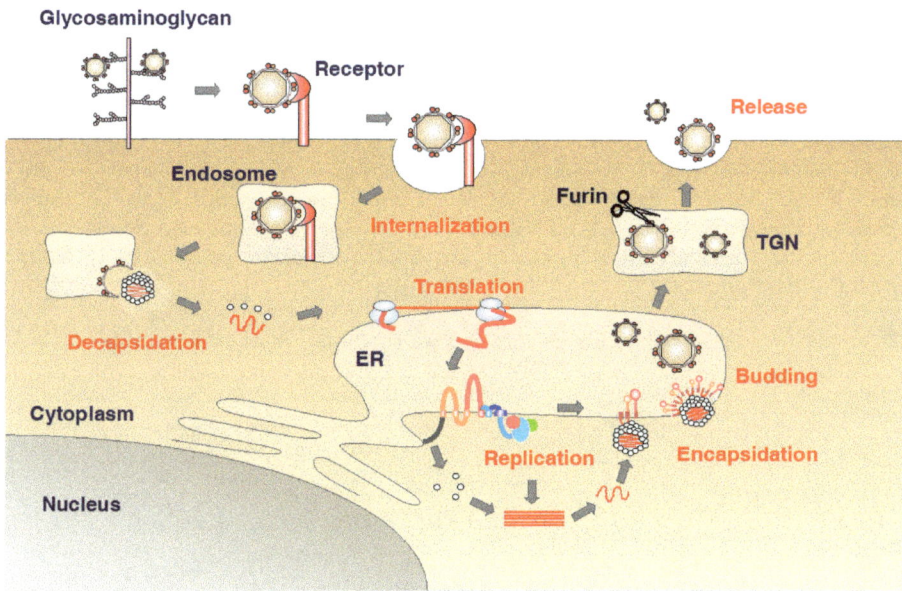

Figure 3. Life cycle of the flaviviruses. The virus first binds to glycosaminoglycan (GAG) and then to specific receptors, and enters the cell by endocytosis. The *Flavivirus* genome is a single-stranded positive-sense RNA molecule that acts directly as an mRNA, which is translated into a polyprotein (Figure 1). Nonstructural proteins form a replication complex that replicates the viral RNA on the ER. Capsid protein molecules incorporate the viral RNA, and the immature virus buds from the ER together with prM and E. The immature protein is processed in the trans-Golgi network, with the removal of prM by furin. Mature virus is subsequently released by exocytosis.

Viral particles primarily interact with glycosaminoglycans, then bind to specific receptors, and are internalized into cells via endocytosis. Under the acidic conditions in the endosome, the capsid releases into cytoplasm after fusion of the viral membrane with the cellular membrane, and the viral RNA is directly translated into a precursor polyprotein. The VRC is formed on the endoplasmic reticulum (ER). The immature viral particles bearing prM and E bud into ER rumen after nucleocapsid formation with viral RNA. PrM is cleaved in the trans-Golgi network by the cellular protease furin to form the mature particles. The mature virions are released by exocytosis.

2. Apoptosis during Flavivirus-Infection

2.1. Pro-Survival and Pro Apoptotic Activity of Viral Proteins

Viral proteins have been shown to regulate apoptosis, as summarized in Table 1. Netsawang et al. showed that nuclear localization of DENV capsid protein is required for its interaction with Fas death domain associated protein xx (DAXX) and the induction of apoptosis [30]. The WNV capsid protein also induces apoptosis through the interaction with importin-α and the phosphorylation by protein kinase C [31]. These results suggest that the nuclear localization of the capsid protein facilitates caspase-9-dependent apoptosis. In contrast, the WNV capsid protein stimulates the phosphorylation of AKT to suppress the activation of caspases 3 and 8 [32]. The ectodomain of the flavivirus M protein also induces apoptosis, but the overexpression of BCL2 suppresses the cell death induced by the M protein [33]. The E protein, but not the NS2B-3, of *Langat virus* (another flavivirus) also induces apoptosis [34]. Injection of recombinant DENV-E protein domain III suppressed megakaryopoiesis through activation of apoptosis in its progenitors [35]. A mutation in the WNV NS2A protein, converting alanine 30 to proline (A30P), attenuated viral virulence by an unknown mechanism [36]. The propagation of the mutant A30P virus was similar to that of the wild-type (WT) virus, but the A30P virus showed less severe cytopathic effects. The number of terminal deoxynucleotidyl transferase dUTP nick end labeling (TUNEL)-positive cells was significantly reduced by the infection with the mutant WNV [37], suggesting that NS2A is involved in WNV-induced apoptosis and pathogenesis. The protease activity of NS3 in JEV, DENV, and WNV induces apoptosis through the activation of caspase 3 or caspase 8. NS2B, a cofactor of NS3, is required for the induction of NS3-induced apoptosis [38–41] (Figure 4A).

Table 1. Pro-apoptotic or pro-survival activity of viral proteins.

Viral Protein	Virus	Tested Cell Lines	Function	Reference
Core	DENV	HepG2	Pro-apoptotic	[30]
	WNV	BHK	Pro-apoptotic	[31]
		A549, HEL/18	Pro-survival	[32]
M	DENV, JEV, WNV, Yellow fever virus (YFV)	HeLa, HepG2, COS-7	Pro-apoptotic	[33]
E	Langat virus	Vero, Neuro-2a	Pro-apoptotic	[34]
	DENV	Progenitor cells of megakaryocytes	Pro-apoptotic	[35]
NS2A	WNV	A549, L929, Vero	Pro-apoptotic	[36,37]
NS2B and NS3	JEV, DENV, WNV	TE671, BHK	Pro-apoptotic	[38]

(A)

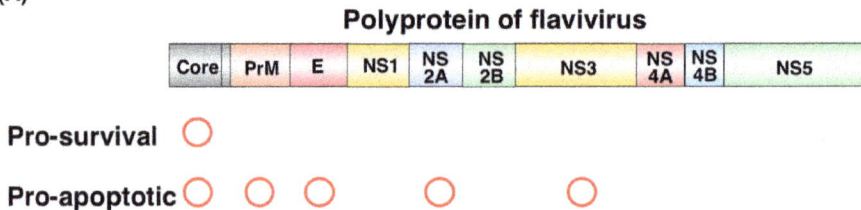

Figure 4. *Cont.*

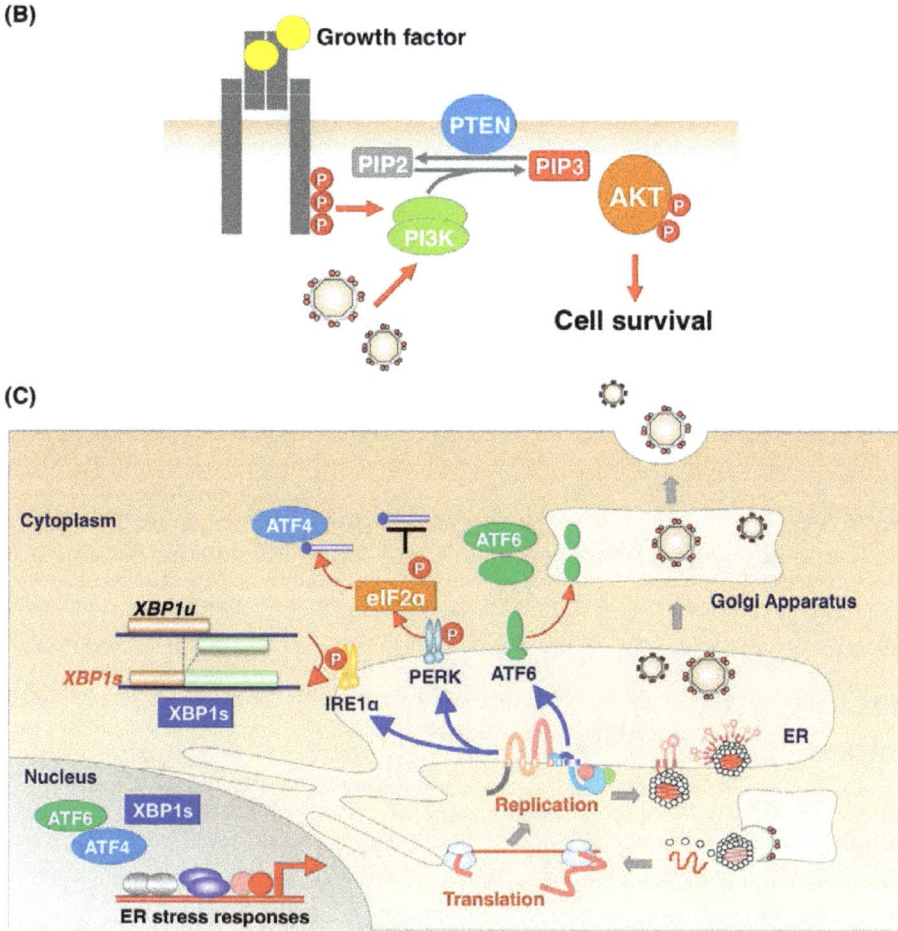

Figure 4. *Flavivirus* infection regulates apoptosis. (**A**) Regulation of viral proteins. Several viral proteins function as proapoptotic proteins; (**B**) AKT/PI3K signaling pathway. In general, growth factors, such as insulin, bind to their receptors to induce receptor phosphorylation, leading to the activation of PI3K, which stimulates the phosphorylation of AKT. Phosphorylation of AKT promotes cellular events, including the suppression of proapoptotic proteins or the promotion of cell survival. Core protein of WNV or infection by JEV or DENV activates the phosphorylation of AKT by activating PI3K to block caspase-dependent apoptosis; (**C**) ER stress induction. The accumulation of unfolded proteins induces ER stress, which stimulates three molecules, IRE1α, PERK, and ATF6. IRE1α is a serine/threonine protein kinase with endonuclease activity. ER stress induces the autophosphorylation of IRE1α to catalyze the excision of XBP1 mRNA. The phosphorylation of PERK, a protein kinase, is induced by ER stress and p-PERK phosphorylates eIF2α to suppress protein translation and to induce the expression of ATF4. ATF6 is also activated by ER stress, translocated to the Golgi, and activated by cleavage of the site-2 proteinase. XBP1, ATF4, and ATF6 are transcription factors to result in response to ER stress. Thus, *Flavivirus* infection stimulates ER stress in many ways.

2.2. Pro-Apoptotic or Pro-Survival Activity during Viral Infection

It is important for us to understand the consequences of regulation of apoptosis during viral infection, and not only the viral protein itself. *Flavivirus* infection has been shown to induce or protect apoptosis as summarized in Table 2. DENV isolated from human patients induced apoptosis in the mouse neuroblastoma cell line, Neuro 2a [42], endothelial cells [43,44], liver cancer cell lines, such as HepG2 [45,46] and Hep3B [47], and human monocyte-derived dendritic cells (Mo-DC) [48,49]. DENV also induced TP53 expression and mitochondrial permeabilization through TP53 [50]. WNV also induced BAX-dependent apoptosis in Neuro 2a cells and K562 cells [51]. Furthermore, apoptosis was induced in neurons derived from embryonic stem (ES) cells during WNV infection [52]. The AKT/PI3K pathway controls cell growth, survival, and energy consumption (Figure 4B). Lee et al. showed that infection with JEV or DENV upregulated the phosphorylation of AKT to activate PI3K/AKT signaling, resulting in the blockage of caspase-mediated cell death [53]. Treatment with LY294002, a PI3K inhibitor, accelerated virus-induced apoptosis [53]. The accumulation of unfolded proteins in the ER triggers the unfolded protein stress response, leading to ER stress [54]. ER stress induces the upregulation of chaperones to refold the unfolded proteins, the ER-associated degradation of any unfolded proteins, and the death of unnecessary cells. *Flavivirus* infection stimulates the ER stress response. JEV, WNV, and DENV infections activate inositol-requiring protein 1α (IRE1α), protein kinase RNA-like ER kinase (PERK), or activating transcription factor 6 (ATF6) [46,55–58]. The activation of IRE1α, PERK1, and ATF6 causes the mRNA encoding unspliced X box-binding protein 1 (XBP1u) to be processed, the expression of ATF4, and the cleavage of ATF6 in the Golgi apparatus, respectively. The ER stress induced by JEV and DENV has been shown to activate ER-associated protein degradation to protect the infected cells from cell death [59]. However, other research groups have shown that the induction of ER stress is associated with DENV pathogenesis [56] (Figure 4C). These discrepancies may be attributable to the different extent of ER stress induced in different cell lines or by different viral strains. Further research, using animal models or gene-edited mice, is required to understand the physiological consequences of the induction of ER stress by flaviviruses. WNV infection also upregulates cellular microRNA Hs_154, to induce apoptosis by targeting antiapoptotic proteins, including the coamplified CCCTC-binding factor (CTCF) and epidermal growth factor receptor (EGFR) and the overexpressed ECOP and VOPP1 proteins. The inhibition of Hs_154 inhibited apoptosis during WNV infection [60].

Table 2. Pro-apoptotic or pro-survival activity of flavivirus infection.

Virus	Tested Cell	Function	Reference
DENV	Neuro 2a	Pro-apoptotic	[42]
DENV	HUVEC EA.hy926	Pro-apoptotic	[43]
DENV	HMEC-1	Pro-apoptotic	[44]
DENV	HepG2	Pro-apoptotic	[45–47]
DENV	Hep3B	Pro-apoptotic	[47]
DENV	Monocyte derived dendritic cells (Mo-DC)	Pro-apoptotic	[48,49]
DENV	Huh7, BHK, Vero	Pro-apoptotic	[50]
WNV	Neuro 2a, K562	Pro-apoptotic	[51]
WNV	ES cells derived neuron	Pro-apoptotic	[52]
JEV, DENV	N18, A549, BHK	Pro-survival	[53]
WNV	SK-N-MC, MEF, HEK293T Primary rat hippocampal neuron	Pro-apoptotic	[55]
DENV	2fTGH, MEF	Pro-apoptotic	[56]
JEV	BHK	Pro-apoptotic	[57]
JEV	Neuro 2a	Pro-apoptotic	[58]
WNV	MEF	Pro-apoptotic	[59]

2.3. Physiological Significance of Apoptosis Signaling during Flavivirus Infection

As described above, many studies have shown that viral infection or the expression of viral proteins induces various types of apoptosis, with the activation of caspases. However, it is important to understand the physiological consequences of apoptotic regulation during viral infection and especially whether the induction or suppression of apoptosis is regulated by host factors to clear infected cells or by the virus to achieve its efficient propagation in vivo. In a mouse model, a deficiency of caspase 3 reduced the disease symptoms and mortality rate during WNV infection, whereas viral propagation was similar in the knockdown and WT mice. This suggests that the lack of caspase 3 activity impaired cell death, suppressing the viral burden and reducing neuronal injury [61]. Not only the extrinsic pathway, but also the mitochondria-mediated apoptotic pathway, leads to the activation of effector caspases, such as caspase 3. A recent study suggested that the blockage of effector caspases caused adverse effects, such as the cytosolic DNA sensor cyclic GMP-AMP synthase (cGAS)/stimulator of interferon gene (STING)-mediated induction of type I interferon (IFN) through mitochondrial DNA (mtDNA) [62,63].

The same research group also reported the functions of the TNF family of ligands, TNF-related apoptosis-inducing ligand (TRAIL) and Fas-ligand (FASL), during WNV infection. The pathogenicity of WNV was enhanced in FASL-deficient (gld) mice or TRAIL$^{-/-}$ mice after their subcutaneous infection [64,65]. There were significantly more viral particles in the brains of the gene-deficient mice than in the WT mice, whereas viral propagation in the brain after intracranial infection with WNV did not differ between the two groups. The CD8$^+$ T cells in both mouse groups showed impaired clearance of the WNV-infected cells. These findings strongly suggest that death-receptor-mediated apoptosis is required for the clearance of WNV-infected cells.

3. Conclusions and Future Directions

The regulation of apoptosis by RNA viruses, including flaviviruses, seems to be more complex than its regulation by DNA viruses. Lessons from the DNA viruses suggest that viruses inhibit apoptosis to achieve efficient propagation in vivo, but apoptosis seems to be induced in virus-infected cells by a host response to clear the infected cells in vitro. Recent advances in gene-editing technologies, such as CRISPR/Cas9, allow us to generate gene-modified mice, including knockout and knockin mice. Therefore, using these technologies, future studies should extend our understanding of the physiological importance of apoptosis during viral infection in vivo. In terms of antiviral therapies, apoptotic cells were detected within 4 h in CLL patients treated with ABT-199 in vitro experiments and the rapid clearance of apoptotic cells was observed at 6–24 h in vivo [66,67]. Therefore, the control of apoptosis by BH3 mimetics can efficiently remove apoptotic cells, including viral-infected cells, from the human body.

Acknowledgments: The authors are grateful to Minako Tomiyama and Junko Higuchi for secretarial work. This work was supported by the Program for Basic and Clinical Research on Hepatitis from Japan Agency for Medical Research and development (AMED) (17fk0210206h0002, 17fk0210305h0003, 17fk0210210h0002, 17fk0210209h0502, 17fk0210304h0003), the Ministry of Education, Culture, Sports, Science, and Technology (MEXT) of Japan (16H06432, 16H06429, 16K21723, 15H04736, 16K19139).

Author Contributions: T.O. drafted the manuscript. T.S., S.K., M.T., J.H., Y.Mi. and Y.Ma. edited and corrected the manuscript.

Conflicts of Interest: The authors declare no conflict of interest.

References

1. Czabotar, P.E.; Lessene, G.; Strasser, A.; Adams, J.M. Control of apoptosis by the BCL-2 protein family: Implications for physiology and therapy. *Nat. Rev. Mol. Cell Biol.* **2014**, *15*, 49–63. [CrossRef] [PubMed]
2. Tait, S.W.G.; Green, D.R. Mitochondria and cell death: Outer membrane permeabilization and beyond. *Nat. Rev. Mol. Cell Biol.* **2010**, *11*, 621–632. [CrossRef] [PubMed]

3. Youle, R.J.; Strasser, A. The BCL-2 protein family: Opposing activities that mediate cell death. *Nat. Rev. Mol. Cell Biol.* **2008**, *9*, 47–59. [CrossRef] [PubMed]

4. Willis, S.N.; Fletcher, J.I.; Kaufmann, T.; van Delft, M.F.; Chen, L.; Czabotar, P.E.; Ierino, H.; Lee, E.F.; Fairlie, W.D.; Bouillet, P.; et al. Apoptosis initiated when BH3 ligands engage multiple BCL-2 homologs, not BAX or BAK. *Science* **2007**, *315*, 856–859. [CrossRef] [PubMed]

5. Adams, J.M.; Cory, S. The Bcl-2 apoptotic switch in cancer development and therapy. *Oncogene* **2007**, *26*, 1324–1337. [CrossRef] [PubMed]

6. Bouillet, P.; O'Reilly, L.A. CD95, BIM and T cell homeostasis. *Nat. Rev. Immunol.* **2009**, *9*, 514–519. [CrossRef] [PubMed]

7. Kaufmann, T.; Schlipf, S.; Sanz, J.; Neubert, K.; Stein, R.; Borner, C. Characterization of the signal that directs BCL-x(L), but not BCL-2, to the mitochondrial outer membrane. *J. Cell Biol.* **2003**, *160*, 53–64. [CrossRef] [PubMed]

8. Schütze, S.; Tchikov, V.; Schneider-Brachert, W. Regulation of TNFR1 and CD95 signaling by receptor compartmentalization. *Nat. Rev. Mol. Cell Biol.* **2008**, *9*, 655–662. [CrossRef]

9. Oltersdorf, T.; Elmore, S.W.; Shoemaker, A.R.; Armstrong, R.C.; Augeri, D.J.; Belli, B.A.; Bruncko, M.; Deckwerth, T.L.; Dinges, J.; Hajduk, P.J.; et al. An inhibitor of BCL-2 family proteins induces regression of solid tumours. *Nature* **2005**, *435*, 677–681. [CrossRef] [PubMed]

10. Ashkenazi, A.; Fairbrother, W.J.; Leverson, J.D.; Souers, A.J. From basic apoptosis discoveries to advanced selective BCL-2 family inhibitors. *Nat. Rev. Drug Discov.* **2017**, *16*, 273–284. [CrossRef] [PubMed]

11. Souers, A.J.; Leverson, J.D.; Boghaert, E.R.; Ackler, S.L.; Catron, N.D.; Chen, J.; Dayton, B.D.; Ding, H.; Enschede, S.H.; Fairbrother, W.J.; et al. ABT-199, a potent and selective BCL-2 inhibitor, achieves antitumor activity while sparing platelets. *Nat. Med.* **2013**, *19*, 202–208. [CrossRef] [PubMed]

12. Leverson, J.D.; Phillips, D.C.; Mitten, M.J.; Boghaert, E.R.; Diaz, D.; Tahir, S.K.; Belmont, L.D.; Nimmer, P.; Xiao, Y.; Ma, X.M.; et al. Exploiting selective BCL-2 family inhibitors to dissect cell survival dependencies and define improved strategies for cancer therapy. *Sci. Transl. Med.* **2015**, *7*, 279ra40. [CrossRef] [PubMed]

13. Roberts, A.W.; Huang, D. Targeting BCL2 with BH3 mimetics: Basic science and clinical application of venetoclax in chronic lymphocytic leukemia and related B cell malignancies. *Clin. Pharmacol. Ther.* **2016**, *101*, 89–98. [CrossRef] [PubMed]

14. Cuconati, A. Viral homologs of BCL-2: Role of apoptosis in the regulation of virus infection. *Genes Dev.* **2002**, *16*, 2465–2478. [CrossRef] [PubMed]

15. Campion, E.M.; Hakimjavadi, R.; Loughran, S.T.; Phelan, S.; Smith, S.M.; D'Souza, B.N.; Tierney, R.J.; Bell, A.I.; Cahill, P.A.; Walls, D. Repression of the proapoptotic cellular BIK/NBK gene by Epstein-Barr virus antagonizes transforming growth factor β1-induced B-cell apoptosis. *J. Virol.* **2014**, *88*, 5001–5013. [CrossRef] [PubMed]

16. Yee, J.; White, R.E.; Anderton, E.; Allday, M.J. Latent Epstein-Barr virus can inhibit apoptosis in B cells by blocking the induction of NOXA expression. *PLoS ONE* **2011**, *6*, 28506–28515. [CrossRef] [PubMed]

17. Shinozaki-Ushiku, A.; Kunita, A.; Isogai, M.; Hibiya, T.; Ushiku, T.; Takada, K.; Fukayama, M. Profiling of virus-encoded microRNAs in Epstein-Barr virus-associated gastric carcinoma and their roles in gastric carcinogenesis. *J. Virol.* **2015**, *89*, 5581–5591. [CrossRef] [PubMed]

18. Choy, E.Y.-W.; Siu, K.-L.; Kok, K.-H.; Lung, R.W.-M.; Tsang, C.M.; To, K.-F.; Kwong, D.L.-W.; Tsao, S.W.; Jin, D.-Y. An Epstein-Barr virus-encoded microRNA targets PUMA to promote host cell survival. *J. Exp. Med.* **2008**, *205*, 2551–2560. [CrossRef] [PubMed]

19. Kvansakul, M.; van Delft, M.F.; Lee, E.F.; Gulbis, J.M.; Fairlie, W.D.; Huang, D.C.S.; Colman, P.M. A structural viral mimic of prosurvival Bcl-2: A pivotal role for sequestering proapoptotic BAX and BAK. *Mol. Cell* **2007**, *25*, 933–942. [CrossRef] [PubMed]

20. Okamoto, T.; Campbell, S.; Mehta, N.; Thibault, J.; Colman, P.M.; Barry, M.; Huang, D.C.S.; Kvansakul, M. Sheeppox virus SPPV14 encodes a BCL-2-like cell death inhibitor that counters a distinct set of mammalian proapoptotic proteins. *J. Virol.* **2012**, *86*, 11501–11511. [CrossRef]

21. Zhou, Q.; Krebs, J.F.; Snipas, S.J.; Price, A.; Alnemri, E.S.; Tomaselli, K.J.; Salvesen, G.S. Interaction of the baculovirus anti-apoptotic protein p35 with caspases. Specificity, kinetics, and characterization of the caspase/p35 complex. *Biochemistry* **1998**, *37*, 10757–10765. [CrossRef]

22. Bertin, J.; Armstrong, R.C.; Ottilie, S.; Martin, D.A.; Wang, Y.; Banks, S.; Wang, G.H.; Senkevich, T.G.; Alnemri, E.S.; Moss, B.; et al. Death effector domain-containing herpesvirus and poxvirus proteins inhibit both Fas- and TNFR1-induced apoptosis. *Proc. Natl. Acad. Sci. USA* **1997**, *94*, 1172–1176. [CrossRef] [PubMed]

23. Stürzl, M.; Hohenadl, C.; Zietz, C. Expression of K13/v-FLIP gene of human herpesvirus 8 and apoptosis in Kaposi's sarcoma spindle cells. *J. Natl. Cancer Inst.* **1999**, *91*, 1725–1733. [CrossRef]

24. Thome, M.; Schneider, P.; Hofmann, K.; Fickenscher, H.; Meinl, E.; Neipel, F.; Mattmann, C.; Burns, K.; Bodmer, J.L.; Schröter, M.; et al. Viral FLICE-inhibitory proteins (FLIPs) prevent apoptosis induced by death receptors. *Nature* **1997**, *386*, 517–521. [CrossRef]

25. Hu, F.Q.; Smith, C.A.; Pickup, D.J. Cowpox virus contains two copies of an early gene encoding a soluble secreted form of the type II TNF receptor. *Virology* **1994**, *204*, 343–356. [CrossRef] [PubMed]

26. Gould, E.A.; Solomon, T. Pathogenic flaviviruses. *Lancet* **2008**, *371*, 500–509. [CrossRef]

27. Mackenzie, J.S.; Gubler, D.J.; Petersen, L.R. Emerging flaviviruses: The spread and resurgence of Japanese encephalitis, West Nile and dengue viruses. *Nat. Med.* **2004**, *10*, 98–109. [CrossRef] [PubMed]

28. Fernandez-Garcia, M.-D.; Mazzon, M.; Jacobs, M.; Amara, A. Pathogenesis of flavivirus infections: Using and abusing the host cell. *Cell Host Microbe* **2009**, *5*, 318–328. [CrossRef] [PubMed]

29. Mukhopadhyay, S.; Kuhn, R.J.; Rossmann, M.G. A structural perspective of the flavivirus life cycle. *Nat. Rev. Microbiol.* **2005**, *3*, 13–22. [CrossRef] [PubMed]

30. Netsawang, J.; Noisakran, S.; Puttikhunt, C.; Kasinrerk, W.; Wongwiwat, W.; Malasit, P.; Yenchitsomanus, P.-T.; Limjindaporn, T. Nuclear localization of dengue virus capsid protein is required for DAXX interaction and apoptosis. *Virus Res.* **2010**, *147*, 275–283. [CrossRef] [PubMed]

31. Bhuvanakantham, R.; Cheong, Y.K.; Ng, M.-L. West Nile virus capsid protein interaction with importin and HDM2 protein is regulated by protein kinase C-mediated phosphorylation. *Microbes Infect.* **2010**, *12*, 615–625. [CrossRef] [PubMed]

32. Urbanowski, M.D.; Hobman, T.C. The West Nile virus capsid protein blocks apoptosis through a phosphatidylinositol 3-kinase-dependent mechanism. *J. Virol.* **2012**, *87*, 872–881. [CrossRef] [PubMed]

33. Catteau, A. Dengue virus M protein contains a proapoptotic sequence referred to as ApoptoM. *J. Gen. Virol.* **2003**, *84*, 2781–2793. [CrossRef] [PubMed]

34. Prikhod'ko, G.G.; Prikhod'ko, E.A.; Cohen, J.I.; Pletnev, A.G. Infection with Langat flavivirus or expression of the envelope protein induces apoptotic cell death. *Virology* **2001**, *286*, 328–335. [CrossRef] [PubMed]

35. Lin, G.-L.; Chang, H.-H.; Lien, T.-S.; Chen, P.-K.; Chan, H.; Su, M.-T.; Liao, C.-Y.; Sun, D.-S. Suppressive effect of dengue virus envelope protein domain III on megakaryopoiesis. *Virulence* **2017**, *40*, 1–13. [CrossRef] [PubMed]

36. Liu, W.J.; Wang, X.J.; Clark, D.C.; Lobigs, M.; Hall, R.A.; Khromykh, A.A. A single amino acid substitution in the West Nile virus nonstructural protein NS2A disables its ability to inhibit α/β interferon induction and attenuates virus virulence in mice. *J. Virol.* **2006**, *80*, 2396–2404. [CrossRef]

37. Melian, E.B.; Edmonds, J.H.; Nagasaki, T.K.; Hinzman, E.; Floden, N.; Khromykh, A.A. West Nile virus NS2A protein facilitates virus-induced apoptosis independently of interferon response. *J. Gen. Virol.* **2013**, *94*, 308–313. [CrossRef] [PubMed]

38. Yang, T.-C.; Shiu, S.-L.; Chuang, P.-H.; Lin, Y.-J.; Wan, L.; Lan, Y.-C.; Lin, C.-W. Japanese encephalitis virus NS2B-NS3 protease induces caspase 3 activation and mitochondria-mediated apoptosis in human medulloblastoma cells. *Virus Res.* **2009**, *143*, 77–85. [CrossRef]

39. Matusan, A.E.; Kelley, P.G.; Pryor, M.J.; Whisstock, J.C.; Davidson, A.D.; Wright, P.J. Mutagenesis of the dengue virus type 2 NS3 proteinase and the production of growth-restricted virus. *J. Gen. Virol.* **2001**, *82*, 1647–1656. [CrossRef] [PubMed]

40. Shafee, N. Dengue virus type 2 NS3 protease and NS2B-NS3 protease precursor induce apoptosis. *J. Gen. Virol.* **2003**, *84*, 2191–2195. [CrossRef] [PubMed]

41. Ramanathan, M.P.; Chambers, J.A.; Pankhong, P.; Chattergoon, M.; Attatippaholkun, W.; Dang, K.; Shah, N.; Weiner, D.B. Host cell killing by the West Nile virus NS2B-NS3 proteolytic complex: NS3 alone is sufficient to recruit caspase-8-based apoptotic pathway. *Virology* **2006**, *345*, 56–72. [CrossRef] [PubMed]

42. Despres, P.; Flamand, M.; Ceccaldi, P.E.; Deubel, V. Human isolates of dengue type 1 virus induce apoptosis in mouse neuroblastoma cells. *J. Virol.* **1996**, *70*, 4090–4096. [PubMed]

43. Huang, J.; Li, Y.; Qi, Y.; Zhang, Y.; Zhang, L.; Wang, Z.; Zhang, X.; Gui, L. Coordinated regulation of autophagy and apoptosis determines endothelial cell fate during Dengue virus type 2 infection. *Mol. Cell. Biochem.* **2014**, *397*, 157–165. [CrossRef] [PubMed]

44. Vásquez Ochoa, M.; García Cordero, J.; Gutiérrez Castañeda, B.; Santos Argumedo, L.; Villegas Sepúlveda, N.; Cedillo Barrón, L. A clinical isolate of dengue virus and its proteins induce apoptosis in HMEC-1 cells: A possible implication in pathogenesis. *Arch. Virol.* **2009**, *154*, 919–928. [CrossRef] [PubMed]

45. El-Bacha, T.; Midlej, V.; Pereira da Silva, A.P.; Silva da Costa, L.; Benchimol, M.; Galina, A.; Da Poian, A.T. Mitochondrial and bioenergetic dysfunction in human hepatic cells infected with dengue 2 virus. *Biochim. Biophys. Acta (BBA)* **2007**, *1772*, 1158–1166. [CrossRef] [PubMed]

46. Thepparit, C.; Khakpoor, A.; Khongwichit, S.; Wikan, N.; Fongsaran, C.; Chingsuwanrote, P.; Panraksa, P.; Smith, D.R. Dengue 2 infection of HepG2 liver cells results in endoplasmic reticulum stress and induction of multiple pathways of cell death. *BMC Res. Notes* **2013**, *6*, 372. [CrossRef] [PubMed]

47. Thongtan, T.; Panyim, S.; Smith, D.R. Apoptosis in dengue virus infected liver cell lines HepG2 and Hep3B. *J. Med. Virol.* **2004**, *72*, 436–444. [CrossRef] [PubMed]

48. Olagnier, D.; Peri, S.; Steel, C.; van Montfoort, N.; Chiang, C.; Beljanski, V.; Slifker, M.; He, Z.; Nichols, C.N.; Lin, R.; et al. Cellular oxidative stress response controls the antiviral and apoptotic programs in Dengue virus-infected dendritic cells. *PLoS Pathog.* **2014**, *10*, e1004566. [CrossRef] [PubMed]

49. Silveira, G.F.; Meyer, F.; Delfraro, A.; Mosimann, A.L.P.; Coluchi, N.; Vasquez, C.; Probst, C.M.; Bafica, A.; Bordignon, J.; Santos, C.N.D.D. Dengue virus type 3 isolated from a fatal case with visceral complications induces enhanced proinflammatory responses and apoptosis of human dendritic cells. *J. Virol.* **2011**, *85*, 5374–5383. [CrossRef] [PubMed]

50. Nasirudeen, A.M.A.; Wang, L.; Liu, D.X. Induction of p53-dependent and mitochondria-mediated cell death pathway by Dengue virus infection of human and animal cells. *Microbes Infect.* **2008**, *10*, 1124–1132. [CrossRef] [PubMed]

51. Parquet, M.D.; Kumatori, A.; Hasebe, F.; Morita, K.; Igarashi, A. West Nile virus-induced BAX-dependent apoptosis. *FEBS Lett.* **2001**, *500*, 17–24. [CrossRef]

52. Shrestha, B.; Gottlieb, D.; Diamond, M.S. Infection and injury of neurons by West Nile encephalitis virus. *J. Virol.* **2003**, *77*, 13203–13213. [CrossRef] [PubMed]

53. Lee, C.J.; Liao, C.L.; Lin, Y.L. Flavivirus activates phosphatidylinositol 3-kinase signaling to block caspase-dependent apoptotic cell death at the early stage of virus infection. *J. Virol.* **2005**, *79*, 8388–8399. [CrossRef] [PubMed]

54. Hetz, C. The unfolded protein response: Controlling cell fate decisions under ER stress and beyond. *Nat. Rev. Mol. Cell Biol.* **2012**, *13*, 89–102. [CrossRef] [PubMed]

55. Medigeshi, G.R.; Lancaster, A.M.; Hirsch, A.J.; Briese, T.; Lipkin, W.I.; Defilippis, V.; Früh, K.; Mason, P.W.; Nikolich-Zugich, J.; Nelson, J.A. West Nile virus infection activates the unfolded protein response, leading to CHOP induction and apoptosis. *J. Virol.* **2007**, *81*, 10849–10860. [CrossRef]

56. Pena, J.; Harris, E. Dengue virus modulates the unfolded protein response in a time-dependent manner. *J. Biol. Chem.* **2011**, *286*, 14226–14236. [CrossRef] [PubMed]

57. Huang, M.; Xu, A.; Wu, X.; Zhang, Y.; Guo, Y.; Guo, F.; Pan, Z.; Kong, L. Japanese encephalitis virus induces apoptosis by the IRE1/JNK pathway of ER stress response in BHK-21 cells. *Arch. Virol.* **2015**, *161*, 699–703. [CrossRef] [PubMed]

58. Bhattacharyya, S.; Sen, U.; Vrati, S. Regulated IRE1-dependent decay pathway is activated during Japanese encephalitis virus-induced unfolded protein response and benefits viral replication. *J. Gen. Virol.* **2013**, *95*, 71–79. [CrossRef]

59. Ambrose, R.L.; Mackenzie, J.M. ATF6 signaling is required for efficient West Nile virus replication by promoting cell survival and inhibition of innate immune responses. *J. Virol.* **2013**, *87*, 2206–2214. [CrossRef]

60. Smith, J.L.; Grey, F.E.; Uhrlaub, J.L.; Nikolich-Zugich, J.; Hirsch, A.J. Induction of the cellular microRNA, Hs_154, by West Nile virus contributes to virus-mediated apoptosis through repression of antiapoptotic factors. *J. Virol.* **2012**, *86*, 5278–5287. [CrossRef] [PubMed]

61. Samuel, M.A.; Morrey, J.D.; Diamond, M.S. Caspase 3-dependent cell death of neurons contributes to the pathogenesis of West Nile virus encephalitis. *J. Virol.* **2007**, *81*, 2614–2623. [CrossRef] [PubMed]

62. White, M.J.; McArthur, K.; Metcalf, D.; Lane, R.M.; Cambier, J.C.; Herold, M.J.; van Delft, M.F.; Bedoui, S.; Lessene, G.; Ritchie, M.E.; et al. Apoptotic caspases suppress mtDNA-induced STING-mediated type I IFN production. *Cell* **2014**, *159*, 1549–1562. [CrossRef] [PubMed]

63. Rongvaux, A.; Jackson, R.; Harman, C.C.D.; Li, T.; West, A.P.; de Zoete, M.R.; Wu, Y.; Yordy, B.; Lakhani, S.A.; Kuan, C.-Y.; et al. Apoptotic caspases prevent the induction of type I interferons by mitochondrial DNA. *Cell* **2014**, *159*, 1563–1577. [CrossRef] [PubMed]

64. Shrestha, B.; Diamond, M.S. Fas ligand interactions contribute to CD8+ T-cell-mediated control of West Nile virus infection in the central nervous system. *J. Virol.* **2007**, *81*, 11749–11757. [CrossRef] [PubMed]

65. Shrestha, B.; Pinto, A.K.; Green, S.; Bosch, I.; Diamond, M.S. CD8+ T cells use TRAIL to restrict West Nile virus pathogenesis by controlling infection in neurons. *J. Virol.* **2012**, *86*, 8937–8948. [CrossRef] [PubMed]

66. Anderson, M.A.; Deng, J.; Seymour, J.F.; Tam, C.; Kim, S.Y.; Fein, J.; Yu, L.; Brown, J.R.; Westerman, D.; Si, E.G.; et al. The BCL2 selective inhibitor venetoclax induces rapid onset apoptosis of CLL cells in patients via a TP53-independent mechanism. *Blood* **2016**, *127*, 3215–3224. [CrossRef] [PubMed]

67. Roberts, A.W.; Davids, M.S.; Pagel, J.M.; Kahl, B.S.; Puvvada, S.D.; Gerecitano, J.F.; Kipps, T.J.; Anderson, M.A.; Brown, J.R.; Gressick, L.; et al. Targeting BCL2 with venetoclax in relapsed chronic lymphocytic leukemia. *N. Engl. J. Med.* **2016**, *374*, 311–322. [CrossRef] [PubMed]

Review

Host and Viral Factors in HIV-Mediated Bystander Apoptosis

Himanshu Garg * and Anjali Joshi *

Center of Emphasis in Infectious Diseases, Department of Biomedical Sciences, Texas Tech University Health Sciences Center, 5001 El Paso Dr., El Paso, TX 79905, USA
* Correspondence: himanshu.garg@ttuhsc.edu (H.G.); anjali.joshi@ttuhsc.edu (A.J.);
  Tel.: +1-915-215-4271 (H.G.); +1-915-215-4263 (A.J.); Fax: +1-915-783-1253 (H.G. & A.J.)

Academic Editor: Marc Kvansakul
Received: 7 July 2017; Accepted: 16 August 2017; Published: 22 August 2017

Abstract: Human immunodeficiency virus (HIV) infections lead to a progressive loss of CD4 T cells primarily via the process of apoptosis. With a limited number of infected cells and vastly disproportionate apoptosis in HIV infected patients, it is believed that apoptosis of uninfected bystander cells plays a significant role in this process. Disease progression in HIV infected individuals is highly variable suggesting that both host and viral factors may influence HIV mediated apoptosis. Amongst the viral factors, the role of Envelope (Env) glycoprotein in bystander apoptosis is well documented. Recent evidence on the variability in apoptosis induction by primary patient derived Envs underscores the role of Env glycoprotein in HIV disease. Amongst the host factors, the role of C-C Chemokine Receptor type 5 (CCR5), a coreceptor for HIV Env, is also becoming increasingly evident. Polymorphisms in the *CCR5* gene and promoter affect CCR5 cell surface expression and correlate with both apoptosis and CD4 loss. Finally, chronic immune activation in HIV infections induces multiple defects in the immune system and has recently been shown to accelerate HIV Env mediated CD4 apoptosis. Consequently, those factors that affect CCR5 expression and/or immune activation in turn indirectly regulate HIV mediated apoptosis making this phenomenon both complex and multifactorial. This review explores the complex role of various host and viral factors in determining HIV mediated bystander apoptosis.

Keywords: HIV; AIDS; apoptosis; bystander; Env; CCR5; immune activation; fusion; hemifusion; gp41

1. Introduction

Human immunodeficiency virus (HIV) infection in humans leads to a progressive loss of CD4 T cells culminating in immunodeficiency. While HIV is known to selectively infect CD4 cells, the mechanism of CD4 T cell loss is more complex than virus infection alone. Mounting evidence suggests that both and host and viral factors play a significant role in determining CD4 T cell loss. Recent studies have started to unravel this complex interplay and a better picture is emerging, which can to a great extent explain the selectivity of CD4 loss as well as the variations in disease progression. Fundamental to HIV pathogenesis is the phenomenon of apoptosis that is believed to be a major pathway in CD4 loss. This review looks at the various host and viral factors that play a role in regulating CD4 apoptosis in HIV infections.

Apoptosis in HIV infections: The phenomenon of apoptosis in HIV infections has been observed from the earliest days of HIV research [1–3]. Support for a role of apoptosis in CD4 T cell loss in HIV infections also comes from simian immunodeficiency virus (SIV) infections in non-human primates. Pathogenic SIV infection in macaques is characterized by increased apoptosis, which is remarkably absent in non-pathogenic SIV infection in natural host [4–7]. In the SIV/HIV chimeric SHIV virus infection in macaques, a correlation between apoptosis and CD4 loss is also evident [8,9].

Multiple studies have reported increased apoptosis in peripheral blood mononuclear cells (PBMCs) from HIV infected individuals that correlates with CD4 decline [10–12]. Although the involvement of apoptosis in CD4 decline in HIV infections is widely accepted, the mechanism of apoptosis induction remains debated.

Role of virus replication in apoptosis: Immunopathogenic features of HIV infection, including CD4 loss in some cases correlate poorly with viremia [13–15]. Furthermore, high levels of viremia in natural SIV infections fails to induce CD4 apoptosis and subsequent CD4 loss [4]. These observations argue against a role of viremia in HIV mediated apoptosis. On the other hand, a role of viremia in CD4 apoptosis is supported by the observation that suppression of virus replication with highly active antiretroviral therapy (HAART) can reverse both CD4 loss as well as T cells apoptosis in the peripheral blood [16–18]. Similarly, untreated HIV infections have been shown to be associated with high levels of apoptosis [11]. Recent evidence from our lab demonstrates that a number of immunopathological features of HIV infection including CD4 apoptosis correlates with viremia [10]. The success of HAART in delaying acquired immunodeficiency syndrome (AIDS) onset and increasing life expectancy in HIV infected individuals is probably the strongest evidence for a role of viremia in HIV pathology including CD4 T cell decline [19,20]. At the same time, this process is complex and cases where viremia fails to mediate CD4 apoptosis/decline; other host and viral factors are likely involved.

Bystander Apoptosis: The number of HIV infected cells in patients is relatively low and cannot solely account for the loss of CD4 cells in vivo. Hence, it is believed that the loss of CD4 cells during HIV infection is due to the process of bystander apoptosis induction [3,21,22]. Early studies by Finkel et al. [23] demonstrated that the majority of cells undergoing apoptosis during HIV infection are not infected but in close proximity to infected cells. A role of direct infection in loss of CD4 cells is also refuted by SIV infection in natural host where high levels of infection and viremia do not result in AIDS development [24]. Hence, bystander apoptosis is believed to be one of the major causes of CD4 loss leading to AIDS [3,22,25]. In vitro infection of cell lines and PBMCs with HIV-1 also shows that apoptosis is largely restricted to uninfected bystander cells [26–28]. Furthermore, in pathogenic SIV models [5,6] and HIV infection in humanized mouse models [26,29], apoptosis in bystander cells has been observed.

2. Viral Factors

2.1. Env Glycoprotein

The role of Envelope (Env) glycoprotein in mediating bystander apoptosis has been extensively studied [21,30,31]. The involvement of Env glycoprotein in this process is supported by three major observations. Firstly, as cell death in HIV infection outnumbers the infected cell population, a role of bystander T cell apoptosis in progression to AIDS has been proposed. Secondly, as the depletion of immune cells is largely restricted to CD4 T cells and as the Env glycoprotein binds directly to CD4, it likely plays a role in CD4 T cell death. Finally, Env glycoprotein is expressed on the surface of infected cells and has been shown to interact with bystander cells expressing CD4 and a coreceptor CXCR4/CCR5 to mediate apoptosis [32–36]. Thus, the Env glycoprotein is believed to be a major player in HIV induced cell death.

The structural features of HIV Env are fundamental to its ability to mediate bystander apoptosis. The Env glycoprotein of HIV is arranged on the surface of the virus and virus-infected cells as a hetero-trimer. Each monomer is composed of a receptor-binding gp120 unit and a gp41 transmembrane unit that mediates fusion of viral and cellular membrane [37]. The sequence of events that lead to fusion catalyzed by HIV-1 Env are initiated by the binding of gp120 subunit to CD4 and a coreceptor, either CXCR4 or CCR5, on CD4 cells. This binding of HIV gp120 to CD4 triggers several conformational changes in gp120 that result in exposure of coreceptor binding site and the N-terminal and C-terminal heptad repeat (HR) regions of gp41 [38]. Subsequently, interaction of the HR domains in a leucine zipper like fashion facilitates effector and target membranes to come in close proximity resulting in fusion [39]. Although the primary function of HIV Env glycoprotein is to facilitate viral entry, its role in bystander apoptosis is becoming increasingly evident [40,41].

2.2. Mechanism of HIV Env-Mediated Apoptosis: Role of gp120 and gp41

The gp120 subunit of HIV Env binds to CD4 and a coreceptor making it a likely candidate to mediate apoptosis. In fact, early studies showed that inhibition of gp120 binding to CD4 or the coreceptor eliminates Env mediated bystander apoptosis [42]. However, in later studies, it was observed that, while gp120 binding is required for this process, it is the gp41 subunit that mediates fusion of membranes and plays a critical role in this process [35,36,43,44]. Studies by several groups have found that membrane hemifusion between HIV-1 Env expressing cells and CD4 bystander cells correlates with Env mediated bystander apoptosis [35,36,44]. Hemifusion is a process that involves transient interaction of cellular membranes characterized by mixing of the outer leaflets of the bilayers without progression to fusion pore formation [45]. Although hemifusion has been extensively studied in influenza, it has been demonstrated by multiple independent groups in HIV as well [35,44,46–48]. It is hypothesized that membrane damage mediated during this hemifusion process in part mediates bystander apoptosis; a phenomenon also referred to as the "kiss of death" [31].

Further evidence for a role of HIV Env mediated fusion in CD4 loss can be found in clinical studies where presence of syncytia inducing (SI) viruses has been associated with poor prognosis and a rapid CD4 decline [49,50]. The SI phenotype is not only associated with increased pathogenesis [51] but also CXCR4 tropism [52] of viruses. In severe combined immunodeficiency humanized (SCID-hu) mouse model, SI phenotype and CXCR4 tropism is linked to CD4 loss as certain CCR5 tropic lab strains fail to induce CD4 decline [53]. The correlation of HIV mediated CD4 loss with the fusogenic activity of Env glycoprotein is also supported by animal models such as the chimeric SHIV infection in rhesus macaques [54–56] and HIV infection in humanized mice [26,29].

Although gp41 mediated fusion plays an important role in bystander apoptosis, the role of gp120 subunit of Env in HIV mediated bystander apoptosis cannot be ignored. As the initial contact between bystander cells and infected cells is made via the gp120 subunit, the phenotypic characteristics of gp120 determine the subsequent steps in Env mediated fusion and therefore bystander apoptosis. Increased affinity of gp120 for CD4 receptor and/or coreceptor can influence both Env mediated fusion as well as apoptosis [28,57–59]. We have also found a negative correlation between potential N-glycosylation sites (PNGS) and bystander apoptosis inducing phenotype among primary patient derived Envs [60]. Thus, the phenotypic characteristic of Env glycoprotein is complex and a consequence of genotypic characteristics of both gp120 and gp41 subunits.

2.3. HIV Env Glycoprotein Variability and Evolution

HIV Env glycoprotein is probably the most variable protein found in nature [61] and evolves throughout the course of infection within individuals [62]. One of the interesting phenomenon in HIV Env evolution is switching of coreceptor usage from CCR5 tropic (R5) early viruses to CXCR4 tropic (X4) viruses during late stages of the disease [49–51]. This phenotypic change in virus often precedes a rapid decline in CD4 cells suggesting an increased pathogenic potential of X4 viruses [51,63,64]. However, coreceptor switching is not a requirement for HIV mediated bystander apoptosis and it has been shown that both X4 and R5 tropic Envs are capable of inducing apoptosis in bystander cells given the appropriate coreceptor is expressed [43,65,66]. Furthermore, late stage AIDS associated R5 tropic viruses have been shown to be more pathogenic with increased replication and cytopathic effect [67,68]. These AIDS associated R5 viruses have been shown to be more fusogenic [69] and this phenotypic difference has been mapped to the glycosylation site at position Asn 362 [58]. We have found that in vitro adaptation of HIV to low levels of CCR5 results in evolution of virus to higher CCR5 affinity and increased bystander apoptosis in cell expressing low CCR5 levels [66]. An in vitro analysis of primary CCR5 tropic Envs for bystander apoptosis induction also shows that patient derived Envs vary considerably with respect to apoptosis phenotype and several genetic signatures correlated with apoptosis, including low levels of PNGS [60]. Recently, we have found that the apoptosis inducing potential (AIP) of primary Envs from patients correlates with CD4 decline and CD4:CD8 ratio in patients [10]. These studies provide strong evidence for a role of HIV Env in bystander apoptosis as well as disease progression.

2.4. Signaling via HIV Env Glycoprotein

The apoptotic signaling mediated by HIV Env has been studied extensively and activation of classical apoptosis markers such as caspase-3, mitochondrial depolarization [70,71] and reactive oxygen species (ROS) production [36] has been observed during the process. However, a role of apoptosis signaling ligands such as Fas and tumor necrosis factor (TNF) has not been found in Env mediated apoptosis, suggesting that it constitutes a unique signaling pathway [36,72,73]. As the gp120 subunit of HIV Env binds to CD4 and CXCR4/CCR5, signaling via either of these receptors could be important for Env mediated apoptosis. However, eliminating CD4 signaling via cytoplasmic tail truncation and CXCR4 signaling via G protein inhibitors fails to inhibit Env mediated apoptosis [34,35] suggesting that signaling via either of these receptors is not required for this phenomenon. The inhibition of HIV Env mediated apoptosis by gp41 inhibitors such as T20 (enfuvirtide) and C34 peptide indicates that the signaling may be initiated by gp41 catalyzed events such as hemifusion of membranes [21,35,74]. Moreover, although the activation of caspase-3 and mitochondrial depolarization has been seen in HIV Env mediated apoptosis [35,36,74], the specific target that initiates this process, perhaps in the plasma membrane remains undetermined.

Current evidence largely suggests that HIV-1 Env mediated apoptosis in bystander cells involves the mitochondria and is associated with ROS production [75], a global increase in cell metabolism and increase in mitochondrial fission. ROS generation mediated by Env can be inhibited by the CXCR4 antagonist AMD-3100 although it is not clear if ROS generation is a consequence of signaling upon Env binding to CXCR4 or as a result of Env mediated oxidative stress. Similarly, HIV Env is known to trigger upregulation of the stress protein family of chaperones such as Heat shock protein 70 (Hsp70). Both mitochondrial and cytoplasmic forms of Hsp70 are upregulated after contact with Env which is also upregulated in HIV infected cells functioning as an innate immunity factor [76]. Molina et al. [77] showed that this defensive response is initiated in uninfected cells after Env–CXCR4 interaction. Although heat shock proteins can both inhibit and enhance apoptosis via different pathways, reduction in cellular expression of HSp60 increases Env mediated bystander apoptosis [70]. While other proteins including the proteasome- and ubiquitin-related proteins have been known to be involved in the process, further investigation is required to delineate the precise pathway involved in HIV Env mediated bystander cell death, especially in physiologically relevant CD4 T cells.

2.5. Targeting HIV gp41 Can Alter Bystander Apoptosis

The inhibition of HIV Env mediated bystander apoptosis by gp41 fusion inhibitors opens the door for targeting gp41 not only to inhibit HIV infection but also bystander apoptosis [78]. In this context, it has been reported that Enfuvirtide therapy may have beneficial effects in patients by inhibiting gp41 mediated cell death [79]. However, similar to other anti-retrovirals, Enfuvirtide come with the caveat of resistance development. Interestingly, resistance against enfuvirtide in many cases comes at the expense of alteration in the Env fusogenic properties [80]. Thus, theoretically, methods that alter HIV gp41 mediated fusion could have therapeutic benefits via reducing bystander apoptosis. In support of this hypothesis, a clinical study by Aquaro et al. [81] found that certain resistant viruses emerging during enfuvirtide therapy were associated with CD4 increase in patients even after virological failure. These mutations were localized in the gp41 HR1 region that is known to regulate gp41 fusion activity [80]. Subsequently, Melby et al. [82] reported that mutations at position V38 in gp41 are associated with increase in CD4 recovery in enfuvirtide treated patients after virological failure. Further evidence for a role of gp41 in this process was shown by Cunyat et al. who found that the presence of V38A in combination with N140I mutation in gp41 was associated with reduced

HIV associated cytopathic phenotype [48]. In vitro mutagenesis studies by our lab later confirmed that mutations at the V38 position, especially V38E, reduces bystander apoptosis activity in vitro [83]. The non-pathogenic nature of gp41 mutant V38E was not limited to in vitro studies but was also observed in humanized mice, where infection with V38E mutant resulted in slower CD4 decline accompanied with a lack of bystander apoptosis compared to infection with wild type virus [26]. Collectively, these findings support a strategy for targeting HIV gp41 to limit bystander apoptosis.

2.6. Fas, TNF/TRAIL and Other Viral Proteins in Bystander Apoptosis

Although a role of tumor necrosis factor receptor 1 (TNFR1) and Fas (CD95) pathways has not been found in Env mediated apoptosis, there is evidence that HIV infected cells show greater susceptibility to Fas induced apoptosis. Both membrane bound and soluble Fas are upregulated in HIV infected patients and correlate with disease progression [84]. Macrophages are believed to play a major role in cell death via this pathway as TNF expressed on the surface of activated macrophages can induce apoptosis in bystander T cells via TRAIL (TNF Related Apoptosis Inducing Ligand)–DR5 (Death Receptor) and Fas–Fas ligand interactions in a major histocompatibility complex (MHC) unrestricted manner [85,86]. Further impairment in cell viability has also been observed due to other viral proteins such as Nef and Vpr that mimic the biological effects of TNF, while gp120 and Vpu can exacerbate the pro apoptotic effects of TNF [87]. It has been shown that CD4 cross linking via gp120 activates the CD95/CD95L pathway [88] and Nef expressing T cells upregulate CD95L that mediates apoptosis in CD95 expressing bystander T cells [89]. Furthermore, Tat protein secreted from infected cells is capable of upregulating CD95L expression in uninfected cells, enhancing susceptibility to apoptosis via this pathway [90]. Another member of the TNF family that has been implicated in HIV induced bystander apoptosis in vivo is TRAIL/Apo2 Ligand (APO2L) as T cells from HIV+ patients are more prone to TRAIL mediated cell death [91]. Finally, Tat protein can induce upregulation of TRAIL in macrophages that can lead to apoptosis in bystander T cells [92].

2.7. Effects of HIV Env on Cell Types Other Than CD4 T Cells

While HIV is a disease affecting CD4 T cells, many other cell types also suffer the consequences of T helper cell dysfunction either directly or indirectly. These include cells of the neuronal lineage, thymocytes, CD34+ stem cells, B cells and cells of the monocytic lineage. With regards to CD34+ stem cells, both stimulatory and inhibitory effects of HIV Env on uninfected CD34+ progenitor cells have been observed. The stimulatory effects manifest as increased myeloid colony formation via indirect effects of gp160 through production of cytokines such as granulocyte monocyte-colony stimulating Factor (GM–CSF) [93]. On the other hand, the HIV gp120 has been shown to inhibit hematological colony formation via TNF-α section (from mononuclear cells) which is a potent hematopoiesis inhibitor [94]. Interestingly, HIV is also shown to induce cytopathic effects in thymocytes via induction of bystander apoptosis [95] along with aberrant positive and negative selection [96]. With regards to B cells, other than affecting all the functions that require CD4 help, HIV Env is known to stimulate B cells to differentiate into antibody secreting cells [97]. Similarly, gp120 binding to CD4 leads to reduced expression of co-stimulatory molecules such as B7 in macrophages derived from HIV infected individuals [98,99]. Finally, the effects of HIV infection on the nervous system cannot be overstated with the prevalence of HIV associated neurocognitive disorders (HAND) on the rise with increasing life span of HIV patients with access to HAART. Primarily, the effects are the result of viral proteins such as gp120 that are shed form HIV infected microglia resulting in neuronal apoptosis via caspase 3 activation, release of inflammatory cytokines and increase in permeability of the blood–brain barrier [100].

3. Host Factors

3.1. CCR5: Role of CCR5 in HIV Disease

The discovery of CCR5 and CXCR4 as co-receptors for HIV infection was a major breakthrough in HIV research [101]. While HIV variants can use CXCR4 for viral entry, most early viral isolates have been found to be R5 tropic [102,103]. This is most likely due to their better transmission potential and abundance of CCR5 expressing cells in the mucosal tissues, the natural sites for establishment of HIV infection [102–105]. The predominance of CCR5 as the major coreceptor for HIV has led to the development of multiple CCR5 antagonists as drug candidates [106]. The therapeutic potential of CCR5 antagonist Maraviroc in suppressing virus replication and virus-related pathologies in patients harboring CCR5 tropic viruses has been well documented [107]. Furthermore, the natural variations in CCR5 expression in humans, as a result of gene and promoter polymorphisms, have been linked to HIV disease progression [108,109].

3.2. CCR5 Polymorphisms

The role of CCR5 polymorphisms in HIV disease was first realized with the observation that patients with the CCR5Δ32 heterozygous genotype, while susceptible to infection, progress slower to AIDS than wild type (WT) genotype [110,111]. While CCR5Δ32 homozygous individuals are resistant to CCR5 tropic virus infection due to lack of functional CCR5 on the cell surface [112,113], the slower disease progression in the CCR5Δ32 heterozygous patients [111] was attributed to lower CCR5 levels on the cell surface [114]. Other factors affecting cell surface CCR5 levels in the host are several single nucleotide polymorphisms (SNPs) in the CCR5 promoter region [108,115]. Seven major SNPs in the CCR5 promoter make up eight different promoter haplotypes (HHA–HHG) and haplotypes such as HHC have been associated with slower disease progression [109,116]. Recent evidence suggests that CCR5 expression is further influenced by epigenetic factors along with activation status of T cell cells playing a role in this process [117]. While CCR5 levels have clearly been documented to play a role in HIV pathogenesis, the mechanism behind the phenomenon is only recently becoming clear.

3.3. CCR5 Env Interaction in HIV-Mediated Apoptosis

Two main processes via which CCR5 levels may regulate disease progression is by modulating virus infection and bystander apoptosis [65,118,119]. Virus replication and fusion mediated by different R5 tropic HIV isolates was shown to be dependent on CCR5 levels in HeLa cells expressing different levels of CCR5 [118]. Moreover, in CD4 T cell lines expressing varying levels of CCR5, bystander apoptosis mediated by the Env glycoprotein was dependent on CCR5 expression and the fusion capacity of the viral Env [65]. Interestingly, in the SCID-hu model of HIV pathogenesis, reconstitution of mice with thymic grafts from CCR5Δ32 heterozygous individuals supported virus replication without CD4 T cell depletion [67]. Our studies show that cells expressing low levels of CCR5 can support long-term HIV replication in the absence of bystander apoptosis [65,66]. Collectively, these data suggest that a different threshold of CCR5 level may be required for bystander apoptosis compared to virus replication.

Recently, Env CCR5 interactions were also found to be important for bystander apoptosis induction via a dual tropic HIV isolate, with the process being reduced by CCR5 inhibitors or mutations in the Env glycoprotein that abrogate CCR5 interaction [29]. Thus, the process of bystander apoptosis in the larger context of HIV pathogenesis seems to rely on two key aspects, the Env fusogencity and host CCR5 expression levels [29,60]. Patients harboring highly fusogenic Envs would likely be efficient in using low levels of CCR5 while patients with less fusogenic Envs would require high CCR5 levels for bystander apoptosis induction. Consistent with this idea, we have found that in the presence of low CCR5 levels in vitro, HIV evolves over time to acquire changes that help utilize low CCR5 levels accompanied by increase in bystander apoptosis [66].

3.4. CCR5 in Primate Models of SIV Infection

A role of CCR5 levels in SIV infections is strengthened by the observations that despite high levels of virus replication, African green monkeys (AGMs) and sooty mangabeys (SMs) do not progress to AIDS in nature. A paucity of CD4CCR5+ T cells in different tissues of five different primate species, where natural SIV infection leads to a non-pathogenic infection, indicates that CCR5 expression plays a significant role in SIV infections [120]. Furthermore, in mandrills (MNDs), lower number of CD4CCR5+ T cells in the mucosal surfaces has been linked to reduced transmission of SIVmnd from mother to infant [121]. In a recent study, Paiardini et al. [122] observed that reduced CCR5 expression in the CD4 central memory T cell compartment in SMs limits SIV infection and progression to AIDS. In the same study, CD4 T cell activation failed to upregulate CCR5 in SMs thereby protecting cells against SIV infection. Recently, an SIV such as phenotype has been reported in non-progressing HIV infected children characterized by low CCR5 expression and reduced immune activation [123].

4. Immune Activation

4.1. Immune Activation in HIV Disease

Chronic immune activation is a distinctive hallmark of pathogenic HIV infections [124–126]. Furthermore, activation of both CD4 and CD8 T cells as determined by expression of activation markers such as Ki67, HLA-DR, CD25 and CD38 [127,128] has been associated with HIV disease progression. Some studies have found immune activation to be a better predictor of disease progression than plasma viremia [13–15,129]. Furthermore, pathogenic HIV and SIV infections can be differentiated from non-pathogenic SIV infections in natural hosts by lack of the immune activation in the latter [4,126,130,131]. Correlation between immune activation and CD4 decline has also been reported in humanized mouse models of HIV infection [132]. The mechanism of immune activation in pathogenic HIV infection involves several factors including virus replication, loss of Th17 cells, disruption of intestinal barrier, leakage of gut microbes in peripheral circulation and stimulation of toll-like receptor (TLR) pathways [130,133,134].

4.2. Viremia and Immune Activation

The mechanism of immune activation in HIV infection remains controversial. There is strong evidence for the involvement of viremia in immune activation as both CD4 and CD8 T cell activation is higher in viremic patients [10,135–137]. A reduction in immune activation after initiation of HAART therapy [138–141] and association with residual viremia in patients unable to suppress virus replication [142] is also supportive of a role of viremia in immune activation. Further evidence can be found in in vitro infection of lymph node histocultures with HIV-1 that results in activation of CD4 and CD8 T cells characterized by upregulation of CD25 and HLA-DR [143]. An involvement of plasmacytoid dendritic cells in detecting virus via TLRs and mediating immune activation via Interferon-1 (IFN-1) secretion has recently been demonstrated [144,145]. Furthermore, a recent study by Cheng et al. showed that IFN-1 is involved in CD4 loss via apoptosis in humanized mouse model of HIV infection [146].

4.3. Microbial Translocation, Th17 Depletion and Immune Activation

The mechanism of immune activation in HIV disease remains debated and microbial translocation has been proposed [147–149]. Brenchley et al. demonstrated that increased levels of lipopolysaccharide (LPS) [150] and bacterial DNA [151] in HIV infected individuals correlates with immune activation. The correlation of microbial translocation with HIV disease has been reported by several independent groups [152–155]. A specific loss of CD4 T cells in the gastrointestinal tract is a hallmark of primary HIV-1/SIV infection [156–159] and likely responsible for the impaired mucosal immunity and microbial translocation [160]. Recent studies suggest that a specific loss of gut associated Th17 subset of CD4 T cells may be behind this phenomenon [161].

Th17 are Interleukin (IL)-17 producing subset of CD4 T helper cells found on mucosal surfaces that maintain intestinal barrier integrity [162]. Both HIV infection and pathogenic SIV infection in rhesus macaques are characterized by a loss of Th17 cells in the gut while in natural SIV infections this subset is preserved [161,163,164] As Th17 cells are involved in microbial clearance in the gut, a depletion of Th17 cells is linked directly to microbial translocation and immune activation in HIV infections [165]. This subset of CD4 T cells are highly susceptible to HIV infection [122] due to the expression of CCR5 at least in the gut-associated lymphoid tissue (GALT) [166]. While the mechanism of Th17 cell depletion has not been studied directly, the susceptibility of these cells to HIV infection suggests that these cells may also be susceptible to HIV mediated apoptosis. In fact, initiation of HAART at early stages in HIV infection can preserve Th17 cell function and reverse HIV associated immune activation [141]. Furthermore, strategies aimed at restoring this population have shown clinical benefits in animal models [167].

4.4. Toll-Like Receptors in Immune Activation

The innate immune system recognizes pathogens via a family of TLRs that modulate adaptive immune response especially to chronic infections such as HIV [168,169]. Recognition of pathogens by TLRs results in production of inflammatory cytokines making this pathway important for immune activation [170]. The correlation between immune activation and viremia suggests that TLR family of innate sensors, specifically TLR7 that senses viral RNA and TLR9 that senses unmethylated CpG viral DNA may be involved in HIV mediated immune activation [145,171]. The sensing of viral nucleic acids by TLR7 and TLR9 in plasmacytoid dendritic cells and subsequent IFN-1 production [144] has been proposed as the mechanism behind this phenomenon. O'Brien et al. have recently demonstrated that interaction of HIV Env with CD4 on plasmacytoid dendritic cells is key to dendritic cell stimulation and IFN production [172]. Furthermore, in vitro stimulation of PBMCs with TLR ligands mediates activation in CD4 and CD8 T cells [173]. Besides recognizing viral CpG DNA, TLR9 is also stimulated by CpG DNA from bacteria. Incidentally, translocation of gut microbes and plasma levels of bacterial DNA are both increased in HIV patients and correlate with immune activation [150,151]. A role of TLRs in HIV induced immune activation is also supported by the reduction in immune activation by TLR7 and TLR9 signaling inhibitors such as chloroquine and hydroxychloroquine [174,175]. How immune activation ties into CD4 apoptosis in HIV infection is an area of active research and recent studies are uncovering the relationship between these two phenomena.

4.5. Immune Activation and Apoptosis

Immune activation per se can induce apoptosis in cells via the process of activation-induced cell death (AICD) and has been proposed as a mechanism of CD4 loss in HIV infections [176]. However, recent studies are less supportive of this pathway for two reasons. Firstly, high levels of immune activation is also seen in CD8 T cells in HIV infected patients but the loss is largely limited to CD4 cells [10]. Secondly, experimental induction of immune activation in vivo in rhesus macaques using LPS [177] and in humanized mice [178] fails to cause a specific loss of CD4 T cells or alter the CD4:CD8 ratio. At the same time, induction of immune activation in natural SIV infection in AGMs can lead to partial CD4 loss in an otherwise non-pathogenic infection [179]. Recent studies by our group have found that activation of PBMCs in vitro enhances the susceptibility of CD4 T cells to HIV-1 Env mediated bystander apoptosis and alters CD4:CD8 ratio similar to that observed in HIV infections [10]. These findings support the hypothesis that while bystander apoptosis is largely mediated by HIV-1 Env, this process can be significantly enhanced if the cells are activated. The mechanism by which immune activation increases susceptibility of CD4 T cells to HIV Env mediated apoptosis remains to be determined. One possibility is that immune activation causes upregulation of coreceptors CXCR4 and CCR5 on cells [180,181] that not only enhances virus replication but could also potentially accelerate Env mediated apoptosis.

5. Other Pathways of Cell Death

5.1. Role of Autophagy in HIV-Mediated Cell Death

Although programmed cell death (apoptosis) is considered to be the key mechanism via which HIV causes CD4 T cells death, other cell death pathways such as autophagy [182,183] have also been proposed as mediators of T cell decline. Autophagy or type II programmed cell death is a catabolic process by which cellular cytoplasmic components and organelles are delivered to the lysosomes for degradation with the objective of establishing homeostasis after stress related stimuli [184]. Autophagy is characterized by the formation of membrane bound compartments that engulf cytoplasmic material, involves the autophagy-related (Atg) group of proteins [185] and has been described in several bacterial and viral infections including HIV [182,186–188]. Studies in HIV induced autophagy have found a role of the Env glycoprotein in this process via interaction with the co-receptor CXCR4 [187]. Interestingly, cells infected with HIV or those expressing the Env glycoprotein on the surface [186,187] induce autophagy in uninfected bystander CD4 T cells characterized by accumulation of Beclin 1 [187]. The process required the presence of CD4 and CXCR4 on the target cells surface, was independent of CD4 and CXCR4 signaling and could be inhibited by drugs that block autophagy such as bafilomycin A1 or siRNAs specific for *Beclin 1* and *Atg7* genes [187]. As CD4/CXCR4 signaling was not required for HIV induced autophagy, later studies identified the role of HIV gp41 in this process as fusion inhibitors (T20 and C34) or gp41 mutations (V2E) [189] inhibited Env mediated autophagy. As the mechanism of autophagy induction by HIV Env glycoprotein is similar to apoptosis, combined with the extensive cross talk between these pathways [190,191], it is plausible that apoptosis and autophagy may both play a role in CD4 T cell loss.

5.2. Role of Pyroptosis in HIV-Mediated Cell Death

Recent studies have suggested a role of the pro-inflammatory cell death pathway called pyroptosis [192] in HIV mediated bystander cell death. Studies by Doitsh et al. demonstrated that cell death in majority of bystander CD4 T cells is due to abortive infection of non-permissive resting CD4 T cells where there is accumulation of incomplete reverse transcription products [193,194]. These incomplete transcripts are detected by the cellular IFl16 DNA sensor to activate a pro inflammatory and pro apoptotic response characterized by activation of caspase-1 [195]. Activation of caspace-1 in quiescent T cells leads to pyroptosis, a form of programmed cell death marked by activation of caspase-1 rather than caspase-3 and release of pro-inflammatory cytokines such as IL-1 beta [196]. It has been speculated that this mechanism does not aid in clearing virus infection but rather creates a vicious cycle of inflammation by attracting new permissive cells to the site of infection. Thus, targeting caspase-1 via inhibitors such as VX-765 was suggested as a safe and viable approach to reduce HIV induced CD4 T cell death [193]. Recent studies from the same group suggest that cell to cell contact between infected and uninfected cells was essential for this form of cell death as cell free virus failed to induce pyroptosis underscoring the importance of the virological synapse in HIV pathogenesis [197]. Although pyroptosis has been suggested as an alternate pathway of cell death in HIV infection the studies are based on ex vivo human lymphoid aggregate culture model. Currently there is limited in vivo data from primate or humanized mouse model to suggest that this pathway is active in pathogenic HIV/SIV infections in vivo. In fact, a recent study by Cheng et al. failed to detect caspase-1 activation in humanized mouse model of HIV infection while apoptosis and caspase-3 activation were readily detected [146].

6. Model of HIV-Mediated Bystander Apoptosis

6.1. Detailed Model of Host and Viral Factors in HIV-Mediated Bystander Apoptosis

Apoptosis mediated by HIV infections is more complex than previously thought. A role of both host and viral factors in this phenomenon is becoming increasingly evident. Based on recent evidence we are proposing a detailed model of HIV mediated bystander apoptosis (Figure 1).

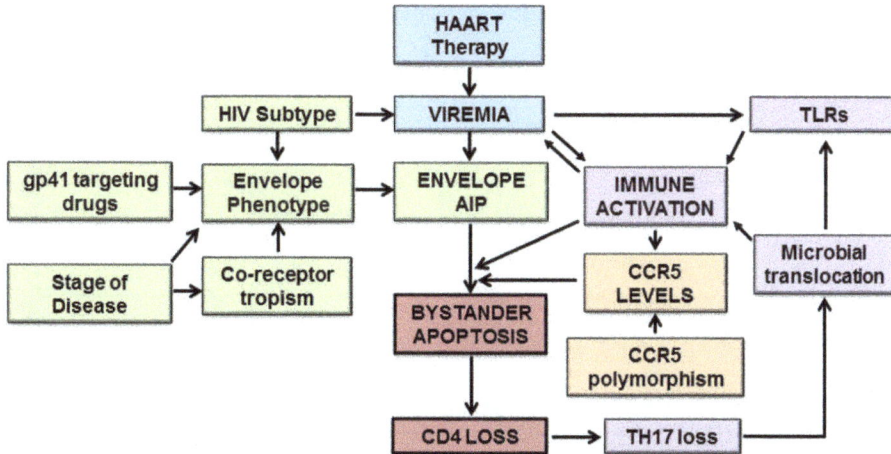

Figure 1. Model of host and viral factors in human immunodeficiency virus (HIV)-mediated bystander apoptosis. HIV mediated bystander apoptosis and CD4 decline can be attributed to both host and viral factors. Fundamental to this process is active virus replication (viremia) as suppressing virus replication via highly active anti-retroviral therapy (HAART) suppresses the major immunopathological variables of the disease including CD4 apoptosis and immune activation. The phenotype of the Envelope (Env) glycoprotein is another major determinant of HIV pathogenesis as the Env apoptosis inducing potential (AIP) correlates with CD4 loss. Other genotypic and phenotypic variations in Env like coreceptor usage, virus subtype, fusogenic activity of Env, etc. have been associated with disease. The binding of Env to CCR5 is also fundamental to HIV induced bystander apoptosis and variations in CCR5 levels can influence this phenomenon. Immune activation is a key immunopathological feature of HIV infections that correlates with CD4 decline and can be affected via multiple pathways as shown by arrows. TLR: toll-like receptor).

Fundamental to this process is active virus replication which forms the primary focus of the model. The suppression of major immunopathological variables including apoptosis with successful HAART suggests that viremia plays a key role in this process. Besides active virus replication, the phenotype of the Env glycoprotein is the other major determinant of HIV pathogenesis. Recent evidence from our lab suggests that variability in HIV Env glycoprotein in terms of apoptosis inducing potential (AIP) correlates with CD4 loss [10]. Multiple genotypic and phenotypic variations in Env have been associated with disease including coreceptor usage, virus subtype, fusogenic activity of Env, N-glycosylation sites and coreceptor binding [60,66]. The binding of Env to CCR5 is also fundamental to HIV pathogenesis and bystander apoptosis and variations in CCR5 levels as a result of CCR5 gene and promoter polymorphisms can influence this phenomenon [65,109]. Immune activation is a key immunopathological feature of HIV infections that correlates with CD4 decline. Recent evidence from our lab shows that in vitro activated PBMCs are more susceptible to HIV Env mediated bystander apoptosis [10]. These new findings help tie the role of immune activation with Env mediated apoptosis completing the complex network of major factors in HIV associated apoptosis. We also include in our model factors that influence immune activation, including a role of virus mediated TLR activation [145] as well as stimulation of TLRs via microbial translocation [150]. The translocation of microbes from the gut may be further facilitated by a specific loss of Th17 subset of CD4 T cells in the gut [161] possibly as a consequence of HIV-mediated apoptosis (Figure 1).

6.2. Host and Viral Factors Determining Differential CD4 Loss in HIV Infections

Based on our model above, we can speculate several factors that may influence differential CD4 apoptosis and rate of progression in HIV patients. A list of these factors is given in Table 1.

Table 1. Host and viral factors determining differential CD4 loss in HIV infections.

Factors Limiting Bystander Apoptosis	Factors Enhancing Bystander Apoptosis
➢ Poor virus replication	➢ High virus replication
➢ Low AIP phenotype	➢ High AIP phenotype
➢ Low CCR5 levels	➢ High CCR5 levels
➢ Low immune activation	➢ High immune activation
➢ Low Env CCR5 binding affinity	➢ High Env CCR5 binding affinity
➢ Less fusogenic Env	➢ Highly fusogenic Env

One could envision that either individually or a combination of these factors would determine different scenarios that could constitute pathogenic or non-pathogenic infections. For example, the non-pathogenic SIV infections in natural host are likely a consequence of low immune activation combined with low CCR5 expression [4,122]. In support of this scenario, experimental induction of immune activation in AGMs infected with SIV increases virus replication and CD4 T cells depletion [179]. Similarly, long term non-progressors may have low levels of CCR5 or low viremia or a combination of both [198]. A recent study in non-progressing HIV-infected children found that low levels of CCR5 combined with low immune activation is likely behind the lack of disease progression in this unique cohort [123]. In viremic non-progressors, limited immune activation [199] and limited infection of central memory cells combined with impaired viral fitness has been linked to non-progression [200]. On the other spectrum, high viremia combined with highly pathogenic phenotype of Env results is progression to AIDS as seen in several studies [10,56,201–203].

7. Conclusions

Recent findings have provided researchers with a better picture of the host and viral factors involved in HIV mediated bystander apoptosis. It is clear that this phenomenon is a complex interplay between the virus and the host. The contribution of each of these factors varies between individuals as well as viral variants. Strategies that limit immune activation, alter CCR5 levels or target Env phenotype like gp41 inhibitors remain attractive approaches to circumvent HIV pathogenesis and warrant further investigation. However, from currently available data, it is apparent that the onus of HIV mediated bystander apoptosis is largely on virus replication and phenotype of the virus. This suggests that suppressing virus replication is perhaps the best strategy to limit HIV-mediated bystander apoptosis for now.

Acknowledgments: This work was supported in part by the National Institutes of Health Grant AI116240-01A1 (to H.G.).

Conflicts of Interest: The authors declare no conflict of interest.

References

1. Pantaleo, G.; Fauci, A.S. Apoptosis in HIV infection. *Nat. Med.* **1995**, *1*, 118–120. [CrossRef] [PubMed]
2. Oyaizu, N.; Pahwa, S. Role of apoptosis in HIV disease pathogenesis. *J. Clin. Immunol.* **1995**, *15*, 217–231. [CrossRef] [PubMed]

3. Gougeon, M.; Montagnier, L. Apoptosis in AIDS. *Science* **1993**, *260*, 1269–1270. [CrossRef] [PubMed]
4. Silvestri, G.; Sodora, D.; Koup, R.; Paiardini, M.; O'Neil, S.; McClure, H.; Staprans, S.; Feinberg, M. Nonpathogenic SIV infection of sooty mangabeys is characterized by limited bystander immunopathology despite chronic high-level viremia. *Immunity* **2003**, *18*, 441–452. [CrossRef]
5. Meythaler, M.; Martinot, A.; Wang, Z.; Pryputniewicz, S.; Kasheta, M.; Ling, B.; Marx, P.A.; O'Neil, S.; Kaur, A. Differential CD4+ T-lymphocyte apoptosis and bystander T-cell activation in rhesus macaques and sooty mangabeys during acute simian immunodeficiency virus infection. *J. Virol.* **2009**, *83*, 572–583. [CrossRef] [PubMed]
6. Meythaler, M.; Pryputniewicz, S.; Kaur, A. Kinetics of T lymphocyte apoptosis and the cellular immune response in SIVmac239-infected rhesus macaques. *J. Med. Primatol.* **2008**, *37*, 33–45. [CrossRef] [PubMed]
7. Cumont, M.C.; Diop, O.; Vaslin, B.; Elbim, C.; Viollet, L.; Monceaux, V.; Lay, S.; Silvestri, G.; Le Grand, R.; Muller-Trutwin, M.; et al. Early divergence in lymphoid tissue apoptosis between pathogenic and nonpathogenic simian immunodeficiency virus infections of nonhuman primates. *J. Virol.* **2008**, *82*, 1175–1184. [CrossRef] [PubMed]
8. Reinberger, S.; Spring, M.; Nisslein, T.; Stahl-Hennig, C.; Hunsmann, G.; Dittmer, U. Kinetics of lymphocyte apoptosis in macaques infected with different simian immunodeficiency viruses or simian/human immunodeficiency hybrid viruses. *Clin. Immunol.* **1999**, *90*, 141–146. [CrossRef] [PubMed]
9. Iida, T.; Ichimura, H.; Shimada, T.; Ibuki, K.; Ui, M.; Tamaru, K.; Kuwata, T.; Yonehara, S.; Imanishi, J.; Hayami, M. Role of apoptosis induction in both peripheral lymph nodes and thymus in progressive loss of CD4+ cells in SHIV-infected macaques. *AIDS Res. Hum. Retrovir.* **2000**, *16*, 9–18. [CrossRef] [PubMed]
10. Joshi, A.; Sedano, M.; Beauchamp, B.; Punke, E.B.; Mulla, Z.D.; Meza, A.; Alozie, O.K.; Mukherjee, D.; Garg, H. HIV-1 Env Glycoprotein phenotype along with immune activation determines CD4 T cell loss in HIV patients. *J. Immunol.* **2016**, *196*, 1768–1779. [CrossRef] [PubMed]
11. Sternfeld, T.; Tischleder, A.; Schuster, M.; Bogner, J.R. Mitochondrial membrane potential and apoptosis of blood mononuclear cells in untreated HIV-1 infected patients. *HIV Med.* **2009**, *10*, 512–519. [CrossRef] [PubMed]
12. Oyaizu, N.; McCloskey, T.W.; Coronesi, M.; Chirmule, N.; Kalyanaraman, V.S.; Pahwa, S. Accelerated apoptosis in peripheral blood mononuclear cells (PBMCs) from human immunodeficiency virus type-1 infected patients and in CD4 cross-linked PBMCs from normal individuals. *Blood* **1993**, *82*, 3392–3400. [PubMed]
13. Leng, Q.; Borkow, G.; Weisman, Z.; Stein, M.; Kalinkovich, A.; Bentwich, Z. Immune activation correlates better than HIV plasma viral load with CD4 T-cell decline during HIV infection. *J. Acquir. Immune Defic. Syndr.* **2001**, *27*, 389–397. [CrossRef] [PubMed]
14. Sousa, A.; Carneiro, J.; Meier-Schellersheim, M.; Grossman, Z.; Victorino, R. CD4 T cell depletion is linked directly to immune activation in the pathogenesis of HIV-1 and HIV-2 but only indirectly to the viral load. *J. Immunol.* **2002**, *169*, 3400–3406. [CrossRef] [PubMed]
15. Resino, S.; Seoane, E.; Gutiérrez, M.; León, J.; Muñoz-Fernández, M. CD4(+) T-cell immunodeficiency is more dependent on immune activation than viral load in HIV-infected children on highly active antiretroviral therapy. *J. Acquir. Immune Defic. Syndr.* **2006**, *42*, 269–276. [CrossRef] [PubMed]
16. Chavan, S.J.; Tamma, S.L.; Kaplan, M.; Gersten, M.; Pahwa, S.G. Reduction in T cell apoptosis in patients with HIV disease following antiretroviral therapy. *Clin. Immunol.* **1999**, *93*, 24–33. [CrossRef] [PubMed]
17. Roger, P.M.; Breittmayer, J.P.; Arlotto, C.; Pugliese, P.; Pradier, C.; Bernard-Pomier, G.; Dellamonica, P.; Bernard, A. Highly active anti-retroviral therapy (HAART) is associated with a lower level of CD4+ T cell apoptosis in HIV-infected patients. *Clin. Exp. Immunol.* **1999**, *118*, 412–416. [CrossRef] [PubMed]
18. Roger, P.M.; Breittmayer, J.P.; Durant, J.; Sanderson, F.; Ceppi, C.; Brignone, C.; Cua, E.; Clevenbergh, P.; Fuzibet, J.G.; Pesce, A.; et al. Early CD4(+) T cell recovery in human immunodeficiency virus-infected patients receiving effective therapy is related to a down-regulation of apoptosis and not to proliferation. *J. Infect. Dis.* **2002**, *185*, 463–470. [CrossRef] [PubMed]
19. Mocroft, A.; Ledergerber, B.; Katlama, C.; Kirk, O.; Reiss, P.; d'Arminio Monforte, A.; Knysz, B.; Dietrich, M.; Phillips, A.N.; Lundgren, J.D. Decline in the AIDS and death rates in the EuroSIDA study: An observational study. *Lancet* **2003**, *362*, 22–29. [CrossRef]

20. Palella, F.J., Jr.; Delaney, K.M.; Moorman, A.C.; Loveless, M.O.; Fuhrer, J.; Satten, G.A.; Aschman, D.J.; Holmberg, S.D. Declining morbidity and mortality among patients with advanced human immunodeficiency virus infection. HIV Outpatient Study Investigators. *N. Engl. J. Med.* **1998**, *338*, 853–860. [CrossRef] [PubMed]

21. Garg, H.; Blumenthal, R. Role of HIV Gp41 mediated fusion/hemifusion in bystander apoptosis. *Cell. Mol. Life Sci.* **2008**, *65*, 3134–3144. [CrossRef] [PubMed]

22. Finkel, T.; Banda, N. Indirect mechanisms of HIV pathogenesis: How does HIV kill T cells? *Curr. Opin. Immunol.* **1994**, *6*, 605–615. [CrossRef]

23. Finkel, T.; Tudor-Williams, G.; Banda, N.; Cotton, M.; Curiel, T.; Monks, C.; Baba, T.; Ruprecht, R.; Kupfer, A. Apoptosis occurs predominantly in bystander cells and not in productively infected cells of HIV- and SIV-infected lymph nodes. *Nat. Med.* **1995**, *1*, 129–134. [CrossRef] [PubMed]

24. Gordon, S.; Dunham, R.; Engram, J.; Estes, J.; Wang, Z.; Klatt, N.; Paiardini, M.; Pandrea, I.; Apetrei, C.; Sodora, D.; et al. Short-lived infected cells support virus replication in sooty mangabeys naturally infected with simian immunodeficiency virus: Implications for AIDS pathogenesis. *J. Virol.* **2008**, *82*, 3725–3735. [CrossRef] [PubMed]

25. Gougeon, M.; Colizzi, V.; Dalgleish, A.; Montagnier, L. New concepts in AIDS pathogenesis. *AIDS Res. Hum. Retrovir.* **1993**, *9*, 287–289. [CrossRef] [PubMed]

26. Garg, H.; Joshi, A.; Ye, C.; Shankar, P.; Manjunath, N. Single amino acid change in gp41 region of HIV-1 alters bystander apoptosis and CD4 decline in humanized mice. *Virol. J.* **2011**, *8*, 34. [CrossRef] [PubMed]

27. Holm, G.H.; Gabuzda, D. Distinct mechanisms of CD4+ and CD8+ T-cell activation and bystander apoptosis induced by human immunodeficiency virus type 1 virions. *J. Virol.* **2005**, *79*, 6299–6311. [CrossRef] [PubMed]

28. Holm, G.H.; Zhang, C.; Gorry, P.R.; Peden, K.; Schols, D.; De Clercq, E.; Gabuzda, D. Apoptosis of bystander T cells induced by human immunodeficiency virus type 1 with increased envelope/receptor affinity and coreceptor binding site exposure. *J. Virol.* **2004**, *78*, 4541–4551. [CrossRef] [PubMed]

29. Tsao, L.C.; Guo, H.; Jeffrey, J.; Hoxie, J.A.; Su, L. CCR5 interaction with HIV-1 Env contributes to Env-induced depletion of CD4 T cells In Vitro and In Vivo. *Retrovirology* **2016**, *13*, 22. [CrossRef] [PubMed]

30. Ahr, B.; Robert-Hebmann, V.; Devaux, C.; Biard-Piechaczyk, M. Apoptosis of uninfected cells induced by HIV envelope glycoproteins. *Retrovirology* **2004**, *1*, 12. [CrossRef] [PubMed]

31. Perfettini, J.; Castedo, M.; Roumier, T.; Andreau, K.; Nardacci, R.; Piacentini, M.; Kroemer, G. Mechanisms of apoptosis induction by the HIV-1 envelope. *Cell Death Differ.* **2005**, *12*, 916–923. [CrossRef] [PubMed]

32. Laurent-Crawford, A.G.; Coccia, E.; Krust, B.; Hovanessian, A.G. Membrane-expressed HIV envelope glycoprotein heterodimer is a powerful inducer of cell death in uninfected CD4+ target cells. *Res. Virol.* **1995**, *146*, 5–17. [CrossRef]

33. Laurent-Crawford, A.G.; Krust, B.; Riviere, Y.; Desgranges, C.; Muller, S.; Kieny, M.P.; Dauguet, C.; Hovanessian, A.G. Membrane expression of HIV envelope glycoproteins triggers apoptosis in CD4 cells. *AIDS Res. Hum. Retrovir.* **1993**, *9*, 761–773. [CrossRef] [PubMed]

34. Biard-Piechaczyk, M.; Robert-Hebmann, V.; Richard, V.; Roland, J.; Hipskind, R.; Devaux, C. Caspase-dependent apoptosis of cells expressing the chemokine receptor CXCR4 is induced by cell membrane-associated human immunodeficiency virus type 1 envelope glycoprotein (gp120). *Virology* **2000**, *268*, 329–344. [CrossRef] [PubMed]

35. Blanco, J.; Barretina, J.; Ferri, K.; Jacotot, E.; Gutiérrez, A.; Armand-Ugón, M.; Cabrera, C.; Kroemer, G.; Clotet, B.; Esté, J. Cell-surface-expressed HIV-1 envelope induces the death of CD4 T cells during GP41-mediated hemifusion-like events. *Virology* **2003**, *305*, 318–329. [CrossRef] [PubMed]

36. Garg, H.; Blumenthal, R. HIV gp41-induced apoptosis is mediated by caspase-3-dependent mitochondrial depolarization, which is inhibited by HIV protease inhibitor nelfinavir. *J. Leukoc. Biol.* **2006**, *79*, 351–362. [CrossRef] [PubMed]

37. Wyatt, R.; Sodroski, J. The HIV-1 envelope glycoproteins: Fusogens, antigens, and immunogens. *Science* **1998**, *280*, 1884–1888. [CrossRef] [PubMed]

38. Gallo, S.; Finnegan, C.; Viard, M.; Raviv, Y.; Dimitrov, A.; Rawat, S.; Puri, A.; Durell, S.; Blumenthal, R. The HIV Env-mediated fusion reaction. *Biochim. Biophys. Acta* **2003**, *1614*, 36–50. [CrossRef]

39. Wild, C.; Dubay, J.; Greenwell, T.; Baird, T.J.; Oas, T.; McDanal, C.; Hunter, E.; Matthews, T. Propensity for a leucine zipper-like domain of human immunodeficiency virus type 1 gp41 to form oligomers correlates with a role in virus-induced fusion rather than assembly of the glycoprotein complex. *Proc. Natl. Acad. Sci. USA* **1994**, *91*, 12676–12680. [CrossRef] [PubMed]

40. Andreau, K.; Perfettini, J.; Castedo, M.; Métivier, D.; Scott, V.; Pierron, G.; Kroemer, G. Contagious apoptosis facilitated by the HIV-1 envelope: Fusion-induced cell-to-cell transmission of a lethal signal. *J. Cell Sci.* **2004**, *117*, 5643–5653. [CrossRef] [PubMed]

41. Garg, H.; Mohl, J.; Joshi, A. HIV-1 induced bystander apoptosis. *Viruses* **2012**, *4*, 3020–3043. [CrossRef] [PubMed]

42. Biard-Piechaczyk, M.; Robert-Hebmann, V.; Roland, J.; Coudronniere, N.; Devaux, C. Role of CXCR4 in HIV-1-induced apoptosis of cells with a CD4+, CXCR4+ phenotype. *Immunol. Lett.* **1999**, *70*, 1–3. [CrossRef]

43. Blanco, J.; Barretina, J.; Clotet, B.; Este, J.A. R5 HIV gp120-mediated cellular contacts induce the death of single CCR5-expressing CD4 T cells by a gp41-dependent mechanism. *J. Leukoc. Biol.* **2004**, *76*, 804–811. [CrossRef] [PubMed]

44. Garg, H.; Joshi, A.; Freed, E.; Blumenthal, R. Site-specific mutations in HIV-1 gp41 reveal a correlation between HIV-1-mediated bystander apoptosis and fusion/hemifusion. *J. Biol. Chem.* **2007**, *282*, 16899–16906. [CrossRef] [PubMed]

45. Chernomordik, L.; Kozlov, M. Membrane hemifusion: Crossing a chasm in two leaps. *Cell* **2005**, *123*, 375–382. [CrossRef] [PubMed]

46. Symeonides, M.; Lambele, M.; Roy, N.H.; Thali, M. Evidence showing that tetraspanins inhibit HIV-1-induced cell-cell fusion at a post-hemifusion stage. *Viruses* **2014**, *6*, 1078–1090. [CrossRef] [PubMed]

47. Bar, S.; Alizon, M. Role of the ectodomain of the gp41 transmembrane envelope protein of human immunodeficiency virus type 1 in late steps of the membrane fusion process. *J. Virol.* **2004**, *78*, 811–820. [CrossRef] [PubMed]

48. Cunyat, F.; Marfil, S.; Garcia, E.; Svicher, V.; Perez-Alvarez, N.; Curriu, M.; Perno, C.F.; Clotet, B.; Blanco, J.; Cabrera, C. The HR2 polymorphism N140I in the HIV-1 gp41 combined with the HR1 V38A mutation is associated with a less cytopathic phenotype. *Retrovirology* **2012**, *9*, 15. [CrossRef] [PubMed]

49. Koot, M.; van't Wout, A.; Kootstra, N.; de Goede, R.; Tersmette, M.; Schuitemaker, H. Relation between changes in cellular load, evolution of viral phenotype, and the clonal composition of virus populations in the course of human immunodeficiency virus type 1 infection. *J. Infect. Dis.* **1996**, *173*, 349–354. [CrossRef] [PubMed]

50. Spijkerman, I.; de Wolf, F.; Langendam, M.; Schuitemaker, H.; Coutinho, R. Emergence of syncytium-inducing human immunodeficiency virus type 1 variants coincides with a transient increase in viral RNA level and is an independent predictor for progression to AIDS. *J. Infect. Dis.* **1998**, *178*, 397–403. [CrossRef] [PubMed]

51. Schuitemaker, H.; Koot, M.; Kootstra, N.; Dercksen, M.; de Goede, R.; van Steenwijk, R.; Lange, J.; Schattenkerk, J.; Miedema, F.; Tersmette, M. Biological phenotype of human immunodeficiency virus type 1 clones at different stages of infection: Progression of disease is associated with a shift from monocytotropic to T-cell-tropic virus population. *J. Virol.* **1992**, *66*, 1354–1360. [PubMed]

52. Van Rij, R.P.; Blaak, H.; Visser, J.; Brouwer, M.; Rientsma, R.; Broersen, S.; de Roda Husman, A.M.; Schuitemaker, H. Differential coreceptor expression allows for independent evolution of non-syncytium-inducing and syncytium-inducing HIV-1. *J. Clin. Investig.* **2000**, *106*, 1039–1052. [CrossRef] [PubMed]

53. Kaneshima, H.; Su, L.; Bonyhadi, M.; Connor, R.; Ho, D.; McCune, J. Rapid-high, syncytium-inducing isolates of human immunodeficiency virus type 1 induce cytopathicity in the human thymus of the SCID-hu mouse. *J. Virol.* **1994**, *68*, 8188–8192. [PubMed]

54. Etemad-Moghadam, B.; Rhone, D.; Steenbeke, T.; Sun, Y.; Manola, J.; Gelman, R.; Fanton, J.W.; Racz, P.; Tenner-Racz, K.; Axthelm, M.K.; et al. Understanding the basis of CD4(+) T-cell depletion in macaques infected by a simian-human immunodeficiency virus. *Vaccine* **2002**, *20*, 1934–1937. [CrossRef]

55. Etemad-Moghadam, B.; Rhone, D.; Steenbeke, T.; Sun, Y.; Manola, J.; Gelman, R.; Fanton, J.W.; Racz, P.; Tenner-Racz, K.; Axthelm, M.K.; et al. Membrane-fusing capacity of the human immunodeficiency virus envelope proteins determines the efficiency of CD+ T-cell depletion in macaques infected by a simian-human immunodeficiency virus. *J. Virol.* **2001**, *75*, 5646–5655. [CrossRef] [PubMed]

56. Etemad-Moghadam, B.; Sun, Y.; Nicholson, E.K.; Fernandes, M.; Liou, K.; Gomila, R.; Lee, J.; Sodroski, J. Envelope glycoprotein determinants of increased fusogenicity in a pathogenic simian-human immunodeficiency virus (SHIV-KB9) passaged in vivo. *J. Virol.* **2000**, *74*, 4433–4440. [CrossRef] [PubMed]

57. Meissner, E.G.; Coffield, V.M.; Su, L. Thymic pathogenicity of an HIV-1 envelope is associated with increased CXCR4 binding efficiency and V5-gp41-dependent activity, but not V1/V2-associated CD4 binding efficiency and viral entry. *Virology* **2005**, *336*, 184–197. [CrossRef] [PubMed]

58. Sterjovski, J.; Churchill, M.; Ellett, A.; Gray, L.; Roche, M.; Dunfee, R.; Purcell, D.; Saksena, N.; Wang, B.; Sonza, S.; et al. Asn 362 in gp120 contributes to enhanced fusogenicity by CCR5-restricted HIV-1 envelope glycoprotein variants from patients with AIDS. *Retrovirology* **2007**, *4*, 89. [CrossRef] [PubMed]

59. Meissner, E.G.; Zhang, L.; Jiang, S.; Su, L. Fusion-induced apoptosis contributes to thymocyte depletion by a pathogenic human immunodeficiency virus type 1 envelope in the human thymus. *J. Virol.* **2006**, *80*, 11019–11030. [CrossRef] [PubMed]

60. Joshi, A.; Lee, R.T.; Mohl, J.; Sedano, M.; Khong, W.X.; Ng, O.T.; Maurer-Stroh, S.; Garg, H. Genetic signatures of HIV-1 envelope-mediated bystander apoptosis. *J. Biol. Chem.* **2014**, *289*, 2497–2514. [CrossRef] [PubMed]

61. Korber, B.; Gaschen, B.; Yusim, K.; Thakallapally, R.; Kesmir, C.; Detours, V. Evolutionary and immunological implications of contemporary HIV-1 variation. *Br. Med. Bull.* **2001**, *58*, 19–42. [CrossRef] [PubMed]

62. Curlin, M.E.; Zioni, R.; Hawes, S.E.; Liu, Y.; Deng, W.; Gottlieb, G.S.; Zhu, T.; Mullins, J.I. HIV-1 envelope subregion length variation during disease progression. *PLoS Pathog.* **2010**, *6*, e1001228. [CrossRef] [PubMed]

63. Camerini, D.; Su, H.P.; Gamez-Torre, G.; Johnson, M.L.; Zack, J.A.; Chen, I.S. Human immunodeficiency virus type 1 pathogenesis in SCID-hu mice correlates with syncytium-inducing phenotype and viral replication. *J. Virol.* **2000**, *74*, 3196–3204. [CrossRef] [PubMed]

64. Waters, L.; Mandalia, S.; Randell, P.; Wildfire, A.; Gazzard, B.; Moyle, G. The impact of HIV tropism on decreases in CD4 cell count, clinical progression, and subsequent response to a first antiretroviral therapy regimen. *Clin. Infect. Dis.* **2008**, *46*, 1617–1623. [CrossRef] [PubMed]

65. Joshi, A.; Nyakeriga, A.M.; Ravi, R.; Garg, H. HIV ENV glycoprotein-mediated bystander apoptosis depends on expression of the CCR5 co-receptor at the cell surface and ENV fusogenic activity. *J. Biol. Chem.* **2011**, *286*, 36404–36413. [CrossRef] [PubMed]

66. Garg, H.; Lee, R.T.; Maurer-Stroh, S.; Joshi, A. HIV-1 adaptation to low levels of CCR5 results in V3 and V2 loop changes that increase envelope pathogenicity, CCR5 affinity and decrease susceptibility to Maraviroc. *Virology* **2016**, *493*, 86–99. [CrossRef] [PubMed]

67. Scoggins, R.M.; Taylor, J.R., Jr.; Patrie, J.; van't Wout, A.B.; Schuitemaker, H.; Camerini, D. Pathogenesis of primary R5 human immunodeficiency virus type 1 clones in SCID-hu mice. *J. Virol.* **2000**, *74*, 3205–3216. [CrossRef] [PubMed]

68. Choudhary, S.K.; Choudhary, N.R.; Kimbrell, K.C.; Colasanti, J.; Ziogas, A.; Kwa, D.; Schuitemaker, H.; Camerini, D. R5 human immunodeficiency virus type 1 infection of fetal thymic organ culture induces cytokine and CCR5 expression. *J. Virol.* **2005**, *79*, 458–471. [CrossRef] [PubMed]

69. Olivieri, K.; Scoggins, R.; Bor, Y.; Matthews, A.; Mark, D.; Taylor, J.J.; Chernauskas, D.; Hammarskjöld, M.; Rekosh, D.; Camerini, D. The envelope gene is a cytopathic determinant of CCR5 tropic HIV-1. *Virology* **2007**, *358*, 23–38. [CrossRef] [PubMed]

70. Roggero, R.; Robert-Hebmann, V.; Harrington, S.; Roland, J.; Vergne, L.; Jaleco, S.; Devaux, C.; Biard-Piechaczyk, M. Binding of human immunodeficiency virus type 1 gp120 to CXCR4 induces mitochondrial transmembrane depolarization and cytochrome c-mediated apoptosis independently of Fas signaling. *J. Virol.* **2001**, *75*, 7637–7650. [CrossRef] [PubMed]

71. Ferri, K.F.; Jacotot, E.; Blanco, J.; Este, J.A.; Kroemer, G. Mitochondrial control of cell death induced by HIV-1-encoded proteins. *Ann. N. Y. Acad. Sci.* **2000**, *926*, 149–164. [CrossRef] [PubMed]

72. Ohnimus, H.; Heinkelein, M.; Jassoy, C. Apoptotic cell death upon contact of CD4+ T lymphocytes with HIV glycoprotein-expressing cells is mediated by caspases but bypasses CD95 (Fas/Apo-1) and TNF receptor 1. *J. Immunol.* **1997**, *159*, 5246–5252. [PubMed]

73. Gandhi, R.; Chen, B.; Straus, S.; Dale, J.; Lenardo, M.; Baltimore, D. HIV-1 directly kills CD4+ T cells by a Fas-independent mechanism. *J. Exp. Med.* **1998**, *187*, 1113–1122. [CrossRef] [PubMed]

74. Wang, X.M.; Nadeau, P.E.; Lo, Y.T.; Mergia, A. Caveolin-1 modulates HIV-1 envelope-induced bystander apoptosis through gp41. *J. Virol.* **2010**, *84*, 6515–6526. [CrossRef] [PubMed]

75. Li, N.; Ragheb, K.; Lawler, G.; Sturgis, J.; Rajwa, B.; Melendez, J.A.; Robinson, J.P. Mitochondrial complex I inhibitor rotenone induces apoptosis through enhancing mitochondrial reactive oxygen species production. *J. Biol. Chem.* **2003**, *278*, 8516–8525. [CrossRef] [PubMed]

76. Wainberg, Z.; Oliveira, M.; Lerner, S.; Tao, Y.; Brenner, B.G. Modulation of stress protein (hsp27 and hsp70) expression in CD4+ lymphocytic cells following acute infection with human immunodeficiency virus type-1. *Virology* **1997**, *233*, 364–373. [CrossRef] [PubMed]

77. Molina, L.; Grimaldi, M.; Robert-Hebmann, V.; Espert, L.; Varbanov, M.; Devaux, C.; Granier, C.; Biard-Piechaczyk, M. Proteomic analysis of the cellular responses induced in uninfected immune cells by cell-expressed X4 HIV-1 envelope. *Proteomics* **2007**, *7*, 3116–3130. [CrossRef] [PubMed]

78. Garg, H.; Viard, M.; Jacobs, A.; Blumenthal, R. Targeting HIV-1 gp41-induced fusion and pathogenesis for anti-viral therapy. *Curr. Top. Med. Chem.* **2011**, *11*, 2947–2958. [CrossRef] [PubMed]

79. Barretina, J.; Blanco, J.; Bonjoch, A.; Llano, A.; Clotet, B.; Esté, J. Immunological and virological study of enfuvirtide-treated HIV-positive patients. *AIDS* **2004**, *18*, 1673–1682. [CrossRef] [PubMed]

80. Reeves, J.; Lee, F.; Miamidian, J.; Jabara, C.; Juntilla, M.; Doms, R. Enfuvirtide resistance mutations: Impact on human immunodeficiency virus envelope function, entry inhibitor sensitivity, and virus neutralization. *J. Virol.* **2005**, *79*, 4991–4999. [CrossRef] [PubMed]

81. Aquaro, S.; D'Arrigo, R.; Svicher, V.; Perri, G.; Caputo, S.; Visco-Comandini, U.; Santoro, M.; Bertoli, A.; Mazzotta, F.; Bonora, S.; et al. Specific mutations in HIV-1 gp41 are associated with immunological success in HIV-1-infected patients receiving enfuvirtide treatment. *J. Antimicrob. Chemother.* **2006**, *58*, 714–722. [CrossRef] [PubMed]

82. Melby, T.; Despirito, M.; Demasi, R.; Heilek, G.; Thommes, J.; Greenberg, M.; Graham, N. Association between specific enfuvirtide resistance mutations and CD4 cell response during enfuvirtide-based therapy. *AIDS* **2007**, *21*, 2537–2539. [CrossRef] [PubMed]

83. Garg, H.; Joshi, A.; Blumenthal, R. Altered bystander apoptosis induction and pathogenesis of enfuvirtide-resistant HIV type 1 Env mutants. *AIDS Res. Hum. Retrovir.* **2009**, *25*, 811–817. [CrossRef] [PubMed]

84. Mitra, D.; Steiner, M.; Lynch, D.H.; Staiano-Coico, L.; Laurence, J. HIV-1 upregulates Fas ligand expression in CD4+ T cells In Vitro and In Vivo: Association with Fas-mediated apoptosis and modulation by aurintricarboxylic acid. *Immunology* **1996**, *87*, 581–585. [CrossRef] [PubMed]

85. Badley, A.D.; Dockrell, D.; Simpson, M.; Schut, R.; Lynch, D.H.; Leibson, P.; Paya, C.V. Macrophage-dependent apoptosis of CD4+ T lymphocytes from HIV-infected individuals is mediated by FasL and tumor necrosis factor. *J. Exp. Med.* **1997**, *185*, 55–64. [CrossRef] [PubMed]

86. Yang, Y.; Tikhonov, I.; Ruckwardt, T.J.; Djavani, M.; Zapata, J.C.; Pauza, C.D.; Salvato, M.S. Monocytes treated with human immunodeficiency virus Tat kill uninfected CD4(+) cells by a tumor necrosis factor-related apoptosis-induced ligand-mediated mechanism. *J. Virol.* **2003**, *77*, 6700–6708. [CrossRef] [PubMed]

87. Herbein, G.; Khan, K.A. Is HIV infection a TNF receptor signalling-driven disease? *Trends Immunol.* **2008**, *29*, 61–67. [CrossRef] [PubMed]

88. Banda, N.K.; Bernier, J.; Kurahara, D.K.; Kurrle, R.; Haigwood, N.; Sekaly, R.P.; Finkel, T.H. Crosslinking CD4 by human immunodeficiency virus gp120 primes T cells for activation-induced apoptosis. *J. Exp. Med.* **1992**, *176*, 1099–1106. [CrossRef] [PubMed]

89. Zauli, G.; Gibellini, D.; Secchiero, P.; Dutartre, H.; Olive, D.; Capitani, S.; Collette, Y. Human immunodeficiency virus type 1 Nef protein sensitizes CD4(+) T lymphoid cells to apoptosis via functional upregulation of the CD95/CD95 ligand pathway. *Blood* **1999**, *93*, 1000–1010. [PubMed]

90. Westendorp, M.O.; Frank, R.; Ochsenbauer, C.; Stricker, K.; Dhein, J.; Walczak, H.; Debatin, K.M.; Krammer, P.H. Sensitization of T cells to CD95-mediated apoptosis by HIV-1 Tat and gp120. *Nature* **1995**, *375*, 497–500. [CrossRef] [PubMed]

91. Jeremias, I.; Herr, I.; Boehler, T.; Debatin, K.M. TRAIL/Apo-2-ligand-induced apoptosis in human T cells. *Eur. J. Immunol.* **1998**, *28*, 143–152. [CrossRef]

92. Zhang, M.; Li, X.; Pang, X.; Ding, L.; Wood, O.; Clouse, K.; Hewlett, I.; Dayton, A.I. Identification of a potential HIV-induced source of bystander-mediated apoptosis in T cells: Upregulation of trail in primary human macrophages by HIV-1 tat. *J. Biomed. Sci.* **2001**, *8*, 290–296. [CrossRef] [PubMed]

93. Than, S.; Oyaizu, N.; Pahwa, R.N.; Kalyanaraman, V.S.; Pahwa, S. Effect of human immunodeficiency virus type-1 envelope glycoprotein gp160 on cytokine production from cord-blood T cells. *Blood* **1994**, *84*, 184–188. [PubMed]

94. Chirmule, N.; Pahwa, S. Envelope glycoproteins of human immunodeficiency virus type 1: Profound influences on immune functions. *Microbiol. Rev.* **1996**, *60*, 386–406. [PubMed]

95. Su, L.; Kaneshima, H.; Bonyhadi, M.; Salimi, S.; Kraft, D.; Rabin, L.; McCune, J.M. HIV-1-induced thymocyte depletion is associated with indirect cytopathogenicity and infection of progenitor cells in vivo. *Immunity* **1995**, *2*, 25–36. [CrossRef]

96. Robey, E.; Axel, R. CD4: Collaborator in immune recognition and HIV infection. *Cell* **1990**, *60*, 697–700. [CrossRef]

97. Chirmule, N.; Oyaizu, N.; Kalyanaraman, V.S.; Pahwa, S. Inhibition of normal B-cell function by human immunodeficiency virus envelope glycoprotein, gp120. *Blood* **1992**, *79*, 1245–1254. [PubMed]

98. Chirmule, N.; McCloskey, T.W.; Hu, R.; Kalyanaraman, V.S.; Pahwa, S. HIV gp120 inhibits T cell activation by interfering with expression of costimulatory molecules CD40 ligand and CD80 (B71). *J. Immunol.* **1995**, *155*, 917–924. [PubMed]

99. Meyaard, L.; Schuitemaker, H.; Miedema, F. T-cell dysfunction in HIV infection: Anergy due to defective antigen-presenting cell function? *Immunol. Today* **1993**, *14*, 161–164. [CrossRef]

100. Silverstein, P.S.; Shah, A.; Weemhoff, J.; Kumar, S.; Singh, D.P.; Kumar, A. HIV-1 gp120 and drugs of abuse: Interactions in the central nervous system. *Curr. HIV Res.* **2012**, *10*, 369–383. [CrossRef] [PubMed]

101. Berger, E.A.; Murphy, P.M.; Farber, J.M. Chemokine receptors as HIV-1 coreceptors: Roles in viral entry, tropism, and disease. *Annu. Rev. Immunol.* **1999**, *17*, 657–700. [CrossRef] [PubMed]

102. Keele, B.F.; Giorgi, E.E.; Salazar-Gonzalez, J.F.; Decker, J.M.; Pham, K.T.; Salazar, M.G.; Sun, C.; Grayson, T.; Wang, S.; Li, H.; et al. Identification and characterization of transmitted and early founder virus envelopes in primary HIV-1 infection. *Proc. Natl. Acad. Sci. USA* **2008**, *105*, 7552–7557. [CrossRef] [PubMed]

103. Van't Wout, A.B.; Kootstra, N.A.; Mulder-Kampinga, G.A.; Albrecht-van Lent, N.; Scherpbier, H.J.; Veenstra, J.; Boer, K.; Coutinho, R.A.; Miedema, F.; Schuitemaker, H. Macrophage-tropic variants initiate human immunodeficiency virus type 1 infection after sexual, parenteral, and vertical transmission. *J. Clin. Investig.* **1994**, *94*, 2060–2067. [CrossRef] [PubMed]

104. Salazar-Gonzalez, J.F.; Bailes, E.; Pham, K.T.; Salazar, M.G.; Guffey, M.B.; Keele, B.F.; Derdeyn, C.A.; Farmer, P.; Hunter, E.; Allen, S.; et al. Deciphering human immunodeficiency virus type 1 transmission and early envelope diversification by single-genome amplification and sequencing. *J. Virol.* **2008**, *82*, 3952–3970. [CrossRef] [PubMed]

105. Shaw, G.M.; Hunter, E. HIV transmission. *Cold Spring Harb. Perspect. Med.* **2012**, *2*, a006965. [CrossRef] [PubMed]

106. Boesecke, C.; Pett, S.L. Clinical studies with chemokine receptor-5 (CCR5)-inhibitors. *Curr. Opin. HIV AIDS* **2012**, *7*, 456–462. [CrossRef] [PubMed]

107. Woollard, S.M.; Kanmogne, G.D. Maraviroc: A review of its use in HIV infection and beyond. *Drug Des. Dev. Ther.* **2015**, *9*, 5447–5468.

108. Mummidi, S.; Ahuja, S.S.; Gonzalez, E.; Anderson, S.A.; Santiago, E.N.; Stephan, K.T.; Craig, F.E.; O'Connell, P.; Tryon, V.; Clark, R.A.; et al. Genealogy of the CCR5 locus and chemokine system gene variants associated with altered rates of HIV-1 disease progression. *Nat. Med.* **1998**, *4*, 786–793. [CrossRef] [PubMed]

109. Joshi, A.; Punke, E.B.; Sedano, M.; Beauchamp, B.; Patel, R.; Hossenlopp, C.; Alozie, O.K.; Gupta, J.; Mukherjee, D.; Garg, H. CCR5 promoter activity correlates with HIV disease progression by regulating CCR5 cell surface expression and CD4 T cell apoptosis. *Sci. Rep.* **2017**, *7*, 232. [CrossRef] [PubMed]

110. Dean, M.; Carrington, M.; Winkler, C.; Huttley, G.A.; Smith, M.W.; Allikmets, R.; Goedert, J.J.; Buchbinder, S.P.; Vittinghoff, E.; Gomperts, E.; et al. Genetic restriction of HIV-1 infection and progression to AIDS by a deletion allele of the CKR5 structural gene. Hemophilia Growth and Development Study, Multicenter AIDS Cohort Study, Multicenter Hemophilia Cohort Study, San Francisco City Cohort, ALIVE Study. *Science* **1996**, *273*, 1856–1862. [PubMed]

111. Huang, Y.; Paxton, W.A.; Wolinsky, S.M.; Neumann, A.U.; Zhang, L.; He, T.; Kang, S.; Ceradini, D.; Jin, Z.; Yazdanbakhsh, K.; et al. The role of a mutant CCR5 allele in HIV-1 transmission and disease progression. *Nat. Med.* **1996**, *2*, 1240–1243. [CrossRef] [PubMed]

112. Liu, R.; Paxton, W.; Choe, S.; Ceradini, D.; Martin, S.; Horuk, R.; MacDonald, M.; Stuhlmann, H.; Koup, R.; Landau, N. Homozygous defect in HIV-1 coreceptor accounts for resistance of some multiply-exposed individuals to HIV-1 infection. *Cell* **1996**, *86*, 367–377. [CrossRef]

113. Samson, M.; Libert, F.; Doranz, B.J.; Rucker, J.; Liesnard, C.; Farber, C.M.; Saragosti, S.; Lapoumeroulie, C.; Cognaux, J.; Forceille, C.; et al. Resistance to HIV-1 infection in caucasian individuals bearing mutant alleles of the CCR-5 chemokine receptor gene. *Nature* **1996**, *382*, 722–725. [CrossRef] [PubMed]

114. Wu, L.; Paxton, W.A.; Kassam, N.; Ruffing, N.; Rottman, J.B.; Sullivan, N.; Choe, H.; Sodroski, J.; Newman, W.; Koup, R.A.; et al. CCR5 levels and expression pattern correlate with infectability by macrophage-tropic HIV-1, in vitro. *J. Exp. Med.* **1997**, *185*, 1681–1691. [CrossRef] [PubMed]

115. Martin, M.P.; Dean, M.; Smith, M.W.; Winkler, C.; Gerrard, B.; Michael, N.L.; Lee, B.; Doms, R.W.; Margolick, J.; Buchbinder, S.; et al. Genetic acceleration of AIDS progression by a promoter variant of CCR5. *Science* **1998**, *282*, 1907–1911. [CrossRef] [PubMed]

116. Gonzalez, E.; Bamshad, M.; Sato, N.; Mummidi, S.; Dhanda, R.; Catano, G.; Cabrera, S.; McBride, M.; Cao, X.H.; Merrill, G.; et al. Race-specific HIV-1 disease-modifying effects associated with CCR5 haplotypes. *Proc. Natl. Acad. Sci. USA* **1999**, *96*, 12004–12009. [CrossRef] [PubMed]

117. Gornalusse, G.G.; Mummidi, S.; Gaitan, A.A.; Jimenez, F.; Ramsuran, V.; Picton, A.; Rogers, K.; Manoharan, M.S.; Avadhanam, N.; Murthy, K.K.; et al. Epigenetic mechanisms, T-cell activation, and CCR5 genetics interact to regulate T-cell expression of CCR5, the major HIV-1 coreceptor. *Proc. Natl. Acad. Sci. USA* **2015**, *112*, E4762–E4771. [CrossRef] [PubMed]

118. Platt, E.; Wehrly, K.; Kuhmann, S.; Chesebro, B.; Kabat, D. Effects of CCR5 and CD4 cell surface concentrations on infections by macrophagetropic isolates of human immunodeficiency virus type 1. *J. Virol.* **1998**, *72*, 2855–2864. [PubMed]

119. Reeves, J.; Gallo, S.; Ahmad, N.; Miamidian, J.; Harvey, P.; Sharron, M.; Pohlmann, S.; Sfakianos, J.; Derdeyn, C.; Blumenthal, R.; et al. Sensitivity of HIV-1 to entry inhibitors correlates with envelope/coreceptor affinity, receptor density, and fusion kinetics. *Proc. Natl. Acad. Sci. USA* **2002**, *99*, 16249–16254. [CrossRef] [PubMed]

120. Pandrea, I.; Apetrei, C.; Gordon, S.; Barbercheck, J.; Dufour, J.; Bohm, R.; Sumpter, B.; Roques, P.; Marx, P.A.; Hirsch, V.M.; et al. Paucity of CD4+ CCR5+ T cells is a typical feature of natural SIV hosts. *Blood* **2007**, *109*, 1069–1076. [CrossRef] [PubMed]

121. Pandrea, I.; Onanga, R.; Souquiere, S.; Mouinga-Ondeme, A.; Bourry, O.; Makuwa, M.; Rouquet, P.; Silvestri, G.; Simon, F.; Roques, P.; et al. Paucity of CD4+ CCR5+ T cells may prevent transmission of simian immunodeficiency virus in natural nonhuman primate hosts by breast-feeding. *J. Virol.* **2008**, *82*, 5501–5509. [CrossRef] [PubMed]

122. Paiardini, M.; Cervasi, B.; Reyes-Aviles, E.; Micci, L.; Ortiz, A.M.; Chahroudi, A.; Vinton, C.; Gordon, S.N.; Bosinger, S.E.; Francella, N.; et al. Low levels of SIV infection in sooty mangabey central memory CD(4)(+) T cells are associated with limited CCR5 expression. *Nat. Med.* **2011**, *17*, 830–836. [CrossRef] [PubMed]

123. Muenchhoff, M.; Adland, E.; Karimanzira, O.; Crowther, C.; Pace, M.; Csala, A.; Leitman, E.; Moonsamy, A.; McGregor, C.; Hurst, J.; et al. Nonprogressing HIV-infected children share fundamental immunological features of nonpathogenic SIV infection. *Sci. Transl. Med.* **2016**, *8*, 358ra125. [CrossRef] [PubMed]

124. Grossman, Z.; Meier-Schellersheim, M.; Paul, W.; Picker, L. Pathogenesis of HIV infection: What the virus spares is as important as what it destroys. *Nat. Med.* **2006**, *12*, 289–295. [CrossRef] [PubMed]

125. Grossman, Z.; Meier-Schellersheim, M.; Sousa, A.; Victorino, R.; Paul, W. CD4+ T-cell depletion in HIV infection: Are we closer to understanding the cause? *Nat. Med.* **2002**, *8*, 319–323. [CrossRef] [PubMed]

126. Bentwich, Z.; Kalinkovich, A.; Weisman, Z.; Grossman, Z. Immune activation in the context of HIV infection. *Clin. Exp. Immunol.* **1998**, *111*, 1–2. [CrossRef] [PubMed]

127. Al-Harthi, L.; MaWhinney, S.; Connick, E.; Schooley, R.; Forster, J.; Benson, C.; Thompson, M.; Judson, F.; Palella, F.; Landay, A. Immunophenotypic alterations in acute and early HIV infection. *Clin. Immunol.* **2007**, *125*, 299–308. [CrossRef] [PubMed]

128. Biancotto, A.; Grivel, J.; Iglehart, S.; Vanpouille, C.; Lisco, A.; Sieg, S.; Debernardo, R.; Garate, K.; Rodriguez, B.; Margolis, L.; et al. Abnormal activation and cytokine spectra in lymph nodes of people chronically infected with HIV-1. *Blood* **2007**, *109*, 4272–4279. [CrossRef] [PubMed]

129. Gougeon, M. T cell apoptosis as a consequence of chronic activation of the immune system in HIV infection. *Adv. Exp. Med. Biol.* **1995**, *374*, 121–127. [PubMed]

130. Sodora, D.; Silvestri, G. Immune activation and AIDS pathogenesis. *AIDS* **2008**, *22*, 439–446. [CrossRef] [PubMed]

131. Estes, J.; Gordon, S.; Zeng, M.; Chahroudi, A.; Dunham, R.; Staprans, S.; Reilly, C.; Silvestri, G.; Haase, A. Early resolution of acute immune activation and induction of PD-1 in SIV-infected sooty mangabeys distinguishes nonpathogenic from pathogenic infection in rhesus macaques. *J. Immunol.* **2008**, *180*, 6798–6807. [CrossRef] [PubMed]

132. Brainard, D.M.; Seung, E.; Frahm, N.; Cariappa, A.; Bailey, C.C.; Hart, W.K.; Shin, H.S.; Brooks, S.F.; Knight, H.L.; Eichbaum, Q.; et al. Induction of robust cellular and humoral virus-specific adaptive immune responses in human immunodeficiency virus-infected humanized BLT mice. *J. Virol.* **2009**, *83*, 7305–7321. [CrossRef] [PubMed]

133. Paiardini, M.; Muller-Trutwin, M. HIV-associated chronic immune activation. *Immunol. Rev.* **2013**, *254*, 78–101. [CrossRef] [PubMed]

134. Ploquin, M.J.; Silvestri, G.; Muller-Trutwin, M. Immune activation in HIV infection: What can the natural hosts of simian immunodeficiency virus teach us? *Curr. Opin. HIV AIDS* **2016**, *11*, 201–208. [CrossRef] [PubMed]

135. Hatano, H.; Jain, V.; Hunt, P.W.; Lee, T.H.; Sinclair, E.; Do, T.D.; Hoh, R.; Martin, J.N.; McCune, J.M.; Hecht, F.; et al. Cell-based measures of viral persistence are associated with immune activation and programmed cell death protein 1 (PD-1)-expressing CD4+ T cells. *J. Infect. Dis.* **2013**, *208*, 50–56. [CrossRef] [PubMed]

136. Cockerham, L.R.; Siliciano, J.D.; Sinclair, E.; O'Doherty, U.; Palmer, S.; Yukl, S.A.; Strain, M.C.; Chomont, N.; Hecht, F.M.; Siliciano, R.F.; et al. CD4+ and CD8+ T cell activation are associated with HIV DNA in resting CD4+ T cells. *PLoS ONE* **2014**, *9*, e110731. [CrossRef] [PubMed]

137. Zheng, L.; Taiwo, B.; Gandhi, R.T.; Hunt, P.W.; Collier, A.C.; Flexner, C.; Bosch, R.J. Factors associated with CD8+ T-cell activation in HIV-1-infected patients on long-term antiretroviral therapy. *J. Acquir. Immune Defic. Syndr.* **2014**, *67*, 153–160. [CrossRef] [PubMed]

138. Hazenberg, M.D.; Stuart, J.W.; Otto, S.A.; Borleffs, J.C.; Boucher, C.A.; de Boer, R.J.; Miedema, F.; Hamann, D. T-cell division in human immunodeficiency virus (HIV)-1 infection is mainly due to immune activation: A longitudinal analysis in patients before and during highly active antiretroviral therapy (HAART). *Blood* **2000**, *95*, 249–255. [PubMed]

139. Hunt, P.W.; Martin, J.N.; Sinclair, E.; Bredt, B.; Hagos, E.; Lampiris, H.; Deeks, S.G. T cell activation is associated with lower CD4+ T cell gains in human immunodeficiency virus-infected patients with sustained viral suppression during antiretroviral therapy. *J. Infect. Dis.* **2003**, *187*, 1534–1543. [CrossRef] [PubMed]

140. Mohri, H.; Perelson, A.S.; Tung, K.; Ribeiro, R.M.; Ramratnam, B.; Markowitz, M.; Kost, R.; Hurley, A.; Weinberger, L.; Cesar, D.; et al. Increased turnover of T lymphocytes in HIV-1 infection and its reduction by antiretroviral therapy. *J. Exp. Med.* **2001**, *194*, 1277–1287. [CrossRef] [PubMed]

141. Schuetz, A.; Deleage, C.; Sereti, I.; Rerknimitr, R.; Phanuphak, N.; Phuang-Ngern, Y.; Estes, J.D.; Sandler, N.G.; Sukhumvittaya, S.; Marovich, M.; et al. Initiation of ART during early acute HIV infection preserves mucosal Th17 function and reverses HIV-related immune activation. *PLoS Pathog.* **2014**, *10*, e1004543. [CrossRef] [PubMed]

142. Buzon, M.J.; Massanella, M.; Llibre, J.M.; Esteve, A.; Dahl, V.; Puertas, M.C.; Gatell, J.M.; Domingo, P.; Paredes, R.; Sharkey, M.; et al. HIV-1 replication and immune dynamics are affected by raltegravir intensification of HAART-suppressed subjects. *Nat. Med.* **2010**, *16*, 460–465. [CrossRef] [PubMed]

143. Biancotto, A.; Iglehart, S.; Vanpouille, C.; Condack, C.; Lisco, A.; Ruecker, E.; Hirsch, I.; Margolis, L.; Grivel, J. HIV-1 induced activation of CD4+ T cells creates new targets for HIV-1 infection in human lymphoid tissue ex vivo. *Blood* **2008**, *111*, 699–704. [CrossRef] [PubMed]

144. Li, G.; Cheng, M.; Nunoya, J.; Cheng, L.; Guo, H.; Yu, H.; Liu, Y.J.; Su, L.; Zhang, L. Plasmacytoid dendritic cells suppress HIV-1 replication but contribute to HIV-1 induced immunopathogenesis in humanized mice. *PLoS Pathog.* **2014**, *10*, e1004291. [CrossRef] [PubMed]

145. Beignon, A.S.; McKenna, K.; Skoberne, M.; Manches, O.; DaSilva, I.; Kavanagh, D.G.; Larsson, M.; Gorelick, R.J.; Lifson, J.D.; Bhardwaj, N. Endocytosis of HIV-1 activates plasmacytoid dendritic cells via Toll-like receptor-viral RNA interactions. *J. Clin. Investig.* **2005**, *115*, 3265–3275. [CrossRef] [PubMed]

146. Cheng, L.; Yu, H.; Li, G.; Li, F.; Ma, J.; Li, J.; Chi, L.; Zhang, L.; Su, L. Type I interferons suppress viral replication but contribute to T cell depletion and dysfunction during chronic HIV-1 infection. *JCI Insight* **2017**, *2*, 94366. [CrossRef] [PubMed]

147. Douek, D. HIV disease progression: Immune activation, microbes, and a leaky gut. *Top. HIV Med.* **2007**, *15*, 114–117. [PubMed]

148. Tincati, C.; Douek, D.C.; Marchetti, G. Gut barrier structure, mucosal immunity and intestinal microbiota in the pathogenesis and treatment of HIV infection. *AIDS Res. Ther.* **2016**, *13*, 19. [CrossRef] [PubMed]

149. Marchetti, G.; Tincati, C.; Silvestri, G. Microbial translocation in the pathogenesis of HIV infection and AIDS. *Clin. Microbiol. Rev.* **2013**, *26*, 2–18. [CrossRef] [PubMed]

150. Brenchley, J.; Price, D.; Schacker, T.; Asher, T.; Silvestri, G.; Rao, S.; Kazzaz, Z.; Bornstein, E.; Lambotte, O.; Altmann, D.; et al. Microbial translocation is a cause of systemic immune activation in chronic HIV infection. *Nat. Med.* **2006**, *12*, 1365–1371. [CrossRef] [PubMed]

151. Jiang, W.; Lederman, M.M.; Hunt, P.; Sieg, S.F.; Haley, K.; Rodriguez, B.; Landay, A.; Martin, J.; Sinclair, E.; Asher, A.I.; et al. Plasma levels of bacterial DNA correlate with immune activation and the magnitude of immune restoration in persons with antiretroviral-treated HIV infection. *J. Infect. Dis.* **2009**, *199*, 1177–1185. [CrossRef] [PubMed]

152. Marchetti, G.; Bellistri, G.M.; Borghi, E.; Tincati, C.; Ferramosca, S.; La Francesca, M.; Morace, G.; Gori, A.; Monforte, A.D. Microbial translocation is associated with sustained failure in CD4+ T-cell reconstitution in HIV-infected patients on long-term highly active antiretroviral therapy. *AIDS* **2008**, *22*, 2035–2038. [CrossRef] [PubMed]

153. Merlini, E.; Bai, F.; Bellistri, G.M.; Tincati, C.; d'Arminio Monforte, A.; Marchetti, G. Evidence for polymicrobic flora translocating in peripheral blood of HIV-infected patients with poor immune response to antiretroviral therapy. *PLoS ONE* **2011**, *6*, e18580. [CrossRef] [PubMed]

154. Bukh, A.R.; Melchjorsen, J.; Offersen, R.; Jensen, J.M.; Toft, L.; Stovring, H.; Ostergaard, L.; Tolstrup, M.; Sogaard, O.S. Endotoxemia is associated with altered innate and adaptive immune responses in untreated HIV-1 infected individuals. *PLoS ONE* **2011**, *6*, e21275. [CrossRef] [PubMed]

155. Nowroozalizadeh, S.; Mansson, F.; da Silva, Z.; Repits, J.; Dabo, B.; Pereira, C.; Biague, A.; Albert, J.; Nielsen, J.; Aaby, P.; et al. Microbial translocation correlates with the severity of both HIV-1 and HIV-2 infections. *J. Infect. Dis.* **2010**, *201*, 1150–1154. [CrossRef] [PubMed]

156. Guadalupe, M.; Reay, E.; Sankaran, S.; Prindiville, T.; Flamm, J.; McNeil, A.; Dandekar, S. Severe CD4+ T-cell depletion in gut lymphoid tissue during primary human immunodeficiency virus type 1 infection and substantial delay in restoration following highly active antiretroviral therapy. *J. Virol.* **2003**, *77*, 11708–11717. [CrossRef] [PubMed]

157. Mehandru, S.; Poles, M.A.; Tenner-Racz, K.; Horowitz, A.; Hurley, A.; Hogan, C.; Boden, D.; Racz, P.; Markowitz, M. Primary HIV-1 infection is associated with preferential depletion of CD4+ T lymphocytes from effector sites in the gastrointestinal tract. *J. Exp. Med.* **2004**, *200*, 761–770. [CrossRef] [PubMed]

158. Mehandru, S.; Poles, M.A.; Tenner-Racz, K.; Manuelli, V.; Jean-Pierre, P.; Lopez, P.; Shet, A.; Low, A.; Mohri, H.; Boden, D.; et al. Mechanisms of gastrointestinal CD4+ T-cell depletion during acute and early human immunodeficiency virus type 1 infection. *J. Virol.* **2007**, *81*, 599–612. [CrossRef] [PubMed]

159. Li, Q.; Duan, L.; Estes, J.D.; Ma, Z.M.; Rourke, T.; Wang, Y.; Reilly, C.; Carlis, J.; Miller, C.J.; Haase, A.T. Peak SIV replication in resting memory CD4+ T cells depletes gut lamina propria CD4+ T cells. *Nature* **2005**, *434*, 1148–1152. [CrossRef] [PubMed]

160. Sankaran, S.; George, M.D.; Reay, E.; Guadalupe, M.; Flamm, J.; Prindiville, T.; Dandekar, S. Rapid onset of intestinal epithelial barrier dysfunction in primary human immunodeficiency virus infection is driven by an imbalance between immune response and mucosal repair and regeneration. *J. Virol.* **2008**, *82*, 538–545. [CrossRef] [PubMed]

161. Brenchley, J.M.; Paiardini, M.; Knox, K.S.; Asher, A.I.; Cervasi, B.; Asher, T.E.; Scheinberg, P.; Price, D.A.; Hage, C.A.; Kholi, L.M.; et al. Differential Th17 CD4 T-cell depletion in pathogenic and nonpathogenic lentiviral infections. *Blood* **2008**, *112*, 2826–2835. [CrossRef] [PubMed]

162. Stockinger, B.; Omenetti, S. The dichotomous nature of T helper 17 cells. *Nat. Rev. Immunol.* **2017**. [CrossRef] [PubMed]

163. Favre, D.; Lederer, S.; Kanwar, B.; Ma, Z.M.; Proll, S.; Kasakow, Z.; Mold, J.; Swainson, L.; Barbour, J.D.; Baskin, C.R.; et al. Critical loss of the balance between Th17 and T regulatory cell populations in pathogenic SIV infection. *PLoS Pathog.* **2009**, *5*, e1000295. [CrossRef] [PubMed]

164. Raffatellu, M.; Santos, R.L.; Verhoeven, D.E.; George, M.D.; Wilson, R.P.; Winter, S.E.; Godinez, I.; Sankaran, S.; Paixao, T.A.; Gordon, M.A.; et al. Simian immunodeficiency virus-induced mucosal interleukin-17 deficiency promotes *Salmonella* dissemination from the gut. *Nat. Med.* **2008**, *14*, 421–428. [CrossRef] [PubMed]

165. Mudd, J.C.; Brenchley, J.M. Gut mucosal barrier dysfunction, microbial dysbiosis, and their role in HIV-1 disease progression. *J. Infect. Dis.* **2016**, *214*, S58–S66. [CrossRef] [PubMed]

166. Sato, W.; Aranami, T.; Yamamura, T. Cutting edge: Human Th17 cells are identified as bearing CCR2+CCR5-phenotype. *J. Immunol.* **2007**, *178*, 7525–7529. [CrossRef] [PubMed]

167. Ortiz, A.M.; Klase, Z.A.; DiNapoli, S.R.; Vujkovic-Cvijin, I.; Carmack, K.; Perkins, M.R.; Calantone, N.; Vinton, C.L.; Riddick, N.E.; Gallagher, J.; et al. IL-21 and probiotic therapy improve Th17 frequencies, microbial translocation, and microbiome in ARV-treated, SIV-infected macaques. *Mucosal Immunol.* **2016**, *9*, 458–467. [CrossRef] [PubMed]

168. Akira, S.; Takeda, K.; Kaisho, T. Toll-like receptors: Critical proteins linking innate and acquired immunity. *Nat. Immunol.* **2001**, *2*, 675–680. [CrossRef] [PubMed]

169. Iwasaki, A.; Medzhitov, R. Toll-like receptor control of the adaptive immune responses. *Nat. Immunol.* **2004**, *5*, 987–995. [CrossRef] [PubMed]

170. Takeda, K.; Kaisho, T.; Akira, S. Toll-like receptors. *Annu. Rev. Immunol.* **2003**, *21*, 335–376. [CrossRef] [PubMed]

171. Mandl, J.N.; Barry, A.P.; Vanderford, T.H.; Kozyr, N.; Chavan, R.; Klucking, S.; Barrat, F.J.; Coffman, R.L.; Staprans, S.I.; Feinberg, M.B. Divergent TLR7 and TLR9 signaling and type I interferon production distinguish pathogenic and nonpathogenic AIDS virus infections. *Nat. Med.* **2008**, *14*, 1077–1087. [CrossRef] [PubMed]

172. O'Brien, M.; Manches, O.; Wilen, C.; Gopal, R.; Huq, R.; Wu, V.; Sunseri, N.; Bhardwaj, N. CD4 receptor is a key determinant of divergent HIV-1 sensing by plasmacytoid dendritic cells. *PLoS Pathog.* **2016**, *12*, e1005553. [CrossRef] [PubMed]

173. Funderburg, N.; Luciano, A.A.; Jiang, W.; Rodriguez, B.; Sieg, S.F.; Lederman, M.M. Toll-like receptor ligands induce human T cell activation and death, a model for HIV pathogenesis. *PLoS ONE* **2008**, *3*, e1915. [CrossRef] [PubMed]

174. Murray, S.M.; Down, C.M.; Boulware, D.R.; Stauffer, W.M.; Cavert, W.P.; Schacker, T.W.; Brenchley, J.M.; Douek, D.C. Reduction of immune activation with chloroquine therapy during chronic HIV infection. *J. Virol.* **2010**, *84*, 12082–12086. [CrossRef] [PubMed]

175. Piconi, S.; Parisotto, S.; Rizzardini, G.; Passerini, S.; Terzi, R.; Argenteri, B.; Meraviglia, P.; Capetti, A.; Biasin, M.; Trabattoni, D.; et al. Hydroxychloroquine drastically reduces immune activation in HIV-infected, antiretroviral therapy-treated immunologic nonresponders. *Blood* **2011**, *118*, 3263–3272. [CrossRef] [PubMed]

176. Ameisen, J.C.; Capron, A. Cell dysfunction and depletion in AIDS: The programmed cell death hypothesis. *Immunol. Today* **1991**, *12*, 102–105. [CrossRef]

177. Bao, R.; Zhuang, K.; Liu, J.; Wu, J.; Li, J.; Wang, X.; Ho, W.Z. Lipopolysaccharide induces immune activation and SIV replication in rhesus macaques of Chinese origin. *PLoS ONE* **2014**, *9*, e98636. [CrossRef] [PubMed]

178. Hofer, U.; Schlaepfer, E.; Baenziger, S.; Nischang, M.; Regenass, S.; Schwendener, R.; Kempf, W.; Nadal, D.; Speck, R.F. Inadequate clearance of translocated bacterial products in HIV-infected humanized mice. *PLoS Pathog.* **2010**, *6*, e1000867. [CrossRef] [PubMed]

179. Pandrea, I.; Gaufin, T.; Brenchley, J.M.; Gautam, R.; Monjure, C.; Gautam, A.; Coleman, C.; Lackner, A.A.; Ribeiro, R.M.; Douek, D.C.; et al. Cutting edge: Experimentally induced immune activation in natural hosts of simian immunodeficiency virus induces significant increases in viral replication and CD4+ T cell depletion. *J. Immunol.* **2008**, *181*, 6687–6691. [CrossRef] [PubMed]

180. Koning, F.; Otto, S.; Hazenberg, M.; Dekker, L.; Prins, M.; Miedema, F.; Schuitemaker, H. Low-level CD4+ T cell activation is associated with low susceptibility to HIV-1 infection. *J. Immunol.* **2005**, *175*, 6117–6122. [CrossRef] [PubMed]

181. Juffermans, N.P.; Paxton, W.A.; Dekkers, P.E.; Verbon, A.; de Jonge, E.; Speelman, P.; van Deventer, S.J.; van der Poll, T. Up-regulation of HIV coreceptors CXCR4 and CCR5 on CD4(+) T cells during human endotoxemia and after stimulation with (myco)bacterial antigens: The role of cytokines. *Blood* **2000**, *96*, 2649–2654. [PubMed]

182. Espert, L.; Codogno, P.; Biard-Piechaczyk, M. What is the role of autophagy in HIV-1 infection? *Autophagy* **2008**, *4*, 273–275. [CrossRef] [PubMed]

183. Espert, L.; Denizot, M.; Grimaldi, M.; Robert-Hebmann, V.; Gay, B.; Varbanov, M.; Codogno, P.; Biard-Piechaczyk, M. Autophagy and CD4+ T lymphocyte destruction by HIV-1. *Autophagy* **2007**, *3*, 32–34. [CrossRef] [PubMed]

184. Yoshimori, T. Autophagy: A regulated bulk degradation process inside cells. *Biochem. Biophys. Res. Commun.* **2004**, *313*, 453–458. [CrossRef] [PubMed]

185. Ohsumi, Y. Molecular dissection of autophagy: Two ubiquitin-like systems. *Nat. Rev. Mol. Cell Biol.* **2001**, *2*, 211–216. [CrossRef] [PubMed]

186. Espert, L.; Biard-Piechaczyk, M. Autophagy in HIV-induced T cell death. *Curr. Top. Microbiol. Immunol.* **2009**, *335*, 307–321. [PubMed]

187. Espert, L.; Denizot, M.; Grimaldi, M.; Robert-Hebmann, V.; Gay, B.; Varbanov, M.; Codogno, P.; Biard-Piechaczyk, M. Autophagy is involved in T cell death after binding of HIV-1 envelope proteins to CXCR4. *J. Clin. Investig.* **2006**, *116*, 2161–2172. [CrossRef] [PubMed]

188. Kirkegaard, K.; Taylor, M.P.; Jackson, W.T. Cellular autophagy: Surrender, avoidance and subversion by microorganisms. *Nat. Rev. Microbiol.* **2004**, *2*, 301–314. [CrossRef] [PubMed]

189. Denizot, M.; Varbanov, M.; Espert, L.; Robert-Hebmann, V.; Sagnier, S.; Garcia, E.; Curriu, M.; Mamoun, R.; Blanco, J.; Biard-Piechaczyk, M. HIV-1 gp41 fusogenic function triggers autophagy in uninfected cells. *Autophagy* **2008**, *4*, 998–1008. [CrossRef] [PubMed]

190. Booth, L.A.; Tavallai, S.; Hamed, H.A.; Cruickshanks, N.; Dent, P. The role of cell signalling in the crosstalk between autophagy and apoptosis. *Cell. Signal.* **2014**, *26*, 549–555. [CrossRef] [PubMed]

191. Djavaheri-Mergny, M.; Maiuri, M.C.; Kroemer, G. Cross talk between apoptosis and autophagy by caspase-mediated cleavage of Beclin 1. *Oncogene* **2010**, *29*, 1717–1719. [CrossRef] [PubMed]

192. Doitsh, G.; Greene, W.C. Dissecting how CD4 T cells are lost during HIV infection. *Cell Host Microbe* **2016**, *19*, 280–291. [CrossRef] [PubMed]

193. Doitsh, G.; Galloway, N.L.; Geng, X.; Yang, Z.; Monroe, K.M.; Zepeda, O.; Hunt, P.W.; Hatano, H.; Sowinski, S.; Munoz-Arias, I.; et al. Cell death by pyroptosis drives CD4 T-cell depletion in HIV-1 infection. *Nature* **2014**, *505*, 509–514. [CrossRef] [PubMed]

194. Doitsh, G.; Cavrois, M.; Lassen, K.G.; Zepeda, O.; Yang, Z.; Santiago, M.L.; Hebbeler, A.M.; Greene, W.C. Abortive HIV infection mediates CD4 T cell depletion and inflammation in human lymphoid tissue. *Cell* **2010**, *143*, 789–801. [CrossRef] [PubMed]

195. Monroe, K.M.; Yang, Z.; Johnson, J.R.; Geng, X.; Doitsh, G.; Krogan, N.J.; Greene, W.C. IFI16 DNA sensor is required for death of lymphoid CD4 T cells abortively infected with HIV. *Science* **2014**, *343*, 428–432. [CrossRef] [PubMed]

196. Fink, S.L.; Cookson, B.T. Apoptosis, pyroptosis, and necrosis: Mechanistic description of dead and dying eukaryotic cells. *Infect. Immun.* **2005**, *73*, 1907–1916. [CrossRef] [PubMed]

197. Galloway, N.L.; Doitsh, G.; Monroe, K.M.; Yang, Z.; Munoz-Arias, I.; Levy, D.N.; Greene, W.C. Cell-to-Cell Transmission of HIV-1 Is Required to Trigger Pyroptotic Death of Lymphoid-Tissue-Derived CD4 T Cells. *Cell Rep.* **2015**, *12*, 1555–1563. [CrossRef] [PubMed]

198. Poropatich, K.; Sullivan, D.J., Jr. Human immunodeficiency virus type 1 long-term non-progressors: The viral, genetic and immunological basis for disease non-progression. *J. Gen. Virol.* **2011**, *92*, 247–268. [CrossRef] [PubMed]

199. Choudhary, S.K.; Vrisekoop, N.; Jansen, C.A.; Otto, S.A.; Schuitemaker, H.; Miedema, F.; Camerini, D. Low immune activation despite high levels of pathogenic human immunodeficiency virus type 1 results in long-term asymptomatic disease. *J. Virol.* **2007**, *81*, 8838–8842. [CrossRef] [PubMed]

200. Klatt, N.R.; Bosinger, S.E.; Peck, M.; Richert-Spuhler, L.E.; Heigele, A.; Gile, J.P.; Patel, N.; Taaffe, J.; Julg, B.; Camerini, D.; et al. Limited HIV infection of central memory and stem cell memory CD4+ T cells is associated with lack of progression in viremic individuals. *PLoS Pathog.* **2014**, *10*, e1004345. [CrossRef] [PubMed]

201. Sivaraman, V.; Zhang, L.; Meissner, E.G.; Jeffrey, J.L.; Su, L. The heptad repeat 2 domain is a major determinant for enhanced human immunodeficiency virus type 1 (HIV-1) fusion and pathogenicity of a highly pathogenic HIV-1 Env. *J. Virol.* **2009**, *83*, 11715–11725. [CrossRef] [PubMed]

202. Dalmau, J.; Rotger, M.; Erkizia, I.; Rauch, A.; Reche, P.; Pino, M.; Esteve, A.; Palou, E.; Brander, C.; Paredes, R.; et al. Highly pathogenic adapted HIV-1 strains limit host immunity and dictate rapid disease progression. *AIDS* **2014**, *28*, 1261–1272. [CrossRef] [PubMed]

203. Si, Z.; Gorry, P.; Babcock, G.; Owens, C.M.; Cayabyab, M.; Phan, N.; Sodroski, J. Envelope glycoprotein determinants of increased entry in a pathogenic simian-human immunodeficiency virus (SHIV-HXBc2P 3.2) passaged in monkeys. *AIDS Res. Hum. Retrovir.* **2004**, *20*, 163–173. [CrossRef] [PubMed]

Review

Interference of Apoptosis by Hepatitis B Virus

Shaoli Lin and Yan-Jin Zhang *

Molecular Virology Laboratory, VA-MD College of Veterinary Medicine and Maryland Pathogen
Research Institute, University of Maryland, College Park, MD 20742, USA; lsl1990@umd.edu
* Correspondence: zhangyj@umd.edu; Tel.: +1-(301)-314-6596

Academic Editor: Marc Kvansakul
Received: 30 June 2017; Accepted: 10 August 2017; Published: 18 August 2017

Abstract: Hepatitis B virus (HBV) causes liver diseases that have been a consistent problem for human health, leading to more than one million deaths every year worldwide. A large proportion of hepatocellular carcinoma (HCC) cases across the world are closely associated with chronic HBV infection. Apoptosis is a programmed cell death and is frequently altered in cancer development. HBV infection interferes with the apoptosis signaling to promote HCC progression and viral proliferation. The HBV-mediated alteration of apoptosis is achieved via interference with cellular signaling pathways and regulation of epigenetics. HBV X protein (HBX) plays a major role in the interference of apoptosis. There are conflicting reports on the HBV interference of apoptosis with the majority showing inhibition of and the rest reporting induction of apoptosis. In this review, we described recent studies on the mechanisms of the HBV interference with the apoptosis signaling during the virus infection and provided perspective.

Keywords: hepatitis B virus (HBV); apoptosis; hepatocellular carcinoma (HCC); X protein

1. Introduction

Hepatitis B virus (HBV) is an enveloped DNA virus with a reverse transcription phase, belonging to the *Orthohepadnavirus* genus, the *Hepadnaviridae* family [1,2]. The genome of HBV is only 3.2 kb, containing four overlapping open reading frames (ORFs). The four ORFs encode seven viral proteins: pre-S1, pre-S2, S, pre-C, C, viral polymerase, and HBV X protein (HBX). There are four regulatory elements in the genome: enhancer II/basal core promoter, pre-S1 promoter, pre-S2/S promoter, and enhancer I/X promoter. The core protein and the viral polymerase are translated from the pre-genomic RNA (pgRNA), while the regulatory HBX protein and the three envelope proteins are encoded by the subgenomic RNAs [1,2]. The HBV virions attach to host cells through heparan sulfate proteoglycans or the hepatocyte-specific pre-S1 receptor, sodium taurocholate cotransporting polypeptide (NTCP) [3]. The virions enter the cells by endocytosis or fusion of the viral envelope at the plasma membrane. Once entering the cells, the viral nucleocapsid containing the partially double-stranded DNA, known as the relaxed circular DNA (rcDNA), would be released into the cytoplasm and transported into the nucleus [2]. The plus strand of the rcDNA is repaired and completed by the viral polymerase in the nucleus to generate the covalently closed circular DNA (cccDNA), which is transcribed into RNAs for the viral replication. The HBV DNA can be integrated into the host genome and the integration is commonly seen in patients with hepatocellular carcinoma (HCC).

HBV infection causes both acute and chronic liver diseases, and accounts for most of the chronic liver diseases globally, affecting over 240 million people worldwide [4]. More than one million individuals die from cirrhosis and liver cancer caused by the chronic HBV infection each year [5]. HBV infection is predominantly prevalent in Asian countries, such as China, Japan, Taiwan, and Korea [6]. In the USA, HBV infection is most common among Asians [7]. Up to now, ten genotypes (A–J) of HBV have been identified [8]. Genotypes A and D are ubiquitous but prevalent in Europe and Africa, while genotypes B and C are confined in Asia and Oceania. Other genotypes (E–J) are occasionally observed in some Asian countries. The virus can be transmitted through blood, semen, and body fluid, or from mother to baby at birth [9,10]. For some people, hepatitis B is an acute or short-term illness; but for others, it can become a long-term, chronic infection [11,12]. This indicates that HBV infection is just a trigger for liver diseases and HCC development, which are possibly the consequence of a complex interplay of many factors including host immune response. There is a large proportion of HBV-inactive carriers, who present little virus replication, normal alanine aminotransferase (ATL) level and minimal liver inflammation [13]. However, some of them may undergo HBV reactivation when treated with immunosuppressive drugs or suffer a higher risk of hepatocellular carcinoma after excessive alcohol consumption [14,15]. Despite an effective vaccine for prevention, there is still no known cure for existing HBV infection.

The frequent integration (>70%) of HBV DNA into the cell genome contributes to the instability of the chromosome, interruption of key cellular pathways and mutation of some pro-cancer genes [16]. HCC progression is found to associate with the patient ages and the HBV genotypes. The HBV genotype A mainly causes acute liver disease in clinical settings. An epidemiological investigation in China shows that, in young HCC patients (<30 years old), the HBV B2 is predominant, and a breakpoint in chromosome 8q24 located between *c-Myc* and plasmacytoma variant translocation 1 (*PVT1*) is more frequently found than in older patients. HBV integration into this site leads to the overexpression of *c-Myc* and *PVT1*, and consequent HCC progression [17]. Aside from the *HBV B2* genotype, the *HBV C* genotype also accounts for a large number of clinical HCC cases [8].

Chronic HBV infection is often accompanied by HCC. Among all the cancer cases caused by the infectious agents, 19.2% are attributed to HBV, while 7.8% are caused by hepatitis C virus (HCV) [18]. Some tumor diseases or HCC are frequently accompanied by defective apoptosis. In clinically histochemical staining of human carcinoma tissues, the Fas-expression is much lower than their corresponding non-carcinoma tissues, in both frequency and amount. Moreover, the apoptotic cell percentage is lower in the Fas-defective tissues [19]. During virus infection, the host cells take some protective measures such as cell death to prevent virus replication or dissemination [20]. To survive in the host cells, the viruses have evolved various mechanisms to modulate the apoptosis signaling during infection. They can promote the cell apoptosis and fission to facilitate the virus dissemination, or antagonize the apoptosis to gain time to proliferate in the infected cells. For example, HCV accelerates the cell apoptosis by the activation of caspase 3 and the release of cytochrome C, whereas Myxoma virus produces viral B-cell lymphoma 2 (BCL-2) to prevent the cell apoptosis [21]. In addition, some oncogenes are also upregulated during virus infection, such as HBV and avian leukemia virus [22,23].

HBX, approximately 17.4 kDa, is a viral protein with multiple functions. This protein can be translated in host cells from the integrated HBV genome even in the absence of complete virus replication cycle [24,25]. Thus far, the function of X protein is the most widely studied among all the HBV proteins. The small regulatory protein is implicated to play a major role in HCC progression [26–28]. It is able to not only suppress the DNA repair machinery [29,30] but also affect the DNA methylation of the host cells. These activities might contribute to the HCC progression [31,32]. HBX has also been demonstrated to have interplay with non-coding RNA (ncRNA) and various signaling pathways to regulate the host cell activities. Since the cell apoptosis is highly correlated with the HCC progress, understanding how the virus interferes with the apoptotic process may shed light on the mechanism of HCC formation and facilitate the development of anti-tumor therapeutics. Here, we summarize recent studies on the mechanisms of HBV interference of the apoptosis.

2. Apoptosis

Apoptosis is a programmed cell death, which is highly organized and acts as a protective strategy for healthy organisms to maintain the homeostasis [33]. The apoptosis plays a vital role in the innate and adaptive immune responses. The morphology of apoptotic cells is featured by cell shrinkage, membrane blebbing and the formation of apoptotic bodies [34–36]. Apoptotic cells do not release their cellular contents to the surroundings before being phagocytosed in vivo. The predominant pathways of the apoptosis are mainly composed of the extrinsic pathway (death receptor pathway) and the intrinsic pathway (mitochondrial pathway) [37].

The extrinsic pathway is mostly triggered by the external stimuli such as Fas, tumor necrosis factor-α (TNF-α), TRAIL (TNF-related apoptosis-inducing ligand), APO3L and APO2L [37]. Binding of the extracellular ligands to their receptors on the cell surface leads to the recruitment of adaptor proteins to transmit the intracellular signals via the caspase cascades. The recruited caspase 8422 and Fas-associated protein with death domain (FADD) forms an oligomeric death-inducing signaling complex (DISC), leading to the cleavage and activation of caspase 8. The activated caspase 8 then cleaves and activates the effector caspase 3/7, which are able to cleave a broad spectrum of cellular targets, such as receptor-interacting protein (RIP), X-linked inhibitor of apoptosis protein (X-IAP), signal transducer and activator of transcription-1 (STAT1), topoisomerase I, vimentin, retinoblastoma (Rb), and lamin B, consequently resulting in the cell death [38–40]. For the TNF receptor (TNFR), upon activation, TNFR recruits adaptor proteins TNFR type 1-associated DEATH domain protein (TRADD), Fas-associated protein with death domain (FADD), receptor-interacting serine/threonine-protein kinase 1 (RIPK1), cellular inhibitors of apoptosis (cIAP), TNF receptor associated factors 2 (TRAF2), and TRAF5. These proteins form the complex I (Figure 1), which leads to initiation of the canonical nuclear factor kappa-light-chain-enhancer of activated B cells (NF-κB) pathway [41]. In the complex I, IAPs can exert an anti-apoptotic role in the cells, for instance, cIAP1/2 interact with the second mitochondrial activator of caspases (SMAC) and sequester it from the XIAP, and the released XIAP inhibits caspases and apoptosis [42,43]. However, the anti-apoptotic role of cIAP1 can be counteracted by the SMAC mimetics [44]. Under certain circumstances, such as loss of cIAP, a secondary death promoting complex, termed as the complex II, can be formed. The complex II is composed of TRADD, FADD, RIPK1, caspase 8, and is capable of inducing apoptosis or necroptosis [45–47]. Notably, the complex II-initiated apoptosis can be prevented by cellular FLICE-inhibitory protein (cFLIP) through the inhibition of caspase 8 activity [48]. The expression of cFLIP is mediated by the NF-κB signaling.

The intrinsic pathway is initiated by the internal stimuli, such as DNA damage, endoplasmic reticulum (ER) stress, hypoxia and metabolic stress [37]. In healthy cells, the BCL-2 sequesters its proapoptotic counterparts, BAX (BCL-2-associated X protein), BAK (BCL-2 homologous antagonist/killer), and BCL-2 homology domain 3 (BH3)-only proteins into inactive complexes. Under cell stress, BH3-only proteins are activated, followed by the release of BAK/BAX and the engagement of the homo-oligomerization of the two proteins. The self-association of BAK/BAX is inclined to form a lipidic pore in the outer membrane (OM) of mitochondria by inserting α-helices 5 and 6 of the dimer into the OM, resulting in mitochondrial outer membrane permeabilization (MOMP) [49,50]. The MOMP allows inner mitochondrial proteins, such as apoptosis inducing factor (AIF), SMAC and cytochrome C, to be released into the cytosol. While SMAC exerts its pro-apoptotic role through binding with cIAPs, cytochrome C interacts with apoptotic protease activating factor 1 (APAF1) to facilitate the formation of the apoptosome. Once formed, the apoptosome will then recruit and activate the pro-caspase 9. The cleaved pro-caspase 9 becomes active and then activates the caspase 3/7, culminating in the cell apoptosis [51,52].

In addition to the caspase-dependent intrinsic pathway, the AIF protein can induce apoptosis by triggering chromatin condensation and DNA fragmentation, independent of caspase activation. After being released from the mitochondria, AIF ends up in the nucleus where it signals the cell to chromosome condensation and DNA fragmentation by Ca^{2+} and Mg^{2+}-dependent endonucleases [53,54].

The extrinsic pathway and the intrinsic pathway have some crosstalk via protein BH3 interacting-domain death agonist (BID). The activated caspase 8 is demonstrated to cleave BID, which belongs to the "BH3-domain-only" subset of the BCL-2 family. The truncated BID (tBID) triggers BAK or BAX homo-oligomerization and consequently MOMP [55]. In the process of apoptosis, the level of reactive oxygen species (ROS) plays an important role in deciding the cell fate. It interferes with both the extrinsic and the intrinsic pathways. Low-level ROS promotes the cell survival signaling, while toxic level ROS induces cell apoptosis [56]. The cancer cells usually have higher level ROS and are more capable to scavenge excessive ROS than the normal cells [57]. The higher level of ROS enhances cell proliferation through inducing abnormal cell growth caused by genetic mutation, enhanced autophagy, or activation of various signaling pathways, such as the phosphatidylinositol-4,5-bisphosphate 3-kinase-protein kinase B (PI3K-Akt) pathway, NF-κB, and protein kinase D (PKD) pathway. The toxic level of ROS leads to the apoptotic death of cancer cells.

Apart from the two extensively studied pathways, there are the physiological pathway, the perforin/granzyme pathway, and the pathological apoptosis pathway. In order to maintain the homeostasis of the human body, numerous cells need to be sacrificed every day to balance the metabolism. For instance, during the development of immune system, most lymphocytes lacking B cell receptor (BCR) or T cell receptor (TCR) will be eliminated by selection process [58]. Some diseases also cause excessive apoptosis in human tissue. A classic example is that the human immunodeficiency virus (HIV) Tat protein increases the Fas expression of CD4$^+$ T cells, increasing the possibility of T cell elimination [59]. The perforin/granzyme B pathway is one of the mechanisms that the cytotoxic T lymphocytes (CTL) and natural killer (NK) cells utilize to kill their target cells. The perforin forms poly-perforin pores on the target cell membrane, inducing the osmotic instability and allowing granzyme B to pass through the cell membrane to cause cell lysis. This mechanism can lead to cell death in the absence or presence of the activated caspases [60].

3. Hepatitis B Virus and Apoptosis

Plenty of studies have been performed to determine the relationship of HBV infection and apoptosis, but the results are still contradictory. The majority of the papers showed that HBV or HBX could inhibit the cellular apoptosis, thereby facilitating the virus proliferation and promoting the HCC progression [61–63]. Various studies have been done to define the balance among the HBV proliferation, apoptosis and HCC. For long-term persistence in the host cells, HBV may inhibit cell death by either activating oncogenes or disrupting signaling pathways, thereby promoting the HCC progression. HBV can also inhibit apoptosis and promote HCC development through the upregulation of some pro-growth proteins, such as cationic amino acid transporter 1 (CAT-1) [64]. In some clinical cases, the CTL response is also relatively weak in chronic HBV patients, culminating in apoptosis of a smaller proportion of infected hepatocytes [65,66]. In chronical HBV-infected mice, cIAPs restrict the TNF-mediated HBV elimination as well as HBV-infected hepatocytes death, while the inhibition of cIAPs by SMAC mimetics or silence of cIAPs boosts HBV clearance in the presence of TNF and HBV-specific CD4$^+$ T cells [67,68]. On the other hand, HBV induction of apoptosis is also reported in some papers [69–74]. The reason for the discrepant results of HBV effect on apoptosis is not known, but possibly due to the different experimental conditions or the HBV genotypes used in the different laboratories.

3.1. Inhibition of Apoptosis by Hepatitis B Virus

Among the HBV proteins, HBX is the most frequently reported one to be associated with the inhibition of apoptosis and the activation of HCC progression. It may block apoptosis through the sequestration of cytoplasmic p53, activation of PI3K-Akt pathway, inhibition of death receptor mediated apoptotic pathway, activation of NF-κB signaling pathway, inhibition of mitochondrial apoptotic pathway, as well as interplay with ncRNA [75–80]. The mechanisms of HBX inhibition of apoptosis are summarized below (Figure 1).

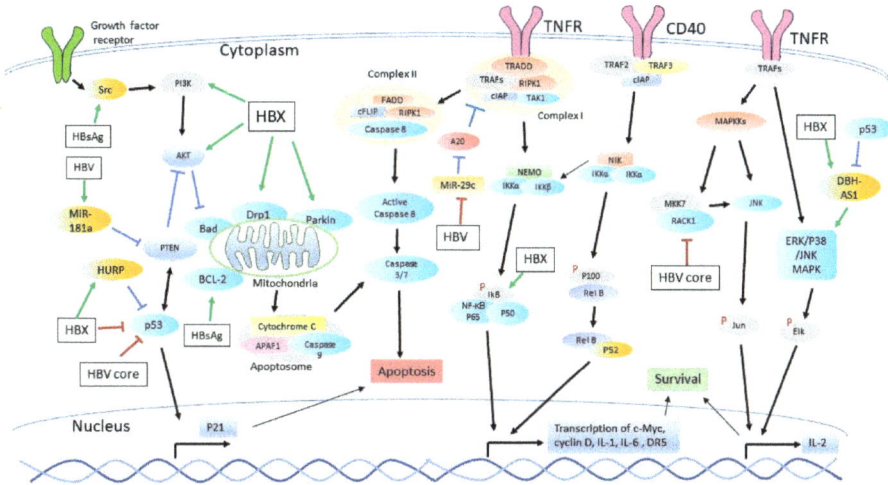

Figure 1. Inhibition of apoptosis by hepatitis B Virus (HBV) infection. Hepatitis B Virus X protein (HBX) and HBV core inhibit p53-mediated apoptosis. HBX activates the phosphatidylinositol-4,5-bisphosphate 3-kinase-protein kinase B (PI3K-Akt) pathway to inhibit apoptosis via the upregulation of PI3K and the induction of Akt phosphorylation. HBX inhibits the intrinsic apoptotic pathway by recruitment of Drp-1 and Parkin to the mitochondria for mitochondrial fission and mitophagy. The activation of Akt also prevents translocation of BAD to the mitochondria, thereby preventing apoptosis. HBX can activate the nuclear factor kappa-light-chain-enhancer of activated B cells (NF-κB) signaling via the degradation of IκB. In the MAPK-JNK pathway, HBV can attenuate the function of the kinase that activates JNK. HBV can downregulate apoptosis by either elevation of anti-apoptotic ncRNAs such as MiR-181a, or decrease of pro-apoptotic ncRNAs, such as MiR-29c. Green arrows next to the white boxes (HBV or HBV proteins) denote the activation of apoptosis. Red bars to the white boxes stand for the inhibition of apoptosis. Drp-1: Dynamin-1-like protein. BAD: BCL-2-associated death promoter. MAPK: MAP kinase. JNK: c-Jun N-terminal kinases. ncRNA: Non-coding RNA. MiR-181a/29c: MicroRNA 181a/29c.

3.1.1. Sequestration of p53 Signaling

P53, a tumor suppressor, is not only a transcription factor that regulates the expression of a variety of genes but also induces apoptosis [81,82]. A Japanese research group shows that by expressing HBX in HepG2 and HLF cells with Cre/Lox system, apoptosis is induced independently of p53 [83]. In contrast, HBX is demonstrated to be capable of abolishing p53-induced apoptosis by binding it (Figure 1) [75,84]. Mutations of nucleotides, A1762T and G1764A, of HBX are frequently reported in HBV isolated from chronically infected patients. The mutant HBX promotes the replication of HBV (subtype *adw*2), resulting in a higher viral load in hepatoma cells [85,86]. In HepG2.2.15 cells, the mutant HBX binds to p53 and blocks its downstream gene transcription, while the wild type HBX (subtype *ayw*) only binds to p53 without affecting the p53-mediated transcription [87]. In addition, the restriction of p53 signaling by HBX may vary in different cell types. In primary human hepatocytes, the wild type HBX (genotypes *ayw* and *adr*) is able to quench p53 in the cytoplasm, and the C terminal portion of HBX is responsible for the sequestration [88,89]. In contrast, in HepG2 and Hep3B tumor cells, HBX is pro-apoptotic and enhances the nuclear translocation of p53 through the activation of ATM kinase, which phosphorylates p53. However, in a later study, the inhibition of p53 by HBX could only be achieved when p53 is expressed at a relatively high level in the cells (HepG2, Hep3B and NIH/3T3 cells) [90]. The hepatoma upregulated protein (HURP), a cellular oncogene that mediates the degradation of p53, is upregulated by HBX in HCC, and consequently, inhibits the cisplatin-induced apoptosis [62]. Although the HBX and p53 interaction is reported, transduction of HBX in transgenic mice indicates that this interaction is not sufficient for tumor formation [91]. Therefore, the HBV-mediated inhibition of apoptosis through the interference of p53 may be dependent on cell types, experimental models, HBX structure and p53 level.

3.1.2. Activation of PI3K Pathway

In many cancers, the PI3K-Akt pathway is overactive, thus reducing apoptosis and allowing cell proliferation (Figure 1) [92]. Akt is normally considered as an anti-apoptotic protein through antagonizing pro-apoptotic proteins or facilitating the induction of other anti-apoptotic proteins [93–96]. In HBX-overexpressed Hep3B cells and 293T cells, Akt is activated to phosphorylate IκB kinase (IKKα) and promote its nuclear translocation, which is found to promote cell migration and invasion [97]. In Chang liver cells, HBX activates the PI3K-Akt, leading to the phosphorylation and blockade of BCL-2-associated death promoter (BAD), a pro-apoptotic protein inducing mitochondrial permeability transition pore (MPTP). Consequently, the cytochrome C release and apoptosis are prevented [98]. In a human placental trophoblastic cell line, HBX inhibits apoptosis via the elevation of PI3K expression to strengthen the activity of the PI3K-Akt pathway [99]. In addition, the transforming growth factor beta (TGF-β)-induced apoptosis in Hep3B cells can be rescued by the HBX-activated PI3K-Akt pathway [76]. A previous study shows that the anti-apoptotic effect of HBX is dependent on its isoforms [100]. The HBX isoform that contains the Akt phosphorylation site at Ser31 functions as an anti-apoptotic protein. This isoform can be phosphorylated by Akt and in turn activate the PI3K-Akt pathway. In contrast, the isoform that does not contain the Akt phosphorylation site plays an opposite function in apoptosis [100].

3.1.3. Inhibition of the Death Receptor-Mediated Apoptotic Pathway

In the extrinsic apoptotic pathway, HBX potently inhibits the caspase 3 activity [79,101]. HBX has been shown to inhibit the Fas-induced apoptosis, and this process is independent of p53 [79]. In this study, HBX transfection rate in primary hepatocytes is significantly enhanced from 5% to 80% by co-expressing HBX and enhanced green fluorescence protein (EGFP). Simultaneously, HBX inhibited the activation of caspase 8 and 3 and the release of cytochrome C. The HBX expression is associated with the upregulation of SAPK/JNK signaling, and furthermore, the [26]RXRXXS motif of HBX is essential for the SAPK upregulation and the inhibition of Fas-mediated cell killing. HBX induces the activation of NF-κB signaling via the degradation of inhibitor of kappa B (IκB), which also contributes to the inhibition of Fas-induced apoptosis (Figure 1) [63,102].

3.1.4. The Activation of NF-κB Pathway

NF-κB is generally regarded as a positive regulator of cell growth [103–105]. The NF-κB signaling consists of the canonical and the non-canonical NF-κB signaling pathways (Figure 1). The canonical NF-κB signaling is initiated through receptors such as toll-like receptors (TLRs), tumor necrosis factor receptor (TNFR), T-cell receptor (TCR) or B cell receptor (BCR). The receptor-mediated activation of transforming growth factor beta-activated kinase 1 (TAK1) phosphorylates IKK complex, which consequently degrades IκB, leading to the release of NF-κB heterodimer (p65/p50) into the nucleus. The non-canonical pathway is triggered by a signaling from a subset of TNFR members, such as B cell activating factor receptor (BAFFR), CD40, lymphotoxin β-receptor (LTβT) and receptor activator for nuclear factor κB (RANK). Through the activation of NF-kappa-B-inducing kinase (NIK) and IKKα, p100 is processed into the active p52, which forms a heterodimer with *RelB*. The subsequent nuclear translocation of the RelB/p52 results in a persistent stimulation of the pathway [106]. Moreover, the accumulation of NIK is reported to activate the canonical NF-κB pathway through the enhancement of IKK complex activity [107]. In the process, the anti-apoptotic protein IAPs can both positively and negatively regulate the canonical or the non-canonical NF-κB signaling. For instance, upon engagement of TNFR (see Section 2), the cIAP in complex I promotes the ubiquitination of RIPK1, which leads to the activation of TAK1 [108]. In addition, the ubiquitinated RIPK1 can prevent apoptosis by suppressing the formation of complex II and the activation of caspase 8. In contrast, IAPs also exert an inhibitory role in the NF-κB pathways. For example, the basal level of NIK activation is very low due to the control of upstream TRAF3-TRAF2-cIAP complex, and cIAP1/2 degrades NIK by ubiquitinating the protein, while the loss of any component of the complex leads to an accumulation of NIK and the activation of both NF-κB signaling pathways [107,109–111].

The NF-κB signaling is constitutively activated in many cancers, and the NF-κB activation contributes to tumorigenesis [112,113]. There are several mechanisms that NF-κB antagonizes cell death. First, the NF-κB activation leads to an elevation of anti-apoptotic genes. Secondly, NF-κB induces the production of immune response cytokines, such as TNF-α, IL-1 (Interleukin-1), IL-6, and IL-8. Moreover, NF-κB induces the expression of some pro-oncogenic genes, such as *cyclin D1*, *c-Myc*, and cIAPs [114]. In addition, the NF-κB signaling contributes to tumor progression by facilitating epithelial to mesenchymal transition and metastasis, as well as aiding the vascularization of tumors via the upregulation of vascular endothelial growth factor (VEGF) [115–117]. The activation of NF-κB signaling increases the stability of HBX protein [118]. Several studies have demonstrated that HBV infection leads to the activation of NF-κB, followed by the inhibition of apoptosis and increase of cell progression (Figure 1) [119–124]. HBX induces the activation of NF-κB by degrading IκB [119,120]. IκB is responsible for sequestering NF-κB in the cytoplasm. Once IκB is phosphorylated and degraded, the NF-κB heterodimer translocates into the nucleus and initiates the transcription of downstream genes. To examine the correlation between NF-κB activation and apoptosis, IκB-SR, an isoform that cannot be phosphorylated, is introduced into NIH/3T3 cells. The cotransfection of IκB-SR and HBX results in increased apoptosis [121]. In the presence of IκB-SR, HBX overexpression induces MPTP. Notably, in the context of HBV replication, HBX activates NF-κB and inhibits the cytochrome C release from the mitochondria. However, when the NF-κB activity is inhibited, the HBX in the context of HBV replication could induce apoptosis through MPTP. This study indicates that, depending on the status of NF-κB activity, HBX can be either pro- or anti-apoptotic [122]. In the HBV-positive cell line, HepG2.2.15, the cIAP1 and cIAP2 are expressed much higher than in HepG2, indicating that HBV replication might boost the anti-apoptotic proteins [78]. In addition, the activation of NF-κB is found to initiate enhanced transcription of both anti-apoptotic genes such as gp96, survivin, p21 and the pro-apoptotic genes such as death receptor 5 (*DR5*) [123]. The expression of the anti-apoptotic genes may be responsible for the multidrug resistance of HBX-transfected HepG2 cells [124]. On the other hand, the upregulation of death domain receptor accounts for the increased sensitivity of cells to apoptotic stimuli. In different cell lines, the regulation effect of HBX also varies [125]. To illustrate this point, two HBX-expressing stable cell lines were established, namely, Huh-7-X and CHANG-X. The mRNA levels of p21, p27, and TGF-β are drastically downregulated in Huh-7-X stable cells but have a minimum change in CHANG-X stable cells [125]. Collectively, the NF-κB signaling is important not only in the innate immune system but also in the release of cell stress and the promotion of hepatocytes growth, as well as in the regulation of apoptosis upon HBV infection.

3.1.5. Inhibition of the Mitochondria-Mediated Apoptotic Pathway

Sequence analysis from tumor tissues and para-tumor tissues of 47 patients shows a combination of mutations (10Ala/Arg and 144Ser/Arg) exists in HBX with high frequency [61]. HBX harboring these two mutations reduces BAX expression and inhibits apoptosis in HepG2 cells. HBX is also able to inhibit serum-starvation induced mitochondrial apoptosis via the activation of autophagy, which is featured by increased microtubule-associated proteins 1A/1B light chain 3B (LC3II) and Beclin-1 [126,127]. As a core component of PI3K-III complex, Beclin-1 plays an important role in autophagy and cell death [128], while the interaction between BCL-2 and Beclin-1 does not counteract the anti-apoptotic role of BCL-2 [129]. HBX can sequester AIF, a caspase-independent protein in the intrinsic pathway, in the cytoplasm, resulting in the prevention of DNA fragmentation and apoptosis [130]. In Huh-7 cells, HBV and HBX can disrupt mitochondrial dynamics by inducing the translocation of dynamin-related protein Drp-1 to the mitochondria and the subsequent mitochondrial fission [77]. Parkin, an E3 ligase, is also translocated to the mitochondria and associated with the mitophagosome triggered by HBV/HBX. Parkin expression is upregulated in the presence of HBV/HBX. The enhanced Parkin level promotes the mitophagy, which attenuates apoptosis. Silencing of Parkin induces the mitochondrial apoptotic signaling. Thus, HBV promotes aberrant mitochondrial dynamics to protect cells from apoptosis in HepAD38 cells [77]. In chronic HBV-infected patients, the mitochondrial polarization in CD8$^+$ T cells is impaired, and a higher level of ROS is detected in chronic patients in comparison with healthy individuals [131]. Further, the restoration of mitochondrial function via mitochondria-targeted antioxidants reactivates the exhausted T cells and helps with HBV clearance in the chronic HBV patients [131].

3.1.6. Interference of Apoptosis through ncRNA

NcRNA accounts for 90% of genomic RNA, and it can be divided into the long non-coding RNA (lncRNA) and the microRNA (miRNA or MiR hereafter) [132]. Recently, increasing studies show that ncRNA has essential biological functions such as modulating cell proliferation, cell cycle, apoptosis, invasion and metastasis in cancers [133]. The interaction of HBV and ncRNA has also been widely studied. HBV and HBX inhibit the cell apoptosis by the interference of ncRNA. This is supported by the observation that HBV or HBX-transfected HepG2 cells have significantly upregulated MiR-181a and decreased PTEN, a tumor suppressor protein. PTEN inhibits PI3K-Akt and protects p53 by attenuating the mouse double minute 2 homolog (Mdm2) translocation into the nucleus (Figure 1) [134]. Upregulation of miR-181a suppresses PTEN expression, and inhibition of miR-181a abolishes the inhibitory effect of HBX on PTEN protein [135]. A novel lncRNA DBH-AS1 has been shown to activate ERK/p38/JNK MAPK (extracellular signal-regulated kinases/p38/ c-Jun N-terminal kinases mitogen-activated protein kinase) signaling and promote cell proliferation. HBX promotes the generation of DBH-AS1, thereby inhibiting serum starvation-induced apoptosis in HCC [136]. MiR-221, promoting cell proliferation by suppression of estrogen receptor-α, is also obviously increased in HBX-transfected HCC cells [137]. Aside from the function of HBV proteins, HBV transcripts in transgenic mice absorb the MiR-15a/16 and increase expression of the anti-apoptotic proteins BCL-2 and Smad7 [138,139].

In HBV-transfected HCC cell lines and clinical tumor tissues, pro-apoptotic microRNA MiR-29c is significantly downregulated [80]. The MiR-29c inhibits cell proliferation through suppressing A20, an E3 ligase negatively regulating NF-κB signaling and TNF-induced apoptosis via downregulating the E3 ligase activity of TRAF2 and TRAF6 (Figure 1) [140]. In HBV-related HCC patients, the MiR-122 and MiR-22 are significantly lower than those in benign liver diseases and non-HBV-related HCC patients, underlying that the miRNAs play vital roles in the HBV-related HCC formation [141]. These data suggested that ncRNA could possibly play an important role in the regulation of cell progression, either positively or negatively, while HBV may interfere with cell apoptosis through the modulation of those ncRNAs. The research progress on the interplay between ncRNA and apoptosis during HBV infection is recently reviewed in detail by Zhang et al. [142].

3.1.7. Other Inhibitory Pathways

In addition to the signaling pathways described above, HBV also inhibits apoptosis by the upregulation of pro-oncogenesis genes or the activation of cell progression pathway. For instance, cell division control protein 42 homolog (CDC42), a member of the Rho GTPase family, is known to facilitate tumorigenesis and cancer progression. It is upregulated in HBX-overexpressed Huh-7 cells, resulting in higher cell proliferation and reduced apoptosis [143]. Manganese superoxide dismutase (MnSOD) is responsible for scavenging superoxide anion and preventing cells from DNA damage. HBV infection increases the expression of MnSOD, which is mediated by HBX protein [144]. Notch signaling and Smad pathway promote cell proliferation, while, in HCC and HTR-8/SVneo cells, HBX expression activates these pathways to suppress apoptosis [145,146].

3.2. Pro-Apoptotic Effect of HBV and HBX

Although the suppression of apoptosis contributes to the progression of carcinogenesis, apoptosis can still be observed in untreated malignant tumors [147]. The apoptosis could be induced by CTL in tumor tissue or could be activated by TNF-α treatment [72,148]. HBV can also activate apoptosis or sensitize host cells to apoptosis induction in in vitro studies, through the direct activation of apoptotic proteins, regulation of Ca^{2+} concentration or the upregulation of cell death receptors [149,150]. Following are signaling pathways that HBV interrupts to induce cell apoptosis (Figure 2).

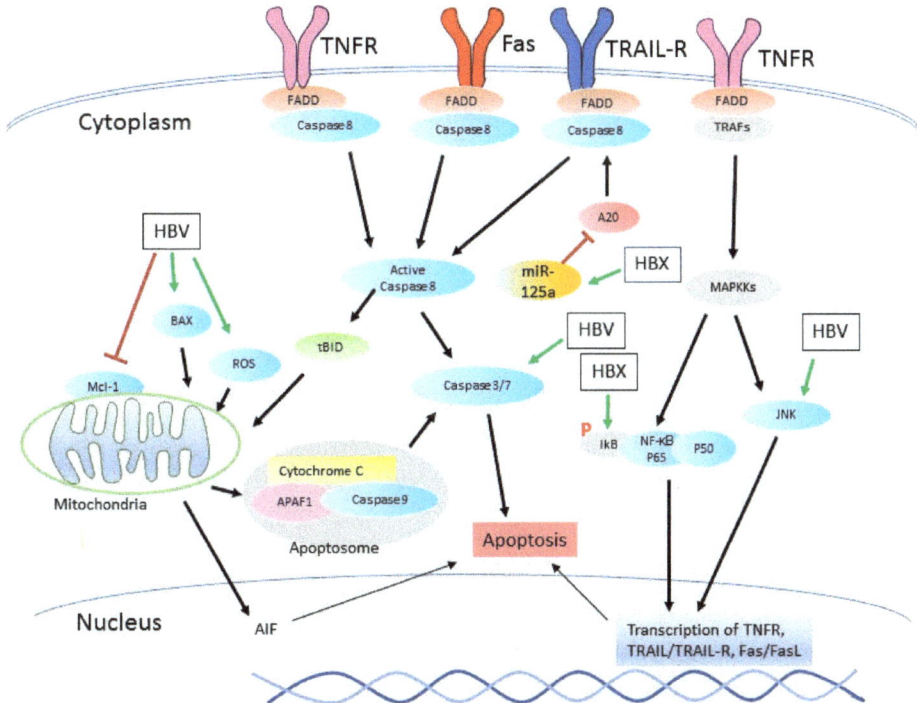

Figure 2. Pro-apoptotic role of HBV. HBV infection causes the activation of NF-κB in hepatoma cells, and subsequent excessive expression of the death-associated receptors, which increase the cell sensitivity to stimuli. In addition, HBV directly induces the cleavage of caspase 3 to activate apoptosis. HBV induces the mitochondrial apoptotic signaling pathway by increasing the BAX expression and the ROS level or downregulating Mcl-1. The BCL-2 homology domain 3 (BH3)-like domain in HBX also plays a role in the induction of apoptosis. Green arrows mean the activation step of apoptosis; Red bars stand for the inhibition step of apoptosis.

3.2.1. Death Receptor-Mediated Signaling Pathways

HBV has been reported to induce apoptosis in liver biopsies of HBV patients [151]. The virus infection in transgenic mice and hepatocytes increases the cell sensibility to TRAIL-induced apoptosis by increasing the expression of BAX (Figure 2) [73]. HBX expression in hepatocytes has the same outcome as HBV infection. In clinical HBV liver samples, the TRAIL expression in chronic hepatitis B samples is the highest in comparison to acute hepatitis B samples, liver cirrhosis, and normal liver samples. This result suggests that TRAIL expression may have some correlation with the extent of liver injury [152]. Further studies show that the death receptor TRAIL-R5 expression is enhanced by HBX in Huh-7 cells through the activation of the NF-κB pathway, which contributes to the increased apoptosis induced by TRAIL [74]. Although A20 is upregulated in HBV-infected HCC cells, liver tissue, and serum of chronic HBV-infected patients [80,153], HBX overexpression in hepatocytes leads to A20 reduction [154]. HBX sensitizes the hepatocytes to the TRAIL-induced apoptosis by repressing the A20 expression and its ubiquitin ligase activity (Figure 2) [154].

In addition to TRAIL, HBX is also able to sensitize cells to the TNF-α-induced apoptosis through the activation of MAP kinase kinase kinase/c-Jun N-terminal kinases (MEKK/JNK) signaling pathway and the nuclear accumulation of N-Myc [72], or by decreasing the expression of Bcl-xL [155].

Another death receptor, Fas, and its ligand FasL are upregulated in rat renal tubular epithelial cells (NRK-52E) transfected with HBX, and this increase is due to the activation of the MLK3-MKK7-JNK pathway [156]. The Fas sensitivity is reconstituted in HBX transgenic mice through the decrease of BCL-2, despite no direct interaction between HBX and BCL-2 family members [70]. Another group demonstrates that HBX activates the p38 MAP kinase and JNK pathways, inducing the transcription of Fas/FasL and TNFR1/TNF-α. The increased expression of death receptors induces the cleavage of pro-caspase 8, with subsequent tBID activation and cytochrome C release [150]. In addition, a recently identified novel ORF of HBV, HBwX that fuses HBX and its upstream 56 amino acid residues, has been shown to sensitize HCC cells to the adriamycin (ADM) and LPS-induced apoptosis [157].

3.2.2. The Mitochondria-Mediated Cell Death

The interaction between HBV and the mitochondria-mediated apoptosis is extensively studied. HBX overexpression contributes to the aggregation of the mitochondria, leading to cell death [158]. In the same study, HBX was found to colocalize with p53, but this association has no correlation with mitochondrial aggregation, suggesting two independent mechanisms in apoptosis induction. HBX binds to BAX in HepG2 cells, leading to enhanced translocation to the mitochondria, followed by loss of mitochondrial membrane potential [71]. Aside from the pathways mentioned above, HBV also sensitizes HL7702 cells to the oxidative stress-induced apoptosis through increasing the opening of MPTP [159]. The Mcl-1, a member of the anti-apoptotic BCL-2 family, is drastically declined during this process (Figure 2) [160].

BH3-like protein is an important initiator of the mitochondrial apoptotic pathway. Both HBX and a spliced viral protein, HBSP, have BH3 domain. Both proteins induce caspase 3 dependent apoptosis in HepG2 cells, while an amino acid mutation in the BH3 domain results in loss of the capability to induce apoptosis [161]. Possibly due to mutations in the BH3 domain of genotypes A and C of HBV, the two genotypes have weaker pro-apoptotic activity than genotype B in HepG2 cells [162]. Further study shows that HBX causes apoptosis in *Caenorhabditis elegans* by targeting BCL-2 homolog protein CED9, through the interaction between the BH3 domain and CED9 [163]. However, a recent structural and biochemistry analysis presents an opposite result, that is, the interaction between BCL-2 and HBX BH3-like domain is much weaker than the canonical BH3 and BCL-2 [164], indicating that the mechanism of HBX BH3 motif interfering with BCL-2 might be different. In addition, HBV has been shown to induce oxidative stress, accompanied by increased ROS level [165].

3.3. *The Roles of the Other HBV Proteins*

Aside from HBX protein, the other HBV proteins also have roles in cell apoptosis, either positive or negative. For example, HBsAg prevents the translocation through interaction with jumping translocation breakpoint protein (JTB), thereby inhibiting cell apoptosis mediated by JTB [166]. The large HBsAg glycoprotein inhibits apoptosis by activating the Src/PI3K/Akt pathway through the activation of Src kinase (Figure 1) [167]. Simultaneously, a prevalent mutant large HBsAg protein with the deletion of amino acids 2 to 55 of the pre-S2 region, enhances the expression of pro-survival BCL-2 proteins; and BCL-2 contributes to 5-fluorouracil resistance in Huh-7 cells [168]. In addition, the HBV core protein inhibits apoptosis in HepG2 cells via the downregulation of Fas, p53 and FasL [169]. Another group shows that the core protein impairs the phosphorylation of mitogen-activated protein kinase kinase 7 (MKK7) by binding to its scaffold protein, RACK1, which consequently down-regulates JNK pathway and sensitizes HepG2 cells to TNF-α-induced apoptosis (Figure 1) [170].

4. Conclusions and Perspective

As an HCC-associated virus, HBV has drawn a lot of attention. To investigate the virus–cell interaction, a series of cell models are used in the experiments [171]. Primary human hepatocytes are the most physiologically relevant in vitro model for HBV infection with the natural viral receptor. However, the constraints of this cell model cannot be ignored, such as the limited span of life, limited sources and disparity from different donors. Huh-7 and HepG2 cell lines are commonly used in the studies, though tumor cells can only partially represent the physiological hepatic functions. To overcome the in vitro virus culture problem, hepatocyte cell lines stably transduced with HBV (HepAD38 and HepG2.2.15) are established as a source of HBV infectious particles [3,172–174]. However, the stable cell lines are not susceptible to HBV infection due to the lack of efficient receptors [175,176]. To obtain an efficient HBV replication, HepG2 cells stably expressing the HBV receptor (NTCP) have been established and might be useful for the basic research of HBV biology [3,174].

Although the HCC development is usually featured by the inhibition of apoptosis, the outcome of cell apoptosis is a result of a complex biological process, as there are different regulations of multiple cell signaling pathways by the various HBV proteins. Even the same apoptotic signaling pathway can be affected towards opposite consequences in the cells, such as the mitochondrial apoptotic pathway. This pathway can be activated by the BAX insertion into the mitochondrial membrane, which is driven by HBV (Figure 2). In contrast, this pathway can be inhibited by the recruitment of Parkin and Drp-1 during HBV infection (Figure 1). Activation of NF-κB signaling is a double-edged sword: promoting the expression of pro-survival genes to facilitate the cell proliferation and upregulating the expression of the death-associated receptor to sensitize cells to apoptotic stimuli. In the cell models with HBX overexpression, one of the potential considerations is that the protein overexpression level should mimic the "physiological level" in the HBV-infected patients, which can be achieved by optimization of transfection or a vector with a mild promoter [177]. In HepG2 cells, the HBX-deficient HBV replication can be rescued even when HBX is expressed at a very low level (beyond the detection limit of Western Blotting) [178]. The HBX functions identified in transiently transfected cells can be further assessed in HBV cell culture models. Therefore, the HBV effect on cell apoptosis varies depending on cellular context, different signaling pathways, HBV genotypes, protein mutations and possibly different clinical stages.

Technically, the cell culture models exhibit certain extent of defect due to its failure to mimic the host microenvironment. The mechanism of interaction between the virus and cell apoptosis should be further investigated under a more comprehensive situation. It is worthy to note that the frequently used cell line in the HBV biology study, Chang cells, are reported to have HeLa cell contamination at least in several clones [179]. Additional caution is needed when using this cell model in further studies.

Although apoptosis induction is widely considered as a positive strategy against cancer, the apoptotic cells are able to promote proliferation of the surrounding tumor cells by affecting the microenvironment [180–182]. In addition, there is no exact study of HBV effect on cell apoptosis in different clinical stages. Since HBV is frequently detected in clinical HCC cancers, it is more relevant that HBV/HBX accelerates cell transformation and facilitates apoptosis inhibition as a long-term effect. In most studies, HBV/HBX is shown to promote the anti-apoptotic proteins or inhibit the function of pro-apoptotic proteins. However, in some experiments, the higher level of ROS or the upregulated death associated receptors in the hepatocytes might be factors for apoptosis induction. In addition, co-effect of HBV/HBX levels in infected patients and the stimulation from the surrounding environment needs to be considered. We postulate there might be certain signaling-competing mechanism during the disease progression under the comprehensive effect of these myriad factors. The HBV levels in patients with acute or chronic HBV infection may be monitored and analyzed for correlation with outcomes in different clinical stages, if possible. Due to the limited host range of the virus, the establishment of an efficient animal model is experimentally important. According to the sequence analysis, duck hepatitis B virus and woodchuck hepatitis virus may be surrogate viruses in the HBV biological study. In addition, transgenic mice are frequently used in the pathogenesis and immune response studies of HBV infection. A more advanced experimental model is much needed to better elucidate the interaction mechanism between HBV and apoptosis.

Acknowledgments: S.L. was partially sponsored by China Scholarship Council. This project was partially funded by an internal fund from the University of Maryland (College Park, MD, USA).

Author Contributions: S.L. wrote the paper. Y.Z. guided the process and edited the paper.

Conflicts of Interest: The authors declare no conflict of interest.

Abbreviations

TNF-α	Tumor necrosis factor alpha
TRAIL	TNF-related apoptosis-inducing ligand
APO2L	APO2 ligand
APO3L	APO3 ligand
BCL-2	B-cell lymphoma 2
BAK	BCL-2 homologous antagonist/killer
FADD	Fas-associated protein with death domain
DISC	death-inducing signaling complex
RIP	receptor-interacting protein
X-IAP	X-linked inhibitor of apoptosis protein
TRADD	TNFR type 1-associated DEATH domain protein
RIPK1	receptor-interacting serine/threonine-protein kinase 1
cIAP	cellular inhibitors of apoptosis
TRAF2/3/5/6	TNF receptor associated factors 2/3/5/6
cFLIP	Cellular FLICE-inhibitory protein
TCR	T cell receptor
BCR	B cell receptor
TAK1	Transforming growth factor beta-activated kinase 1
BAX	BCL-2-associated X protein
BAK	BCL-2 homologous antagonist/killer
APAF-1	apoptotic protease activating factor 1
ROS	reactive oxygen species
BAD	BCL-2-associated death promoter
BID	BH3 interacting-domain death agonist
SMAC	a second mitochondrial activator of caspases
AIF	apoptosis inducing factor
MOMP	mitochondrial outer membrane permeabilization
ncRNA	non-coding RNA
miR	microRNA
lncRNA	long coding RNA
HURP	Hepatoma upregulated protein
MPTP	mitochondrial permeability transition pore
PI3K	Phosphatidylinositol-4,5-bisphosphate 3-kinase
Akt	Protein kinase B
TGF-β	Transforming growth factor beta
SAPK	stress-activated protein kinase
BAFFR	TNF family receptor
LTβR	lymphotoxin β-receptor
RANK	receptor activator for nuclear factor κB

NIK	NF-kappa-B-inducing kinase
IKK	IκB kinase
PTEN	Phosphatase and tensin homolog
gp96	Heat shock protein 90kDa beta member 1
Mcl-1	Induced myeloid leukemia cell differentiation protein
p21	s cyclin-dependent kinase inhibitor 1
p27	Cyclin-dependent kinase inhibitor 1B
DR5	Death receptor 5/TRAIL receptor 2
IL-1/-6/-8	Interleukin-1/-6/-8
ER	endoplasmic reticulum
Mdm2	Mouse double minute 2 homolog
LC3 II	Microtubule-associated proteins 1A/1B light chain 3B
STAT3	Signal transducer and activator of transcription 3
CDC42	Cell division control protein 42 homolog
EGFP	Enhanced green fluorescent protein
CAT-1	Cationic amino acid transporter 1
Smad7	Mothers against decapentaplegic homolog 7
MLK3	Mitogen-activated protein kinase kinase kinase 3
MKK7	Dual specificity mitogen-activated protein kinase kinase 7
ERK	extracellular signal-regulated kinases
JNK	c-Jun N-terminal kinases
MEKK	MAP kinase kinase kinase
RACK	Receptor for activated C-kinase
Drp-1	Dynamin-1-like protein
Elk	ETS domain-containing protein
MAPKKs	MAP kinase cascades
NTCP	sodium taurocholate cotransporting polypeptide

References

1. Urban, S.; Schulze, A.; Dandri, M.; Petersen, J. The replication cycle of hepatitis B virus. *J. Hepatol.* **2010**, *52*, 282–284. [CrossRef] [PubMed]
2. Nassal, M. Hepatitis B viruses: Reverse transcription a different way. *Virus Res.* **2008**, *134*, 235–249. [CrossRef] [PubMed]
3. Yan, H.; Zhong, G.; Xu, G.; He, W.; Jing, Z.; Gao, Z.; Huang, Y.; Qi, Y.; Peng, B.; Wang, H.; et al. Sodium taurocholate cotransporting polypeptide is a functional receptor for human hepatitis B and D virus. *Elife* **2012**, *1*, e00049. [CrossRef] [PubMed]
4. Ott, J.J.; Stevens, G.A.; Groeger, J.; Wiersma, S.T. Global epidemiology of hepatitis B virus infection: New estimates of age-specific HBsAg seroprevalence and endemicity. *Vaccine* **2012**, *30*, 2212–2219. [CrossRef] [PubMed]
5. Revill, P.; Testoni, B.; Locarnini, S.; Zoulim, F. Global strategies are required to cure and eliminate HBV infection. *Nat. Rev. Gastroenterol. Hepatol.* **2016**, *13*, 239–248. [CrossRef] [PubMed]
6. Wang, M.; Xi, D.; Ning, Q. Virus-induced hepatocellular carcinoma with special emphasis on HBV. *Hepatol. Int.* **2017**, *11*, 171–180. [CrossRef] [PubMed]
7. Kim, H.S.; Rotundo, L.; Yang, J.D.; Kim, D.; Kothari, N.; Feurdean, M.; Ruhl, C.; Unalp-Arida, A. Racial/ethnic disparities in the prevalence and awareness of Hepatitis B virus infection and immunity in the United States. *J. Viral Hepat.* **2017**, 1–15. [CrossRef] [PubMed]
8. Tian, Q.; Jia, J. Hepatitis B virus genotypes: Epidemiological and clinical relevance in Asia. *Hepatol. Int.* **2016**, *10*, 854–860. [CrossRef] [PubMed]
9. Inoue, T.; Tanaka, Y. Hepatitis B virus and its sexually transmitted infection—An update. *Microb. Cell* **2016**, *3*, 420–437. [CrossRef] [PubMed]
10. Li, Z.; Hou, X.; Cao, G. Is mother-to-infant transmission the most important factor for persistent HBV infection? *Emerg. Microb. Infect.* **2015**, *4*, e30. [CrossRef] [PubMed]

11. Blumberg, B.S. The discovery of the hepatitis B virus and the invention of the vaccine: A scientific memoir. *J. Gastroenterol. Hepatol.* **2002**, *17*, S502–S503. [CrossRef] [PubMed]

12. Schweitzer, I.L.; Dunn, A.E.; Peters, R.L.; Spears, R.L. Viral hepatitis b in neonates and infants. *Am. J. Med.* **1973**, *55*, 762–771. [CrossRef]

13. Pita, I.; Horta-Vale, A.M.; Cardoso, H.; Macedo, G. Hepatitis B inactive carriers: An overlooked population? *GE Port. J. Gastroenterol.* **2014**, *21*, 241–249. [CrossRef]

14. Xuan, D.; Yu, Y.; Shao, L.; Wang, J.; Zhang, W.; Zou, H. Hepatitis reactivation in patients with rheumatic diseases after immunosuppressive therapy—A report of long-term follow-up of serial cases and literature review. *Clin. Rheumatol.* **2014**, *33*, 577–586. [CrossRef] [PubMed]

15. Chen, J.D.; Yang, H.I.; Iloeje, U.H.; You, S.L.; Lu, S.N.; Wang, L.Y.; Su, J.; Sun, C.A.; Liaw, Y.F.; Chen, C.J. Carriers of inactive hepatitis B virus are still at risk for hepatocellular carcinoma and liver-related death. *Gastroenterology* **2010**, *138*, 1747–1754. [CrossRef] [PubMed]

16. Zhao, L.H.; Liu, X.; Yan, H.X.; Li, W.Y.; Zeng, X.; Yang, Y.; Zhao, J.; Liu, S.P.; Zhuang, X.H.; Lin, C.; et al. Genomic and oncogenic preference of HBV integration in hepatocellular carcinoma. *Nat. Commun.* **2016**, *7*, 12992. [CrossRef] [PubMed]

17. Yan, H.; Yang, Y.; Zhang, L.; Tang, G.; Wang, Y.; Xue, G.; Zhou, W.; Sun, S. Characterization of the genotype and integration patterns of hepatitis B virus in early- and late-onset hepatocellular carcinoma. *Hepatology* **2015**, *61*, 1821–1831. [CrossRef] [PubMed]

18. Plummer, M.; de Martel, C.; Vignat, J.; Ferlay, J.; Bray, F.; Franceschi, S. Global burden of cancers attributable to infections in 2012: A synthetic analysis. *Lancet. Glob. Health* **2016**, *4*, e609–e616. [CrossRef]

19. Higaki, K.; Yano, H.; Kojiro, M. Fas antigen expression and its relationship with apoptosis in human hepatocellular carcinoma and noncancerous tissues. *Am. J. Pathol.* **1996**, *149*, 429–437. [PubMed]

20. Barber, G.N. Host defense, viruses and apoptosis. *Cell Death Differ.* **2001**, *8*, 113–126. [CrossRef] [PubMed]

21. Galluzzi, L.; Brenner, C.; Morselli, E.; Touat, Z.; Kroemer, G. Viral control of mitochondrial apoptosis. *PLoS Pathog.* **2008**, *4*, e1000018. [CrossRef] [PubMed]

22. He, P.; Zhang, D.; Li, H.; Yang, X.; Li, D.T.; Zhai, Y.Z.; Ma, L.; Feng, G.H. Hepatitis B virus X protein modulates apoptosis in human renal proximal tubular epithelial cells by activating the JAK2/STAT3 signaling pathway. *Int. J. Mol. Med.* **2013**, *31*, 1017–1029. [PubMed]

23. Clurman, B.E.; Hayward, W.S. Multiple proto-oncogene activations in avian leukosis virus-induced lymphomas: Evidence for stage-specific events. *Mol. Cell. Biol.* **1989**, *9*, 2657–2664. [CrossRef] [PubMed]

24. Peng, Z.; Zhang, Y.; Gu, W.; Wang, Z.; Li, D.; Zhang, F.; Qiu, G.; Xie, K. Integration of the hepatitis B virus X fragment in hepatocellular carcinoma and its effects on the expression of multiple molecules: A key to the cell cycle and apoptosis. *Int. J. Oncol.* **2005**, *26*, 467–473. [CrossRef] [PubMed]

25. Hwang, G.Y.; Lin, C.Y.; Huang, L.M.; Wang, Y.H.; Wang, J.C.; Hsu, C.T.; Yang, S.S.; Wu, C.C. Detection of the hepatitis B virus X protein (HBX) antigen and anti-HBX antibodies in cases of human hepatocellular carcinoma. *J. Clin. Microbiol.* **2003**, *41*, 5598–5603. [CrossRef] [PubMed]

26. Feitelson, M.A.; Lee, J. Hepatitis B virus integration, fragile sites, and hepatocarcinogenesis. *Cancer Lett.* **2007**, *252*, 157–170. [CrossRef] [PubMed]

27. Seifer, M.; Hohne, M.; Schaefer, S.; Gerlich, W.H. In vitro tumorigenicity of hepatitis B virus DNA and HBX protein. *J. Hepatol.* **1991**, *13* (Suppl. 4), S61–S65. [CrossRef]

28. Jung, J.K.; Park, S.H.; Jang, K.L. Hepatitis B virus X protein overcomes the growth-inhibitory potential of retinoic acid by downregulating retinoic acid receptor-beta2 expression via DNA methylation. *J. Gen. Virol.* **2010**, *91*, 493–500. [CrossRef] [PubMed]

29. Prost, S.; Ford, J.M.; Taylor, C.; Doig, J.; Harrison, D.J. Hepatitis B X protein inhibits p53-dependent DNA repair in primary mouse hepatocytes. *J. Biol. Chem.* **1998**, *273*, 33327–33332. [CrossRef] [PubMed]

30. Becker, S.A.; Lee, T.H.; Butel, J.S.; Slagle, B.L. Hepatitis B virus X protein interferes with cellular DNA repair. *J. Virol.* **1998**, *72*, 266–272. [PubMed]

31. Geng, M.; Xin, X.; Bi, L.Q.; Zhou, L.T.; Liu, X.H. Molecular mechanism of hepatitis B virus X protein function in hepatocarcinogenesis. *World J. Gastroenterol.* **2015**, *21*, 10732–10738. [CrossRef] [PubMed]

32. Wei, X.; Xiang, T.; Ren, G.; Tan, C.; Liu, R.; Xu, X.; Wu, Z. miR-101 is down-regulated by the hepatitis B virus X protein and induces aberrant DNA methylation by targeting DNA methyltransferase 3A. *Cell Signal.* **2013**, *25*, 439–446. [CrossRef] [PubMed]

33. Guicciardi, M.E.; Gores, G.J. Apoptosis: A mechanism of acute and chronic liver injury. *Gut* **2005**, *54*, 1024–1033. [CrossRef] [PubMed]

34. Wyllie, A.H. Glucocorticoid-induced thymocyte apoptosis is associated with endogenous endonuclease activation. *Nature* **1980**, *284*, 555–556. [CrossRef] [PubMed]

35. Shiokawa, D.; Maruta, H.; Tanuma, S. Inhibitors of poly(ADP-ribose) polymerase suppress nuclear fragmentation and apoptotic-body formation during apoptosis in HL-60 cells. *FEBS Lett.* **1997**, *413*, 99–103. [CrossRef]

36. Kurosaka, K.; Takahashi, M.; Watanabe, N.; Kobayashi, Y. Silent Cleanup of Very Early Apoptotic Cells by Macrophages. *J. Immunol.* **2003**, *171*, 4672–4679. [CrossRef] [PubMed]

37. Elmore, S. Apoptosis: A review of programmed cell death. *Toxicol. Pathol.* **2007**, *35*, 495–516. [CrossRef] [PubMed]

38. Riedl, S.J.; Shi, Y. Molecular mechanisms of caspase regulation during apoptosis. *Nat. Rev. Mol. Cell Biol.* **2004**, *5*, 897–907. [CrossRef] [PubMed]

39. Slee, E.A.; Adrain, C.; Martin, S.J. Executioner caspase-3, -6, and -7 perform distinct, non-redundant roles during the demolition phase of apoptosis. *J. Biol. Chem.* **2001**, *276*, 7320–7326. [CrossRef] [PubMed]

40. Marsters, S.A.; Sheridan, J.P.; Donahue, C.J.; Pitti, R.M.; Gray, C.L.; Goddard, A.D.; Bauer, K.D.; Ashkenazi, A. Apo-3, a new member of the tumor necrosis factor receptor family, contains a death domain and activates apoptosis and NF-kappa B. *Curr. Biol.* **1996**, *6*, 1669–1676. [CrossRef]

41. Micheau, O.; Tschopp, J. Induction of TNF receptor I-mediated apoptosis via two sequential signaling complexes. *Cell* **2003**, *114*, 181–190. [CrossRef]

42. Hu, S.; Yang, X. Cellular inhibitor of apoptosis 1 and 2 are ubiquitin ligases for the apoptosis inducer Smac/DIABLO. *J. Biol. Chem.* **2003**, *278*, 10055–10060. [CrossRef] [PubMed]

43. Lau, R.; Pratt, M.A. The opposing roles of cellular inhibitor of apoptosis proteins in cancer. *ISRN Oncol.* **2012**, *2012*, 928120. [CrossRef] [PubMed]

44. Guicciardi, M.E.; Mott, J.L.; Bronk, S.F.; Kurita, S.; Fingas, C.D.; Gores, G.J. Cellular inhibitor of apoptosis 1 (cIAP-1) degradation by caspase 8 during TNF-related apoptosis-inducing ligand (TRAIL)-induced apoptosis. *Exp. Cell Res.* **2011**, *317*, 107–116. [CrossRef] [PubMed]

45. Varfolomeev, E.; Goncharov, T.; Fedorova, A.V.; Dynek, J.N.; Zobel, K.; Deshayes, K.; Fairbrother, W.J.; Vucic, D. c-IAP1 and c-IAP2 are critical mediators of tumor necrosis factor alpha (TNF-alpha)-induced NF-kappaB activation. *J. Biol. Chem.* **2008**, *283*, 24295–24299. [CrossRef] [PubMed]

46. Mahoney, D.J.; Cheung, H.H.; Mrad, R.L.; Plenchette, S.; Simard, C.; Enwere, E.; Arora, V.; Mak, T.W.; Lacasse, E.C.; Waring, J.; et al. Both cIAP1 and cIAP2 regulate TNFalpha-mediated NF-kappaB activation. *Proc. Natl. Acad. Sci. USA* **2008**, *105*, 11778–11783. [CrossRef] [PubMed]

47. Wang, L.; Du, F.; Wang, X. TNF-alpha induces two distinct caspase-8 activation pathways. *Cell* **2008**, *133*, 693–703. [CrossRef] [PubMed]

48. Micheau, O.; Lens, S.; Gaide, O.; Alevizopoulos, K.; Tschopp, J. NF-kappaB signals induce the expression of c-FLIP. *Mol. Cell Biol.* **2001**, *21*, 5299–5305. [CrossRef] [PubMed]

49. Westphal, D.; Dewson, G.; Czabotar, P.E.; Kluck, R.M. Molecular biology of Bax and Bak activation and action. *Biochim. Biophys. Acta* **2011**, *1813*, 521–531. [CrossRef] [PubMed]

50. Wei, M.C.; Zong, W.X.; Cheng, E.H.; Lindsten, T.; Panoutsakopoulou, V.; Ross, A.J.; Roth, K.A.; MacGregor, G.R.; Thompson, C.B.; Korsmeyer, S.J. Proapoptotic BAX and BAK: A requisite gateway to mitochondrial dysfunction and death. *Science* **2001**, *292*, 727–730. [CrossRef] [PubMed]

51. Ichim, G.; Tait, S.W. A fate worse than death: Apoptosis as an oncogenic process. *Nat. Rev. Cancer* **2016**, *16*, 539–548. [CrossRef] [PubMed]

52. Tait, S.W.; Green, D.R. Mitochondria and cell death: Outer membrane permeabilization and beyond. *Nat. Rev. Mol. Cell Biol.* **2010**, *11*, 621–632. [CrossRef] [PubMed]

53. Cande, C.; Vahsen, N.; Garrido, C.; Kroemer, G. Apoptosis-inducing factor (AIF): Caspase-independent after all. *Cell Death Differ.* **2004**, *11*, 591–595. [CrossRef] [PubMed]

54. Bortner, C.D.; Oldenburg, N.B.E.; Cidlowski, J.A. The Role of DNA Fragmentation in Apoptosis. *Trends Cell Biol.* **1995**, *5*, 21–26. [CrossRef]

55. Li, H.; Zhu, H.; Xu, C.J.; Yuan, J. Cleavage of BID by caspase 8 mediates the mitochondrial damage in the Fas pathway of apoptosis. *Cell* **1998**, *94*, 491–501. [CrossRef]

56. Redza-Dutordoir, M.; Averill-Bates, D.A. Activation of apoptosis signalling pathways by reactive oxygen species. *Biochim. Biophys. Acta* **2016**, *1863*, 2977–2992. [CrossRef] [PubMed]

57. Moloney, J.N.; Cotter, T.G. ROS signalling in the biology of cancer. *Semin. Cell Dev. Biol.* **2017**. [CrossRef] [PubMed]

58. Owen, J.J.; Jenkinson, E.J. Apoptosis and T-cell repertoire selection in the thymus. *Ann. N. Y. Acad. Sci.* **1992**, *663*, 305–310. [CrossRef] [PubMed]

59. Li, C.J.; Friedman, D.J.; Wang, C.; Metelev, V.; Pardee, A.B. Induction of apoptosis in uninfected lymphocytes by HIV-1 Tat protein. *Science* **1995**, *268*, 429–431. [CrossRef] [PubMed]

60. Trapani, J.A.; Smyth, M.J. Functional significance of the perforin/granzyme cell death pathway. *Nat. Rev. Immunol.* **2002**, *2*, 735–747. [CrossRef] [PubMed]

61. Shi, Y.; Wang, J.; Wang, Y.; Wang, A.; Guo, H.; Wei, F.; Mehta, S.R.; Espitia, S.; Smith, D.M.; Liu, L.; et al. A novel mutant 10Ala/Arg together with mutant 144Ser/Arg of hepatitis B virus X protein involved in hepatitis B virus-related hepatocarcinogenesis in HepG2 cell lines. *Cancer Lett.* **2016**, *371*, 285–291. [CrossRef] [PubMed]

62. Chao, C.C. Inhibition of apoptosis by oncogenic hepatitis B virus X protein: Implications for the treatment of hepatocellular carcinoma. *World J. Hepatol.* **2016**, *8*, 1061–1066. [CrossRef] [PubMed]

63. Yun, C.; Um, H.R.; Jin, Y.H.; Wang, J.H.; Lee, M.O.; Park, S.; Lee, J.H.; Cho, H. NF-kappaB activation by hepatitis B virus X (HBX) protein shifts the cellular fate toward survival. *Cancer Lett.* **2002**, *184*, 97–104. [CrossRef]

64. Dai, R.; Peng, F.; Xiao, X.; Gong, X.; Jiang, Y.; Zhang, M.; Tian, Y.; Xu, Y.; Ma, J.; Li, M.; et al. Hepatitis B virus X protein-induced upregulation of CAT-1 stimulates proliferation and inhibits apoptosis in hepatocellular carcinoma cells. *Oncotarget* **2017**. [CrossRef]

65. Kondo, Y.; Kobayashi, K.; Asabe, S.; Shiina, M.; Niitsuma, H.; Ueno, Y.; Kobayashi, T.; Shimosegawa, T. Vigorous response of cytotoxic T lymphocytes associated with systemic activation of CD8 T lymphocytes in fulminant hepatitis B. *Liver Int.* **2004**, *24*, 561–567. [CrossRef] [PubMed]

66. Gogoi, D.; Borkakoty, B.; Biswas, D.; Mahanta, J. Activation and Exhaustion of Adaptive Immune Cells in Hepatitis B Infection. *Viral Immunol.* **2015**, *28*, 348–353. [CrossRef] [PubMed]

67. Ebert, G.; Preston, S.; Allison, C.; Cooney, J.; Toe, J.G.; Stutz, M.D.; Ojaimi, S.; Scott, H.W.; Baschuk, N.; Nachbur, U.; et al. Cellular inhibitor of apoptosis proteins prevent clearance of hepatitis B virus. *Proc. Natl. Acad. Sci. USA* **2015**, *112*, 5797–5802. [CrossRef] [PubMed]

68. Ebert, G.; Allison, C.; Preston, S.; Cooney, J.; Toe, J.G.; Stutz, M.D.; Ojaimi, S.; Baschuk, N.; Nachbur, U.; Torresi, J.; et al. Eliminating hepatitis B by antagonizing cellular inhibitors of apoptosis. *Proc. Natl. Acad. Sci. USA* **2015**, *112*, 5803–5808. [CrossRef] [PubMed]

69. Kim, H.; Lee, H.; Yun, Y. X-gene product of hepatitis B virus induces apoptosis in liver cells. *J. Biol. Chem.* **1998**, *273*, 381–385. [CrossRef] [PubMed]

70. Terradillos, O.; de La Coste, A.; Pollicino, T.; Neuveut, C.; Sitterlin, D.; Lecoeur, H.; Gougeon, M.L.; Kahn, A.; Buendia, M.A. The hepatitis B virus X protein abrogates Bcl-2-mediated protection against Fas apoptosis in the liver. *Oncogene* **2002**, *21*, 377–386. [CrossRef] [PubMed]

71. Kim, H.J.; Kim, S.Y.; Kim, J.; Lee, H.; Choi, M.; Kim, J.K.; Ahn, J.K. Hepatitis B virus X protein induces apoptosis by enhancing translocation of Bax to mitochondria. *IUBMB Life* **2008**, *60*, 473–480. [CrossRef] [PubMed]

72. Su, F.; Schneider, R.J. Hepatitis B virus HBX protein sensitizes cells to apoptotic killing by tumor necrosis factor alpha. *Proc. Natl. Acad. Sci. USA* **1997**, *94*, 8744–8749. [CrossRef] [PubMed]

73. Liang, X.; Liu, Y.; Zhang, Q.; Gao, L.; Han, L.; Ma, C.; Zhang, L.; Chen, Y.H.; Sun, W. Hepatitis B Virus Sensitizes Hepatocytes to TRAIL-Induced Apoptosis through Bax. *J. Immunol.* **2006**, *178*, 503–510. [CrossRef]

74. Kong, F.Y.; You, H.J.; Zhao, J.J.; Liu, W.; Hu, L.; Luo, W.Y.; Hu, W.; Tang, R.X.; Zheng, K.Y. The enhanced expression of death receptor 5 (DR5) mediated by HBV X protein through NF-kappaB pathway is associated with cell apoptosis induced by (TNF-alpha related apoptosis inducing ligand) TRAIL in hepatoma cells. *Virol. J.* **2015**, *12*, 192. [CrossRef] [PubMed]

75. Wang, X.W.; Gibson, M.K.; Vermeulen, W.; Yeh, H.; Forrester, K.; Sturzbecher, H.W.; Hoeijmakers, J.H.; Harris, C.C. Abrogation of p53-induced apoptosis by the hepatitis B virus X gene. *Cancer Res.* **1995**, *55*, 6012–6016. [PubMed]

76. Shih, W.L.; Kuo, M.L.; Chuang, S.E.; Cheng, A.L.; Doong, S.L. Hepatitis B virus X protein inhibits transforming growth factor-beta-induced apoptosis through the activation of phosphatidylinositol 3-kinase pathway. *J. Biol. Chem.* **2000**, *275*, 25858–25864. [CrossRef] [PubMed]

77. Kim, S.J.; Khan, M.; Quan, J.; Till, A.; Subramani, S.; Siddiqui, A. Hepatitis B virus disrupts mitochondrial dynamics: Induces fission and mitophagy to attenuate apoptosis. *PLoS Pathog.* **2013**, *9*, e1003722. [CrossRef] [PubMed]

78. Lu, X.; Lee, M.; Tran, T.; Block, T. High level expression of apoptosis inhibitor in hepatoma cell line expressing Hepatitis B virus. *Int. J. Med. Sci.* **2005**, *2*, 30–35. [CrossRef] [PubMed]

79. Diao, J.; Khine, A.A.; Sarangi, F.; Hsu, E.; Iorio, C.; Tibbles, L.A.; Woodgett, J.R.; Penninger, J.; Richardson, C.D. X protein of hepatitis B virus inhibits Fas-mediated apoptosis and is associated with up-regulation of the SAPK/JNK pathway. *J. Biol. Chem.* **2001**, *276*, 8328–8340. [CrossRef] [PubMed]

80. Wang, C.M.; Wang, Y.; Fan, C.G.; Xu, F.F.; Sun, W.S.; Liu, Y.G.; Jia, J.H. miR-29c targets TNFAIP3, inhibits cell proliferation and induces apoptosis in hepatitis B virus-related hepatocellular carcinoma. *Biochem. Biophys. Res. Commun.* **2011**, *411*, 586–592. [CrossRef] [PubMed]

81. Amaral, J.D.; Xavier, J.M.; Steer, C.J.; Rodrigues, C.M. The role of p53 in apoptosis. *Discov. Med.* **2010**, *9*, 145–152. [PubMed]

82. Vousden, K.H.; Lane, D.P. p53 in health and disease. *Nat. Rev. Mol. Cell. Biol.* **2007**, *8*, 275–283. [CrossRef] [PubMed]

83. Shintani, Y.; Yotsuyanagi, H.; Moriya, K.; Fujie, H.; Tsutsumi, T.; Kanegae, Y.; Kimura, S.; Saito, I.; Koike, K. Induction of apoptosis after switch-on of the hepatitis B virus X gene mediated by the Cre/loxP recombination system. *J. Gen. Virol.* **1999**, *80*, 3257–3265. [CrossRef] [PubMed]

84. Feitelson, M.A.; Zhu, M.; Duan, L.X.; London, W.T. Hepatitis B x antigen and p53 are associated in vitro and in liver tissues from patients with primary hepatocellular carcinoma. *Oncogene* **1993**, *8*, 1109–1117. [PubMed]

85. Buckwold, V.E.; Xu, Z.; Chen, M.; Yen, T.S.; Ou, J.H. Effects of a naturally occurring mutation in the hepatitis B virus basal core promoter on precore gene expression and viral replication. *J. Virol.* **1996**, *70*, 5845–5851. [PubMed]

86. Buckwold, V.E.; Xu, Z.; Yen, T.S.; Ou, J.H. Effects of a frequent double-nucleotide basal core promoter mutation and its putative single-nucleotide precursor mutations on hepatitis B virus gene expression and replication. *J. Gen. Virol.* **1997**, *78*, 2055–2065. [CrossRef] [PubMed]

87. Iyer, S.; Groopman, J.D. Interaction of Mutant Hepatitis B X Protein With p53 Tumor Suppressor Protein Affects Both Transcription and Cell Survival. *Mol. Carcinogen.* **2011**, *50*, 972–980. [CrossRef] [PubMed]

88. Elmore, L.W.; Hancock, A.R.; Chang, S.F.; Wang, X.W.; Chang, S.; Callahan, C.P.; Geller, D.A.; Will, H.; Harris, C.C. Hepatitis B virus X protein and p53 tumor suppressor interactions in the modulation of apoptosis. *Proc. Natl. Acad. Sci. USA* **1997**, *94*, 14707–14712. [CrossRef] [PubMed]

89. Knoll, S.; Furst, K.; Thomas, S.; Villanueva Baselga, S.; Stoll, A.; Schaefer, S.; Putzer, B.M. Dissection of cell context-dependent interactions between HBX and p53 family members in regulation of apoptosis: A role for HBV-induced HCC. *Cell Cycle* **2011**, *10*, 3554–3565. [CrossRef] [PubMed]

90. Ahn, J.Y.; Jung, E.Y.; Kwun, H.J.; Lee, C.W.; Sung, Y.C.; Jang, K.L. Dual effects of hepatitis B virus X protein on the regulation of cell-cycle control depending on the status of cellular p53. *J. Gen. Virol.* **2002**, *83*, 2765–2772. [CrossRef] [PubMed]

91. Keng, V.W.; Tschida, B.R.; Bell, J.B.; Largaespada, D.A. Modeling hepatitis B virus X-induced hepatocellular carcinoma in mice with the Sleeping Beauty transposon system. *Hepatology* **2011**, *53*, 781–790. [CrossRef] [PubMed]

92. Zhao, H.F.; Wang, J.; Shao, W.; Wu, C.P.; Chen, Z.P.; To, S.T.; Li, W.P. Recent advances in the use of PI3K inhibitors for glioblastoma multiforme: Current preclinical and clinical development. *Mol. Cancer* **2017**, *16*, 100. [CrossRef] [PubMed]

93. Franke, T.F.; Hornik, C.P.; Segev, L.; Shostak, G.A.; Sugimoto, C. PI3K/Akt and apoptosis: Size matters. *Oncogene* **2003**, *22*, 8983–8998. [CrossRef] [PubMed]

94. Khwaja, A. Akt is more than just a Bad kinase. *Nature* **1999**, *401*, 33–34. [CrossRef] [PubMed]

95. Datta, S.R.; Dudek, H.; Tao, X.; Masters, S.; Fu, H.; Gotoh, Y.; Greenberg, M.E. Akt phosphorylation of BAD couples survival signals to the cell-intrinsic death machinery. *Cell* **1997**, *91*, 231–241. [CrossRef]

96. Kane, L.P.; Shapiro, V.S.; Stokoe, D.; Weiss, A. Induction of NF-kappaB by the Akt/PKB kinase. *Curr. Biol.* **1999**, *9*, 601–604. [CrossRef]

97. Huang, W.C.; Chen, W.S.; Chen, Y.J.; Wang, L.Y.; Hsu, S.C.; Chen, C.C.; Hung, M.C. Hepatitis B virus X protein induces IKKalpha nuclear translocation via Akt-dependent phosphorylation to promote the motility of hepatocarcinoma cells. *J. Cell. Physiol.* **2012**, *227*, 1446–1454. [CrossRef] [PubMed]

98. Lee, Y.I.; Kang-Park, S.; Do, S.I.; Lee, Y.I. The hepatitis B virus-X protein activates a phosphatidylinositol 3-kinase-dependent survival signaling cascade. *J. Biol. Chem.* **2001**, *276*, 16969–16977. [CrossRef] [PubMed]

99. Wang, W.; Shi, Y.; Bai, G.; Tang, Y.; Yuan, Y.; Zhang, T.; Li, C. HBXAg suppresses apoptosis of human placental trophoblastic cell lines via activation of the PI3K/Akt pathway. *Cell Biol. Int.* **2016**, *40*, 708–715. [CrossRef] [PubMed]

100. Lee, W.P.; Lan, K.H.; Li, C.P.; Chao, Y.; Lin, H.C.; Lee, S.D. Pro-apoptotic or anti-apoptotic property of X protein of hepatitis B virus is determined by phosphorylation at Ser31 by Akt. *Arch. Biochem. Biophys.* **2012**, *528*, 156–162. [CrossRef] [PubMed]

101. Gottlob, K.; Fulco, M.; Levrero, M.; Graessmann, A. The hepatitis B virus HBX protein inhibits caspase 3 activity. *J. Biol. Chem.* **1998**, *273*, 33347–33353. [CrossRef] [PubMed]

102. Pan, J.; Duan, L.X.; Sun, B.S.; Feitelson, M.A. Hepatitis B virus X protein protects against anti-Fas-mediated apoptosis in human liver cells by inducing NF-kappa B. *J. Gen. Virol.* **2001**, *82*, 171–182. [CrossRef] [PubMed]

103. Doi, T.S.; Takahashi, T.; Taguchi, O.; Azuma, T.; Obata, Y. NF-kappa B RelA-deficient lymphocytes: Normal development of T cells and B cells, impaired production of IgA and IgG1 and reduced proliferative responses. *J. Exp. Med.* **1997**, *185*, 953–961. [CrossRef] [PubMed]

104. Franzoso, G.; Carlson, L.; Xing, L.; Poljak, L.; Shores, E.W.; Brown, K.D.; Leonardi, A.; Tran, T.; Boyce, B.F.; Siebenlist, U. Requirement for NF-kappaB in osteoclast and B-cell development. *Genes Dev.* **1997**, *11*, 3482–3496. [CrossRef] [PubMed]

105. Kontgen, F.; Grumont, R.J.; Strasser, A.; Metcalf, D.; Li, R.; Tarlinton, D.; Gerondakis, S. Mice lacking the c-rel proto-oncogene exhibit defects in lymphocyte proliferation, humoral immunity, and interleukin-2 expression. *Genes Dev.* **1995**, *9*, 1965–1977. [CrossRef] [PubMed]

106. Sun, S.C. Non-canonical NF-kappaB signaling pathway. *Cell Res.* **2011**, *21*, 71–85. [CrossRef] [PubMed]

107. Zarnegar, B.; Yamazaki, S.; He, J.Q.; Cheng, G. Control of canonical NF-kappaB activation through the NIK-IKK complex pathway. *Proc. Natl. Acad. Sci. USA* **2008**, *105*, 3503–3508. [CrossRef] [PubMed]

108. Wang, C.; Deng, L.; Hong, M.; Akkaraju, G.R.; Inoue, J.; Chen, Z.J. TAK1 is a ubiquitin-dependent kinase of MKK and IKK. *Nature* **2001**, *412*, 346–351. [CrossRef] [PubMed]

109. Gardam, S.; Sierro, F.; Basten, A.; Mackay, F.; Brink, R. TRAF2 and TRAF3 signal adapters act cooperatively to control the maturation and survival signals delivered to B cells by the BAFF receptor. *Immunity* **2008**, *28*, 391–401. [CrossRef] [PubMed]

110. He, J.Q.; Zarnegar, B.; Oganesyan, G.; Saha, S.K.; Yamazaki, S.; Doyle, S.E.; Dempsey, P.W.; Cheng, G. Rescue of TRAF3-null mice by p100 NF-kappa B deficiency. *J. Exp. Med.* **2006**, *203*, 2413–2418. [CrossRef] [PubMed]

111. Xie, P.; Stunz, L.L.; Larison, K.D.; Yang, B.; Bishop, G.A. Tumor necrosis factor receptor-associated factor 3 is a critical regulator of B cell homeostasis in secondary lymphoid organs. *Immunity* **2007**, *27*, 253–267. [CrossRef] [PubMed]

112. Dejardin, E.; Bonizzi, G.; Bellahcene, A.; Castronovo, V.; Merville, M.P.; Bours, V. Highly-expressed p100/p52 (NFKB2) sequesters other NF-kappa B-related proteins in the cytoplasm of human breast cancer cells. *Oncogene* **1995**, *11*, 1835–1841. [PubMed]

113. Sovak, M.A.; Bellas, R.E.; Kim, D.W.; Zanieski, G.J.; Rogers, A.E.; Traish, A.M.; Sonenshein, G.E. Aberrant nuclear factor-kappaB/Rel expression and the pathogenesis of breast cancer. *J. Clin. Invest.* **1997**, *100*, 2952–2960. [CrossRef] [PubMed]

114. Larosa, F.A.; Pierce, J.W.; Sonenshein, G.E. Differential Regulation of the C-Myc Oncogene Promoter by the Nf-Kappa-B Rel Family of Transcription Factors. *Mol. Cell. Biol.* **1994**, *14*, 1039–1044. [CrossRef]

115. Huber, M.A.; Azoitei, N.; Baumann, B.; Grünert, S.; Sommer, A.; Pehamberger, H.; Kraut, N.; Beug, H.; Wirth, T. NF-κB is essential for epithelial-mesenchymal transition and metastasis in a model of breast cancer progression. *J. Clin. Invest.* **2004**, *114*, 569–581. [CrossRef] [PubMed]

116. Xie, T.X.; Xia, Z.; Zhang, N.; Gong, W.; Huang, S. Constitutive NF-kappaB activity regulates the expression of VEGF and IL-8 and tumor angiogenesis of human glioblastoma. *Oncol. Rep.* **2010**, *23*, 725–732. [PubMed]

117. Yoshida, A.; Yoshida, S.; Ishibashi, T.; Kuwano, M.; Inomata, H. Suppression of retinal neovascularization by the NF-kappaB inhibitor pyrrolidine dithiocarbamate in mice. *Invest. Ophthalmol. Vis. Sci.* **1999**, *40*, 1624–1629.

118. Shukla, R.; Yue, J.; Siouda, M.; Gheit, T.; Hantz, O.; Merle, P.; Zoulim, F.; Krutovskikh, V.; Tommasino, M.; Sylla, B.S. Proinflammatory cytokine TNF-alpha increases the stability of hepatitis B virus X protein through NF-kappaB signaling. *Carcinogenesis* **2011**, *32*, 978–985. [CrossRef] [PubMed]

119. Su, F.; Schneider, R.J. Hepatitis B virus HBX protein activates transcription factor NF-kappaB by acting on multiple cytoplasmic inhibitors of rel-related proteins. *J. Virol.* **1996**, *70*, 4558–4566. [PubMed]

120. Wang, T.; Wang, Y.; Wu, M.C.; Guan, X.Y.; Yin, Z.F. Activating mechanism of transcriptor NF-kappaB regulated by hepatitis B virus X protein in hepatocellular carcinoma. *World J. Gastroenterol.* **2004**, *10*, 356–360. [PubMed]

121. Su, F.; Theodosis, C.N.; Schneider, R.J. Role of NF-kappaB and myc proteins in apoptosis induced by hepatitis B virus HBX protein. *J. Virol.* **2001**, *75*, 215–225. [CrossRef] [PubMed]

122. Clippinger, A.J.; Gearhart, T.L.; Bouchard, M.J. Hepatitis B virus X protein modulates apoptosis in primary rat hepatocytes by regulating both NF-kappaB and the mitochondrial permeability transition pore. *J. Virol.* **2009**, *83*, 4718–4731. [CrossRef] [PubMed]

123. Feng, C.; Wu, B.; Fan, H.; Li, C.; Meng, S. NF-kappaB-induced gp96 up-regulation promotes hepatocyte growth, cell cycle progression and transition. *Acta Microbiol. Sin.* **2014**, *54*, 1212–1220.

124. Liu, Y.; Lou, G.; Wu, W.; Zheng, M.; Shi, Y.; Zhao, D.; Chen, Z. Involvement of the NF-kappaB pathway in multidrug resistance induced by HBX in a hepatoma cell line. *J. Viral Hepat.* **2011**, *18*, e439–e446. [CrossRef] [PubMed]

125. Yang, C.H.; Cho, M. Hepatitis B virus X gene differentially modulates cell cycle progression and apoptotic protein expression in hepatocyte versus hepatoma cell lines. *J. Viral Hepat.* **2013**, *20*, 50–58. [CrossRef] [PubMed]

126. Mao, Y.; Da, L.; Tang, H.; Yang, J.L.; Lei, Y.R.; Tiollais, P.; Li, T.P.; Zhao, M.J. Hepatitis B virus X protein reduces starvation-induced cell death through activation of autophagy and inhibition of mitochondrial apoptotic pathway. *Biochem. Biophys. Res. Commun.* **2011**, *415*, 68–74. [CrossRef] [PubMed]

127. Zhang, H.T.; Chen, G.G.; Hu, B.G.; Zhang, Z.Y.; Yun, J.P.; He, M.L.; Lai, P.B. Hepatitis B virus X protein induces autophagy via activating death-associated protein kinase. *J. Viral Hepat.* **2014**, *21*, 642–649. [CrossRef] [PubMed]

128. McKnight, N.C.; Zhenyu, Y. Beclin 1, an Essential Component and Master Regulator of PI3K-III in Health and Disease. *Curr. Pathobiol. Rep.* **2013**, *1*, 231–238. [CrossRef] [PubMed]

129. Ciechomska, I.A.; Goemans, G.C.; Skepper, J.N.; Tolkovsky, A.M. Bcl-2 complexed with Beclin-1 maintains full anti-apoptotic function. *Oncogene* **2009**, *28*, 2128–2141. [CrossRef] [PubMed]

130. Liu, H.; Yuan, Y.; Guo, H.; Mitchelson, K.; Zhang, K.; Xie, L.; Qin, W.; Lu, Y.; Wang, J.; Guo, Y.; et al. Hepatitis B virus encoded X protein suppresses apoptosis by inhibition of the caspase-independent pathway. *J. Proteome Res.* **2012**, *11*, 4803–4813. [CrossRef] [PubMed]

131. Cully, M. Viral infections: Reinvigorating exhausted T cells in hepatitis B infection. *Nat. Rev. Drug Discov.* **2017**, *16*, 240. [CrossRef] [PubMed]

132. Li, B.; Carey, M.; Workman, J.L. The role of chromatin during transcription. *Cell* **2007**, *128*, 707–719. [CrossRef] [PubMed]

133. Hayes, J.; Peruzzi, P.P.; Lawler, S. MicroRNAs in cancer: Biomarkers, functions and therapy. *Trends Mol. Med.* **2014**, *20*, 460–469. [CrossRef] [PubMed]

134. Mayo, L.D.; Dixon, J.E.; Durden, D.L.; Tonks, N.K.; Donner, D.B. PTEN protects p53 from Mdm2 and sensitizes cancer cells to chemotherapy. *J. Biol. Chem.* **2002**, *277*, 5484–5489. [CrossRef] [PubMed]

135. Tian, Y.; Xiao, X.Q.; Gong, X.; Peng, F.; Xu, Y.; Jiang, Y.F.; Gong, G.Z. HBX promotes cell proliferation by disturbing the cross-talk between miR-181a and PTEN. *Sci. Rep.* **2017**, *7*, 40089. [CrossRef] [PubMed]

136. Huang, J.L.; Ren, T.Y.; Cao, S.W.; Zheng, S.H.; Hu, X.M.; Hu, Y.W.; Lin, L.; Chen, J.; Zheng, L.; Wang, Q. HBX-related long non-coding RNA DBH-AS1 promotes cell proliferation and survival by activating MAPK signaling in hepatocellular carcinoma. *Oncotarget* **2015**, *6*, 33791–33804. [CrossRef] [PubMed]

137. Chen, J.J.; Tang, Y.S.; Huang, S.F.; Ai, J.G.; Wang, H.X.; Zhang, L.P. HBX protein-induced upregulation of microRNA-221 promotes aberrant proliferation in HBVrelated hepatocellular carcinoma by targeting estrogen receptor-alpha. *Oncol. Rep.* **2015**, *33*, 792–798. [PubMed]

138. Liu, N.; Zhang, J.; Jiao, T.; Li, Z.; Peng, J.; Cui, Z.; Ye, X. Hepatitis B virus inhibits apoptosis of hepatoma cells by sponging the MicroRNA 15a/16 cluster. *J. Virol.* **2013**, *87*, 13370–13378. [CrossRef] [PubMed]

139. Liu, N.N.; Jiao, T.; Huang, Y.; Liu, W.J.; Li, Z.W.; Ye, X. Hepatitis B Virus Regulates Apoptosis and Tumorigenesis through the MicroRNA-15a-Smad7-Transforming Growth Factor Beta Pathway. *J. Virol.* **2015**, *89*, 2739–2749. [CrossRef] [PubMed]

140. Shembade, N.; Ma, A.; Harhaj, E.W. Inhibition of NF-kappaB signaling by A20 through disruption of ubiquitin enzyme complexes. *Science* **2010**, *327*, 1135–1139. [CrossRef] [PubMed]

141. Qiao, D.D.; Yang, J.; Lei, X.F.; Mi, G.L.; Li, S.L.; Li, K.; Xu, C.Q.; Yang, H.L. Expression of microRNA-122 and microRNA-22 in HBV-related liver cancer and the correlation with clinical features. *Eur. Rev. Med. Pharmacol. Sci.* **2017**, *21*, 742–747. [PubMed]

142. Zhang, B.; Han, S.; Feng, B.; Chu, X.; Chen, L.; Wang, R. Hepatitis B virus X protein-mediated non-coding RNA aberrations in the development of human hepatocellular carcinoma. *Exp. Mol. Med.* **2017**, *49*, e293. [CrossRef] [PubMed]

143. Xu, Y.; Qi, Y.; Luo, J.; Yang, J.; Xie, Q.; Deng, C.; Su, N.; Wei, W.; Shi, D.; Xu, F.; et al. Hepatitis B Virus X Protein Stimulates Proliferation, Wound Closure and Inhibits Apoptosis of HuH-7 Cells via CDC42. *Int. J. Mol. Sci.* **2017**, *18*, 586. [CrossRef] [PubMed]

144. Li, L.; Hong, H.H.; Chen, S.P.; Ma, C.Q.; Liu, H.Y.; Yao, Y.C. Activation of AMPK/MnSOD signaling mediates anti-apoptotic effect of hepatitis B virus in hepatoma cells. *World J. Gastroenterol.* **2016**, *22*, 4345–4353. [CrossRef] [PubMed]

145. Kongkavitoon, P.; Tangkijvanich, P.; Hirankarn, N.; Palaga, T. Hepatitis B Virus HBX Activates Notch Signaling via Delta-Like 4/Notch1 in Hepatocellular Carcinoma. *PLoS ONE* **2016**, *11*, e0146696. [CrossRef] [PubMed]

146. Cui, H.; Li, Q.L.; Chen, J.; Na, Q.; Liu, C.X. Hepatitis B virus X protein modifies invasion, proliferation and the inflammatory response in an HTR-8/SVneo cell model. *Oncol. Rep.* **2015**, *34*, 2090–2098. [CrossRef] [PubMed]

147. Kerr, J.F.; Winterford, C.M.; Harmon, B.V. Apoptosis. Its significance in cancer and cancer therapy. *Cancer* **1994**, *73*, 2013–2026. [CrossRef]

148. Curson, C.; Weedon, D. Spontaneous regression in basal cell carcinomas. *J. Cutan. Pathol.* **1979**, *6*, 432–437. [CrossRef] [PubMed]

149. Chami, M.; Ferrari, D.; Nicotera, P.; Paterlini-Brechot, P.; Rizzuto, R. Caspase-dependent alterations of Ca^{2+} signaling in the induction of apoptosis by hepatitis B virus X protein. *J. Biol. Chem.* **2003**, *278*, 31745–31755. [CrossRef] [PubMed]

150. Wang, W.H.; Gregori, G.; Hullinger, R.L.; Andrisani, O.M. Sustained activation of p38 mitogen-activated protein kinase and c-Jun N-terminal kinase pathways by hepatitis B virus X protein mediates apoptosis via induction of Fas/FasL and tumor necrosis factor (TNF) receptor 1/TNF-alpha expression. *Mol. Cell Biol.* **2004**, *24*, 10352–10365. [CrossRef] [PubMed]

151. Yeganeh, B.; Moghadam, A.R.; Alizadeh, J.; Wiechec, E.; Alavian, S.M.; Hashemi, M.; Geramizadeh, B.; Samali, A.; Lankarani, K.B.; Post, M.; et al. Hepatitis B and C virus-induced hepatitis: Apoptosis, autophagy, and unfolded protein response. *World J. Gastroenterol.* **2015**, *21*, 13225–13239. [CrossRef] [PubMed]

152. Liu, F.W.; Wu, D.B.; Chen, E.Q.; Liu, C.; Liu, L.; Chen, S.C.; Gong, D.Y.; Zhao, L.S.; Tang, H.; Zhou, T.Y. Expression of TRAIL in liver tissue from patients with different outcomes of HBV infection. *Clin. Res. Hepatol. Gas.* **2013**, *37*, 269–274. [CrossRef] [PubMed]

153. Xu, H.; Wang, L.; Zheng, P.; Liu, Y.; Zhang, C.; Jiang, K.; Song, H.; Ji, G. Elevated serum A20 is associated with severity of chronic hepatitis B and A20 inhibits NF-kappaB-mediated inflammatory response. *Oncotarget* **2017**, *8*, 38914–38926. [PubMed]

154. Zhang, H.; Huang, C.; Wang, Y.; Lu, Z.; Zhuang, N.; Zhao, D.; He, J.; Shi, L. Hepatitis B Virus X Protein Sensitizes TRAIL-Induced Hepatocyte Apoptosis by Inhibiting the E3 Ubiquitin Ligase A20. *PLoS ONE* **2015**, *10*, e0127329. [CrossRef] [PubMed]

155. Miao, J.; Chen, G.G.; Chun, S.Y.; Lai, P.P.S. Hepatitis B virus X protein induces apoptosis in hepatoma cells through inhibiting Bcl-xL expression. *Cancer Lett.* **2006**, *236*, 115–124. [CrossRef] [PubMed]

156. He, P.; Zhang, B.R.; Liu, D.J.; Bian, X.H.; Li, D.T.; Wang, Y.Q.; Sun, G.P.; Zhou, G.Y. Hepatitis B Virus X Protein Modulates Apoptosis in NRK-52E Cells and Activates Fas/FasL Through the MLK3-MKK7-JNK3 Signaling Pathway. *Cell. Physiol. Biochem.* **2016**, *39*, 1433–1443. [CrossRef] [PubMed]

157. Zhang, Y.; Liu, H.L.; Yi, R.T.; Yan, T.T.; He, Y.L.; Zhao, Y.R.; Liu, J.F. Hepatitis B virus whole-X and X protein play distinct roles in HBV-related hepatocellular carcinoma progression. *J. Exp. Clin. Cancer Res.* **2016**, *35*, 87. [CrossRef] [PubMed]

158. Takada, S.; Shirakata, Y.; Kaneniwa, N.; Koike, K. Association of hepatitis B virus X protein with mitochondria causes mitochondrial aggregation at the nuclear periphery, leading to cell death. *Oncogene* **1999**, *18*, 6965–6973. [CrossRef] [PubMed]

159. Gao, W.Y.; Li, D.; Cai, D.E.; Huang, X.Y.; Zheng, B.Y.; Huang, Y.H.; Chen, Z.X.; Wang, X.Z. Hepatitis B virus X protein sensitizes HL-7702 cells to oxidative stress-induced apoptosis through modulation of the mitochondrial permeability transition pore. *Oncol. Rep.* **2017**, *37*, 48–56. [PubMed]

160. Hu, L.; Chen, L.; Yang, G.; Li, L.; Sun, H.; Chang, Y.; Tu, Q.; Wu, M.; Wang, H. HBX sensitizes cells to oxidative stress-induced apoptosis by accelerating the loss of Mcl-1 protein via caspase-3 cascade. *Mol. Cancer* **2011**, *10*, 43. [CrossRef] [PubMed]

161. Lu, Y.W.; Chen, W.N. Human hepatitis B virus X protein induces apoptosis in HepG2 cells: Role of BH3 domain. *Biochem. Biophys. Res. Commun.* **2005**, *338*, 1551–1556. [CrossRef] [PubMed]

162. Lu, Y.W.; Tan, T.L.; Chan, V.; Chen, W.N. The HBSP gene is expressed during HBV replication, and its coded BH3-containing spliced viral protein induces apoptosis in HepG2 cells. *Biochem. Biophys. Res. Commun.* **2006**, *351*, 64–70. [CrossRef] [PubMed]

163. Geng, X.; Harry, B.L.; Zhou, Q.; Skeen-Gaar, R.R.; Ge, X.; Lee, E.S.; Mitani, S.; Xue, D. Hepatitis B virus X protein targets the Bcl-2 protein CED-9 to induce intracellular Ca^{2+} increase and cell death in Caenorhabditis elegans. *Proc. Natl. Acad. Sci. USA* **2012**, *109*, 18465–18470. [CrossRef] [PubMed]

164. Jiang, T.; Liu, M.; Wu, J.; Shi, Y. Structural and biochemical analysis of Bcl-2 interaction with the hepatitis B virus protein HBX. *Proc. Natl. Acad. Sci. USA* **2016**, *113*, 2074–2079. [CrossRef] [PubMed]

165. Ren, J.H.; Chen, X.; Zhou, L.; Tao, N.N.; Zhou, H.Z.; Liu, B.; Li, W.Y.; Huang, A.L.; Chen, J. Protective Role of Sirtuin3 (SIRT3) in Oxidative Stress Mediated by Hepatitis B Virus X Protein Expression. *PLoS ONE* **2016**, *11*, e0150961. [CrossRef] [PubMed]

166. Guleng, B.; Liu, Y.P.; Ren, J.L. HBsAg Inhibits the Translocation of Jtb Into Mitochondria in HEPG2 Cells and Potentially Plays a Role in HCC Progression. *Gastroenterology* **2012**, *142*, S705. [CrossRef]

167. Liu, H.; Xu, J.; Zhou, L.; Yun, X.; Chen, L.; Wang, S.; Sun, L.; Wen, Y.; Gu, J. Hepatitis B virus large surface antigen promotes liver carcinogenesis by activating the Src/PI3K/Akt pathway. *Cancer Res.* **2011**, *71*, 7547–7557. [CrossRef] [PubMed]

168. Hung, J.H.; Teng, Y.N.; Wang, L.H.; Su, I.J.; Wang, C.C.; Huang, W.; Lee, K.H.; Lu, K.Y.; Wang, L.H. Induction of Bcl-2 expression by hepatitis B virus pre-S2 mutant large surface protein resistance to 5-fluorouracil treatment in Huh-7 cells. *PLoS ONE* **2011**, *6*, e28977. [CrossRef] [PubMed]

169. Liu, W.; Lin, Y.T.; Yan, X.L.; Ding, Y.L.; Wu, Y.L.; Chen, W.N.; Lin, X. Hepatitis B virus core protein inhibits Fas-mediated apoptosis of hepatoma cells via regulation of mFas/FasL and sFas expression. *FASEB J.* **2015**, *29*, 1113–1123. [CrossRef] [PubMed]

170. Jia, B.; Guo, M.; Li, G.; Yu, D.; Zhang, X.; Lan, K.; Deng, Q. Hepatitis B virus core protein sensitizes hepatocytes to tumor necrosis factor-induced apoptosis by suppression of the phosphorylation of mitogen-activated protein kinase kinase 7. *J. Virol.* **2015**, *89*, 2041–2051. [CrossRef] [PubMed]

171. Verrier, E.R.; Colpitts, C.C.; Schuster, C.; Zeisel, M.B.; Baumert, T.F. Cell Culture Models for the Investigation of Hepatitis B and D Virus Infection. *Viruses* **2016**, *8*. [CrossRef] [PubMed]

172. Ladner, S.K.; Otto, M.J.; Barker, C.S.; Zaifert, K.; Wang, G.H.; Guo, J.T.; Seeger, C.; King, R.W. Inducible expression of human hepatitis B virus (HBV) in stably transfected hepatoblastoma cells: A novel system for screening potential inhibitors of HBV replication. *Antimicrob. Agents Chemother.* **1997**, *41*, 1715–1720. [PubMed]

173. Sells, M.A.; Chen, M.L.; Acs, G. Production of hepatitis B virus particles in Hep G2 cells transfected with cloned hepatitis B virus DNA. *Proc. Natl. Acad. Sci. USA* **1987**, *84*, 1005–1009. [CrossRef] [PubMed]

174. Ni, Y.; Lempp, F.A.; Mehrle, S.; Nkongolo, S.; Kaufman, C.; Falth, M.; Stindt, J.; Koniger, C.; Nassal, M.; Kubitz, R.; et al. Hepatitis B and D viruses exploit sodium taurocholate co-transporting polypeptide for species-specific entry into hepatocytes. *Gastroenterology* **2014**, *146*, 1070–1083. [CrossRef] [PubMed]

175. Watashi, K.; Urban, S.; Li, W.; Wakita, T. NTCP and beyond: Opening the door to unveil hepatitis B virus entry. *Int. J. Mol. Sci.* **2014**, *15*, 2892–2905. [CrossRef] [PubMed]

176. Thomas, E.; Liang, T.J. Experimental models of hepatitis B and C—New insights and progress. *Nat. Rev. Gastroenterol. Hepatol.* **2016**, *13*, 362–374. [CrossRef] [PubMed]

177. Chan, C.; Wang, Y.; Chow, P.K.; Chung, A.Y.; Ooi, L.L.; Lee, C.G. Altered binding site selection of p53 transcription cassettes by hepatitis B virus X protein. *Mol. Cell. Biol.* **2013**, *33*, 485–497. [CrossRef] [PubMed]

178. Keasler, V.V.; Hodgson, A.J.; Madden, C.R.; Slagle, B.L. Hepatitis B virus HBX protein localized to the nucleus restores HBX-deficient virus replication in HepG2 cells and in vivo in hydrodynamically-injected mice. *Virology* **2009**, *390*, 122–129. [CrossRef] [PubMed]

179. Gao, Q.; Wang, X.Y.; Zhou, J.; Fan, J. Cell line misidentification: The case of the Chang liver cell line. *Hepatology* **2011**, *54*, 1894–1895. [CrossRef] [PubMed]

180. Ryoo, H.D.; Gorenc, T.; Steller, H. Apoptotic cells can induce compensatory cell proliferation through the JNK and the Wingless signaling pathways. *Dev. Cell* **2004**, *7*, 491–501. [CrossRef] [PubMed]

181. Huh, J.R.; Guo, M.; Hay, B.A. Compensatory proliferation induced by cell death in the Drosophila wing disc requires activity of the apical cell death caspase Dronc in a nonapoptotic role. *Curr. Biol.* **2004**, *14*, 1262–1266. [CrossRef] [PubMed]

182. Perez-Garijo, A.; Martin, F.A.; Morata, G. Caspase inhibition during apoptosis causes abnormal signalling and developmental aberrations in Drosophila. *Development* **2004**, *131*, 5591–5598. [CrossRef] [PubMed]

MDPI

Article

Infectious Bronchitis Virus Infection Induces Apoptosis during Replication in Chicken Macrophage HD11 Cells

Xiaoxiao Han [1,2], Yiming Tian [1,2], Ru Guan [1,2], Wenqian Gao [1,2], Xin Yang [1,2], Long Zhou [1,2] and Hongning Wang [1,2,3,*]

[1] Key Laboratory of Bio-Resources and Eco-Environment, Ministry of Education, College of Life Science, Sichuan University, Chengdu 610064, China; silky1010@163.com (X.H.); tymgood@126.com (Y.T.); 17713568487@163.com (R.G.); 18780204364@163.com (W.G.); yangxin0822@163.com (X.Y.); wszl5918@163.com (L.Z.)

[2] Animal Disease Prevention and Food Safety Key Laboratory of Sichuan Province, Chengdu 610064, China

[3] "985 Project" Science Innovative Platform for Resource and Environment Protection of Southwestern China, Sichuan University, Chengdu 610064, China

* Correspondence: whongning@163.com; Tel.: +86-028-8547-1599

Academic Editor: Marc Kvansakul
Received: 18 April 2017; Accepted: 21 July 2017; Published: 26 July 2017

Abstract: Avian infectious bronchitis has caused huge economic losses in the poultry industry. Previous studies have reported that infectious bronchitis virus (IBV) infection can produce cytopathic effects (CPE) and apoptosis in some mammalian cells and primary cells. However, there is little research on IBV-induced immune cell apoptosis. In this study, chicken macrophage HD11 cells were established as a cellular model that is permissive to IBV infection. Then, IBV-induced apoptosis was observed through a cell viability assay, morphological changes, and flow cytometry. The activity of caspases, the inhibitory efficacy of caspase-inhibitors and the expression of apoptotic genes further suggested the activation of apoptosis through both intrinsic and extrinsic pathways in IBV-infected HD11 cells. Additionally, ammonium chloride (NH_4Cl) pretreated HD11 cells blocked IBV from entering cells and inhibited IBV-induced apoptosis. UV-inactivated IBV also lost the ability of apoptosis induction. IBV replication was increased by blocking caspase activation. This study presents a chicken macrophage cell line that will enable further analysis of IBV infection and offers novel insights into the mechanisms of IBV-induced apoptosis in immune cells.

Keywords: IBV infection; chicken macrophage; apoptosis; caspase; virus replication

1. Introduction

Infectious bronchitis virus (IBV) can cause avian infectious bronchitis, an acute and highly infectious disease of chicken. IBV is a member of the family *Coronaviridae* and genus *Coronavirus*. It is a single stranded positive sense, enveloped RNA virus 27–32 kb in length [1,2]. Like some other members of the coronavirus family, IBV mainly causes upper-respiratory tract disease. IBV is characterized by nephritis, proventriculitis and reduction in both laying rate and egg quality in infected chickens [3]. Vaccination is an effective prevention measure, but IBV's ability to mutate has decreased vaccine protection [4–6]. In order to develop better prevention and control measures, the interactions between host and IBV needs to be better studied. Almost all wild-type IBV strains are only able to proliferate in embryonated chicken eggs or primary chicken embryo kidney cells. The Beaudette strains were used previously to study the resistance of IBV to the antiviral state induced by type I interferon (IFN) [7], induction of apoptosis through endoplasmic reticulum stress in Vero cells by IBV infection [8] and activate autophagy by IBV nonstructural protein (NSP) 6 [9]. However, there have been limited studies of the interactions between IBV infection and immune cell apoptosis. Here, the Beaudette strain was used to study the mechanism of IBV infection.

Apoptosis is a form of programmed cell death that results from the activation of intracellular self-destruction biochemical programs [10]. The activation of caspases (a family of cysteine protease) is a significant regulatory event in the apoptosis process [11]. Caspase cascades are triggered by both extrinsic and intrinsic signals to mediate the cell apoptosis [12]. The relationship between cell apoptosis and virus infection is complex. Cell apoptosis induced by virus may cause tissue damage, especially in the immune and nervous systems, suggesting that apoptosis is a pathogenic mechanism in virus-induced disease. At the same time, apoptosis of infected cells can directly interfere with viral replication, and immune cells can engulf apoptotic cells to prevent inflammation [13,14].

Previous studies demonstrated that IBV induced apoptosis in cultured mammalian cells and primary cells [6,15,16]. However, there is limited information about the apoptosis signaling pathways induced by IBV infection in immune cells. Some studies indicated that IBV can transform certain elements of the innate immune system to promote secondary bacterial infections, and macrophages, as an important component of the innate immune system, may play a role in this process [17]. A nephropathogenic IBV strain (B1648) can productively replicate in blood monocytic cells, and the infected cells may act as carrier cells to play a crucial role in cell-associated viremia and the dissemination of virus to the internal organs [18]. Some viruses have been shown to induce apoptosis in macrophage, like human immunodeficiency virus (HIV)-1 [19], Chikungunya virus (CHIKV) [20] and influenza virus [21]. Therefore, additional study is required to investigate the functional roles of macrophages in IBV infection to help understand the mechanistic details of immune responses during virus infections [22].

In this report, we used chicken macrophage HD11 cells considered an accurate representation of primary avian macrophages [23,24]. The HD11 cells were identified and characterized as a novel model that is permissive to IBV infection. The molecular and morphological variations in IBV-infected cells revealed that cell apoptosis was induced by IBV infection and appeared to activate caspase-8 by the Fas/Fas ligand (FasL)-mediated signaling pathway and to activate caspase-9 by the B-cell lymphoma 2 (Bcl-2) family-mediated signaling pathway. Apoptosis required viral replication in IBV-infected cells.

2. Materials and Methods

2.1. Cells and Virus

The chicken macrophage HD11 cells were kindly provided by Prof. Xin-An Jiao (Jiangsu Key Laboratory of Zoonosis, Yang Zhou University, Yang Zhou, Jiangsu Province, China). HD11 cells were cultured in Dulbecco's modified Eagle's medium (DMEM) (HyClone, Logan, UT, USA) supplemented with 10% fetal bovine serum (FBS) (Gemini Bio-Products, West Sacramento, CA, USA), 100 U/mL penicillin and 100 μg/mL streptomycin (HyClone, Logan, UT, USA) at pH 7.2 and were kept at 37 °C with 5% CO_2. The Vero cell-adapted IBV Beaudette strain (p65) [25] used in the current study was kindly provided by Prof. Shi-Qi Sun (State Key Laboratory of Veterinary Etiological Biology, Lanzhou Veterinary Research Institute, Chinese Academy of Agricultural Sciences, Gansu Province, China). Traditional IBV strain M41, vaccine IBV strain H120 and virulent IBV strain SABIK2 [26] were housed in our laboratory (viruses were propagated in specific pathogen free (SPF) 10 days old embryonated chicken eggs). Susceptibility of HD11 cells to IBVs was measured by morphological changes, growth curves using 50% tissue culture infective doses ($TCID_{50}$) and indirect immunofluorescence assay (IFA).

2.2. Virus Titration and Growth Kinetics

Here, $TCID_{50}$ were applied to HD11 cells to quantitate virus titers as described previously [27]. HD11 cells were cultured in 96-well plates, and ten-fold dilutions of the virus were prepared in DMEM supplemented with 2% FBS. Cultured cells were infected with the virus and then observed daily for cytopathic effects (CPE). In order to evaluate the virus growth kinetics in HD11 cells, the cells were infected with IBV at 10 multiplicity of infection (MOI). The infected cells were collected at the indicated time points and analyzed using $TCID_{50}$ assay.

2.3. Cell Viability Assay

A Cell Counting Kit-8 (CCK-8) assay (Beyotime, Haimen, Jiangsu Province, China) was utilized to identify the viability of cells as described previously [28]. HD11 cells were cultured in 96-well plates and infected with IBV at different MOI (0.1, 0.5, 1, 5, and 10 MOI) for specific lengths of times. In parallel, a negative control was set up. The cells were incubated with 10 μL/well CCK-8 solution (Beyotime, Haimen, Jiangsu Province, China) and allowed to react for 2 h at 37 °C. The absorbance was detected using a microplate reader (model 680, Bio-Rad, Hercules, CA, USA) at 450 nm. The negative control was set at 100%, and the treated samples were calculated according to the following formula: Survival rate (%) = optical density (OD) of the treated cells/OD of the negative control × 100.

2.4. Morphological Analysis

HD11 cells were pre-incubated in 96-well plates and infected with IBV at 10 MOI. To assess apoptosis, the condensed and fragmented nuclei were observed using Hoechst 33342 staining (KeyGEN Biotech, Nanjing, Jiangsu Province, China). At the specified time points, the cells were immobilized with 4% paraformaldehyde (KeyGEN Biotech, Nanjing, Jiangsu Province, China) for 30 min and then incubated with Hoechst 33342 (KeyGEN Biotech) in the dark for 15 min. The typical apoptotic morphological changes were observed using a fluorescence microscope (Olympus IX71, Olympus, Tokyo, Japan) with UV excitation at 350 nm.

2.5. Indirect Immunofluorescence Assay

HD11 cells were grown overnight to 75% density in 96-well plates and were then infected with IBV at an MOI of 10. After the appearance of typical CPE, the cells were immobilized with 4% paraformaldehyde for 30 min. A mouse polyclonal antibody against the IBV nucleocapsid (N) protein (1:200 dilution, prepared in our laboratory) was added, followed by incubation for 1 h at 37 °C. Next, the cells were treated with a fluorescein isothiocyanate (FITC)-conjugated goat anti-mouse IgG (1:200, TransGen Biotech, Beijing, China) for 1 h at 37 °C. The specimens were viewed with an Olympus IX71 fluorescence microscope (Olympus) with the appropriate excitation and emission wavelengths for FITC (490 nm and 525 nm, respectively).

2.6. Flow Cytometry

To identify the apoptotic rate, the percentage of cells undergoing apoptosis was determined by an annexin V-FITC apoptosis detection kit (Absin, Shanghai, China). HD11 cells were cultured in 6-well plates and infected with IBV at 10 MOI. Cells were harvested and washed three times with phosphate-buffered saline (PBS) at the indicated times. The cells were centrifuged at 500 g for 5 min and then suspended in 500 μL of binding buffer containing 5 μL FITC-conjugated annexin V antibody and 5 μL propidium iodide (PI). The mixture was incubated at room temperature for 15 min in the dark. The cells were detected by flow cytometer (BD Biosciences, San Jose, CA, USA) within an hour.

2.7. Caspase Activity Assay

The activities of caspase-3, -8, and -9 were detected by colorimetric assay kit (KeyGEN Biotech). The cells were incubated with lysis buffer, and the concentrations of protein were detected by bicinchoninic acid (BCA) protein assay reagent (Vazyme Biotech, Nanjing, Jiangsu Province, China). The protein (200 μg/sample) was treated with caspase-3, -8, and -9 substrate for each sample at 37 °C for 4 h. Samples were read by a microplate reader (model 680, Bio-Rad) at 405 nm.

2.8. Quantitative Real-Time Polymerase Chain Reaction Analysis

Total RNA was isolated using TRIzol agent (Invitrogen, Carlsbad, CA, USA), and each RNA sample was reverse-transcribed to complementary DNA (cDNA) by PrimeScript™ RT Reagent Kit (Takara, Dalian, Liaoning Province, China). cDNA was used for quantitative real-time polymerase chain reaction (qRT-PCR) analysis. The sets of primer pairs of apoptotic regulating genes are listed in Table 1 [29]. For qRT-PCR reactions, the 25 μL reaction mixture included 2 μL cDNA, 12.5 μL SYBR Premix Ex TaqTM II (Takara), 1.0 μL of forward and 1.0 μL of reverse primer and 8.5 μL RNAase-free water (Takara). Reaction conditions were 95 °C for 3 min followed by 44 cycles of 95 °C for 10 s, the specific melting temperature (Tm) of a primer pair for 30 s, and then 95 °C for 10 s, and 72 °C for 10 s, using a Bio-Rad IQ5 Thermal Cycler (Bio-Rad). β-actin was selected as a reference gene. The expression fold changes were calculated using the $2^{-\Delta\Delta CT}$ method [30].

Table 1. Sequences of chicken primer pairs used for quantitative real-time polymerase chain reaction (qRT-PCR).

Gene	Forward Primer (5′–3′)	Reverse Primer (5′–3′)
Fas	TCCACCTGCTCCTCGTCATT	GTGCAGTGTGTGTGGGAACT
FasL	GGCATTCAGTACCGTGACCA	CCGGAAGAGCACATTGGAGT
Bax	GGTGACAGGGATCGTCACAG	TAGGCCAGGAACAGGGTGAA
Bcl-2	TGTTTCTCAAACCAGACACCAA	CAGTAGGCACCTGTGAGATCG
β-actin	TGCTGTGTTCCCATCTATCG	TTGGTGACAATACCGTGTTCA

2.9. Statistical Analysis

All data are expressed as the mean ± standard error of the mean (SEM) from three independent experiments performed in triplicate. The statistical analyses were conducted using Student's *t*-test in GraphPad Prism version 5 (GraphPad Software, San Diego, CA, USA). A *p*-value < 0.05 was considered significant, and a *p*-value < 0.01 was considered highly significant.

3. Results

3.1. Viral Infection in HD11 Cells

To determine whether IBVs replicate in the chicken macrophage cell line, four IBV strains (Beaudette, M41, H120 and SABIK2 strains) were utilized in this study to test the infective processes in HD11 cells. For the M41, H120 and SABIK2 strains, the infected cells were blindly passaged five times, and CPE was not observed. The growth curve using $TCID_{50}$ and IFA showed that replication of the three IBV strains in HD11 cells was severely restricted, and no significant replication was observed. However, the HD11 cells were highly permissive for the propagation of the attenuated IBV Beaudette strain. The results showed that HD11 cells could be infected by IBV Beaudette in passage one.

First, morphological changes of IBV Beaudette-infected HD11 cells were observed. After infection with IBV Beaudette at 10 MOI, CPE appeared in HD11 cells at 24 h post-infection (h.p.i.) and were evident at 36 h.p.i. when compared with the mock infection (Figure 1A). The normal HD11 cells could also be re-infected with the culture supernatant from the virally infected cells. The susceptibility of HD11 cells to IBV Beaudette infection was evaluated by growth kinetics. The growth kinetics of the virus were observed upon infection at an MOI of 10 in HD11 cells. The virus titers increased until reaching the maximal level of $10^{6.875}$ $TCID_{50}/mL$ (Figure 1B). IBV Beaudette replication in HD11 cells was also studied by performing an immunofluorescence assay. The production of FITC-stained virus was observed by 24 h.p.i. In contrast, mock-infected HD11 cells showed no fluorescence (Figure 1C).

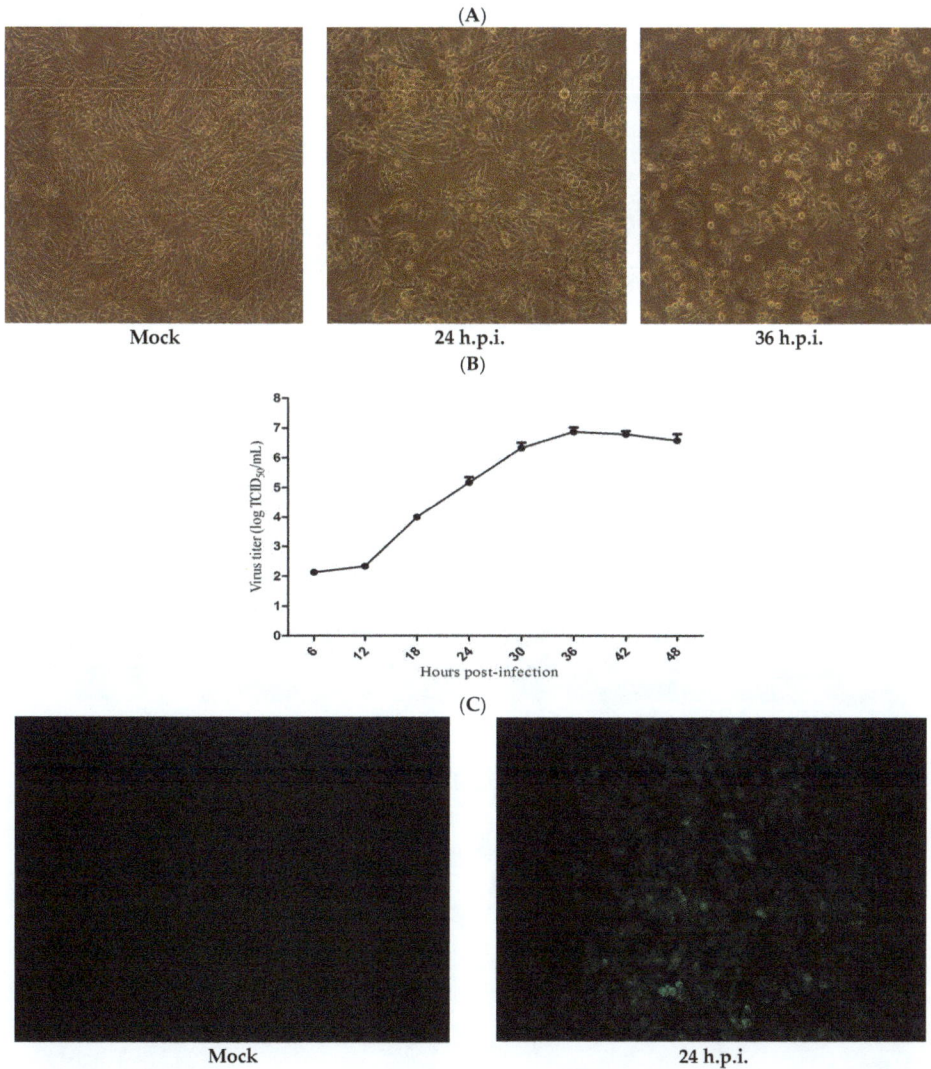

Figure 1. Susceptibility of HD11 to infectious bronchitis virus (IBV) Beaudette infection. (**A**) Cytopathic effects (CPE) observed upon IBV Beaudette infection by 24 and 36 h post-infection (h.p.i.), and mock-infected HD11 cells displayed no CPE. (**B**) Growth kinetics of IBV Beaudette in HD11 cells infected at 10 multiplicity of infection (MOI). Data are shown as the mean ± standard error of the mean (SEM) of three independent experiments. (**C**) Production of IBV Beaudette (fluorescein isothiocyanate, FITC) could be observed at 24 h.p.i., and mock-infected HD11 cells showed no fluorescence.

3.2. IBV Beaudette Induces Apoptosis in HD11 Cells

Infection of HD11 cells with IBV Beaudette caused cell death in a time- and dose- dependent manner, as tested by CCK-8 assay. (Figure 2A). The infected cells showed chromatin condensation and nuclear fragmentation. After 36 h of infection, large amounts of apoptotic bodies were observed in HD11 cells (Figure 2B). The rate of apoptotic cells was measured by flow cytometry. The rate of apoptosis significantly increased at 12 h.p.i. in virus-infected cells when compared with the mock-infected cells (Figure 2C). These findings indicated that apoptosis was induced by IBV Beaudette infection in HD11 cells.

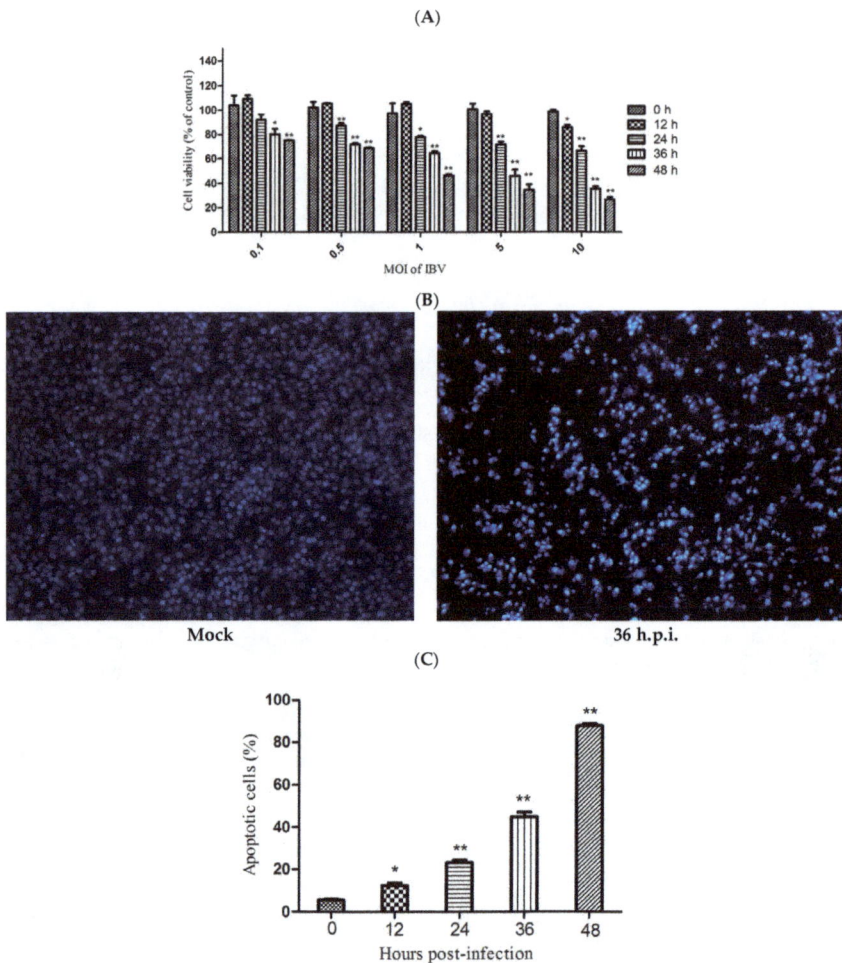

(A)

(B)

Mock   36 h.p.i.

(C)

Figure 2. IBV Beaudette induces apoptosis in HD11 cells. (**A**) The role of IBV Beaudette in cell viability. HD11 cells were infected with IBV Beaudette at different MOIs and detected at the indicated times. The Cell Counting Kit-8 (CCK-8) assay was used to measure cell viability. The data are shown as the mean ± SEM of three independent experiments. * $p < 0.05$, ** $p < 0.01$ versus control group (0 h). (**B**) Morphological changes. IBV Beaudette-infected cells were observed with condensed chromatin and nuclear fragmentation under fluorescence microscopy followed by Hoechst 33342 staining. (**C**) The apoptotic rate of cells. IBV Beaudette-infected cells (10 MOI) were subjected to flow cytometry at different times. Data are shown as the mean ± SEM of three independent experiments. * $p < 0.05$, ** $p < 0.01$ versus control group (0 h).

3.3. IBV Beaudette Triggers Apoptosis by Induction of Caspase Activity

Activation of the caspase proteinases is a significant event in the occurrence of apoptosis. The activity of caspases that play important roles in the activation of the apoptosis pathway was investigated in this study. When HD11 cells were infected with IBV Beaudette at 10 MOI, the levels of caspase-3, -8, and -9 were significantly increased from 8 h.p.i. and then increased further over time (Figure 3A). To further identify the function of caspase-3, -8, and -9 in the apoptotic pathway, we measured the viability of infected-cells incubated with specific inhibitors of caspase-3, -8, and -9 (Z-DEVD-FMK, Z-IETD-FMK, and Z-LEHD-FMK; KeyGEN Biotech, Nanjing, Jiangsu Province, China). The data revealed that cell viability was significantly increased by the specific inhibition of caspase-3, -8, and -9 (Figure 3B). To confirm the function of caspase-8 and caspase-9 to activate caspase-3, the inhibitory efficacy of the caspase-8 or caspase-9 inhibitors on caspase-3 activity was also determined. When cells were pretreated with the caspase-8 or caspase-9 inhibitor, the activity of caspase-3 was significantly decreased in cells, and more significantly decreased when the two inhibitors were added together (Figure 3C). These results revealed that caspase-3 activation and IBV Beaudette-induced apoptosis may be triggered via both extrinsic and intrinsic pathways.

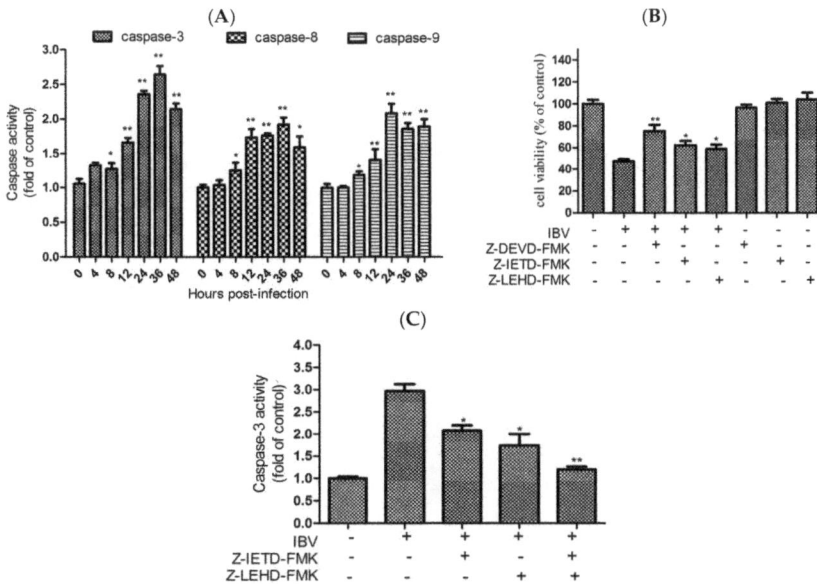

Figure 3. Effects of IBV Beaudette infection on caspases in HD11 cells. (**A**) The activity of caspases in IBV Beaudette-infected cells. The caspases -3, -8 and -9 activity in HD11 cells infected with IBV at 10 MOI at the designed times were determined. The data are shown as the mean $\pm$ SEM, * $p < 0.05$, ** $p < 0.01$ versus the control group (0 h). (**B**) Role of caspase inhibitors in cell viability. Cell viability was determined by CCK-8 assay: 20 μM of each caspase inhibitor was utilized to pre-treat cells for 2 h. Then, the treated and untreated cells were both infected with IBV Beaudette at an 10 MOI for 36 h. The data are shown as the mean $\pm$ SEM, * $p < 0.05$, ** $p < 0.01$ versus IBV infection alone. (**C**) The effect of initiator caspase-8 or -9 on the activation of caspase-3: 20 μM of each caspase inhibitor was utilized to pretreat cells for 2 h. Then, the treated and untreated cells were infected with IBV at 10 MOI for 36 h. Caspase-3 activity was detected using a colorimetric assay kit. Data are shown as the mean $\pm$ SEM, * $p < 0.05$, ** $p < 0.01$ versus virus infection alone.

3.4. Regulation of IBV Beaudette-Inducted Apoptosis Is Regulated by the Fas/FasL and Bcl-2 Family Members

Caspase-8 has an important effect on apoptosis that is mediated by Fas/FasL. The activity of caspase-8 was increased in the IBV Beaudette-infected HD11 cells. This implied that apoptosis is induced by IBV Beaudette infection through the Fas/FasL pathway. To investigate this further, the expression levels of Fas and FasL were detected in IBV Beaudette-infected HD11 cells by qRT-PCR. The data revealed increased gene expression of Fas and FasL over time (Figure 4A). Furthermore, the members of the Bcl-2 family are generally distributed on the surface of mitochondria, and their activation may regulate the intrinsic apoptosis pathway. To test this, the expression levels of Bcl-2 and Bcl-2- associated X protein (Bax) were quantified by qRT-PCR in IBV Beaudette-infected HD11 cells. The results showed the mRNA levels of Bcl-2 were obviously downregulated from 24 h.p.i. and declined over time. Conversely, the mRNA levels of Bax were upregulated from 4 h.p.i. and continuously increased until 48 h.p.i (Figure 4B). Moreover, the activation of caspase-9 was partly inhibited in IBV Beaudette-infected cells pretreated with the inhibitor of caspase-8 (Figure 4C). This suggested that caspase-9 activation was affected by the blocking of caspase-8 activity. Taken together, these findings suggested that the Fas/FasL-mediated signal contributes to the activation of caspase-8. Additionally, Bcl-2 and Bax might play important roles in regulating the activation of caspase-9. The activation of caspase-8 can also affect the extrinsic apoptosis pathway in IBV Beaudette-infected cells.

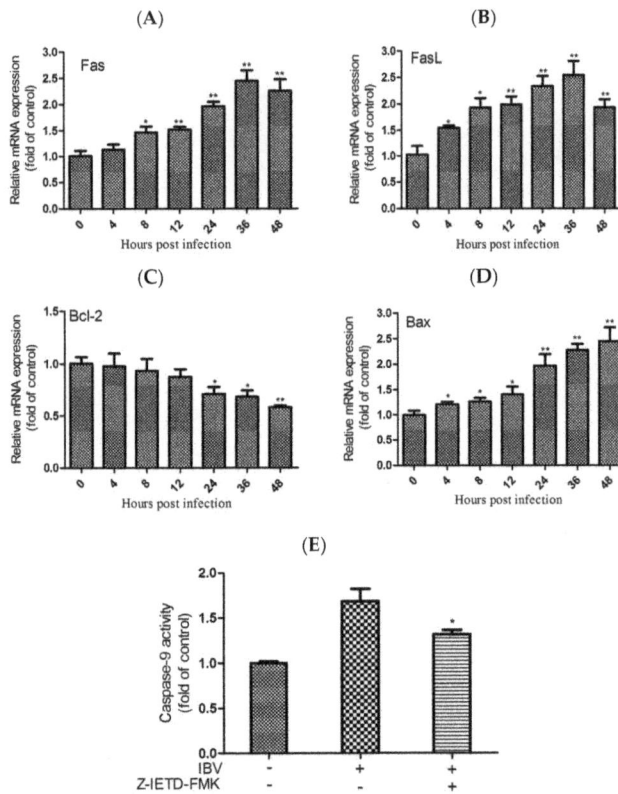

Figure 4. IBV Beaudette-induced apoptosis was regulated by the expression of Fas/Fas ligand (FasL) and the Bcl-2 family. (**A,B**) mRNA expression levels of Fas and FasL were detected in IBV Beaudette-infected cells for the indicated times. The data are shown as the mean ± SEM. * $p < 0.05$, ** $p < 0.01$ versus control. (**C,D**) mRNA expression levels of Bcl-2 and Bax were detected in IBV Beaudette-infected cells for indicated times. The data are shown as the mean ± SEM. * $p < 0.05$, ** $p < 0.01$ versus control. (**E**) The activity of caspase-9 was party blocked by caspase-8 inhibitor. Following incubation with Z-IETD-FMK for 2 h, the cells were infected with IBV Beaudette for 24 h. Caspase-9 activity was detected using a colorimetric assay kit. Data are shown as the mean ± SEM. * $p < 0.05$ versus virus infection alone.

3.5. Cell Apoptosis and Viral Replication are Required Mutually

To determine whether apoptosis plays a pivotal role in inhibition of virus replication, the virus titers of untreated cells or those treated with caspase inhibitors were detected by $TCID_{50}$. The results showed that the caspase-3 inhibitor could increase the titer of IBV Beaudette, but did not show obvious effects on caspase-8 and -9 inhibitor-treated cells (Figure 5A). To test whether the ability of virions to enter cells was crucial to apoptosis, endosomal acidification was blocked by NH_4Cl to prevent the viruses from being released [31]. The virus titer was significantly decreased with NH_4Cl treatment (Figure 5B). Compare with non-treated cells, the rate of apoptotic cells was also decreased in NH_4Cl-treated cells when infected with IBV Beaudette (Figure 5C). Next, the UV-treated virus was used to test whether apoptosis induction required virus replication. When the virus was subjected to UV treatment, the virus titer could not be detected (Figure 5D). Consistently, the rate of apoptotic cells was dramatically decreased in cells infected with UV-inactivated virus, when compared with the UV-untreated virus (Figure 5E). In conclusion, apoptosis induction required viral replication in IBV Beaudette-infected cells.

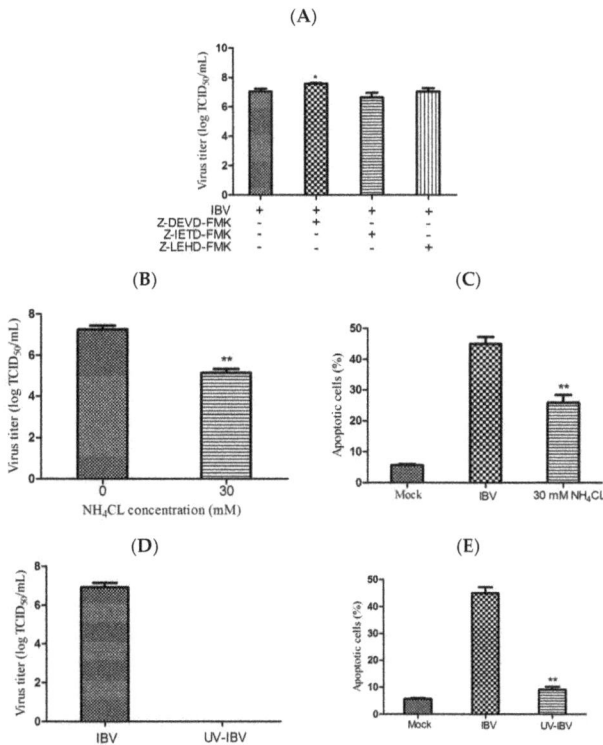

Figure 5. Interplay between cell apoptosis and virus replication. (**A**) Following incubation with inhibitors for 2 h, the cells were infected with IBV Beaudette of 10 MOI for 36 h. After pre-treatment, virus titers were determined as log $TCID_{50}$/mL. Data are shown as the mean ± SEM. * $p < 0.05$ versus virus infection alone. (**B,C**) A 30 µM concentration of NH_4Cl was used to incubate HD11 cells for 2 h before infection, then cells were infected with IBV Beaudette of 10 MOI for 36 h, and the virus titer (**B**) and the rate of apoptotic cells (**C**) were separately tested. Data are shown as the mean ± SEM. ** $p < 0.01$ versus cells infected without NH_4Cl. (**D,E**) UV germicidal light was utilized to inactivate IBV Beaudette for 30 min. UV-inactivated virus infected HD11 cells for 36 h; the virus titer (**D**) and the rate of apoptotic cells (**E**) were measured separately. Data are shown as the mean ± SEM. ** $p < 0.01$ versus UV-untreated IBV Beaudette.

4. Discussion

Compared with other coronaviruses, IBV is not easily adapted to cell culture. Several mammalian cell lines and primary cells have been previously revealed to be permissive to IBV infection. Some strains of IBV can replicate and produce CPE in primary chicken embryo kidney cells. IBV Holte and Beaudette-42 strains can proliferate in the BHK-21 cell line [32], and the Beaudette strain of embryo-culture IBV has adapted to Vero cells [33]. However, previous studies of chicken immune cells and the pathogenesis of IBV focused on primary immune cells separated from the blood or spleen [18,34]. In this study, HD11 cells, a chicken macrophage cell line, were shown to be susceptible to IBV Beaudette. Additionally, the IBV Beaudette-infected cells produced infectious virus progeny with a high virus titer. Morphological assessment of the cells during IBV Beaudette infection showed that CPE can be observed after 24 h.p.i. The virus growth kinetics for HD11 cells also showed peak viral titers at 36 h.p.i. Immunofluorescence was used to identify and analyze virus infection, and strong fluorescence signals were observed in the IBV Beaudette-infected cells. Based on these results, the chicken macrophage HD11 cells will serve as an essential tool for future studies of IBV infection.

Apoptosis is an important part of the antiviral host response. However, some viruses actively trigger this process to facilitate their replication [13]. Infection with coronavirus induced apoptosis in various cell types. Transmissible gastroenteritis virus (TGEV)-induced apoptosis in PK-15 cells was dependent on viral replication [31]. Porcine hemagglutinating encephalomyelitis virus (PHEV) induced apoptosis through a caspase-dependent pathway in PK-15 cells [35]. Severe acute respiratory syndrome (SARS) coronavirus membrane (M) and nucleocapsid (N) proteins can induce apoptosis in HPF cells [36]. According to previous studies, apoptosis occurs in response to IBV infection in Vero cells, DF1 cells and chicken embryo kidney cells [16,37]. This study is the first demonstration that IBV induces apoptosis in chicken macrophage HD11 cells. The IBV Beaudette-infected HD11 cells exhibited typical characteristics of apoptosis including the condensation of the cell nucleus, reduction of cell viability, and an increased rate of apoptotic cells.

Caspases are a family of cysteine-catalyzed proteases that cleave aspartic acids. The triggering of caspase cascades plays indispensable roles in apoptosis and can be induced by many viruses. There are two major signaling pathways that contribute to caspase activation: death receptor and mitochondrial pathways [38]. Some viruses have exhibited cell apoptosis that is mediated by Fas/FasL signaling as a reaction to viral infection, such as hepatitis C virus (HCV) [39] and dengue virus (DENV) [40]. Cell apoptosis can be induced by some viruses by regulating the levels of Bcl-2 family members, such as SARS coronavirus [41] and Epstein–Barr virus (EBV) [42]. Our results showed that the activation of caspase-8 in IBV Beaudette-infected cells was regulated by Fas and FasL. The results also showed that activation of caspase-9 in IBV Beaudette-infected cells was regulated by decreased expression of Bcl-2 and increased expression of Bax. The caspase-3 activation and virus-induced apoptosis might be triggered through both extrinsic and intrinsic pathways. In most cases, cell apoptosis induced by virus is a process of interaction between extrinsic and intrinsic pathways. The activation of caspase-8 was inhibited by Z-IETD-FMK, and the activation of caspase-9 was not completely eliminating by blocking caspase-8 activity, suggesting that the activation of caspase-8 is not the only pathway to activate caspase-9, requiring further research. IBV are known to induce apoptosis through caspase-dependent pathway [15] and intrinsic-dependent pathway regulated by Bcl-2 family proteins [16] in Vero cells. In this study, these two pathways were demonstrated that play an important role in IBV Beaudette-infected HD11 cells. Additionally, this is the first report that the extrinsic pathway regulated by Fas/FasL was activated in IBV-induced apoptosis. Improved knowledge of the mechanisms by which IBV activates the extrinsic and intrinsic apoptotic pathways will help to better understand the pathogenic properties of epidemic IBV strains in the host.

Some viruses can induce cell apoptosis through viral replication. UV-inactivated BHV-1 and TGEV could not induce apoptosis, or the NH_4Cl-pretreated cells prevented the appearance of apoptosis [31,43]. Some other viruses can induce apoptosis without viral replication, such as vaccinia virus and vesicular stomatitis virus [44,45]. Here, infection with UV-inactivated IBV Beaudette or treatment of HD11 cells with NH$_4$Cl reduced virus apoptosis induction, indicating that IBV Beaudette -induced apoptosis in HD11 cells depends on viral replication. This finding is similar to that of a previous study showing that UV-inactivated IBV lost the capacity to induce apoptosis in mammalian cells [16]. We tested whether caspase activation is needed for IBV replication in cells. The finding revealed that treatment with the caspase-3 inhibitor can increase the virus titer in IBV Beaudette-infected cells, suggesting that the caspase inhibitor might increase the survival time of cells to promote replication. From a therapeutic standpoint, available drugs controlling apoptosis could be used to limit IBV spreading [46].

The innate immune response is the first line of defense against viruses, and macrophages are an important component of this system. Some viruses have evolved strategies to induce apoptosis to enhance the production of virus progeny and promote dissemination to neighboring cells with limited host immune/inflammatory responses. The presence of apoptotic cells may also lead to the mobilization and initiation of innate immune defenses [47]. Previous studies have shown that virus-induced apoptosis of macrophage has an important impact on virus infection. Porcine reproductive and respiratory syndrome virus (PRRSV) stimulates anti-apoptotic pathways in macrophages early in infection, and these PRRSV-infected macrophages die by apoptosis late in infection [48]. CHIKV infection induces apoptosis and enhances expression of major histocompatibility complexes (MHCs) and co-stimulatory molecules and interleukin (IL)-6 and monocyte chemoattractant protein (MCP)-1 production in macrophages [20]. However, little is known about IBV-induced immune cell apoptosis. It has been reported that phagocytic cells may play a crucial role in dissemination of virus to the blood circulation and internal organs. Therefore, establishment of this macrophage system of IBV Beaudette infection and determination of the apoptotic mechanism might be proof of principle for IBV infection in the host.

In conclusion, chicken macrophage HD11 cells were established for attenuated IBV strain Beaudette infection. IBV Beaudette induced cell apoptosis through caspase-8 activation mediated by Fas/FasL and caspase-9 activation mediated by Bcl-2/Bax. In addition, IBV Beaudette replication was essential to apoptosis induction, and IBV Beaudette replication increased when caspase activation was blocked. Based on these findings, this study has shown the establishment of a chicken macrophage cell line that will facilitate the further analysis of IBV infection. Additional studies are required to clarify the detailed molecular mechanisms underlying IBV-induced apoptosis.

Acknowledgments: This research was supported by the State Natural Sciences Foundation (NSFC) (31372442), China Agriculture Research System (CARS-40) National System for Layer Production Technology (CARS-40-K09), and the Project for Science and Technology Support Program of Sichuan Province (2014NZ0002, 2016NZ0003).

Author Contributions: X.H. and H.W. conceived and designed the experiments; X.H., R.G. and Y.T. performed the experiments; X.H. and X.Y. analyzed the data; W.G. and L.Z. contributed reagents/materials/analysis tools; X.H. and H.W. wrote the paper. All authors have read and approved the final manuscript.

Conflicts of Interest: The authors declare no conflict of interest.

References

1. Brierley, I.; Boursnell, M.E.; Binns, M.M.; Bilimoria, B.; Blok, V.C.; Brown, T.D.; Inglis, S.C. An efficient ribosomal frame-shifting signal in the polymerase-encoding region of the coronavirus IBV. *EMBO J.* **1987**, *6*, 3779–3785. [PubMed]
2. Cavanagh, D. Coronavirus avian infectious bronchitis virus. *Vet. Res.* **2007**, *38*, 281–297. [CrossRef] [PubMed]
3. Cook, J.K.; Jackwood, M.; Jones, R.C. The long view: 40 Years of infectious bronchitis research. *Avian Pathol. J. WVPA* **2012**, *41*, 239–250. [CrossRef] [PubMed]
4. Hodgson, T.; Casais, R.; Dove, B.; Britton, P.; Cavanagh, D. Recombinant infectious bronchitis coronavirus Beaudette with the spike protein gene of the pathogenic M41 strain remains attenuated but induces protective immunity. *J. Virol.* **2004**, *78*, 13804–13811. [CrossRef] [PubMed]
5. Jordan, B. Vaccination against infectious bronchitis virus: A continuous challenge. *Vet. Microbiol.* **2017**, *206*, 137–143. [CrossRef] [PubMed]

6. Yang, X.; Zhou, Y.; Li, J.; Fu, L.; Ji, G.; Zeng, F.; Zhou, L.; Gao, W.; Wang, H. Recombinant infectious bronchitis virus (IBV) H120 vaccine strain expressing the hemagglutinin-neuraminidase (HN) protein of Newcastle disease virus (NDV) protects chickens against IBV and NDV challenge. *Arch. Virol.* **2016**, *161*, 1209–1216. [CrossRef] [PubMed]

7. Kint, J.; Dickhout, A.; Kutter, J.; Maier, H.J.; Britton, P.; Koumans, J.; Pijlman, G.P.; Fros, J.J.; Wiegertjes, G.F.; Forlenza, M. Infectious bronchitis coronavirus inhibits STAT1 signaling and requires accessory proteins for resistance to type i interferon activity. *J. Virol.* **2015**, *89*, 12047–12057. [CrossRef] [PubMed]

8. Fung, T.S.; Liao, Y.; Liu, D.X. The endoplasmic reticulum stress sensor IRE1α protects cells from apoptosis induced by the coronavirus infectious bronchitis virus. *J. Virol.* **2014**, *88*, 12752–12764. [CrossRef] [PubMed]

9. Cottam, E.M.; Maier, H.J.; Manifava, M.; Vaux, L.C.; Chandra-Schoenfelder, P.; Gerner, W.; Britton, P.; Ktistakis, N.T.; Wileman, T. Coronavirus NSP6 proteins generate autophagosomes from the endoplasmic reticulum via an omegasome intermediate. *Autophagy* **2011**, *7*, 1335–1347. [CrossRef] [PubMed]

10. Elmore, S. Apoptosis: A review of programmed cell death. *Toxicol. Pathol.* **2007**, *35*, 495–516. [CrossRef] [PubMed]

11. Thornberry, N.A.; Lazebnik, Y. Caspases: Enemies within. *Science* **1998**, *281*, 1312–1316. [CrossRef] [PubMed]

12. Fulda, S.; Debatin, K.M. Extrinsic versus intrinsic apoptosis pathways in anticancer chemotherapy. *Oncogene* **2006**, *25*, 4798–4811. [CrossRef] [PubMed]

13. Clarke, P.; Tyler, K.L. Apoptosis in animal models of virus-induced disease. *Nat. Rev. Microbiol.* **2009**, *7*, 144–155. [CrossRef] [PubMed]

14. Hay, S.; Kannourakis, G. A time to kill: Viral manipulation of the cell death program. *J. Gen. Virol.* **2002**, *83*, 1547–1564. [CrossRef] [PubMed]

15. Liu, C.; Xu, H.Y.; Liu, D.X. Induction of caspase-dependent apoptosis in cultured cells by the avian coronavirus infectious bronchitis virus. *J. Virol.* **2001**, *75*, 6402–6409. [CrossRef] [PubMed]

16. Zhong, Y.; Liao, Y.; Fang, S.; Tam, J.P.; Liu, D.X. Up-regulation of Mcl-1 and Bak by coronavirus infection of human, avian and animal cells modulates apoptosis and viral replication. *PLoS ONE* **2012**, *7*, e30191. [CrossRef] [PubMed]

17. Ariaans, M.P.; Matthijs, M.G.; van Haarlem, D.; van de Haar, P.; van Eck, J.H.; Hensen, E.J.; Vervelde, L. The role of phagocytic cells in enhanced susceptibility of broilers to colibacillosis after infectious bronchitis virus infection. *Vet. Immunol. Immunopathol.* **2008**, *123*, 240–250. [CrossRef] [PubMed]

18. Reddy, V.R.; Trus, I.; Desmarets, L.M.; Li, Y.; Theuns, S.; Nauwynck, H.J. Productive replication of nephropathogenic infectious bronchitis virus in peripheral blood monocytic cells, a strategy for viral dissemination and kidney infection in chickens. *Vet. Res.* **2016**, *47*, 70. [CrossRef] [PubMed]

19. Xue, J.; Fu, C.; Cong, Z.; Peng, L.; Peng, Z.; Chen, T.; Wang, W.; Jiang, H.; Wei, Q.; Qin, C. Galectin-3 promotes caspase-independent cell death of HIV-1-infected macrophages. *FEBS J.* **2017**, *284*, 97–113. [CrossRef] [PubMed]

20. Nayak, T.K.; Mamidi, P.; Kumar, A.; Singh, L.P.; Sahoo, S.S.; Chattopadhyay, S.; Chattopadhyay, S. Regulation of viral replication, apoptosis and pro-inflammatory responses by 17-AAG during Chikungunya virus infection in macrophages. *Viruses* **2017**, *9*. [CrossRef] [PubMed]

21. Herold, S.; Steinmueller, M.; von Wulffen, W.; Cakarova, L.; Pinto, R.; Pleschka, S.; Mack, M.; Kuziel, W.A.; Corazza, N.; Brunner, T.; et al. Lung epithelial apoptosis in influenza virus pneumonia: The role of macrophage-expressed TNF-related apoptosis-inducing ligand. *J. Exp. Med.* **2008**, *205*, 3065–3077. [CrossRef] [PubMed]

22. Chhabra, R.; Chantrey, J.; Ganapathy, K. Immune responses to virulent and vaccine strains of infectious bronchitis viruses in chickens. *Viral Immunol.* **2015**, *28*, 478–488. [CrossRef] [PubMed]

23. Ciraci, C.; Lamont, S.J. Avian-specific TLRs and downstream effector responses to CpG-induction in chicken macrophages. *Dev. Comp. Immunol.* **2011**, *35*, 392–398. [CrossRef] [PubMed]

24. He, H.; Genovese, K.J.; Kogut, M.H. Modulation of chicken macrophage effector function by T(H)1/T(H)2 cytokines. *Cytokine* **2011**, *53*, 363–369. [CrossRef] [PubMed]

25. Wei, Y.Q.; Guo, H.C.; Dong, H.; Wang, H.M.; Xu, J.; Sun, D.H.; Fang, S.G.; Cai, X.P.; Liu, D.X.; Sun, S.Q. Development and characterization of a recombinant infectious bronchitis virus expressing the ectodomain region of S1 gene of H120 strain. *Appl. Microbiol. Biotechnol.* **2014**, *98*, 1727–1735. [CrossRef] [PubMed]

26. Wu, X.; Yang, X.; Xu, P.; Zhou, L.; Zhang, Z.; Wang, H. Genome sequence and origin analyses of the recombinant novel IBV virulent isolate SAIBK2. *Virus Genes* **2016**, *52*, 509–520. [CrossRef] [PubMed]

27. Tay, F.P.; Huang, M.; Wang, L.; Yamada, Y.; Liu, D.X. Characterization of cellular furin content as a potential factor determining the susceptibility of cultured human and animal cells to coronavirus infectious bronchitis virus infection. *Virology* **2012**, *433*, 421–430. [CrossRef] [PubMed]

28. Xie, Y.; Liao, J.; Li, M.; Wang, X.; Yang, Y.; Ge, J.; Chen, R.; Chen, H. Impaired cardiac microvascular endothelial cells function induced by Coxsackievirus B3 infection and its potential role in cardiac fibrosis. *Virus Res.* **2012**, *169*, 188–194. [CrossRef] [PubMed]

29. Yuan, S.; Wu, B.; Yu, Z.; Fang, J.; Liang, N.; Zhou, M.; Huang, C.; Peng, X. The mitochondrial and endoplasmic reticulum pathways involved in the apoptosis of bursa of Fabricius cells in broilers exposed to dietary aflatoxin B_1. *Oncotarget* **2016**, *7*, 65295–65306. [CrossRef] [PubMed]

30. Livak, K.J.; Schmittgen, T.D. Analysis of relative gene expression data using real-time quantitative PCR and the $2^{-\Delta\Delta C_T}$ method. *Methods* **2001**, *25*, 402–408. [CrossRef] [PubMed]

31. Ding, L.; Xu, X.; Huang, Y.; Li, Z.; Zhang, K.; Chen, G.; Yu, G.; Wang, Z.; Li, W.; Tong, D. Transmissible gastroenteritis virus infection induces apoptosis through FasL- and mitochondria-mediated pathways. *Vet. Microbiol.* **2012**, *158*, 12–22. [CrossRef] [PubMed]

32. Otsuki, K.; Noro, K.; Yamamoto, H.; Tsubokura, M. Studies on avian infectious bronchitis virus (IBV). II. Propagation of IBV in several cultured cells. *Arch. Virol.* **1979**, *60*, 115–122. [CrossRef] [PubMed]

33. Shen, S.; Law, Y.C.; Liu, D.X. A single amino acid mutation in the spike protein of coronavirus infectious bronchitis virus hampers its maturation and incorporation into virions at the nonpermissive temperature. *Virology* **2004**, *326*, 288–298. [CrossRef] [PubMed]

34. Gurjar, R.S.; Gulley, S.L.; van Ginkel, F.W. Cell-mediated immune responses in the head-associated lymphoid tissues induced to a live attenuated avian coronavirus vaccine. *Dev. Comp. Immunol.* **2013**, *41*, 715–722. [CrossRef] [PubMed]

35. Lan, Y.; Zhao, K.; Wang, G.; Dong, B.; Zhao, J.; Tang, B.; Lu, H.; Gao, W.; Chang, L.; Jin, Z.; et al. Porcine hemagglutinating encephalomyelitis virus induces apoptosis in a porcine kidney cell line via caspase-dependent pathways. *Virus Res.* **2013**, *176*, 292–297. [CrossRef] [PubMed]

36. Zhao, G.; Shi, S.Q.; Yang, Y.; Peng, J.P. M and N proteins of SARS coronavirus induce apoptosis in HPF cells. *Cell Biol. Toxicol.* **2006**, *22*, 313–322. [CrossRef] [PubMed]

37. Chhabra, R.; Kuchipudi, S.V.; Chantrey, J.; Ganapathy, K. Pathogenicity and tissue tropism of infectious bronchitis virus is associated with elevated apoptosis and innate immune responses. *Virology* **2016**, *488*, 232–241. [CrossRef] [PubMed]

38. Marsden, V.S.; O'Connor, L.; O'Reilly, L.A.; Silke, J.; Metcalf, D.; Ekert, P.G.; Huang, D.C.; Cecconi, F.; Kuida, K.; Tomaselli, K.J.; et al. Apoptosis initiated by Bcl-2-regulated caspase activation independently of the cytochrome c/Apaf-1/caspase-9 apoptosome. *Nature* **2002**, *419*, 634–637. [CrossRef] [PubMed]

39. Chen, Z.; Zhu, Y.; Ren, Y.; Tong, Y.; Hua, X.; Zhu, F.; Huang, L.; Liu, Y.; Luo, Y.; Lu, W.; et al. Hepatitis C virus protects human B lymphocytes from Fas-mediated apoptosis via E2-CD81 engagement. *PLoS ONE* **2011**, *6*, e18933. [CrossRef] [PubMed]

40. Liao, H.; Xu, J.; Huang, J. FasL/Fas pathway is involved in dengue virus induced apoptosis of the vascular endothelial cells. *J. Med. Virol.* **2010**, *82*, 1392–1399. [CrossRef] [PubMed]

41. Tan, Y.X.; Tan, T.H.; Lee, M.J.; Tham, P.Y.; Gunalan, V.; Druce, J.; Birch, C.; Catton, M.; Fu, N.Y.; Yu, V.C.; et al. Induction of apoptosis by the severe acute respiratory syndrome coronavirus 7a protein is dependent on its interaction with the Bcl-X_L protein. *J. Virol.* **2007**, *81*, 6346–6355. [CrossRef] [PubMed]

42. Fu, Q.; He, C.; Mao, Z.R. Epstein-Barr virus interactions with the Bcl-2 protein family and apoptosis in human tumor cells. *J. Zhejiang Univ. Sci. B* **2013**, *14*, 8–24. [CrossRef] [PubMed]

43. Xu, X.; Zhang, K.; Huang, Y.; Ding, L.; Chen, G.; Zhang, H.; Tong, D. Bovine herpes virus type 1 induces apoptosis through Fas-dependent and mitochondria-controlled manner in Madin-Darby bovine kidney cells. *Virol. J.* **2012**, *9*, 202. [CrossRef] [PubMed]

44. Pearce, A.F.; Lyles, D.S. Vesicular stomatitis virus induces apoptosis primarily through Bak rather than Bax by inactivating Mcl-1 and Bcl-X_L. *J. Virology* **2009**, *83*, 9102–9112. [CrossRef] [PubMed]

45. Yates, N.L.; Yammani, R.D.; Alexander-Miller, M.A. Dose-dependent lymphocyte apoptosis following respiratory infection with Vaccinia virus. *Virus Res.* **2008**, *137*, 198–205. [CrossRef] [PubMed]

46. Callus, B.A.; Vaux, D.L. Caspase inhibitors: Viral, cellular and chemical. *Cell Death Differ.* **2007**, *14*, 73–78. [CrossRef] [PubMed]

47. Zhong, Y.; Tan, Y.W.; Liu, D.X. Recent progress in studies of arterivirus-and coronavirus-host interactions. *Viruses* **2012**, *4*, 980–1010. [CrossRef] [PubMed]
48. Costers, S.; Lefebvre, D.J.; Delputte, P.L.; Nauwynck, H.J. Porcine reproductive and respiratory syndrome virus modulates apoptosis during replication in alveolar macrophages. *Arch. Virol.* **2008**, *153*, 1453–1465. [CrossRef] [PubMed]

MDPI AG

St. Alban-Anlage 66

4052 Basel, Switzerland

Tel. +41 61 683 77 34

Fax +41 61 302 89 18

http://www.mdpi.com

Viruses Editorial Office

E-mail: viruses@mdpi.com

http://www.mdpi.com/journal/viruses